Unveiling the Universe

Springer
*London
Berlin
Heidelberg
New York
Barcelona
Budapest
Hong Kong
Milan
Paris
Santa Clara
Singapore
Tokyo*

Unveiling the Universe

An Introduction to Astronomy

J.E. van Zyl

With 383 Figures
including 45 Colour Plates

Springer

J.E. van Zyl, BSc, BA(Hons), HED
PO Box 35571, Northcliff 2115, Republic of South Africa

Cover illustration: The front cover shows the Orion Nebula, composited from a mosaic of 15 separate fields taken with NASA's Hubble Space Telescope. C.R. O'Dell (Rice University) and NASA. The back cover also shows Fig. 8.50 (copyright held by NASA); Fig. 10.30; Fig. 14.1 (copyright held by Proff. P.J.E. Peebles, E.J. Roth, M. Seldner and B. Siebers, made from galaxy counts by C.D. Shane and C.A. Wirtanen); and Fig. 6.5.

ISBN 3-540-76023-7 Springer-Verlag Berlin Heidelberg New York

British Library Cataloguing in Publication Data
Van Zyl, Johannes Ebenhaezer
  Unveiling the universe: an introduction to astronomy
  1. Astronomy
  I. Title
  520
ISBN 3540760237

Library of Congress Cataloging-in-Publication Data
Van Zyl, J.E. (Johannes Ebenhaezer), 1913–
  [Ontsluier die heelal. Afrikaans]
  [Unveiling the universe: an introduction to astronomy / J.E. van Zyl.
     p.   cm.
  Includes bibliographical references and index.
  ISBN 3-540-76023-7
  1. Astronomy.   I. Title.
QB43.2.V35613   1996   96-13944
520– –dc20   CIP

Apart from any fair dealing for the purposes of research or private study, or criticism or review, as permitted under the Copyright, Designs and Patents Act 1988, this publication may only be reproduced, stored or transmitted, in any form or by any means, with the prior permission in writing of the publishers, or in the case of reprographic reproduction in accordance with the terms of licences issued by the Copyright Licensing Agency. Enquiries concerning reproduction outside those terms should be sent to the publishers.

© Springer-Verlag London Limited 1996
Printed in Spain

The use of registered names, trademarks, etc. in this publication does not imply, even in the absence of a specific statement, that such names are exempt from the relevant laws and regulations and therefore free for general use.

The publisher makes no representation, express or implied, with regard to the accuracy of the information contained in this book and cannot accept any legal responsibility or liability for any errors or omissions that may be made.

Typeset by EXPO Holdings, Malaysia
Printed and bound by Graficas Zamudio Printek, S.A.L., Zamudio, Spain
58/3830-543210 Printed on acid-free paper

Dedicated to all those who, through their painstaking efforts, have laboured long and hard to unravel the secrets of the Universe – they walk the pages of this book

Preface

Many books on general astronomy have been published in recent years, but this one is exceptional in several respects. It not only provides the complete newcomer to astronomy with a broad picture, covering all aspects – historical, observational, space research methods, cosmology – but it also presents enough more advanced material to enable the really interested student to take matters further.

Astronomy is essentially a mathematical science, but there are many people who are anxious to take more than a passing interest and yet are not equipped to deal with mathematical formulae. In this book, therefore, the mathematical sections are deliberately separated out, so that they can be passed over without destroying the general picture. The result is that the book will be equally useful to beginners, to more advanced readers, and to those who really want to go deeply into the subject – for instance at university level.

The whole text is written with admirable clarity, and there are excellent illustrations, together with extensive appendices which give lists of objects of various types together with more detailed mathematical explanations. All in all, the book may be said to bridge the gap between purely popular works and more advanced treatises; as such it deserves a very wide circulation, and it will undoubtedly run to many future editions.

July 1996   Patrick Moore

Introduction

The purpose of this book is to make the principles of astronomy available to all, so that the general reader can obtain a good background knowledge. The book can serve as a guide to amateur astronomers and as a basis for further study.

The methods and calculations employed by astronomers are explained and reduced to first principles. Actual measurements are used to illustrate the findings and elucidate the explanations.

The reader need take nothing for granted, because the essential mathematics is not by-passed. In order to facilitate the grasp of the astronomy, Appendices are provided, where the school mathematics used, is reduced to first principles. The uninitiated reader would do well to refer to the relevant Appendix where indicated in the text.

Cognate disciplines, such as physics, chemistry, radio-technology, electronics, etc., are not entered into, for fear of making the book too weighty, although these subjects are the foundations on which astronomy is based. Electronics and space technology will certainly continue to enrich astronomy in the future.

It is hoped that this book will help the general reader gain insight into the wonderful world of astronomy.

J.E. van Zyl

Acknowledgements

The author thanks his wife Freda for her assistance with the proofreading and the compilation of the Index; and the following members of the Johannesburg Centre of the Astronomical Society of Southern Africa for their assistance: K.E. Buchmann, M.D. Overbeek, A.S. Hilton, C. Papadopoulos, G.C. Jacobs, G. Pulik.

Alfred Biehler is thanked for doing the digital processing of the eight star charts in Chapter 1 and Figure 13.1. The author expresses his gratitude to the undermentioned astronomers and physicists for their painstaking revision of the text and for their very constructive suggestions. Any remaining shortcomings are entirely due to the author.

Jan Wolterbeek, Senior Lecturer in Astronomy, University of South Africa;
Dr. Chris Koen, South African Astronomical Observatory, Cape Town;
Prof. Lötz Strauss, Professor of Physics, University of Pretoria;
Dr. Chris Engelbrecht, Department of Physics, University of South Africa.

Photographs and Diagrams

The undermentioned Organisations and Persons are thanked for their kind permission to use the images over which they hold the copyrights:

The National Aeronautics and Space Administration (NASA) and the Jet Propulsion Laboratories, for the following figures:

5.10, 5.12, 6.2, 6.3, 6.6, 7.4, 7.8, 7.35, 7.38, 8.2, 8.3, 8.4, 8.5, 8.7, 8.9, 8.10, 8.11, 8.12, 8.18, 8.23, 8.27, 8.28, 8.30, 8.31, 8.32, 8.33, 8.34, 8.35, 8.36, 8.37, 8.40, 8.42, 8.44, 8.46, 8.49, 8.50, 8.51, 8.52, 8.53, 8.56, 8.57, 8.58, 8.59, 8.60, 8.61, 8.62, 8.63, 9.20.

The American National Optical Astronomy Observatories, for the following figures:

7.31, 9.8, 9.12, 9.15, 9.17, 9.18, 9.19, 11.23, 11.24 11.35, 12.4, 12.7, 12.12, 12.13, 12.15, 12.18, 13.2, 13.3, 13.4, 13.7, 13.9, 13.12, 13.13, 13.14, 13.15, 13.19, 13.20, 13.24, 13.25, 13.28.

Dr. Hans Vehrenberg, for the following: 7.13a, 10.33, 11.28, 11.36, 11.37, 12.14, 12.19, 13.10, 13.11.
The British Museum, Copyright: 3.1, 3.3, 3.8, 3.17, 3.18, 3.20, 7.1, 10.4.
Hale Observatories/California Institute of Technology: 4.15, 9.10, 9.11, 9.13, 13.16, 13.22.
Space Telescope Science Institute/NASA: 4.26, 7.16, 7.29.
Lowell Obs. Photos: 3.9, 7.14, 10.29.
Yerkes Obs. Photos: 3.10, 5.1, 10.8
UCO/Lick Observatory Images: 6.4, 6.7.
Dr. C.N. Williams, Bedfordview, Transvaal, South Africa: 7.33, 7.34.
Max Planck Institute: 7.21, 7.30.
C. Papadopoulos (N. Akakios): 7.13b, 12.16.
S A Astronomical Observatory: 7.18, 11.32.
Paul E. Roques (Griffith Obs.): 3.19.
National Museum of American History: 3.22.
Sky and Telescope, photo by Dennis di Cicco: 5.14
Sky and Telescope, photo by Paul Hodge: 10.34

Proff. A.S. Murrell, C.P. Knuckles, Mexico State University: 6.1.
Prof. William M. Sinton, Flagstaff: 7.2.
Official US Navy Photograph: 7.15.
C. Nicollier, NASA Johnson Space Centre: 7.26.
Jack Bennett (Mrs. N. Smith): 7.27.
European Southern Observatory: 7.28.
Dr. J.F. McHone, Arizona State University: 7.36, 7.37.
Astromedia, Milwaukee, USA: 7.39.
Cmdr. H. Hatfield, Kent, UK: 8.14.
U.A. Shatalin, Russian Astronautics: 8.22.
Akira Fujii, Japan: 8.25.
M.D. Overbeek, Edenvale, Transvaal: 9.23.
R.T.A. Innes (J.E. Innes, Perth): 10.5.
J. Bartholdi, Haute Provence Obs.: 10.17.
Dr. R.B. Minton, Martin Marietta: 10.24.
Prof. R.A. Fesen, Dartmouth Coll.: 11.26.
Prof. I. Shelton, Toronto University: 11.31.
Royal Observatory, Edinburgh: 11.33.
B. Baldwin (IAU): 11.34.
Warner Swasey Observatory: 11.38.
Harvard College Observatory: 12.3.
Jason Ware, Frisco, Texas, Internet www page http:galaxy photo.com.: 12.11
Zeiss, Oberkochen, Germany: 13.6.
Arecibo Radio Observatory: 13.26.
National Radio Astronomy Obs.: 13.27.
Dr. S. Mitton, Dr. M. Ryle, Cambridge: 13.29.
Proff. P.J.E. Peebles, E.J. Roth, M. Seldner, B. Siebers, Princeton University: 14.1.

Contents

1 The Visible Night Sky 1
2 The Dawn of Astronomy 13
3 The Formulation of Laws 19
4 Instruments 33
5 Fundamental Measurements 47
6 Surveying the Planets 57
7 New Planets, Comets and Meteors 71
8 Exploring the Solar System 97
9 The Sun 137
10 Stars and Double Stars 157
11 Variable Stars 195
12 The Milky Way Galaxy 221
13 Beyond the Milky Way 237
14 From the Big Bang to the Present 259

Appendices 273
Bibliography 315
Index 317

Chapter 1
The Visible Night Sky

FROM their fairly safe haven among the lakes where the Mediterranean Sea now divides Africa from Europe, our ancestors were able to view the starry sky at night and were enthralled by the brilliant flickering of the stars. They were not entirely safe because their spaceship "Earth" was subject to dangers from storms, lightning, earthquakes, floods and volcanic eruptions – phenomena which unleashed tremendous power.

When the last ice age came to an end about 15 000 years ago, the ice masses, centred on the Earth's poles, had reached the latitude of 45 degrees. As the climate became more and more equable, the ice started melting and the ice caps began to recede. The water from the melted ice flowed into the oceans, so that the sea level gradually started to rise. Eventually, after many centuries, the waters of the Atlantic Ocean had risen so much that they burst through the dry ground connecting Africa to Europe at the position of the present Straits of Gibraltar. In the low-lying area between Africa and Europe, the lakes started to fill and Man therefore, had to move to higher ground. In some areas the inrushing waters dammed up against a ridge and then suddenly overflowed into the low-lying areas beyond, and many people must have drowned.

This disaster was enacted at the very spot where Man's Western Civilisation had taken root. Eventually the whole area between Africa and Europe was filled by the waters of the present Mediterranean Sea. How could Man have known what the scope of the disaster really was, spatially bound as he was to his local area? In his legends, handed down by word of mouth, Man was able to give only a feeble account of the real disaster. He told the story of a great flood.

In this same area the volcanic island Thera (today known as Santorini) suddenly exploded! If this explosion was comparable with the explosion of the volcanic island Krakatoa in 1883 in the East Indies, the bang must have been audible thousands of kilometres away. The tidal wave it caused in the sea flooded the surrounding islands and wiped out the Minoan Civilisation on the island of Crete. In their legends the survivors told of the continent of Atlantis, which vanished under the sea.

FROM time to time men saw a sudden flash among the brilliant jewels of the night sky. Surely this must be a star which had suddenly shot away from its former firm position among all the other immovable stars. It was therefore called a shooting star.

At other times men saw an object with a flaming tail. For a few months it would hang like a sceptre over the Earth – surely an omen that a kingdom was about to fall or that some other great tragedy was imminent. These were the comets.

Probably the most disturbing of all events were those occasions when the Sun suddenly blackened. This had to be a dragon which was trying to swallow it. In those cases when the Sun was completely blacked out, its beautiful corona became visible, completely surrounding it. This must surely be the foam from the dragon's jaws. What could Man do? Well, he had to try and help – so he beat on his drums and blew on his horns and trumpets and rent the air with his screams. And, funnily enough, he was always successful, because, after a few minutes the foam from the dragon's jaws vanished and after an hour or two the Sun was back in all its glory. The dragon had been defeated! Thus Man exercised his powers over the heavenly bodies!

These eclipses of the Sun must have struck terror into the hearts of men since they had only a vague idea of the possible causes of such events.

WITH unpolluted skies above, primitive man had a beautiful view of the starry heavens. Against a pitch black background the twinkling stars provided a magnificent sight, especially during the time of dark Moon. The positions occupied by the stars seemed invariable and they were called the fixed stars. Intently gazing at the heavens, humans naturally connected the stars by imaginary lines, thus forming patterns of animals and of kings and the heroes of their fables. Man gave free rein to his imagination and painted pictures of fabulous creatures among the stars.

The inhabitants of ancient Egypt, Mesopotamia, Babylon and Greece conjured up the gripping Hunting Scene which consisted of four or five constellations visible during the winter months of the Northern Hemisphere, if one looks toward the south (Figure 1.4, page 7). Westwards, there is a V-shaped asterism which men saw as the forehead of Taurus (Latin for Bull). His one eye contains the very bright, reddish-coloured star with the Arabic name Aldebaran, meaning "the follower" (of the Seven Sisters, situated a bit further west).

To the east and slightly south of Taurus is the magnificent group of very bright stars among which Orion, the great hunter, was depicted. His right shoulder has the bright orange-coloured star Betelgeuse (pronounced "Beteldjewz"). Since it is the brightest star in the constellation of Orion, it is called Alpha Orionis. The second brightest star in this constellation, diagonally across from Betelgeuse, in Orion's ankle, is the bright star Rigel (or Beta Orionis). There are three very bright stars in Orion's belt (also known as the Three Kings), namely Mintaka, Alnilam and Alnitak. Alnilam means "string of pearls" and the other two simply mean "belt". Towards the south from Alnilam there is indeed a string of stars, which depict Orion's sword. At the tip of the sword there is the Great Nebula in Orion (M42 for short). There is a great amount of nebulosity (or cloudiness) in the constellation, in which new stars are constantly being formed. The projection of the Earth's equator against the sky – the celestial equator – passes through Mintaka.

With a shield on his left arm and a club in his raised right hand, Orion keeps a wary eye on Taurus.

South-east of Orion we come to Canis Major (the Great Dog) with the brightest of all the stars, Sirius, in his one eye. Sirius is, in fact, the sixth-nearest star to the Earth and is actually 23 times brighter than the Sun. The three constellations, Taurus, Orion and Canis Major, contain many bright stars. Towards the east of Orion and Canis Major, the vague streak of the Milky Way spans the sky from north-west to south-east.

At the feet of Orion, Lepus (the Hare) lies in safety, protected from onslaughts by Canis Major.

Further south, Columba (the Dove), Caelum (graving tool) and Pictor (Painter) are to be found. The second-brightest star in Pictor, Beta Pictoris, was recently found to have a disc of dust around it. This could be material for the forming of planets.

Exactly on the southern horizon lies Dorado (the Goldfish).

To the north of Canis Major we find Monoceros (the Unicorn), where the stars are rather faint. Through a telescope, beautiful nebulae can be seen in this area. Next we come to Canis Minor (the Lesser Dog) with an elongated body and short legs – a prototype of the sausage dog! In the eye of the Lesser Dog is the bright star Procyon, eight times brighter than the Sun. Sirius, Betelgeuse and Procyon form a large triangle – the Winter Triangle for northern viewers.

Further north we have the constellation Gemini (the Twins). The Twins have their arms around each other's shoulders and each has a bright star on his forehead: the two equally bright stars Castor and Pollux. Although Castor is seen by the eye as a single point of light, it actually consists of three pairs of stars.

Next we come to the Lynx and Auriga the Charioteer who has several Kids in his arms. The brightest star in this constellation is Capella (Latin for goat).

To the south of Canis Major lies Puppis (Stern of the Ship) and then Carina (the Keel) which contains Canopus, the second brightest of all the stars. It is very distant, being one thousand and one hundred light years from us. It must therefore be very bright and, indeed, is 150 000 times the brightness of the Sun! North-east of Carina is Vela (the Sails). Formerly these last three constellations constituted one constellation, Argo Navis (the Ship Argo), in which the Argonauts searched for the Golden Fleece. Every ship must have a compass and this is provided by Pyxis, just north of Vela.

Twenty five degrees to the north of Pyxis is the head of Hydra (the Water Snake). The coils of its body twist far to the east. In Hydra there is a great cluster of distant galaxies or stellar systems. To the north is the constellation Cancer (the Crab) with its pair of leering eyes and claws in all directions.

NORTH-EAST of Cancer one finds the stars that are visible during spring in the Northern Hemisphere (Figure 1.3, page 6). Most outstanding are the stars of Ursa Major (the Great Bear). Seven of the bright stars are also depicted as the Plough (the Haywain or the Big Dipper). How did it come to pass that the Bear has such a long tail? Legend has it that Zeus (the chief of the gods according to the Greeks) grabbed the bear by the tail and flung him into the heavens just before a hunter was about to release an arrow from his bow with the Bear in his sights. Any bear that gets thrown so far by the tail will surely have a much elongated tail!

South of the Great Bear is Leo Minor (the Lesser Lion) and then Leo (the Lion) himself, majestic and proud, spreading west to east over 30 degrees. The bright star Regulus (or Alpha Leonis) is situated on the Ecliptic. The Ecliptic is the great circle which marks the path of the Sun against the background of the stars. Because the path of the Moon is inclined by only 5 degrees to the Ecliptic, it often happens that Regulus is occulted when the Moon moves across in front of Regulus.

South of Leo we have Sextans (the Sextant) followed by the coils of Hydra and then Antlia (the Air Pump) bordering on Pyxis and Vela. East of Vela, some of the stars of Centaurus (the Centaur) are just visible on the southern horizon.

Northwards from the coils of Hydra is Crater (the Cup) and Corvus (the Crow) with four bright stars in the form of a cross, which must not be mistaken for the Southern Cross. Next comes Virgo (the Virgin), who can easily be pictured as a young lady, legs stretched out fore and aft and with arms spread out, doing her physical exercises! Virgo is just east of Leo and also encompasses 30 degrees of the Ecliptic. Virgo also has a great cluster of galaxies. Northwards, Coma Berenices (Berenice's Hair), consisting of faint stars, also has a cluster of galaxies, known as the Coma cluster.

Further to the north is Canes Venatici (the Hunting Dogs) which contains the galaxy M51, the first nebula to be discovered as spiral in form. To the east is Boötes (the Herdsman) and next to him Corona Borealis (the Northern Crown). Twenty degrees southwards there are three bright stars which form a triangle with two sides equal in length: Libra (the Scales). The three stars have very euphonious names: Zuben Elgenubi, Zuben Elschemali (the northern and southern claws) and Zuben Elakrab (the claw). These stars therefore actually belong to the neighbouring constellation, Scorpius (the Scorpion) just to the east of Libra. The head of Scorpius and Libra both lie on the Ecliptic.

THE summer stars (Figure 1.2, page 5) lie further east: Lupus (the Wolf), Norma (the Level) and Ara (the Altar). North of these is the magnificent Scorpion, having the very bright red giant star Antares in his heart. The Scorpion's tail makes a beautiful curve in the heavens and ends in the sting. To the north is Ophiuchus (the Serpent Holder) with the head of the Serpent squeezed between him and the Scorpion. The body of the Serpent trails northwards from Ophiuchus towards Hercules, the strong man. It was he who cleaned the stables of King Augeas by diverting the waters of the river Alpheus through them.

North of Hercules, the head of Draco (the Dragon) can be seen. From here to the south-east is Lyra (the Lyre) and the magnificent Cygnus (the Swan), with wings spread wide and long neck stretched out.

To the south lies Vulpecula (the Fox) and Sagitta (the Arrow). Then follow Delphinus (the Dolphin), a very small constellation, and Aquila (the Eagle) diving downwards.

Next to the head of the Serpent is Scutum (the Shield) and southwards the constellation Sagittarius (the Archer), very rich in stars in the middle of the Milky Way. Many nebulae are to be found here. Sagittarius, half horse, half man, aims his arrow at the heart of the Scorpion. Just to the south is Corona Australis (the Southern Crown) and eastwards Capricornus (the Goat) with the curving horns of a gazelle and the tail-end of a fish.

Further south there is Microscopium (the Microscope), Telescopium (the Telescope) and Indus (the Indian).

NEXT follow the stars of the northern autumn (Figure 1.1, page 4): Grus (the Crane) and northwards Piscis Australis (the Southern Fish). Then comes Aquarius (the Water Bearer), on the Ecliptic, and Equuleus (the Foal), followed by Pegasus (the Flying Horse). The rump of Pegasus is demarcated by four bright stars at the corners of an apparent square, 17 by 15 degrees. North of Pegasus is Lacerta (the Lizard) and the young lady Andromeda, chained to a rock waiting to be devoured by Cetus (the Whale). Andromeda's mother, the Queen Cassiopeia, is to the north, and east of her the hero Perseus.

According to legend, the farmers complained to King Cepheus because the Sea Monster Cetus was killing their cattle. King Cepheus visited the Sea Nymphs to ask their advice. The Sea Nymphs were very cross with the royal couple because Queen Cassiopeia had been boasting that she was the most beautiful woman in the land and the Sea Nymphs took a dim view of this. However, they gave King Cepheus this advice: if he really cherished the interests of the farmers, then he must sacrifice his daughter Andromeda to the Sea Monster, by chaining her to a rock by the sea. The Sea Nymphs would then see to it that Cetus would stop slaughtering the farmers' cattle. Cepheus, good democrat that he was, complied with the bidding of the Sea Nymphs. However the hero, Perseus, happened to hear of this. He had just returned from an expedition during which he had slain the Gorgons. With the head of Medusa (one of the Gorgons) in his hand, he mounted the Flying Horse, Pegasus, and made tracks to go and free Andromeda. The hair of the Medusa consisted of serpents and her one eye was the variable star Ra's al Ghoel (Algol, the Demon). Anybody who dared look the Medusa in the eye was immediately turned to stone. When Perseus approached Andromeda, with the Medusa's head in his hand, he realised the danger should Andromeda look into the face of the Medusa. So he gave a long, hard wolf whistle and shouted to Andromeda to look the other way. In those days womenfolk were still obedient towards menfolk and she did look the other way. Perseus and the Sea Monster, Cetus, rapidly came closer. Perseus swung the Medusa's head to

4   The Visible Night Sky

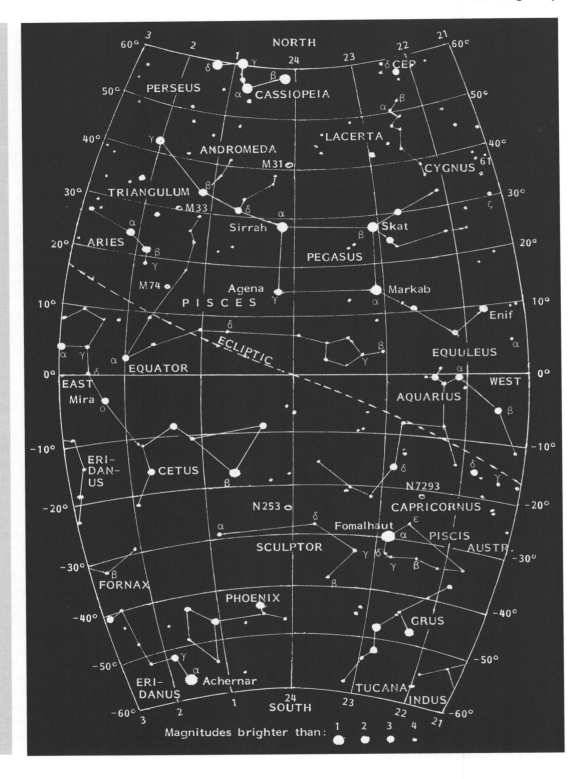

Figure 1.1
The night sky, September to November

The Visible Night Sky

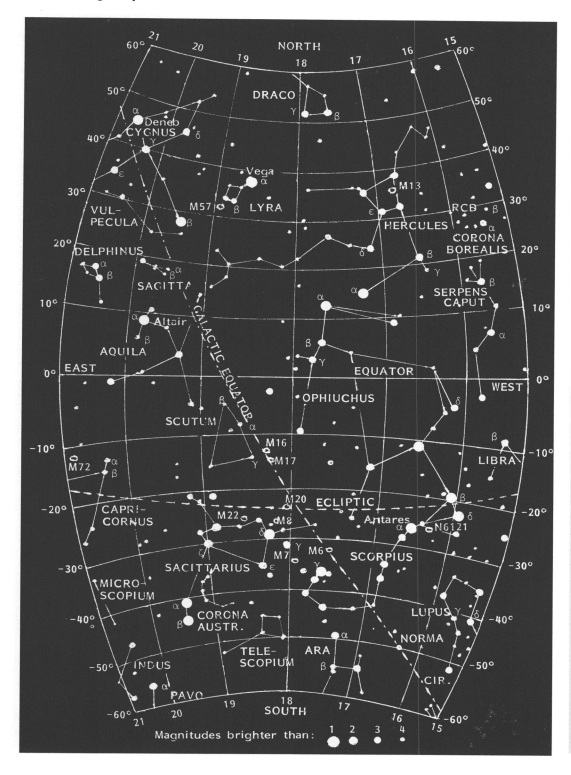

Figure 1.2
The night sky, June to August

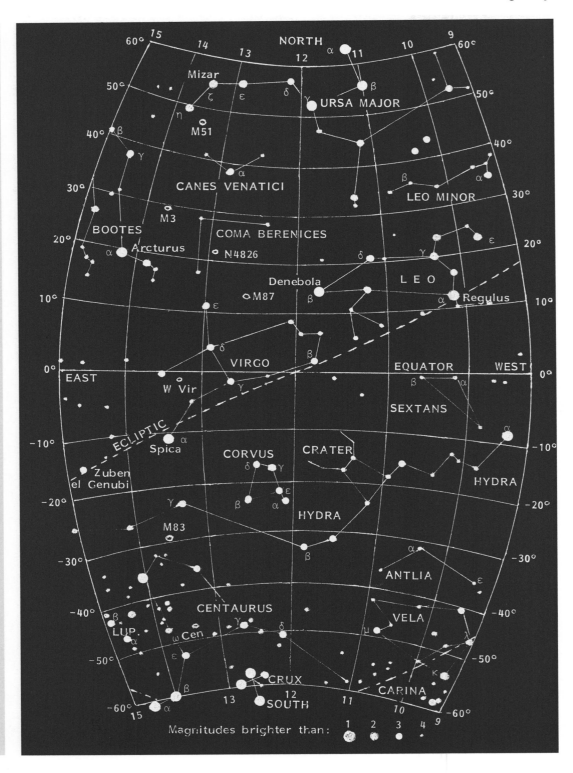

Figure 1.3
The night sky, March to May

The Visible Night Sky

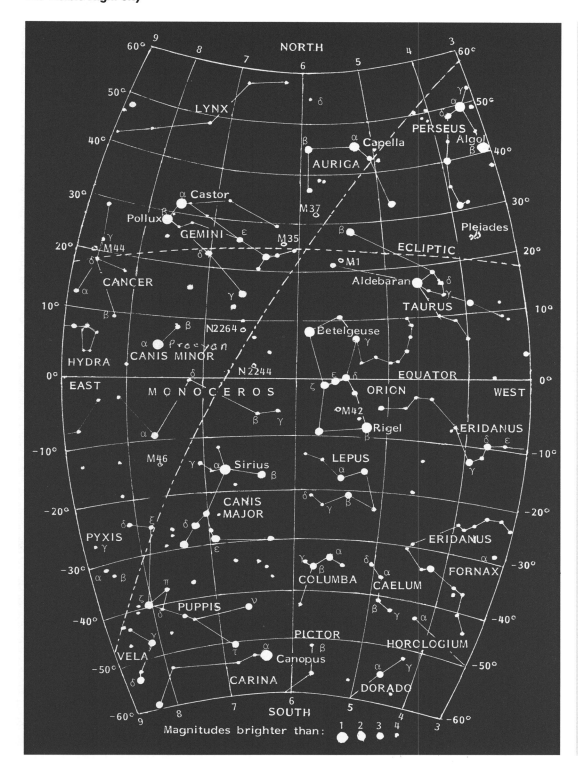

Figure 1.4
The night sky, December to February

and fro before the eyes of Cetus, and there and then Cetus was turned to stone! And the hero Perseus cut Andromeda's chain with his sword and they lived happily ever after!

To the south of Perseus are Triangulum (the Triangle) and Aries (the Ram). Spread over the back of Pegasus are two cords with fishes at their ends: Pisces (the Fishes). Then follows the rump of Cetus (the Whale) and, further south, Sculptor, Fornax (the Furnace) and Phoenix. In this region Reticulum (the Net) and Horologium (the Pendulum Clock) are also located. And winding from south to north is Eridanus (the River).

This completes the cycle, moving from west to east, and we are back at Taurus, our starting point. On the hump of Taurus is a beautiful asterism, the cluster of stars called the Pleiades (or Seven Sisters). Today the naked eye can make out only six stars. What could have happened to the seventh member? The Seven Sisters played an important role in the lives of primitive Man. They noticed that shrubs began sprouting after winters when the Seven Sisters were visible in the east just before dawn. When grains of corn were pushed into the soil at that time of the year, they germinated, grew and produced more corn or wheat. In this way Man first made use of the heavens to serve him as a calendar. He thereby took the first strides towards civilisation. This was a first step towards unveiling the Universe.

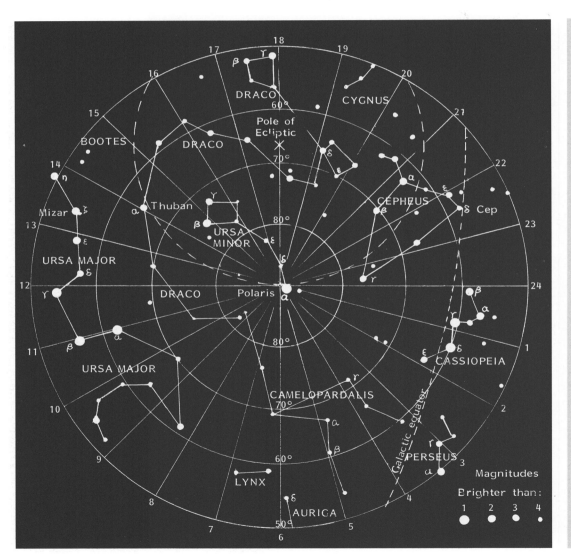

Figure 1.5
North-polar sky, Northern autumn

The Visible Night Sky

THE 360 degrees around the celestial equator are divided into 24 hours of right ascension, beginning with zero at the point where the Ecliptic cuts the equator. At present this point is situated on the western boundary of Pisces, or the eastern boundary of Aquarius. The forehead of Taurus is situated at about $4\frac{1}{2}$ hours right ascension. Taking this point as reference, and turning around so that the observer in the Northern Hemisphere faces north instead of south, we see the stars which are grouped around the projection of the Earth's North Pole.

North of Auriga and Perseus, the straggling stars of Camelopardalis (the Giraffe) spread towards the North Pole. Moving clockwise around the Pole, we find Lynxis, the stars of Ursa Major and the tail end of Draco. If a line is drawn from Beta through Alpha of Ursa Major and projected to the north, we come to the bright star Polaris (the Pole Star) which is situated at 89°02′, and thus within one degree of the North Pole. It therefore serves as a guide in finding direction. It was, however, not always the Pole Star. About 4500 years ago, the star Thuban, midway along the tail of Draco, was very close to the North Pole. This fact was known to the Egyptian priests of those days, in the time of the Pharaoh Cheops. Figure 1.5 shows Thuban on a broken circle which passes through the North Celestial Pole. This circle marks the path that the pole describes in a period of 25 765 years. The circle has a radius of $23\frac{1}{2}$ degrees,

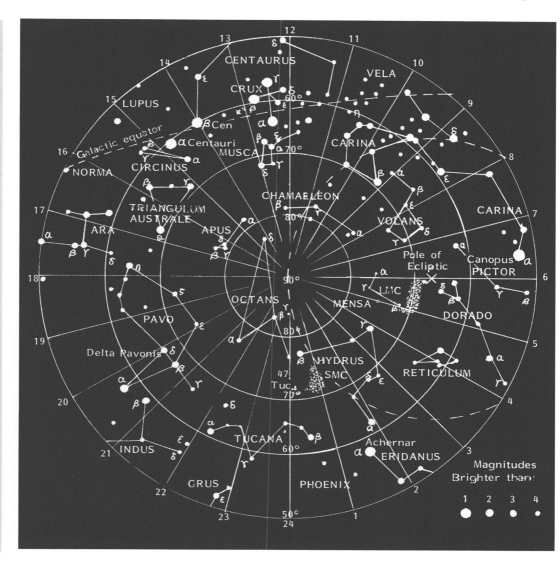

Figure 1.6
South-polar sky, Southern winter

which is the same as the inclination of the Earth's axis to the vertical on the Ecliptic plane in which the Earth revolves around the Sun.

Within the coils of Draco's tail we find Ursa Minor (the Little Bear), with Polaris at the tip of his tail.

Continuing clockwise past the head of Draco, we come to King Cepheus and then we find five or six bright stars in the shape of a flat W, known as Cassiopeia's Chair, on which Queen Cassiopeia sits.

ALTHOUGH the Southern Hemisphere has no pole star, it has many bright circumpolar stars. Starting in the southern winter, June, we see the beautiful Southern Cross standing erect, its long axis pointing to the position of the South Pole, 4½ cross-lengths downwards, past the stars of Musca (the Fly) and the Chameleon (Chamaeleon in Latin). Moving clockwise (Figure 1.6), we come to Carina (the Keel) which has much nebulosity around the star Eta Carinae and also contains the star Canopus (or Alpha Carinae), the second brightest star in the heavens. Nearby is Volans (the Flying Fish) and then Mensa (Table Mountain) and Dorado (the Goldfish). On the boundary between these constellations is the Large Magellanic Cloud, a galaxy outside the Milky Way. Then follows Hydrus (Sea Serpent) and Tucana (the Toucan) which contains the Small Magellanic Cloud. Closer to the Pole is Octans (the Octant), and further round, Pavo (the Peacock). The star Delta Pavonis is very similar to the Sun and has a high probability of having a habitable planet. Next we come to Apus (the Bird of Paradise) and Triangulum Australe (the Southern Triangle), followed by Circinus (the Pair of Compasses). Then comes Centaurus which contains the stars of Crux (Southern Cross), thus completing the circuit of the South Pole.

A list of all the names of the constellations is given in Appendix 2.

WHEN one looks at the starry skies through the night, one notices that the stars move systematically westwards. On average, it takes 12 hours from rising in the east to setting in the west, through 180 degrees. In 12 hours there are 720 minutes. Thus $720 \div 180 = 4$ minutes elapse for every 1 degree that the stars move westwards across the sky. This motion of the stars is due to the rotation of the Earth in the opposite direction, from west to east.

During a year (365.25 days) the Earth revolves once around the Sun, so that the direction of a given star changes by 360 degrees. This is almost 1 degree per day, which is equivalent to 4 minutes of time. A star will thus move westwards by 1 degree every day owing to the Earth's revolution around the Sun, and this will cause it to rise 4 minutes earlier each day. The movement of the stars from east to west proves that the Earth moves from west to east. In one month a star will move 30 degrees westwards. In six months any given constellation will move from the eastern horizon, at sunset, to the western horizon. As the seasons succeed each other – winter, spring, summer and autumn – the constellations of Figure 1.4 are followed in turn by those of Figures 1.3, 1.2 and 1.1. The relative positions of the stars and constellations remain unaltered. That is why the stars are called fixed. The reason their positions do not change is because they are so very far away. By means of a telescope, however, it can be seen that the stars actually do shift.

BESIDES the fixed stars, there are other bodies in the sky: the planets or wanderers. Five planets were known to ancient man: Mercury, Venus, Mars, Jupiter and Saturn. They were seen to move in a very narrow strip of the sky, called the Zodiac or animal strip, because eleven of the twelve constellations in this strip of sky depict animals: Pisces, Aries, Taurus, Gemini, Cancer, Leo, Virgo, Scorpius, Sagittarius, Capricornus and Aquarius, with Libra, the non-

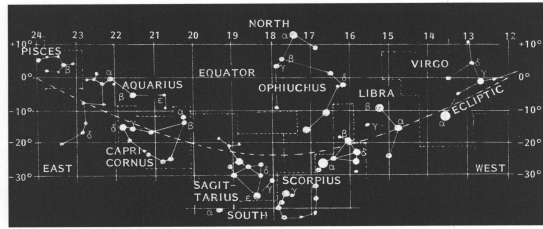

Figure 1.7
Sun's path on Ecliptic (right to left) September to March

The Visible Night Sky

animal, between Virgo and Scorpius (Figures 1.7 and 1.8). Because the tail of Scorpius curls away from the Ecliptic, the 20 degrees between the head of Scorpius and Sagittarius can be filled in by Ophiuchus which can be considered as the thirteenth constellation of the Zodiac.

THE ancient Greeks (and others) thought that the planets must be gods because they did not keep fixed positions in the starry sky. Venus, the brightest, was considered to be the goddess of beauty; Jupiter was the father of the gods; Mars, red in colour, had to be the god of war; Saturn, moving very slowly, was also called Chronos, the god of time; and swiftly-moving Mercury was rightly given the role of messenger of the gods.

The planets were usually seen to move from west to east against the background of the fixed stars, but there were times when they came to a halt, moved westwards for a while, came to a halt again and then moved eastwards again. This was a difficult phenomenon to explain but, of course, the gods could do as they pleased!

The Moon also moves in the narrow belt of the Zodiac because its plane of revolution is inclined at only $5°09'$ to the plane of the Ecliptic. It moves through 13 degrees per day; the area reflecting sunlight continually changes, thus giving rise to the different phases: full moon, half moon or crescent.

The Greeks noticed that the constellation Cancer was just above the western horizon when the Sun set in summer. The Sun must therefore have been in the constellation Gemini which is just west of Cancer. As summer wore on, the next constellation, Leo, sank lower and lower in the west, until it disappeared along with the Sun. Then Virgo followed suit, and so on right around the Zodiac in a period of one year. This showed that the Sun moves eastwards against the stellar background and occurs because the Earth revolves around the Sun. The path of the Sun against the stars is called the Ecliptic. In one year the Sun moves through 360 degrees. The most developed people, the Egyptian priests, took careful note of these phenomena, and they were the first to establish a calendar containing 365 days. They had also noticed that the flooding of the Nile took place annually just after the brightest star, Sirius, made its appearance in the east just before dawn.

DURING one year the Moon moves 13.36 times around the celestial sphere and thus passes the Sun 13 times a year. Both the Sun and the Moon have an angular diameter of very nearly $\frac{1}{2}$ degree. Since the Moon's orbit is inclined to the ecliptic by $5°09'$, the Moon must be precisely on one of the points of intersection of its orbit with the Ecliptic, in order that it may eclipse the Sun. These points are known as the nodes. The Moon need not be so close to a node in order to undergo a lunar eclipse when it passes into the Earth's shadow, because the Earth's shadow cone is considerably larger than that of the Moon. Lunar eclipses are thus more frequent and can be seen from a whole hemisphere of the Earth, whereas a total eclipse of the Sun can be seen only from the narrow strip within the umbra (darkest shadow) of the Moon. This strip is never wider than 269 km. The umbral spot of the Moon's shadow moves swiftly across the Earth's surface so that the longest duration of totality of a solar eclipse is about 7 minutes. The area covered by a partial eclipse is much larger and the duration can be 1 to 2 hours.

It is understandable that a total eclipse must have struck men with terror.

The Babylonians discovered that eclipses repeat every 18 years (actually 6585.32 days), a period called the "Saros". This knowledge enabled them to predict the occurrence of eclipses. The Greek, Thales of Miletus, knew about the

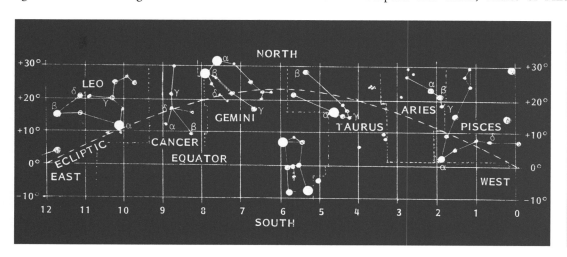

Figure 1.8
Sun's path on Ecliptic (right to left) March to September

Saros. During a war between the Medes and Lydians he predicted that an eclipse of the Sun would take place on a certain day in the year 585 BC. And it did! At that moment a battle was raging, but when the combatants saw the eclipse taking place, the battle ceased abruptly!

All in all, these natural phenomena posed difficult problems. Some very learned men, such as Ptolemy, went off on quite the wrong track. On the other hand, philosophers such as Aristarchus readily realised that the planets move around the Sun, that the Earth is also a planet, and that Earth, Sun and Moon are spherical in shape.

Chapter 2

The Dawn of Astronomy

Aristarchus

The first more or less correct view of the shapes and movements of the Earth, Moon, Sun and Planets, held by Western Man, was probably that of Aristarchus who lived on the Greek island of Samos from 320 to 250 BC. He must have known about the ideas of his countryman, Pythagoras, who preceded him by 250 years. But Pythagoras was more interested in pure geometry and the theory of numbers. By a simple application of geometry, Aristarchus was able to determine the relative sizes and distances of the Sun, Moon and Earth. He accepted the ideas of Anaxagoras (500–428 BC) about the phases of the Moon, namely that they are due to the Moon being spherical in shape. As it revolves around the Earth, the area of its surface which reflects

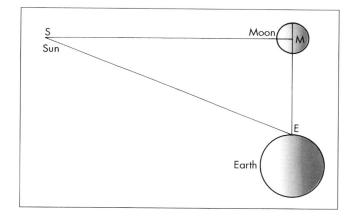

Figure 2.2 Relative distances of the Sun and Moon

sunlight is constantly changing (Figure 2.1). When the Moon is between the Earth and the Sun, position 1, it is invisible for a day or two, shown by a in the figure. After about $3\frac{1}{2}$ days, the Moon moves to position 2 and has the appearance of a thin crescent, b. After 7 days, it is at right angles to the direction of the Sun, at position 3, and has the appearance of a half Moon, c. After 10 to 11 days, it is in position 4 and has a gibbous shape, d. Half a revolution around the Earth brings the Moon to position 5 and it is seen as full Moon, e. Thereafter the phases reverse: gibbous at 6; half Moon at 7; crescent at 8; finally to disappear with the Sun behind it.

Aristarchus realised that the Moon would be at right angles to the direction of the Sun when its phase was that of a half Moon, as shown in Figure 2.2. Thus, if he could measure the angle SEM from his position E on Earth, then he could work out the ratio of the Sun's distance ES to the

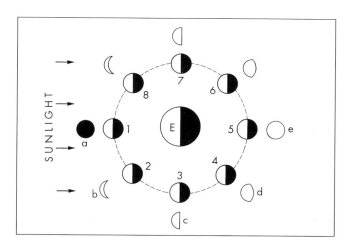

Figure 2.1 Phases of the Moon

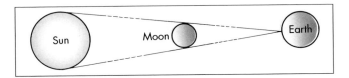

Figure 2.3 Sun and Moon appear equally large when viewed from Earth

Moon's distance EM. It is not known how he actually measured angle SEM. Without a telescope with graduated circles, he could not determine this angle with great accuracy. The value he obtained was 87 degrees. He was then able to draw the triangle to scale and measure the ratio between the lengths ES and EM. He found that ES was 20 times EM. The Sun's distance from the Earth was thus 20 times that of the Moon.

This was a remarkable discovery. Aristarchus had proved that space has depth and that the Sun and Moon do not move against a single crystal sphere, as was believed up to that time.

Aristarchus did not realise that he had tackled a very difficult problem. Lacking a chronometer, he had no way of knowing the exact moment of half Moon, and he could not possibly have obtained an accuracy of 1 degree in measuring the angle. We know today that the angle SEM is equal to $89°51'$. We can therefore calculate the ratio between ES and EM as follows:

$$\frac{ES}{EM} = 1 \div \frac{EM}{ES} = 1 \div \cos(89°51')$$
$$= 1 \div 0.00261799 = 382$$

(See Appendix 1, section 1.1 for definitions of sine (sin), cosine (cos), tangent (tan), etc. and how to find their values with a pocket calculator.)

According to this calculation, the Sun is nearly 400 times further from the Earth than is the Moon. Aristarchus did not have cosine tables, but nevertheless his method was scientifically correct.

Because the Sun is very much further away from the Earth than is the Moon, the Sun must be very much larger than the Moon; however, they appear to be about the same size when viewed from Earth, as shown in Figure 2.3. Aristarchus then posed the question: how does the size of the Earth compare with that of the Moon? It was clear to him that a lunar eclipse takes place when the Moon moves into the shadow of the Earth (Figure 2.4).

When the arc ABC of the Earth's shadow is completed to describe the circle ABCDA, the diameter of this circle will be equal to the Earth's diameter, according to Aristarchus.

Once again he had encountered a difficult problem. Firstly, the Earth's shadow on the Moon is not sharply defined, owing to refraction of the Sun's rays by the Earth's atmosphere. Secondly, the diameter of the Earth's shadow at the distance of the Moon will be less than its actual diameter, since the Sun's rays which graze the surface of the Earth form a cone in space. The value which Aristarchus found for the Earth's diameter, twice that of the Moon, was too small. The actual value is now known to be 3.7 times. According to Aristarchus, the Sun was 20 times larger than the Moon and therefore the Sun had to be 10 times larger than the Earth. In fact, the diameter of the Sun is 109 times the diameter of the Earth.

The importance of Aristarchus' work rested on the fact that man had proved, for the first time, that the Earth, although obviously larger than the Moon, was relatively insignificant compared with the Sun. It was therefore reasonable to assume that the smaller body, the Moon, revolves around the Earth, and this could easily be seen to be so. But it was not as easy to accept the idea that the Earth, being smaller than the Sun, must revolve around the Sun. Archimedes, who was somewhat younger than Aristarchus, mentions in his writings that Aristarchus had held that the Earth revolves around the Sun, along with the five planets.

Aristarchus also correctly believed that the stars remain fixed because they are so infinitely far away from the Earth.

Eratosthenes

Some sixty or seventy years later, Eratosthenes (276–194 BC) made another major step forward. For most of his long life he was librarian at Alexandria. Besides being a philosopher and scientist, he was also a very good athlete!

Eratosthenes discovered that when the Sun was furthest

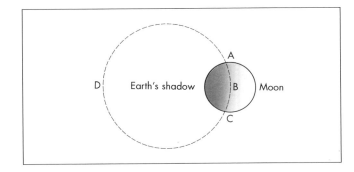

Figure 2.4 Earth's shadow eclipses the Moon

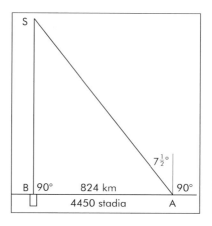

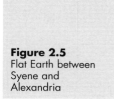

Figure 2.5 Flat Earth between Syene and Alexandria

north, on the day of the solstice in mid-summer, it was vertically above the town of Syene (the present Assuan) on the Nile, so it cast no shadows on the water in a well at that place when it was mid-day. At the same time a pole planted vertically at Alexandria, 4450 stadia to the north of Syene, cast a shadow which made an angle of $7\frac{1}{2}$ degrees with the vertical. On all other days, the walls of the well at Syene formed shadows on the water, and the shadow of the pole at Alexandria made an angle greater than $7\frac{1}{2}$ degrees.

It is not known exactly how Eratosthenes measured the distance of 4450 stadia (= 824 km) but it must be borne in mind that he is known as the founder of geography.

We can place ourselves in Eratosthenes' shoes and reason as follows: when the Sun, S (Figure 2.5) is vertically above Syene B, the angle SBA = 90 degrees, where A is Alexandria, 4450 stadia away. This is so because the Earth is flat, as everybody can see. At the same time, the Sun's rays make an angle of $7\frac{1}{2}$ degrees with the top of a vertical pole at Alexandria. Therefore the angle SAB had to be $82\frac{1}{2}$ degrees.

We can then easily construct the triangle SBA, where S is the Sun, B Syene and A Alexandria, 4450 stadia from B. We would then have been able to say: distance SB divided by distance BA is the tangent of the angle $82\frac{1}{2}$ degrees (see Appendix 1, section 1.1). Now, tan $82\frac{1}{2}$ degrees = 7.596, therefore SB ÷ BA = 7.596; hence SB = BA × 7.596 = 4450 × 7.596 = 33 800 stadia. We should then have been able to say triumphantly: "You see, the Sun is very far above the Earth, exactly as Aristarchus had said!" In those days, very few people would have had an idea of what a distance of 33 800 stadia really meant.

But Eratosthenes did not fall into the trap of assuming that the Earth is flat. He had sufficient proof to assume that the Earth is spherical, and he argued as follows: the Sun is so far away that its rays can be considered as being parallel when they reach Earth. Therefore, although the Sun's rays are vertical at B, the angle of $7\frac{1}{2}$ degrees between the rays and the vertical pole at Alexandria occurs because the surface of the Earth is curved (Figure 2.6). This angle of $7\frac{1}{2}$ degrees will therefore be equal to angle BOA at the centre of the Earth, since lines S_1O and S_2A are parallel. Arc AB of the Earth's surface (= 824 km) subtends an angle of $7\frac{1}{2}$ degrees at the Earth's centre. In the 360 degrees around the Earth's circumference, there are exactly $48 \times 7\frac{1}{2}$ degrees. Therefore the circumference of the Earth must be $48 \times 4450 = 213\,600$ stadia or 39 552 km. Today's most accurate value for the circumference through the North Pole is 40 009 km! Eratosthenes was out by only 457 km, an error of only 1.1%.

Actually, he was very lucky that the angle was $7\frac{1}{2}$ degrees and thus easily measurable. If he had found the angle to be 7.4 degrees, his answer would have been 40 086 km. The important point is, however, that his method was scientifically correct. If he had determined the angle between the Sun's rays and the vertical at a third point, he would have had proof positive.

ERATOSTHENES also succeeded in measuring the angle between the Earth's equator and the Ecliptic, the plane in which the Earth revolves around the Sun. His value was very close to the correct value of $23\frac{1}{2}$ degrees.

It is this inclination of the equator to the Ecliptic that causes the seasons. If the angle between the equator and the Ecliptic were zero, there could be no seasons because the Earth's axis, around which it rotates, would then be at right angles to the Sun's rays. There would then be a hot strip on

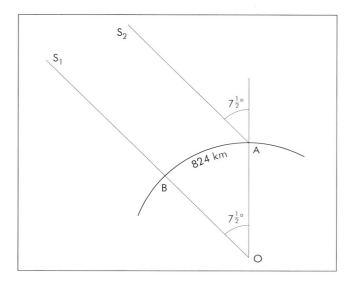

Figure 2.6 Arc BA subtends an angle of $7\frac{1}{2}°$ at Earth's centre O

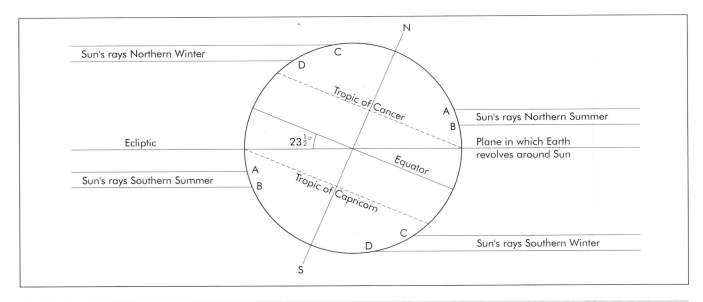

Figure 2.7 The seasons, caused by the inclination of the equator to the Ecliptic

both sides of the equator, with subsequent strips becoming progressively cooler as they were further removed from the equator. These same temperature differences would hold throughout the year.

Figure 2.7 shows bundles of rays of sunshine of equal width falling on the surface of the Earth at latitudes of, say, 40 degrees north and south. During summer, the bundle heats an area represented by AB, and during winter a larger area represented by CD. That is why it is colder in the winter and warmer in the summer.

During the northern summer, the Sun reaches a greatest displacement of $23\frac{1}{2}$ degrees when it is vertically above the Tropic of Cancer; this happens on June 21. In the southern summer, the Sun moves to a position vertically above the Tropic of Capricorn, on 22 December. The tropics are $23\frac{1}{2}$ degrees from the equator.

THE ancient Greeks knew that the Moon takes $29\frac{1}{2}$ days between one new Moon and the next – that is, when it is in the same position relative to the Sun and Earth as before. To move once around the sky and to come back to the same position with regard to the stars took only $27\frac{1}{3}$ days. The first period is called the Moon's synodic period, and the latter its sidereal period. The Greeks later found that the sidereal period of Mars was 687 days, of Jupiter 12 years and of Saturn 30 years. These three planets therefore had to be much further away than the Moon. Venus and Mercury, on the other hand, did not move right across the sky, but oscillated to and fro about the Sun. Venus never moved more than 47 degrees from the Sun and Mercury never more than 28 degrees. It was very difficult to explain these phenomena. A good guess was that they were nearer to the Sun than the Earth. With the passage of time, even ages, the positions of the stars remained unaltered. They were thus much further from the Earth than the planets.

Hipparchus

Antique Greek astronomy reached its pinnacle with Hipparchus, who lived on the island of Rhodes in the second century BC. His interest in astronomy was fired by the sudden appearance of a bright "new star". This spurred him on to compile a star atlas, in which he registered 1000 stars. He very painstakingly measured the positions of the stars to an accuracy of one-fifteenth of a degree.

He found that the positions of the stars on his charts differed from those on older charts by small amounts in a systematic way. He deduced the brilliant idea that this was due to a slow westward movement of the point where the equator intersects the Ecliptic – a movement of 40 arc seconds per year. This in its turn was caused by a slow westward gyration of the axis around which the Earth spins. Such a gyration would describe a circle centred on the pole of the Ecliptic and having a radius of $23\frac{1}{2}$ degrees, which is equal to the inclination of the Earth's axis from the vertical

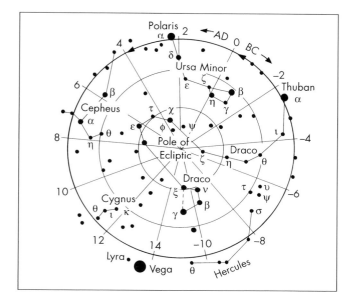

Figure 2.8 Precession of the Earth's axis of rotation

spinning top, except that the axis of a spinning top gyrates forwards while the Earth's axis gyrates backwards. This is because the spinning top is subject to the downward pull of the Earth's gravity, while the spinning Earth is subject to the sideways pull of the gravities of the Moon and the Sun in the plane of the Ecliptic (Figure 2.9).

Because the Earth's equatorial diameter is greater than its polar diameter, there is more mass in the bulge at the equator, which is tilted to the Ecliptic plane by $23\frac{1}{2}$ degrees. Thus the gravitational forces of the Moon and Sun tend to pull this excess mass towards the plane of the Ecliptic. In order to conserve its angular momentum, the Earth's axis gyrates backwards, towards the west, a movement called precession. Such movement can be demonstrated by means of a gyroscope. The effect of precession is that the point where the equator cuts the Ecliptic, the First Point of Aries, or Gamma, γ, moves westwards. In the days of Hipparchus, this point was situated on the western boundary of the constellation Aries. In the 2000 years that have elapsed since then, this point has shifted westwards through the constellation of Pisces, and is now on the boundary of Aquarius.

on the Ecliptic. At 40 arc seconds per year, the circle would be completed in 32 400 years. We now know that the precession of the Equinox (the point where the equator and Ecliptic intersect) is actually 50.3 arc seconds per year. The Earth's axis of rotation will thus describe a full circle in 25 765 years. This circle is shown in Figure 2.8.

In the period between 2800 and 2400 years ago (in the days of the 4th Dynasty of the Pharaohs), the Earth's axis of rotation pointed close to the star Thuban (Alpha Draconis), which was then the Pole Star.

The Earth's axis of rotation behaves like that of a

HIPPARCHUS measured the length of the day to an accuracy of six minutes. He found that the length of the summer half year in the northern hemisphere, from 21 March to 23 September (the equinoxes), was $7\frac{1}{2}$ days longer than the winter half year from September to March.

By studying the positions of the stars that became visible just after sunset, he found a variation in the amounts that the stars shifted. This enabled him to plot the positions of the Sun against the celestial sphere (Figure 2.10). M is the equinox in March and S in September; J is the solstice in June and D in December. O is the Earth, from where the observations were made.

The Sun thus took $7\frac{1}{2}$ days longer to move from M through J to S than it took from S through D to M. This

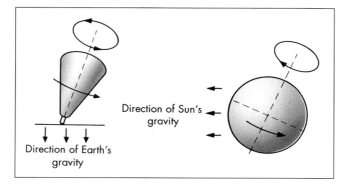

Figure 2.9 Precession of axes of spinning top and Earth

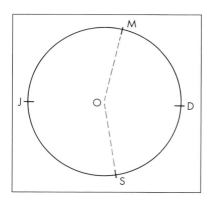

Figure 2.10 Eccentricity of the Sun's positions during the year

meant that the eccentricity of the Sun's positions was $7.5 \div 365.25 = 0.02$.

Today we know that this eccentricity is due to the eccentricity of the Earth's orbit around the Sun. The orbit is almost a circle, but not quite. The eccentricity is now known to be 0.017, so that the length of the northern half year is 6.25 days longer in the summer than in the winter. Hipparchus' measurements gave results very close to today's accepted values.

Hipparchus also classified the stars according to their brightnesses. The brightest stars were classified as magnitude 1, and the faintest that the eye could detect as magnitude 6, which was about 100 times fainter than magnitude 1. This scale has formed the basis of the modern magnitude scale. Hipparchus' work, including his catalogue, also formed the main body of the Almagest, the sole authoritarian work on astronomy over the next 1600 years.

Hipparchus additionally observed that the eccentricity of the Moon's orbit around the Earth varies periodically.

For 1600 years after Hipparchus, no noteworthy progress was made in astronomy. The world had to wait until 1543 when Copernicus' book on the Solar System was published, before any real progress was made. Publications such as the Alphonsine Tables, circa 1272, brought planetary positions up to date, but no new ideas were developed. About 1440, Ulugh Beg compiled a new catalogue of stars, working from his observatory in Samarkand, at that time situated in Persia.

As a result of the contributions made by the ancient Greeks, Aristarchus, Eratosthenes and Hipparchus, it would have been possible over 2000 years ago to have made the following statements regarding the Earth and other celestial bodies:

1. The Earth, Sun and Moon are spherical in shape.
2. The Earth spins on its axis and, in so doing, causes day and night.
3. The circumference of the Earth is very nearly 40 000 km.
4. The Earth is larger than the Moon.
5. The Sun is larger than the Earth.
6. The Sun's distance from the Earth is much greater than the Moon's distance.
7. The Earth and planets revolve around the Sun.
8. The orbital planes of the planets lie very close to the Ecliptic.
9. The planets Mars, Jupiter and Saturn are further from the Sun than is the Earth.
10. Venus and Mercury are nearer to the Sun than is the Earth.
11. The stars are much further away from the Earth than are the Moon, Sun and Planets.
12. Space has depth, and the celestial bodies do not revolve around the Earth against a crystal sphere.
13. The Earth's equator is inclined to the plane of the Ecliptic by $23\frac{1}{2}$ degrees.
14. The axis of rotation of the Earth precesses by 40 minutes of arc per year.
15. The Sun's position against the background of the stars shows a measure of eccentricity.
16. The Moon's orbit around the Earth is eccentric.

Chapter 3
The Formulation of Laws

Copernicus

The next step in deciphering the secrets of the Universe was made by Nicolaus Copernicus (1473–1543). Although a priest in the Roman Catholic Church, he spent many years studying the movements of the Sun, Moon and planets against the background of stars.

Copernicus (Figure 3.1) came to the following conclusions:

1. The Earth is spherical. His proof for this was the fact that the north celestial pole rose higher and higher the further one moved northwards; and the opposite held for movement southwards.
2. The Earth spins from west to east, and the atmosphere moves with the surface; if this were not so, there would be a continual wind blowing from east to west. The spin of the Earth has the effect of making the Sun, Moon and stars rise in the east and set in the west, the opposite direction to the Earth's spin.
3. The Earth is not the centre of the Universe; the Sun is the centre, and the Earth and planets revolve in circles around the Sun. The centres of the circular orbits do not all coincide with the centre of the Sun. (This erroneous explanation had been proposed to explain the changes in the speeds with which the planets move against the starry background.)
4. When the outer planets, Mars, Jupiter and Saturn, are in opposition, that is, when their directions are opposite to that of the Sun, they apparently come to rest, then move backwards for a while before coming to rest once more and moving east again. This appears to be the case because the Earth overtakes them on the inside.
5. Since the Earth describes a circle around the Sun, moving from west to east, the stellar background must move in the opposite direction, from east to west – and this is exactly what is observed.
6. When at opposition, the outer planets are much brighter than at other times because they are then nearer to the Earth.
7. Since the retrograde motion of Mars is larger than that of Jupiter, which in its turn is larger than that of Saturn, this implies that Mars is nearer to the Sun than Jupiter, and Jupiter is nearer than Saturn. The sizes of the retrograde loops are effects of perspective.
8. The sidereal periods of the planets, that is, the times they take to return to the same positions relative to the stars, are indications of their distances from the Sun. Mars takes 687 days, Jupiter almost 12 years and Saturn 30 years. Therefore Saturn is the furthest from the Sun, then Jupiter, and Mars is the nearest of the three. The Earth, having a period of 365 days, is nearer to the Sun than is Mars. Mercury and Venus never travel

Figure 3.1
Copernicus

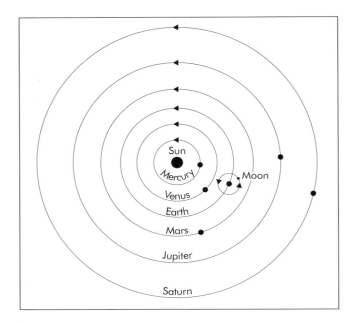

Figure 3.2 The Copernican system of revolving planets

Figure 3.3 Giordano Bruno – burned alive at the stake

completely round the sky, since they move in circles within the circular orbit of the Earth; their sidereal periods are 88 and 225 days respectively. Therefore Mercury must be nearer to the Sun, and Venus further away but still nearer than Earth. Since Mercury and Venus moved in circles within that of the Earth, they should show phases, like the Moon. The fact that this could not be observed by the naked eye did not deter Copernicus from holding that view (see Figure 3.2).

BY 1530 Copernicus had completed his writing, but he continually revised and edited the work. His friend Rhaeticus, who was Professor of Mathematics at Wittenberg, visited him in 1539 and stayed there for two years. In 1541 he published a résumé of Copernicus' work under the title *Narratio Prima* (First Report) which ran to 38 pages. Attacks were launched against it by the Roman Catholic Church as well as by Martin Luther, Melancthon and Calvin.

In the meantime, Copernicus' manuscript was in the hands of Osiander who was a Lutheran churchman. He took it to a publisher, Petrejus at Nurnberg. Aware of the attacks which had been made on the First Report, Osiander added a preface to the effect that the ideas in the book were not to be taken as absolute truth, but merely as a mathematical method for calculating planetary positions. The book, entitled *De Revolutionibus Orbium Coelestium* (On the Revolutions of the Celestial Spheres) was published in 1543, just two months before Copernicus' death on 24 May 1543. Copernicus thus may never have seen the preface – if he had, it would probably have hastened his death!

The opposition of the Church was something that had to be taken into account, considering the fate that befell Giordano Bruno (Figure 3.3), 57 years after the publication of Copernicus' book. Bruno was a very keen supporter of Copernicus' views and he delivered lectures in France, England and Italy. He also stated that the planets were worlds, like the Earth. In 1660 he was charged before the Inquisition and, after a summary trial, was burned alive at the stake!

Copernicus acknowledged the contributions made by Aristarchus and stated that stars show no annual motion due to parallax because they are so very far away from Earth.

Copernicus was the first to give the correct explanation of the fact that Venus and Mercury do not move right round the sky, but oscillate east and west of the Sun through 47 and 28 degrees respectively. This is so because their orbits lie within the Earth's orbit.

Venus was always to be found inside the angle VEW (Figure 3.4): the angles SEV and SEW were never greater than 47 degrees, with a mean of 46.5 degrees. When Venus was in the direction EV it was at its greatest eastern elongation, and when it was in the direction EW it was at its greatest western elongation. These positions prevail in the evenings and mornings respectively. The same explanation holds for Mercury: it is at greatest eastern elongation in the evening when in the direction EM; and at greatest western elongation before dawn when in the direction EN. Mercury is difficult to observe because it is always near to the Sun in direction and its sojourn at elongation is not very long. Venus, on the other hand, has a long period at elongation.

COPERNICUS' explanation of the retrograde motion of the outer planets at and near opposition is illustrated in Figure 3.5. All the planets move in the same direction from west to east around the Sun. Mars, for example, moves from the direction E_1M_1, seen against the sky at position 1, to position 2 along the direction E_2M_2. Here it appears to come to rest, before moving backwards to position 3, as seen along line of sight E_3M_3. It continues to move backwards to 4, seen along line of sight E_4M_4, and then to 5, seen along line of sight E_5M_5, where it appears to come to rest again. From there onwards to position 6 (line of sight E_6M_6), and beyond, the eastward motion continues up to the next opposition.

In the case of Mars, 780 days elapse between one opposition and the next; in Jupiter's case 399 days, and for Saturn 378 days (Figure 3.6). If Mars is in opposition at M_1, it will be in opposition again at M_2 after having moved right round its orbit plus the distance M_1M_2. The Earth will have completed two orbits plus the distance E_1E_m, because 780 days = 2 years plus 49.5 days. Jupiter covers the distance J_1J_2 while the Earth completes one orbit plus the distance E_1E_j. Saturn moves from S_1 to S_2 while the Earth describes one orbit plus distance E_1E_s.

The speeds at which the planets move in their orbits can be calculated only after the shapes and sizes of their orbits and their distances from the Sun are known. The Earth's distance from the Sun is therefore of fundamental importance. The inclinations of the planetary orbits to the plane of the Ecliptic could be determined to a good degree of accuracy, because the path of the Sun against the

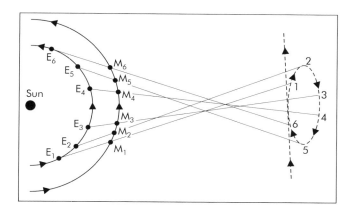

Figure 3.5 Apparent retrograde motion of Mars

background of the stars was well known. Of the five known planets, Mercury's orbit had the greatest angle of inclination to the Ecliptic, 7 degrees. All the other orbits have small angles of inclination: Venus 3°23′, Saturn 2°29′, Mars 1°51′ and Jupiter 1°18′.

Tycho Brahe

The first accurate determinations of planetary positions were made by Tycho Brahe (1546–1601). Brahe did not accept Copernicus' system of planets revolving around the Sun. He held that if the Earth revolved around the Sun, the nearer stars would describe small circles against the

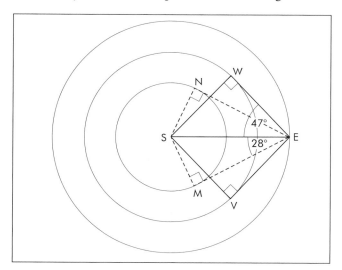

Figure 3.4 Eastern and western elongations of Mercury and Venus

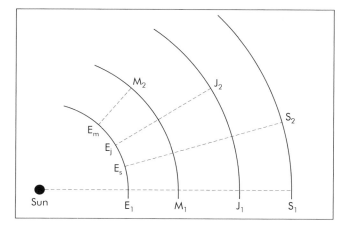

Figure 3.6 Successive oppositions of the outer planets

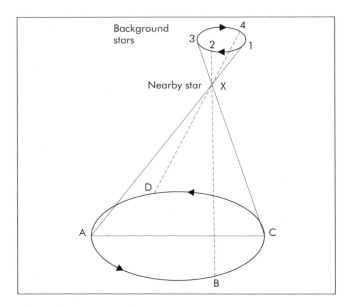

Figure 3.7 Parallax of a nearby star

the movement due to parallax. Copernicus held that the stars would show no effect of parallax because they are so distant. The parallax that a nearby star would show against the background of the distant stars is illustrated in Figure 3.7. When Earth is at A, the nearby star X will be seen at point 1 against the background stars. When Earth is at B, the star will be seen at point 2, and so on: C corresponding to point 3 and D to 4. The circle (or rather, ellipse) described by the star will be a miniature of the Earth's orbit around the Sun. Brahe's argument was correct but his instruments, although the best in existence, were not accurate enough to measure the angle of parallax of the nearer stars. His quadrant could measure to one-tenth of a degree: the line across his quadrant in Figure 3.8 stands at 23.7 degrees. We know today that the diameter of the ellipse formed at the nearest star is only 1.5 arc seconds, which is 240 times smaller than Brahe could measure.

Brahe argued that if Venus and Mercury revolved in orbits within that of the Earth, these planets should show phases, like the Moon, and these could not be seen. According to his system, the Moon, Sun, Mars, Jupiter and Saturn revolved around the Earth, while Mercury and Venus revolved around the Sun. Even if this was the case, Mercury and Venus would still show phases when they came between the Sun and Earth! If only Brahe had had a telescope, he would have been able to see the phases of Venus. Figure 3.9 is a composite by E.C. Slipher of Lowell

background of the more distant stars, that is, they would show the effects of parallax. Parallax can be illustrated by holding the arm outstretched and raising the forefinger. If the finger is looked at alternately with the left and right eyes, it will be seen that the finger appears to jump from left to right and right to left against the background. The greater the distance of the object looked at, the smaller is

 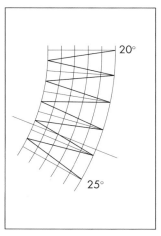

Figure 3.8 Tycho Brahe and his quadrant subdivided into tenths of a degree

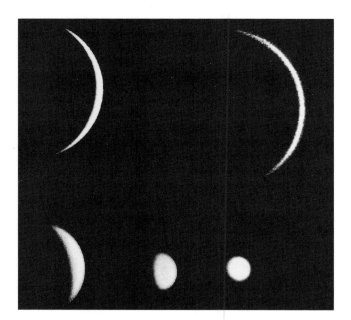

Figure 3.9 The phases of Venus

Observatory. It shows Venus at constant magnifications: bottom right, as a small disc when Venus is on the far side of the Sun; towards the left, becoming apparently larger and larger as it moves towards elongation – firstly a gibbous phase, then a crescent, and finally the top two crescents formed when Venus is almost exactly in line between the Sun and the Earth.

However, Brahe rendered astronomy a great service by the mass of accurate measurements he made of planetary positions, especially of Mars. These enabled Johannes Kepler (Figure 3.10) to formulate the first mathematical laws of planetary motions.

Johannes Kepler

For a year or two Johannes Kepler (1571–1630) worked with Tycho Brahe, after Brahe left Denmark and went to Prague. They had acrimonious arguments about the Copernican system, which Kepler supported wholeheartedly.

When Brahe died in 1601, Kepler saw to it that he obtained Brahe's measurements of planetary positions. By analysing these he was able to prove that the orbit of Mars is not a circle but an ellipse.

Kepler worked out the method for determining the relative distances of the planets from the Sun in terms of the Earth's distance from the Sun.

Using Brahe's measurements, he found that, on average, 105.5 days elapsed between the opposition of Mars when Sun, Earth and Mars are in a straight line, S–E–M (Figure 3.11), and the position of quadrature when the angle Sun, Earth, Mars, SE'M' is a right angle and therefore equal to 90 degrees.

In 105.5 days, while the Earth moves from E to E', it moves through

$$105.5 \div 365.25 \times 360 = 104°$$

While the Earth moves through 104 degrees, Mars moves from M to M', through

$$105.5 \div 687 \times 360 = 55°$$

The value 687 is the length of Mars' year, that is, the time that Mars takes to revolve once around the Sun and to return to the same position relative to the background of the stars. This value, Mars' sidereal period, was well known from the measurements made by Brahe.

The angle E'SM' must therefore be equal to $104 - 55 = 49$ degrees.

Since the sum of the angles of a triangle equal 180 degrees, the angle E'M'S could be found:

$$\text{angle } E'M'S = 180 - (49 + 90) = 41°$$

Figure 3.10 Johannes Kepler

In triangle M'E'S:

$$\sin E'M'S = \frac{E'S}{M'S} \quad \text{(see Appendix 1, section 1.1)}$$

therefore $$M'S = \frac{E'S}{\sin E'M'S} = \frac{E'S}{\sin 41°}$$

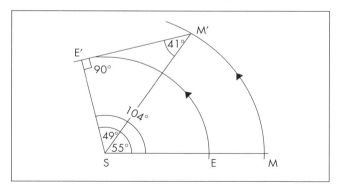

Figure 3.11 Mars in opposition, S–E–M; and at quadrature, angle SE'M'

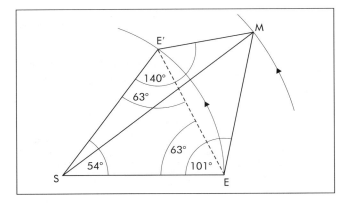

Figure 3.12 Mars seen from Earth before and after one sidereal period of Mars

Because the value of the Earth's distance from the Sun, E'S, was unknown, Kepler set it equal to 1 astronomical unit (1 AU). Thus:

$$M'S = \frac{1}{\sin 41°} = \frac{1}{0.656} = 1.524 \text{ AU}$$

Kepler thus found that Mars was 1.524 times as far from the Sun as the Earth is from the Sun – that was a remarkable achievement. (Ten to twenty years before Kepler tackled this problem, Jean Neper (Napier) invented logarithms and that enabled Kepler to calculate the values of sines, cosines, etc. of angles.)

As a check, Kepler also used the directions Sun–Earth–Mars for two successive positions of Mars after the lapse of one sidereal period of Mars. During that period of time the Earth would have gone twice around the Sun plus the arc E–E' (Figure 3.12).

From the Earth's first position at E, the angle SEM was 101 degrees. From the Earth's second position at E', the angle SE'M was found to be 140 degrees and the angle E'SE was 54 degrees.

Kepler reasoned as follows: considering the Earth's orbit as a circle, which it very nearly is, and which it is if average distances are taken, SE will be equal to SE', so triangle SEE' is isosceles and thus angle SEE' will be equal to angle SE'E. Each of these angles will be equal to

$$\tfrac{1}{2}(180 - 54) = 63°$$

By applying the sine rule (Appendix 1, section 1.2), which states that the sides of a triangle divided by the sines of the opposite angles are equal, Kepler could say:

$$\frac{E E'}{\sin ESE'} = \frac{ES}{\sin SE'E}$$

from which it follows that:

$$\frac{E'E}{\sin 54°} = \frac{1}{\sin 63°}$$

so that

$$E'E = \frac{1 \times \sin 54°}{\sin 63°} = \frac{0.809}{0.891} = 0.908$$

Because angle SEE' = angle SE'E = 63°:
angle E'EM = 101 − 63 = 38°, and
angle EE'M = 140 − 63 = 77°.

$$\therefore \text{ angle E'ME} = 180 - (38 + 77) = 65°$$

In triangle E'EM:

$$\frac{E'M}{\sin E'EM} = \frac{E'E}{\sin E'ME}, \text{ i.e. } \frac{E'M}{\sin 38°} = \frac{E'E}{\sin 65°}$$

Thus

$$E'M = \frac{E'E \sin 38°}{\sin 65°} = \frac{0.908 \times 0.616}{0.906}$$

that is $\quad E'M = 0.617$ AU.

By applying the cosine formula (Appendix 1, section 1.6) to triangle E'SM:

$$SM^2 = E'S^2 + E'M^2 - 2(E'S)(E'M)\cos 140°$$
$$= 1^2 + (0.617)^2 - 2(1)(0.617)(-0.766)$$
$$= 1.3807 + 0.9452 = 2.3259$$

Therefore

$$SM = \sqrt{2.3259} = 1.525 \text{ AU}$$

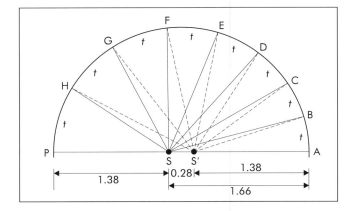

Figure 3.13 The radius vector sweeps out equal areas in equal times, *t*. Speed increases from A to P

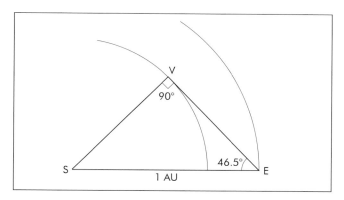

Figure 3.14 Venus at greatest elongation

This value is the same as Kepler obtained from the quadrature calculation. Kepler found, however, that the distance of Mars from the Sun varied between 1.38 and 1.66 AU. This showed, beyond doubt, that the orbit of Mars was not a circle. It had to be an ellipse with an eccentricity equal to $(1.66 - 1.38) \div (1.66 + 1.38)$, i.e. $0.28 \div 3.04 = 0.092$.

When Kepler plotted the directions from the Sun to Mars, S–A, S–B, S–C, etc. (Figure 3.13) having equal time intervals each equal to t, the areas of the triangles BSA, CSB, DSC ... PSH were found to be equal. Angles BSA, CSB ... increase gradually as Mars moves from A, its aphelion, when it is furthest from the Sun, to P, its perihelion, where it is closest to the Sun. This means that the angular distances AB, BC ... increase because the speed of Mars in its orbit gradually increases as it moves from aphelion to perihelion. After passing perihelion, the reverse takes place until Mars reaches aphelion again.

In order to calculate the areas of the triangles, Kepler used the values of the angles BSA, CSB ... and the circular measures AB, BC ... . The area of triangle ABS, for example, is equal to

$\frac{1}{2}(AB)^2 \div \sin BSA$ (Appendix 1, section 1.9)

The finding that the straight line joining the planet to the Sun sweeps out equal areas in equal times is a cardinal property of an ellipse.

When Kepler used a second focus, S′, equally far from the centre O as the first focus, S, he found that the sums of the distances BS + BS′, CS + CS′, etc. are constant. The definition of an ellipse is as follows: an ellipse is the locus of a point which moves in such a way that the sum of its distances from two fixed points, the foci, remains constant. In the case of a circle, the two foci coincide.

The eccentricity of an ellipse is the quotient obtained when the distance between the foci is divided by the major axis. The eccentricity of the orbit of the planet Mars is 0.092.

KEPLER found that the same property held in the cases of the other planets. In 1609, he was therefore able to formulate the first two of his laws of planetary motion:

> ### KEPLER'S FIRST TWO LAWS
> 1. The planets move in orbits which are ellipses having the Sun as one focus.
> 2. The radius vector from the Sun to the planet describes equal areas in equal intervals of time.

To determine the relative distances of the planets Mercury and Venus, Kepler used the angles subtended by the planets and the Sun when the planets were at greatest elongation, so that the angle Earth–Planet–Sun was a right angle (Figure 3.14).

The distance of Venus from the Sun, VS, in terms of the Earth's distance, ES, can be calculated as follows:

$VS \div ES = \sin SEV$ (Appendix 1, section 1.1)

i.e. $VS = ES \times \sin SEV$
$= 1 \times \sin 46.5°$
$= 0.725$ AU

The distance from Venus to the Sun is therefore 0.725 AU. The distance of Mercury was calculated to be 0.387 AU.

As with the case of Mars, Kepler also worked out the relative distances of the inner planets, Mercury and Venus, by measuring the necessary angles subtended by the planets and the Sun with an interval of one sidereal period of the planet (Figure 3.15).

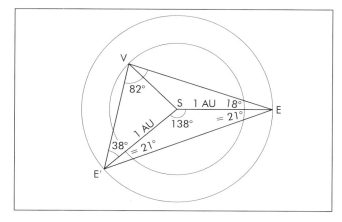

Figure 3.15 Venus as seen before and after one sidereal period of Venus

When the Earth was at E, the angle between Venus, V, and the Sun was 18 degrees. After one sidereal period of Venus, the Earth had moved to E'. Angle VE'S was then 38 degrees; angle E'SE was 138 degrees and angle E'VE 82 degrees. These data were sufficient to work out the relative distance of Venus.

By means of the sine rule applied to triangle ESE', the length of EE' works out at 1.867 AU; and when the sine rule is applied to triangle EVE', the length of EV works out at 1.616 AU. Applying the cosine formula to triangle ESV, the length of VS comes to 0.733 AU, in fair agreement with the value of 0.723 AU which is today's accepted value of the relative distance of Venus from the Sun.

Table 3.1

Planet	Period		Distance
	Days	T (years)	R (AU)
Mercury	88	0.241	0.387
Venus	225	0.616	0.725
Earth	365.25	1.000	1.000
Mars	687	1.881	1.524
Jupiter	4333	11.86	5.2
Saturn	10760	29.46	9.54

THE eccentricity of Venus' orbit was calculated as 0.0068, which makes it the closest to a perfect circle of all the planets. The eccentricity of Mercury's orbit is 0.206. Mercury's aphelion distance from the Sun is therefore $\frac{1}{2}(1 + 0.206) = 0.603$, and its perihelion distance $\frac{1}{2}(1 - 0.206) = 0.397$.

Therefore, at aphelion, Mercury is $0.603 \div 0.397 = 1.5$ times further from the Sun than at perihelion.

Table 3.1 contains the periods of revolution around the Sun, and distances from the Sun of the planets known to Kepler.

The last two columns contain the periods (in years) and the distances from the Sun, taking the Earth's values as equal to 1.

This table revealed the fact that longer periods fall together with greater distances. Kepler wondered whether there could be any linear relationship between periods and distances. For nine years he constantly improved the data and, in the mean time, he had to defend his 69 year old mother who had been charged with witchcraft!

Eventually, in 1618, Kepler published his findings. These showed there was a mathematical relationship between the squares (2nd powers) of the periods and the cubes (3rd powers) of the distances.

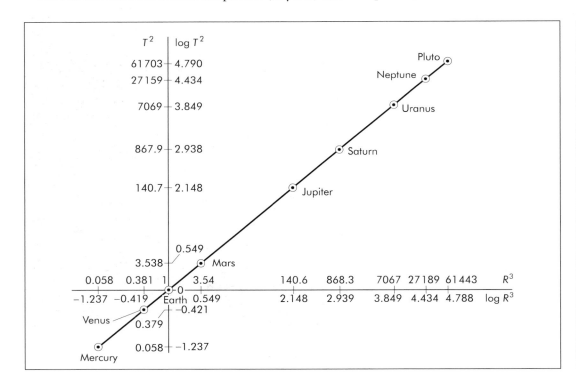

Figure 3.16 Straight line graph is proof of the proportionality between T^2 and R^3

Table 3.2 demonstrates this astonishing discovery. The calculations are performed as follows: the period of Mercury for one revolution around the Sun has a mean value of 0.241 years ($= T$). The square of this, $T^2 = 0.241 \times 0.241 = 0.058$. Mercury's average distance from the Sun is 0.387 AU ($= R$). The cube of this is $0.387 \times 0.387 \times 0.387 = 0.058 = R^3$.

Kepler had discovered that the squares of the periodic times are equal to the cubes of the average distances, namely $T^2 = R^3$, when the periods and distances are expressed in terms of those of the Earth, taken as unity.

Thus $T^2 \div R^3$ is constant and equal to 1. This constancy means that as T^2 becomes larger, R^3 becomes larger in the same proportion. From this, Kepler formulated his third law:

> **KEPLER'S THIRD LAW**
> The squares of the times of revolution of the planets around the Sun ($= T^2$) are proportional to the cubes of their average distances from the Sun ($= R^3$).

This epoch-making discovery became a mighty weapon in the armoury of the astronomers. It also established beyond doubt that the celestial bodies move strictly according to mathematical laws.

The proportionality between T^2 and R^3 can best be illustrated by means of a graph. Because the scale of times and distances is so enormous, the logarithms of these are marked off on the axes so as to bring the graph within bounds (Figure 3.16). The graph is a straight line, with the Earth at the origin because its period and distance from the Sun are taken as equal to 1 and the log of 1 is zero. The straight line shows that T^2 is proportional to R^3.

Later we shall see how Kepler's third law can be used to determine the masses of double stars, simply by measuring their periods of revolution around their common centres of gravity.

Galileo

While Johannes Kepler was making these remarkable discoveries in planetary dynamics, an event of even greater importance had taken place. Galileo Galilei (Figure 3.17) had come into possession of one of the first telescopes. The telescope was probably invented by Hans Lipperscheij (Figure 3.18, *overleaf*).

When Galileo turned his first telescope on the stars, he was overcome by the sight of the multitude of stars in the Milky Way – this hazy strip actually consisted of separate stars.

When he looked at Jupiter, he saw four moons revolving around the planet (Figure 3.19, *overleaf*) – a Solar System in miniature!

Marius von Gunzenhausen has also been credited with discovering Jupiter's four large moons in 1610.

Table 3.2					
Planet	Period T (years)	T^2	Distance R (AU)	R^3	$T^2 \div R^3$
Mercury	0.241	0.058	0.387	0.058	1.000
Venus	0.616	0.379	0.725	0.381	0.995
Earth	1.000	1.000	1.000	1.000	1.000
Mars	1.881	3.538	1.524	3.540	0.999
Jupiter	11.86	140.7	5.200	140.6	1.001
Saturn	29.46	867.9	9.54	868.3	1.000
Uranus	84.08	7069	19.19	7067	1.000
Neptune	164.8	27159	30.07	27189	0.999
Pluto	248.4	61703	39.46	61443	1.000

Figure 3.17 Galileo Galilei (1564–1642)

Figure 3.18 Hans Lipperscheij, inventor of the telescope

Figure 3.19 Jupiter and its four moons, as seen by Galileo

The distances of the moons from the centre of Jupiter in Jupiter radii are those obtained by the great G.D. Cassini in 1693. Kepler's third law can be applied by converting the periods and distances to multiples of that of Io, the innermost moon. We see that, once again, the squares of the periods are proportional to the cubes of the distances (Table 3.3), thus giving Kepler's third law universal credibility.

Galileo also observed the phases of Venus. Out of fear of being ridiculed, he stated this finding in the form of a cryptic riddle.

Galileo noticed that Saturn appeared to have two "ears", which a few years later seemed to have vanished. These were, of course, Saturn's rings which become invisible every fourteen and a half years when the Earth is in the same plane as the rings.

The use of the telescope, initiated by Galileo, started a revolution in astronomy.

In 1633 Galileo was summoned before the Inquisition and forced to recant his teachings, supporting the Copernican system. Mindful of Giordano Bruno's fate in 1600, Galileo was very wise to "recant". He was placed under house arrest and spent the last nine years of his life working out the principles on which Isaac Newton, who was born in the year of Galileo's death, 1642, was able to formulate his laws of motion. When Edmond Halley expressed praise of Newton's grasp of celestial mechanics, Newton replied that he was fortunate in being able to stand on the shoulders of giants – the giants being Kepler and Galileo.

Isaac Newton

Destined to be the discoverer of the law of gravity, Isaac Newton (1642–1727) placed astronomy on a firm mathematical basis. He and G.W. Leibnitz, independently of each other, worked out the first principles of the differential and integral calculus, thus providing the tools which enabled astronomy to reach unprecedented heights: mathematics became the indispensable adjunct of astronomy. The movements of the celestial bodies could henceforth be quantified.

Table 3.3 Kepler's third law applied to Jupiter's moons

Satellite	Period			Distance			
	Hours	$\div 42.5 = T$	T^2	Jupiter radii	$\div 5.667 = R$	R^3	$T^2 \div R^3$
Io	42.5	1.000	1.000	5.667	1.000	1.000	1.000
Europa	85.25	2.006	4.024	9.017	1.591	4.027	0.999
Ganymede	172	4.047	16.378	14.384	2.538	16.348	1.002
Callisto	402	9.459	89.473	25.300	4.464	88.955	1.006

$T^2 \div R^3$ is constant, therefore T^2 is proportional to R^3.

From the results of experiments which Galileo had done on bodies moving on inclined planes, Newton (Figure 3.20) was able to formulate his first law of motion.

> **NEWTON'S FIRST LAW OF MOTION**
> Every body continues in its state of rest, or of uniform motion in a straight line, except so far as it may be compelled by an impressed force to change that state.

Galileo had made the discovery that the velocity of a body, falling under the gravitional force of the Earth, was independent of its mass. The Earth's force of attraction therefore works on every separate particle of the body. This was contrary to the belief held since the time of Aristotle, that a heavier body will fall faster than a lighter body. Galileo said that a hammer and a bird's feather would fall equally fast in a vacuum where there is no air to slow the feather down by buoyancy. James Scott of Apollo 15 demonstrated this by dropping a hammer and the feather of a falcon simultaneously on the surface of the Moon. Television viewers on Earth could see how the hammer and feather drifted down to the surface of the Moon at the same speed!

Because different bodies fall through equal distances in equal times, this means that the acceleration is directly proportional to the applied force of attraction of the Earth. Galileo also found that the acceleration which a body undergoes is inversely proportional to its mass: the greater the mass, the smaller the acceleration; and the smaller the mass, the greater the acceleration. Newton included this idea in his second law of motion.

> **NEWTON'S SECOND LAW OF MOTION**
> Change of motion is proportional to the impressed force, and takes place in the direction of the straight line in which the force acts. The acceleration is inversely proportional to the mass of the body.

Newton expressed this idea mathematically as follows. If a represents the acceleration, F the force and m the mass of the body, then the acceleration, a, is proportional to the force, F, and inversely proportional to the mass, m. That is:

$$a \propto \frac{F}{m}$$

or alternatively:

$$a \propto F \times \frac{1}{m}$$

Newton also found that when one body strikes another so as to make it move, the striking body is slowed down by a reaction.

Figure 3.20 Isaac Newton

> **NEWTON'S THIRD LAW OF MOTION**
> Every action is accompanied by an equal but opposite reaction.

This is the principle on which rocket propulsion acts.

Uniform Circular Motion

Robert Hooke had proposed that circular motion could only take place if a force towards the centre of the circle was acting. This centripetal force would be inversely proportional to the square of the distance from the centre. Newton gave the mathematical proof of this (Figure 3.21).

According to Newton's first law, a body at A will continue its uniform motion with a velocity v, in the direction of the straight line AB, which is the tangent to the circle, unless a force acts upon it.

Now suppose the body moves to C in a time of t seconds

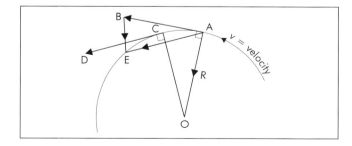

Figure 3.21 Centripetal force. Velocities are represented by lengths of lines

subject to a centripetal force towards the centre of the circle, O. During this lapse of time, the velocity which was represented by the length AB has changed to the velocity represented by CD, which is the tangent to the circle at C. Because the velocity is uniform, then AB and CD are both equal to v.

From A draw AE parallel and equal to CD. To change the velocity AB into the velocity AE, a velocity BE must be added acting in the direction BE. But a change in velocity (distance per second) is the force F which acts on every particle mass. Thus:

$$F = BE \div t$$

If we let t get smaller and smaller, AE will tend more and more to become parallel to AB, and BE will tend more and more to become right-angled to AB. The direction which is at right angles to AB is the direction AO. This means that the force which acts on the body must be acting in the direction AO, towards the centre of the circle. The magnitude of the force is given by:

$$F = BE \div t.$$

In the limiting position, when t is infinitesimally small, arc AC coincides with chord AC, a straight line whose length is infinitesimally small. Length AC will be equal to velocity v multiplied by time t. Namely:

$$AC = vt$$

Because the triangles ABE and OAC are similar (Appendix 1, section 1.4) the sides opposite equal angles are proportional, that is

(Triangle ABE)   (Triangle OAC)

$$\frac{BE}{AB} = \frac{AC}{OA}$$

i.e. $\dfrac{BE}{v} = \dfrac{vt}{R}$ or $\dfrac{BE}{t} = \dfrac{v^2}{R}$

The force which acts on a body of mass m in the direction AO is given by:

$$F = m\frac{BE}{t} = m\frac{v^2}{R}$$

The force $(mv^2) \div R$ is therefore constantly drawing the body towards the centre, O, of the circle. This is the centripetal force.

In the case of circular motion, the time, T, for one revolution is given by:

$$T = \frac{\text{circumference of circle}}{\text{velocity of motion}} = \frac{2\pi R}{v}$$

If R is the radius of a circle, its circumference $= 2\pi R$, where $\pi = 3.14159$ (Appendix 1, section 1.8).

Because $\quad T = \dfrac{2\pi R}{v}, \quad v = \dfrac{2\pi R}{T}$

The centripetal force, F, is given by:

$$F = \frac{mv^2}{R} = \frac{m}{R}\left(\frac{2\pi R}{T}\right)^2 = \frac{4\pi^2 R^2 m}{RT^2}$$

i.e. $\quad F = 4\pi^2 Rm \div T^2$

The ratio of the two forces, F_1 and F_2, of attraction of the Sun on two planets of masses m_1 and m_2 and periods of revolution T_1 and T_2 is given by

$$\frac{F_1}{F_2} = \frac{4\pi^2 R_1 m_1 \div T_1^2}{4\pi^2 R_2 m_2 \div T_2^2}$$

i.e. $\quad \dfrac{F_1}{F_2} = \dfrac{m_1 R_1 T_2^2}{m_2 R_2 T_1^2}$

The supposition of circular orbits is permissible, since the distances R_1 and R_2 in Kepler's third law are average distances from the Sun, as if the planets orbited in circles. Newton then applied Kepler's third law by making the substitution:

$$\frac{T_2^2}{T_1^2} = \frac{R_2^3}{R_1^3} \quad \text{so that} \quad \frac{F_1}{F_2} = \frac{m_1 R_1 R_2^3}{m_2 R_2 R_1^3}$$

therefore $\quad \dfrac{F_1}{F_2} = \dfrac{m_1 R_2^2}{m_2 R_1^2}$

This means that forces F_1 and F_2 are proportional to masses m_1 and m_2, and inversely proportional to the squares of distances R_1 and R_2 (i.e. R_1^2 and R_2^2). This is the law of inverse squares which Newton formulated as follows:

NEWTON'S LAW OF GRAVITATION

The force with which the Sun attracts a planet is proportional to the product of the masses of the Sun and the planet and inversely proportional to the square of the distance between the Sun and the planet, i.e.

$$F \propto \frac{Mm}{R^2}, \quad \text{or as an equation:} \quad F = G\frac{Mm}{R^2}$$

where M is the mass of the Sun, m the mass of the planet and R the average distance of the planet from the Sun. The universal constant of gravitation, G, has been found experimentally to be equal to $(6.672 \pm 0.0041) \times 10^{-11}$ N m^2 kg^{-2}.

Using this value of G, the mass of the Earth has been found to be $(5.9724) \times 10^{24}$ kg and its density 5.515 g cm^{-3}. At great distances, gravity acts as if each mass is concentrated at a point.

The inverse squares of the distances means that the force becomes one-quarter when the distance is doubled: one-

Gravitation

Figure 3.22 Albert Einstein

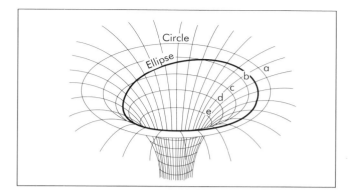

Figure 3.23 Curvature of space–time continuum caused by gravity

ninth when the distance is trebled, etc.

The law of gravitation explained Kepler's Third Law as being based on a centripetal force, inversely proportional to the square of the distance.

Gravity affects all bodies throughout the Universe, while magnetism is strongest in iron. The perturbations that bodies undergo in their orbits are due to the forces of attraction of neighbouring bodies. The Moon experiences greater force from the Sun than from the Earth. The Sun's mass is 330 000 times that of the Earth, but its distance is 389 times the Earth's distance from the Moon. Therefore the ratio of the Sun's force on the Moon to that of the Earth is $333\,000 \div (389)^2 = 2.2$ times.

The idea of a force acting at a distance is difficult to comprehend. Albert Einstein (Figure 3.22) overcame this difficulty in his Theory of Relativity by stating that every concentration of mass causes a bending of the space–time continuum of the Universe. It can be considered as if the world lines of force are bent. Every material body then moves in such a way as always to describe the shortest distance between two points (or rather two events). This is the perpendicular from the first point to the next world line. He showed that these paths in space would be conic sections, such as ellipses, parabolas, hyperbolas or circles. Figure 3.23 is an attempt at depicting the curvature of the space–time continuum, which has three spatial dimensions and one time dimension, on the flat surface of paper, which has only two dimensions. If a body moves in an orbit which is at right angles to the vertical through the centre of the curved space-time continuum, its orbit will be a circle. Five such circles (a, b, c, d and e) have been drawn but, because the drawing is an isometric view, the circles appear to be ellipses. If the plane of the orbit in which a body moves diverges from 90 degrees to the vertical, the orbit will be an ellipse, as shown by the thick curve. This curve is nearer to the centre of attraction on the left side of the sketch. The curve climbs towards the right: the body is then moving uphill towards the right of the sketch and downhill towards the left of the sketch, and it moves respectively slower and faster in its orbit. This is exactly what is observed in the cases of the planets: when moving towards aphelion, they slow down; and towards perihelion, they speed up.

If a curve is tilted more steeply, it may be a parabola or even a hyperbola, in which cases the moving body will not return to its orbit but will move off into space. Some comets have been found to move in orbits that are either parabolas or hyperbolas.

Chapter 4
Instruments

Telescopes

Good progress was made in the development of the telescope during the first half of the 17th century. The first telescopes were refractors, having an objective lens and ocular, or eyepiece, on the optical axis. The principle of the refractor is illustrated in Figure 4.1.

The ray of light A, parallel to the optical axis OFD, is refracted by the objective lens O, so that it passes through the principal focus F. Ray B, through the centre O of the objective lens, goes straight through. The two rays intersect at C, where an inverted image of the object is formed. The ocular can be moved in or out to focus the image so as to obtain an enlarged image EE'. Rays A' and B' are refracted by the ocular so that their backward projections intersect at E.

The magnification is equal to the angle C'DC subtended by the image, divided by the angle C'OC. This is equal to OC' ÷ DC', because the distances DC' and OC' are inversely proportional to the angles C'DC and C'OC. The magnification is thus equal to the distance from the objective to the real image, divided by the distance from the real image to the ocular.

When the telescope is focused on an object infinitely far away, the magnification is equal to the focal length of the objective divided by the focal length of the ocular. The longer the former and the shorter the latter, the greater is the magnification. For observing double stars, a telescope of good resolving power will have an objective lens of long focal length, say 10 metres. If an ocular of focal length 6 mm is used, the magnification will be equal to 10 000 ÷ 6 = 1666 times. If the magnification is too great, the available light rays will be so spread out that the image will become invisible. Therefore it is necessary to use larger objective lenses so as to gather more light. To manufacture large objective lenses is very difficult, and in addition the weight of the glass becomes so great that the lens tends to distort. The largest objective lens ever made is the 100 cm of the Yerkes Observatory, Michigan, USA.

The idea took root that it would be easier to construct curved mirrors to take the place of the objective lens. In 1663 J. Gregory gave a description in his *Optica Promota* of a telescope which would use a parabolic mirror instead of

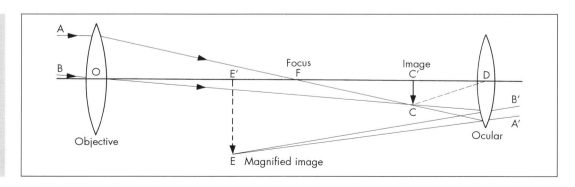

Figure 4.1 The principle of the refracting telescope

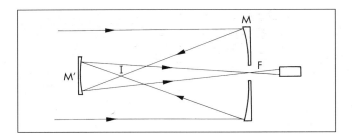

Figure 4.2 The Gregorian telescope

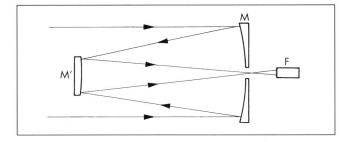

Figure 4.3 Cassegrain telescope

an objective lens (Figure 4.2).

The rays of light from the object are reflected by a parabolic mirror M, to form an image at I. Then the rays are reflected by a secondary parabolic mirror M′ to the final focus of the ocular at F. I and F are the foci of an ellipse of which the small secondary mirror M′ forms part. The rays reflected from the secondary mirror pass through a hole in the centre of the objective mirror on their way to the ocular, where the enlarged image is observed.

An improvement was suggested in 1672 by Cassegrain, whereby the use of a convex secondary mirror, M′, would yield a longer focal length (Figure 4.3). The Cassegrain system is still very popular today.

Besides the renown Newton gained from his law of gravitation, he also built the first reflecting telescope in 1688. He overcame the difficulty of drilling a hole in the centre of the primary mirror by mounting a secondary plane mirror on the optical axis at an angle of 45 degrees, so as to reflect the light rays to the side, out of the telescope tube (Figure 4.4).

One of the most modern systems is the Schmidt–Cassegrain system, which uses a spherical concave primary instead of a parabolic mirror (Figure 4.5), because it is easier to construct a spherical surface than a parabolic one. However, a spherical mirror does not focus rays that are removed from the optical axis at one point, so this is overcome by placing a correcting plate, C, at the centre of curvature of the primary mirror M. A correcting plate of lithium fluoride makes the telescope sensitive to ultra-violet light. A convex secondary mirror, M′, gives the Schmidt–Cassegrain a long focal length, even though the tube is short. The magnification is very good, although the field of view is restricted. This system can also be used with a Newtonian set-up.

Another modern development is the Maksutov system in which all the surfaces are spherical. Figure 4.6 illustrates a

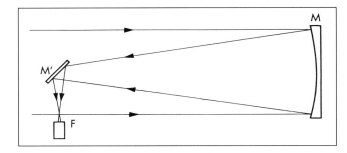

Figure 4.4 Newtonian telescope

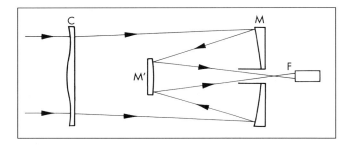

Figure 4.5 Schmidt–Cassegrain telescope

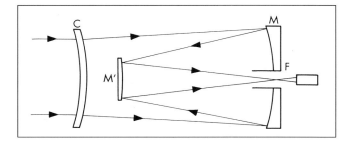

Figure 4.6 Maksutov telescope

Oculars

design by R. Sigler: it has a 15 cm correcting plate, 1.57 cm thick, which is placed −31.5 cm from the 17.5 cm diameter primary mirror. The secondary mirror is 5.2 cm wide. The radii of curvature of the correcting plate are −22.7 and −23.58 cm; that of the primary −100 cm and that of the secondary mirror −64.55 cm. The focal length works out at 121.55 cm. It is therefore an f-7 telescope because the focal length divided by the diameter of the primary mirror (121.55 ÷ 17.5) works out to be 7.

Some Maksutovs have a silvered spot in the middle of the correcting plate and thus dispense with a secondary mirror. The field of view of 2 degrees is four times the diameter of the full Moon. The Maksutov is not subject to any spherical or chromatic aberration, and displays no coma or glow around the image.

Oculars

A single bi-convex lens in an ocular produces a coloured image, because white light is refracted by the lens in such a way that the different frequencies undergo different amounts of bending (Figure 4.7). The tapering edges of a lens are similar to a glass prism. In 1704, in his book on optics, Newton gave an exposition of the spectrum: when white light falls obliquely on a tapered glass surface, it is split into the colours red, orange, yellow, green, blue, indigo and violet. The red component undergoes the least amount of refraction, and the violet most. In a rainbow, the colours are caused by refraction and reflection of light in the fine raindrops which float in the air.

In order to eliminate the colour effect of a single lens, called chromatic aberration, use is made of compound lenses.

A diverging lens (bi-concave or plano-concave) is mounted against a converging (bi-convex) lens, as shown in Figure 4.8. Usually the diverging lens is made of a different kind of glass, such as flint. The nett result must be converging (although Galileo's eyepiece was divergent). The focal lengths of eyepieces are very short being 25, 12 or 6 mm.

Various oculars have been designed, each with a special purpose, such as to eliminate chromatic aberration or astigmatism, or to obtain a flatter field.

The eyepiece devised by C. Huygens (1629–1695) consisted of two plano-convex lenses, having focal lengths in the ratio of 3 : 1 and at a distance apart equal to the focal length of the eye lens (Figure 4.9). Huygens eyepiece gave a

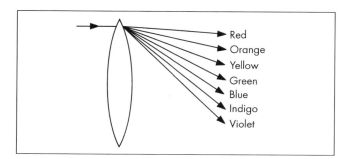

Figure 4.7 Spectrum formed by refraction

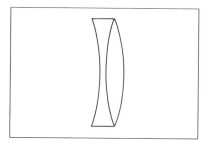

Figure 4.8 Compound lens

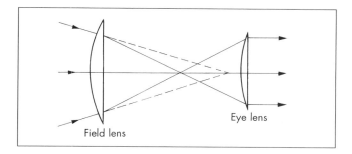

Figure 4.9 Huygens' eyepiece

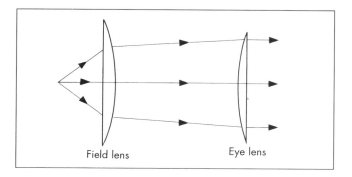

Figure 4.10 Ramsden eyepiece

flat field but did not eliminate chromatic (or colour) aberration.

The Ramsden eyepiece (Figure 4.10, previous page) consists of two plano-convex lenses of equal focal lengths, mounted two-thirds of this distance apart. The eyepiece lessens both chromatic and spherical aberration, but does not eliminate them entirely.

The Kellner eyepiece (Figure 4.11) uses an achromatic eye lens, consisting of a divergent eye lens of flint against a convergent lens of ordinary crown glass. The field lens is plano-convex.

The focal length of the combined field and eye lenses is equal to the distance between the lenses, and equal to the focal length of each lens. This was one of the best of the older oculars: it eliminated chromatic and spherical aberration, but was not free from internal reflections.

One of the best oculars is the orthoscopic eyepiece (Figure 4.12). The field lens consists of two bi-convex lenses, one of which is almost plano-convex, having a bi-concave lens between the two. The eye lens is plano-convex. The system eliminates both chromatic and spherical aberration. The orthoscopic ocular can be used together with a strong objective lens, as strong as f-6, that is, an objective lens having a focal length six times the diameter of the lens.

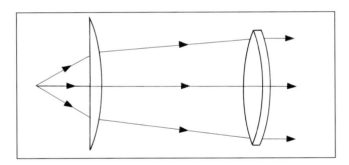

Figure 4.11 Kellner eyepiece

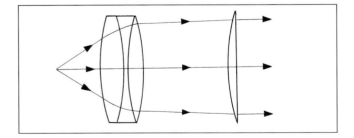

Figure 4.12 Orthoscopic eyepiece

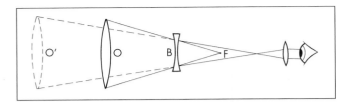

Figure 4.13 Barlow lens increases focal length of objective

The power of the objective lens can be increased by using a Barlow lens as an adjunct to the eyepiece (Figure 4.13). The Barlow lens is diverging and if its magnification is $2\times$, for example, it will double the effective focal length of the objective. The lens, B, is so placed that its distance OB from the objective O is slightly less than the focal length OF. The nett result is that the objective lens, O, is in effect replaced by a new objective lens, O'; the focus, F, is maintained. With an increased focal length, the telescope is then very suitable for observing the planets and for resolving double stars.

The Filar Micrometer

To determine the exact position of a celestial body, the diameter of a planet, or the angular separation between stars, the eyepiece has to be fitted with crosswires in the focal plane of the ocular. One of the crosswires AB (Figure 4.14a) can be moved, say, to A'B' by means of a micrometer screw. This technique was invented in 1644 by W. Gascoigne. The angular separation of A'B' from the fixed wire CD can be read off very accurately on the drum of the micrometer.

By turning the eyepiece as shown in Figure 4.14b, the horizontal wire XY can be set at any angle. When observing

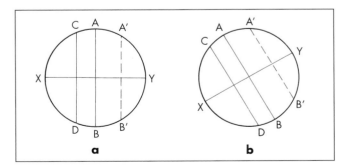

Figure 4.14 The filar micrometer

Telescope Mountings

Figure 4.15 5-metre Hale Telescope, Mount Palomar. The mirror is in the crate above the men on the floor.

double stars, the wire XY is set so as to pass through both stars of the pair. The angle through which the wire has rotated from north, through east, can be read off on the drum of the micrometer.

Many stars are double stars, or binary stars as they are termed. Over periods of years, their separations and position angles alter. Amateur astronomers can render their work professional by fitting micrometer screws to their telescopes.

The reflecting telescope has largely replaced the refractor, because it is easier to construct a mirror than a lens. Through the years, larger and larger mirrors have been constructed. In 1917, the 2.5-metre reflector, the Hooker telescope, came into operation on Mount Wilson. It was the first telescope to enable the diameter of a star to be measured. This was followed in 1948 by the 5-metre Hale telescope on Mount Palomar (Figure 4.15). Later, a 6-metre Russian telescope started operations in the Caucasus Mountains. In recent years, several very large mirrors have been planned, including four 8-metre mirrors which are to operate in concert at the European Southern Observatory in the Andes Mountains.

Telescope Mountings

A telescope must be mounted on a firm base so as to minimise quivering. In order to keep celestial bodies in the field of view, the telescope must be able to turn westwards at the same speed that the Earth spins eastwards, because the spinning of the Earth is the cause of the apparent westward motion of the sky. For this purpose an electric motor, or a stepping motor controlled by a computer, is used.

The axis about which the telescope turns, the polar axis, shown as P–P in Figures 4.16 and 4.17 *overleaf*, must be parallel to the Earth's axis of rotation, and it must therefore point to the celestial pole. The elevation of the celestial pole above the northern or southern horizon is equal to the latitude of the point of observation, that is, the number of degrees north or south of the equator.

In order to be able to point in all directions, the telescope must also turn about a second axis at right angles to the polar axis. This is the declination axis, shown as D–D.

Each axis must have a graduated circle to denote the position of the telescope. The circle attached to the polar axis, the right ascension circle, is subdivided into 24 hours and further into minutes, counting from west to east. The zero must coincide with the point on the celestial sphere where the equator cuts the Ecliptic, this being the First Point of Aries. The declination circle indicates the angle between the equator and the direction in which the

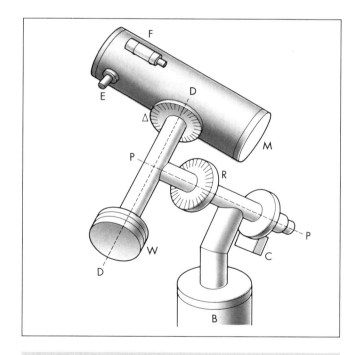

Figure 4.16 Equatorial mounting
B   Massive base   M   Objective mirror
C   Clockdrive   P–P Polar axis
D–D Declination axis   R   Right ascension circle
E   Eyepiece   W   Balance weights
F   Finderscope   Δ   Declination circle

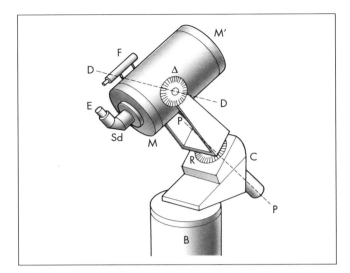

Figure 4.17 Fork mounting
M' Secondary mirror
Sd Star diagonal
Other symbols as for Figure 4.16

telescope is pointing. The positions of celestial bodies are usually given in right ascension and declination.

The Celestial Sphere

The great circles on the celestial sphere, on which the coordinates of celestial bodies are measured, are shown in Figure 4.18, having the observer situated at O, say at 50 degrees north latitude. The observer's horizon is the great circle NWASE, through the north, west, south and east points on the horizon. The observer's meridian is the great circle in the plane of the paper, through the points NPZTQS–P′Z′T′Q′N, where P is the north pole, Z the zenith, vertically overhead and Z′ the nadir, vertically below the observer. The angle NOP equals 50 degrees.

The great circle through the zenith Z and the body X, here represented by the star Gamma Andromedae, meets the horizon at A. The angle between X and A is the elevation of the body.

The angle SB between the meridian and the vertical circle PXRB, through the pole and the body, is the azimuth which is measured eastwards.

The equator is the great circle drawn through the points WγRQEΩQ′W, and is as many degrees from the zenith as is the latitude of the observer. The equator cuts the Ecliptic γTΩT′γ at γ, the First Point of Aries (the Vernal Equinox) and at Ω, the First Point of Libra (the Autumnal Equinox). At the equinoxes, the Sun is vertically above the equator and day and night are equally long. The Ecliptic is inclined to the equator by $23\frac{1}{2}$ degrees; this equals the angle between the direction of the Earth's axis and the vertical on the plane of the Ecliptic. The Ecliptic marks the annual path of the Sun against the background of the stars. The vertical circle through P and the body X cuts the equator at right angles at R; it is also known as the hour circle. The angle γOR is the right ascension of all bodies on the circle PXR. Right ascension is measured eastwards in hours, minutes and seconds from the First Point of Aries γ. In Figure 4.18, the right ascension of the circle PXR (angle α, alpha) is 2 hours, that of the star Gamma Andromedae. The angle ROX (angle δ, delta) is the declination of X, measured from the equator, positive (+) when measured northwards and negative (−) southwards from the equator. In the figure the declination is +42 degrees. A star can be designated by means of its coordinates; for example, if the right ascension is, say, 2 hours, 12 minutes and 45 seconds and the declination 42 degrees, 4 minutes, and 19 seconds, the star can be designated as: 02 12 45 + 42 04 19. The angle between the prime vertical PZQS and the vertical PXRB through the object X, that is, the angle ZPX, is the hour angle, and it is measured westwards from the prime vertical (= SB).

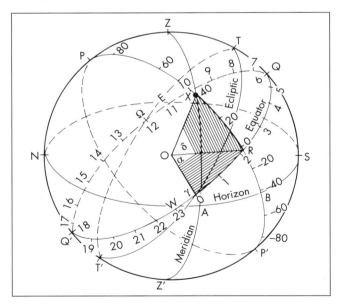

Figure 4.18 Great circles on the celestial sphere
γR = α = Right ascension (in hours)   RX = δ = Declination (in degrees)
SA = Azimuth (in degrees)   AX = Altitude (degrees)

Table 4.1 Standard times for setting right ascension circles

Star	Declination	Right ascension	Sidereal time	Date	Setting time*
Rigel	−08°13′	5 h 14 m	5 h 13 m	Jan 24	20 h 59 m
Procyon	+05°15′	7 h 39 m	7 h 39 m	Mar 2	21 h 00 m
Regulus	+12°01′	10 h 08 m	10 h 08 m	Apr 8	21 h 00 m
Arcturus	+19°14′	14 h 53 m	14 h 51 m	Jun 4	20 h 58 m
Altair	+08°51′	19 h 50 m	19 h 49 m	Sept 3	20 h 59 m

*Add 4 minutes for each degree west of zero longitude.

Setting the Graduated Circles

If the graduated circles are set correctly, a celestial body can be found by setting the telescope according to the coordinates of the body. When Gamma Andromedae is in the centre of the field of the telescope, the declination circle can be clamped to show 42 degrees as the declination. When a star is on the meridian, its right ascension is equal to the sidereal time. The star Procyon, for example, has a right ascension of 7 h 39 m. On 2 March, it is on the meridian at 21 hours standard time for zero longitude on Earth. At that moment the hour angle circle, or right ascension circle, must be clamped so that it reads zero, because sidereal time minus right ascension equals hour angle. Table 4.1 contains the coordinates of five stars that may conveniently be used to set the hour angle circle.

Sidereal Time

The time which the Earth takes to complete one rotation on its axis, relative to the stars as background, is the sidereal day; its length is 23 h 56 m 4.091 s, that is, nearly 4 minutes shorter than the solar day which is 24 hours long. The sidereal time is always equal to the right ascension of celestial bodies on the meridian, the great circle through the north pole and the south point on the horizon. This circle passes through the zenith.

The Measurement of Time

When C. Huygens invented the escapement in 1656, he was able to construct a pendulum clock, which enabled astronomers to measure time accurately. Today, caesium or quartz crystal clocks are used and are accurate to within 1 second over a year.

The Equinoxes

When the Sun is in the First Point of Aries on 21 March or the First Point of Libra on 23 September, the two points where the Ecliptic cuts the equator, day and night, are equally long at all points on the Earth. The Sun then rises in the east point on the horizon and sets in the west point. These dates are known as the equinoxes. On account of precession the First Point of Aries moves backwards, from east to west on the equator. In the 2110 years that have elapsed since the time of Hipparchus, this point has moved through (2110) × (50.3) arc seconds, namely 29 degrees. The annual amount of precession is 50.3 arc seconds. At present the First Point of Aries is located on the border between Pisces and Aquarius, but it is still called the First Point of Aries. Not only does precession affect the values of the right ascensions of bodies but also the declination. Therefore new tables of right ascension and declination are compiled for intervals of 50 years: 1950.0, 2000.0, etc.

By means of computers, right ascentions and declinations are presently calculated on a yearly basis.

The Spectroscope

The importance of the spectroscope in astronomy cannot be over-emphasised. The chromatic aberration of a single bi-convex lens is a problem in the construction of an eyepiece. The similar spectrum, formed when light rays pass through a glass prism, actually carries the secrets of the Universe. The spectroscope (Figure 4.19) consists of a collimator which is a telescope that directs the rays that pass through a narrow slit, into a parallel beam. The parallel rays then fall on to a glass prism, which is the essential part of the spectroscope. In the glass prism the rays of light undergo varying amounts of refraction, depending on the wavelength of the light: red light is refracted the least;

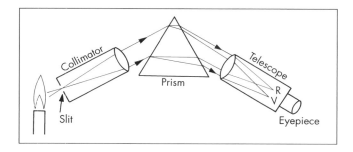

Figure 4.19 The spectroscope

orange a little more; then yellow, green, blue, indigo; and violet is refracted the most.

Crosswires are mounted in the plane R–V of the telescope so that the positions of the images of the collimator slit can be measured precisely in colours varying from red to violet. This series of narrow lines is what constitutes the spectrum. In order to determine the wavelengths of the various coloured lines, a graduated scale of wavelengths is projected into the telescope so that these wavelengths can be read off directly. The wavelengths of the various colours are actually determined by means of interference experiments in which the waves alternately strengthen and eliminate each other.

Red light has the longest wavelength of visible light and therefore the lowest frequency. If λ represents the wavelength and f the frequency of the waves, then the product $f\lambda$ is a constant, equal to c, the speed of light, which is equal to 3×10^8 metres per second. Because c is constant, λ and f are inversely proportional: when λ becomes shorter, f becomes higher; when λ gets longer, f gets lower.

The light waves of longest wavelength therefore form red images of the slit in the form of narrow lines at R, and the shortest waves form violet-coloured lines at V. Between these two points there are line images, coloured orange, yellow, green, blue and indigo.

The same effect is obtained if a very large number of straight lines are ruled by means of a diamond-tipped pen on a sheet of flat glass. This is called a grating and can take the place of the prism. Gratings have between 8000 and 16 000 lines per cm. The lines serve as obstacles around which the rays are diffracted, thus forming a spectrum.

A grating can also be drawn on a flat or concave mirror of speculum which reflects the rays to form a spectrum. Figure 4.20 shows how a grating and an ordinary camera can be used to obtain a spectrum. A collimating mirror is used and the grating is a reflecting one.

H.A. Rowland constructed a grating having 4000 lines per cm on a concave, cylindrical sheet of speculum (an alloy of copper and tin). The sunlight incident on the mirror was reflected by the strips between the lines and formed a spectrum, spread over 15.5 metres. The radius of curvature of the speculum mirror was 6.55 metres. The separations between the lines in the spectrum, formed by the concave mirror, are exactly proportional to the wavelengths of the light.

The Nature of the Spectrum

The direct spectrum of the light from a radiating body consists of bright lines; it is known as an emission spectrum. The positions occupied by the lines depend on the frequencies, or on the wavelengths, of the light rays. Physicists have established that an emission spectrum is formed because the electrons revolving around the nuclei of the atoms jump from outer to inner orbits, becoming nearer to the nuclei. Each jump from one orbit to the next is accompanied by the radiation of a discrete amount of energy at a definite frequency, thus forming an image of the spectroscope slit having a definite position and colour.

When a gas absorbs radiant energy, its electrons jump from inner to outer orbits and with each jump a discrete amount of energy is absorbed. This absorption of energy means that light of that particular frequency is absorbed, thus forming black line images of the slit. This is called an absorption spectrum.

Since the positions of the lines in a spectrum depend on the frequencies of the light rays and since these are determined by the atomic structure, each element will form lines in determinable positions. This means that the spectrum can be used to analyse substances chemically. If table salt, for example, is burned in a gas flame, the sodium in the salt forms two very bright lines, close together in the yellow portion of the spectrum. The wavelengths of these two yellow lines of sodium are 5896 and 5890 (in round numbers). Burning hydrogen gas forms a series of lines, the Balmer series, and also other series.

The spectrum lines are known as Fraunhofer lines, after J. von Fraunhofer who made a systematic study of spectroscopy in 1814. This work was continued by R. Bunsen and G. Kirchhoff: they found that no two elements have the same patterns of lines in the spectrum. Each element has an unique spectrum, different from those of all other elements. Here was a mighty weapon in the hands of astronomers, enabling them to identify elements existing in the Sun and the stars.

Kirchhoff showed that the absorption spectrum of hydrogen, formed when light passes through cold hydrogen gas, is identical to that of the emission spectrum of burning

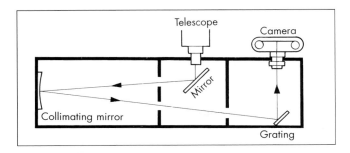

Figure 4.20 Use of a grating and camera to form a spectrum

The Nature of the Spectrum

hydrogen. The electron changes are the same in both cases, except that they occur in different directions.

By 1860, G. Kirchhoff and A. Secchi had made a detailed study of the Sun's spectrum, and they found that most of the lines corresponded with those of terrestrial elements. As time went on, more and more lines were found to correspond in this way. However, in 1868 N. Lockyer found a set of lines in the Sun's spectrum for which no equivalents could be found in terrestrial elements. He therefore ascribed those lines to an unknown element existing in the Sun, which he called helium, after helios, the Greek for sun. Twenty six years later, while W. Ramsay was analysing radioactive ores, he discovered a gas having the same spectrum as helium. Thus helium was discovered on Earth twenty six years after it was discovered in the Sun!

A.J. Angstrom was responsible for devising a system of units of wavelength. The unit is known as the angstrom unit (Å); its length is 10^{-10} metre. On this scale the wavelength of red light is about 6800 Å; orange light, 6200; yellow, 5600; green, 5000; blue, 4200; and violet, 4000 Å. The wavelength of yellow light is therefore 0.000 56 millimetres.

Light of wavelengths longer than 8000 Å and shorter than 3900 is not visible to the human eye; such wavelengths are called infra-red and ultra-violet respectively.

If a prism made of rock salt is used, we find a region, beyond the red end of the spectrum, where a thermometer shows that heat is being received. This phenomenon was discovered by William Herschel. This portion of the spectrum, from 7500 to 10 million Å, that is, from 7.5×10^{-7} up to 1×10^{-3} metres, is the infra-red portion of the spectrum. Wavelengths longer than the deep infra-red constitute microwaves (as used in a microwave oven). Their wavelengths range from 10^{-3} to 2.5 metres. Longer wavelengths than this are FM and television waves (2.5 to 15 metres); the longest are radio waves, 13 to 1500 metres in length (Figure 4.21).

At the other end of the visible spectrum, waves shorter than violet are termed ultra-violet, from 3200 to 100 Å; then follow the X-rays, 10^{-8} to 10^{-13} metres, and the shortest of all known wavelengths, gamma rays, being shorter than 10^{-13} metres.

The shorter the wavelength, the higher is the frequency and the more energetic is the radiation. The total perceptible spectrum thus ranges from 10^{-17} to 1.5×10^3 metres, that is, over 20 powers of 10. The visible spectrum, from 3900 to 8000 Å, is therefore only an infinitesimal portion of the whole electromagnetic spectrum. In Figure 4.21 wavelengths and frequencies are represented on a logarithmic scale, so that each power of 10 stretches over the same distance. A wavelength of 10^{-6} metres is therefore 10 times longer than a wavelength of 10^{-7}.

The wavelengths of the sunlight that permeates the

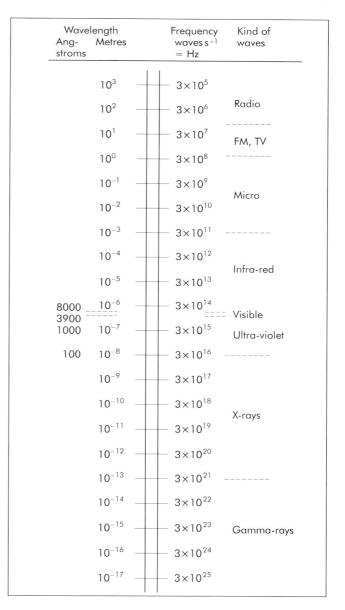

Figure 4.21 The Electromagnetic Spectrum

Earth's atmosphere range from 10^{-7} metres, in the ultra-violet, to 10^{-5}, in the infra-red.

Table 4.2 (*overleaf*) contains the wavelengths of prominent spectral lines of certain elements. The first three entries in the table (*A*, *a* and *B*) contain the telluric lines, these being the lines of oxygen and water in the Earth's

Table 4.2 Fraunhofer lines

Line	Element	Wavelength	Colour
A	oxygen	7594	red
a	water (molecule)	7183	red
B	oxygen	6869.955	red
C	hydrogen–alpha	6562.8	orange
D	sodium	5895.92	yellow
D	sodium	5889.95	yellow
E	calcium	5269.54	yellow
E	iron	5269	yellow
b	magnesium	5183	green
b	magnesium	5172.68	green
b	iron	5169	green
b	magnesium	5167	green
F	hydrogen–beta	4861.32	blue-green
f	hydrogen–gamma	4340.46	blue
G	iron, titanium	4307.90	blue
g	calcium	4227	blue
h	hydrogen–delta	4102	violet
H	calcium	3968.47	violet
K	calcium	3933.66	violet

atmosphere. All spectra of celestial bodies show these three lines, caused by absorption of those wavelengths by the Earth's atmosphere; they are dark lines. The *H* and *K* lines of ionised calcium are very important, especially in determining radial velocities, that is, velocities along the line of sight. They are also very useful in measuring redshifts.

The Spectrum of Hydrogen

Hydrogen plays a very important role in astrophysics. The hydrogen atom is the smallest, lightest and simplest of all atoms. It consists of a nucleus of only one proton, which carries a positive electric charge. The nucleus contains almost all the mass of the atom. When the atom is in the neutral state, the positive charge of the proton in the nucleus is neutralised by the negative charge of the single electron which orbits the nucleus. When the hydrogen atom absorbs energy, the electron jumps from the energy level which it occupies to one of higher energy, such as from level 1 to level 2 or 3, etc. Seven energy levels have been indicated in Figure 4.22. An electron jump from one energy level to the next higher level requires a precise amount of energy, one quantum. This is linked to a certain frequency, which has a certain wavelength. The product, wavelength times frequency, is a constant, equal to the speed of light: $\lambda f = c = 300\,000$ km s^{-1}. Therefore $\lambda = c \div f$.

A frequency of, say, 4.571×10^{14} Hertz (Hz) will have a wavelength

$$\lambda = 3 \times 10^8 \div 4.571 \times 10^{14}$$
$$= 0.6563 \times 10^{8-14} = 0.6563 \times 10^{-6}$$
$$= 6563 \times 10^{-10} \text{ metres}$$
$$= 6563 \text{ Å}$$

When an electron jumps from a higher to a lower energy level, the atom radiates energy of the same wavelength and frequency as it absorbed in the energy-absorbing phase. The amounts of energy absorbed or radiated during jumps to or from energy level 3, for example, are less than those to or from energy level 2, which in their turn, are less than from energy level 1. The latter contains the most energy; the wavelength is the shortest and the frequency the highest. The wavelengths entailed in transitions to or from energy levels 1, 2 and 3 are shown in Figure 4.22.

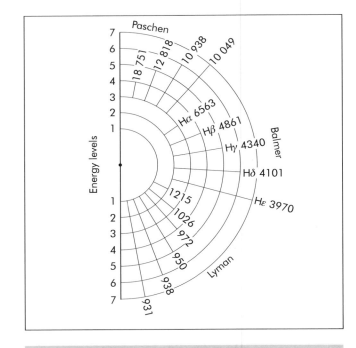

Figure 4.22 Wavelengths of electron transitions in the hydrogen atom

The Doppler Effect

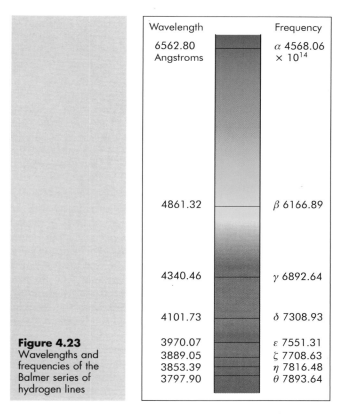

Figure 4.23 Wavelengths and frequencies of the Balmer series of hydrogen lines

The series of spectral lines produced by transitions to or from energy level 1 were discovered by T. Lyman in 1906. The frequencies of these lines range from

$$3 \times 10^8 \div 931 \times 10^{-10} = 3.222 \times 10^{15} \text{ Hz, to}$$
$$3 \times 10^8 \div 1215 \times 10^{-10} = 2.469 \times 10^{15} \text{ Hz}$$

The series of transitions about energy level 2 was discovered in 1885 by J.J. Balmer. The frequencies range from 7.557×10^{14} Hz for the line of wavelength 3970 Å to 4.571×10^{14} Hz for the line of wavelength 6563 Å. The Balmer series, which is in the visible part of the spectrum, is shown in Figure 4.23. The wavelengths and frequencies have been calculated using the most modern value for the speed of light, 299 792.458 km s^{-1}.

The Paschen–Ritz series is in the infra-red. The frequencies range from 2.985×10^{14} to 1.60×10^{14} Hz.

A fourth series, the Pfund–Brackett, is in the far infra-red and has wavelengths of 26 300 to 74 000 Å.

The energy around level 1 is very high because the electron is very strongly bound to the nucleus when in that level, and it requires great force to tear the electron away from the level.

Physics has found that the distances of the energy levels from the nucleus of the atom are in the ratios of the squares of the natural numbers, namely $1^2 = 1$, $2^2 = 4$, $3^2 = 9$, etc. It is not possible for an electron to exist between two energy levels.

The discovery of the fact that each transition is linked to a certain definite amount of energy gave rise to the development of the quantum theory by Max K.E.L. Planck. By using this theory, T. Lyman was able to calculate the possible existence of the Lyman series in the ultra-violet, before it was discovered experimentally. The hallmark of a good theory is that it must be able to make projections which can be tested experimentally. If the test works, it supports the theory; if not, the theory has to be revised or cast aside. This is the accepted basis for the formulation of laws in science.

The Doppler Effect

Besides the use of the spectroscope in chemical analysis, it can also be used to measure radial velocity, that is, velocity along the line of sight. C.J. Doppler and A.H.L. Fizeau showed that the frequency received from a radiating object (whistle of a locomotive or hooter of a motor car) increases when the body is approaching the observer and decreases when it is receding. This means that the wavelength of the radiation which is approaching the observer is shortened, while that which is receding is lengthened. This is called the Doppler effect. In Figure 4.24, the crests of the waves emitted when the body is at points 1, 2 and 3 are shown as the large, medium and small circles, respectively. An observer at A sees the crests as being closer together than they would have been had the body not been moving. Shorter wavelength implies higher frequency. Observer B sees the wavelengths as longer and therefore the frequencies which he receives are lower than they would have been if the radiating body had been at rest. The wavelengths

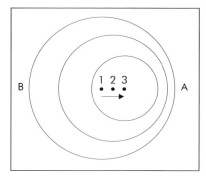

Figure 4.24 Doppler effect

received by B are therefore doubly longer than those received by A. We can consider the waves in A's direction as being compressed and lengthened in B's direction. Fizeau showed that the Doppler effect applies to light as well as sound. The spectroscope will show the increased frequencies of the radiation from a body which is approaching the observer, as a shifting of the spectral lines towards the violet end of the spectrum; and the spectral lines of the radiation from a receding body will be shifted towards the red end of the spectrum. Fizeau proved that the shifts are proportional to the velocities of the moving bodies. In 1888, H.C. Vogel showed that the Doppler effect was also to be found in starlight – that in many cases, the spectral lines of stars were shifted, towards either the violet or the red. This meant that these stars were either coming closer or moving away from the Earth. The stars that showed no blue- or redshift are those that are moving mainly transversely across the sky. By measuring the blue or redshifts, the velocities along the line of sight can be calculated.

Besides revealing the chemical composition of substances in the stars, and the radial velocities of the stars, the spectroscope can also be used to measure temperature, density, angular velocity and the pressure prevailing in a gas, as well as the existence of magnetic fields, as will become clear later.

Some stars are double, and some double stars revolve around their mutual centres of gravity (their barycentre) in planes that are nearly in the line of sight from the Earth. In such cases the spectral lines are not only double because of the two sources present, but move alternately from blue- to redshift, and vice versa. The spectroscope can reveal such double stars (binaries) even if they are so close that the telescope cannot resolve the two members of a pair. Such double stars are called spectroscopic binaries. The periods of binary stars can be measured accurately and, by applying Kepler's third law, the relative masses of the stars can be found, as we shall see later.

Photometers

Before the telescope was invented, the brightnesses of stars had to be estimated visually. The first light meters, or photometers, such as those of Bouguer, Rumford and Bunsen, compared the brightness of two sources of light by casting the light on to a sheet of translucent glass. The brightnesses of the spots of light could then be compared. Joly's photometer contained two rectangular blocks of translucent wax which were tightly pressed together. The light from a standard source of light, such as a candle flame, was directed from one side on to a block of wax, while starlight was sent through the other block. By moving the candle nearer to or further from one block of wax, the illumination of the two blocks could be made equal, so that the brightness of the star could be equated to that of the candle flame at its particular distance from the wax.

Modern photometers make use of the photoelectric effect of light. When light impinges on a substance such as silicon or germanium, electrons are set free. These electrons then generate an electric current of which the strength can be measured. The amount of electrons freed, and therefore the strength of the electric current generated, are proportional to the brightness of the incident light.

A photoelectric photometer has two electrodes: the negative cathode, overlaid with a layer of silicon, and the positive anode, which is kept at a certain electrical potential. The electrodes are contained in a glass bulb, containing an inactive gas such as nitrogen or argon, at a very low pressure. When light falls on the cathode, the electrons freed move across to the anode. The electric current thus generated can be measured very accurately. The strength of the current is proportional to the electron flux, which, in its turn, is proportional to the intensity of the light falling on the cathode.

The most modern photometers consist of photoelectric cells of semi-conductors between two electrical contacts. They are therefore not dependent on maintaining the near vacuum in a glass bulb. Today the brightnesses of stars can be measured correct to three significant figures.

It will pay any amateur astronomer to obtain a good photometer, rather than to go for a large telescope. An amateur can do professional work with a photometer.

The Photomultiplier

In the modern photomultiplier, the cathode consists of antimony–caesium oxide which is deposited on to the reception end of the tube. The electrons set free by the incident light have their electrical potentials repeatedly

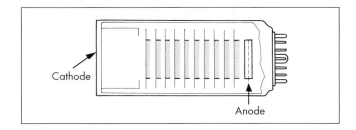

Figure 4.25 Photomultiplier

increased by a series of dynodes consisting of an alloy of antimony and caesium. At each dynode more and more electrons are set free, eventually increasing the electron flux by millions of times. The resultant electric current then registers the brightness of the source. The brightnesses of very faint objects can be determined in this way (Figure 4.25).

The Charge Coupled Device

This is a modern electronic masterpiece. It consists of a silicon chip 10 mm square and about eight-thousandths of a mm thick. The freed electrons are shuttled into compartments by means of electrical potentials towards a registering device. Each light-element (pixel) of the field of view is scanned and the varying electron fluxes are fed into a computer which prints out a picture of the field of view. The instrument is 50 times more sensitive than a photographic plate. In a given period of time, five to ten times as many photographs can be obtained by the charge coupled device as by ordinary photography.

The Telescope Race

Recording devices, such as those just described, have reached their peak and it has once again become necessary to construct telescopes with larger mirrors and to employ new technology in mirror-making, in supporting the mirrors in their crates, and in dealing with the optics.

First in the field was the European Southern Observatory with their 3.8-metre New Technology Telescope at La Silla in the Andes. The mirror is very thin and has 75 supports which are electronically monitored to control the parabolic shape of the mirror.

The Very Large Telescope will have four mirrors, each of 8 metres. Each of these mirrors will be only 20 cm thick and use will be made of several hundred supports, electronically monitored to maintain the perfect parabolic shape of the mirrors. The technique of fibre optics will also be used to canalise the images into one unified focus. Jointly the four mirrors will have an effective width of a 32-metre mirror. A multiple mirror telescope is already in action in Arizona.

The concept of the multiple mirror has been adapted in the W.M. Keck telescope which has been erected on Mauna Kea, Hawaii. The multiple mirror consists of 36 hexagonal mirrors, each 1.8 metres across. Around a central opening of 1.8 metres, there are six hexagons in the first ring of mirrors, 12 in the second ring and 18 in the third ring. The 36 mirrors are also supported by means of active optics. The effective width of the 36 mirrors is 10 metres. The spectrograph to be used with the telescope weighs 4 tons. The project was made possible by a donation of $70 million by the W.M. Keck Foundation. The Foundation has donated another $74.6 million for the construction of a second 10-metre telescope, which is being erected 85 metres from the first. These two telescopes, acting in tandem, will have a light-gathering power eight times that of the 5-metre Hale Telescope of Mount Palomar.

A new technique for casting mirrors has been developed by Roger Angel. While the mirror is being cast, the whole oven is rotated at such a speed that the molten glass will, on cooling, have a parabolic shape. This will eliminate much of the grinding necessary. Mirrors of widths of 6 metres have been successfully cast. The Japanese Observatory has an 8-metre mirror on order. It has already been cast and the telescope ought to be in operation by the year 2000.

The Hubble Space Telescope

The Hubble Space Telescope (Figure 4.26) was launched by the space shuttle Discovery in 1990 and placed into an orbit 610 km above the surface of the Earth. The orbit is inclined by an angle of $28\frac{1}{2}$ degrees to the equator. The diameter of the reflecting mirror is 2.4 metres. The mirror is an f-24. Its

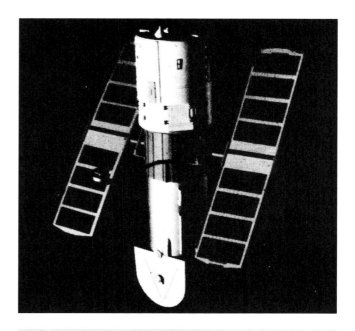

Figure 4.26 The Hubble Space Telescope

field of view is 18 minutes of arc (slightly more than half the Moon's diameter) and its resolving power is 0.05 arc seconds, which is much better than that of any existing telescope. From its vantage point outside the Earth's atmosphere, the HST is very well suited to operate at those wavelengths, such as the ultra-violet and infra-red, which do not penetrate the Earth's atmosphere.

Great was the disappointment among astronomers when it transpired that there was a fault with the main mirror – that it had been polished to a spherical shape instead of parabolic. America's National Aeronautics and Space Administration undertook a repair mission and in December 1993 the space shuttle Endeavour was sent off to repair the optics. The mission was a great success. Not only did the astronauts repair the optics of the main mirror, but also the wide angle and faint object cameras and the protective covers of the magnetometers. The solar cells were replaced and the gyroscopes now keep the telescope jitter-free.

The capabilities of the telescope are as follows:

1. The f-12.9 wide angle camera has a scanning grid of 640 000 pixels in the charge coupled device. The camera also has an f-30 ratio which can produce photographs as good as those made by the Voyager spacecraft when it was 5 days from Jupiter. The camera is sensitive to wavelengths from 1150 Å in the ultra-violet to 11 000 Å in the infra-red. It is provided with 48 filters which can be employed to let a variety of wavelengths through. Objects as faint as magnitude 28 can be detected. This is 40 times fainter than the 5-metre Hale telescope can detect.
2. The faint object camera has ratios of f-48 and f-288 and is sensitive to wavelengths from 1150 Å to 6500 Å. This camera can employ 44 filters.
3. A high-resolution spectrograph can operate in the ultra-violet. The spectrograph has a built-in exposure meter which works automatically.
4. A faint object spectrograph operates in two wave bands, namely 1150 Å to 5500 Å and 1700 Å to 8500 Å, that is, from ultra-violet to deep red.
5. The high-speed photometer can record 100 000 measurements of brightness per second.
6. Fine-tuned sensors can direct the telescope by means of light intensities and can also steady the telescope.
7. The telemetry systems include radio reception and transmission.
8. The two photoelectric solar panels convert sunshine into electricity to drive the instrument and to charge the batteries.

The total mass is 11 600 kg and the total length is 13 metres.

Chapter 5
Fundamental Measurements

Telescopic Discoveries

When the telescope came into use, a nearly endless series of discoveries followed. In one of the unpublished manuscripts of Thomas Harriot, mention is made of the fact that he saw Jupiter's four large moons before Galileo mentioned his discovery in 1610. Harriot also discovered sunspots. Galileo kept his discovery of sunspots secret, out of fear for his life! How could he dare say that the immaculate Sun had blemishes? In 1665 Christiaan Huygens discovered the large moon of Saturn, Titan. Giovanni D. Cassini (1625–1712) discovered a gap in Saturn's rings. He also discovered four of Saturn's moons: Iapetus (1671), Rhea (1672), Dione and Tethys (1684). In addition, he made very careful measurements of the periods of Jupiter's large moons. This enabled Olaus C. Rømer (1644–1710) to make the first determination of the velocity of light (Figure 5.1). Rømer noticed that the times at which the four moons disappeared behind Jupiter became systematically later as the Earth moved

Figure 5.1 Olaus C. Rømer

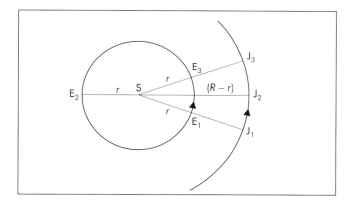

Figure 5.2 Light takes 990 seconds longer to cover the distance J_2E_2 than the distance J_1E_1

away from its position of opposition, E_1 (Figure 5.2), to the other side of the Sun, E_2. As the Earth came back towards opposition, the times became earlier. Rømer correctly ascribed these discrepancies to the time that it takes light to travel across the diameter of the Earth's orbit. This implied that the velocity of light could not be infinite.

When Jupiter was at J_2 and the Earth at E_2, the times of commencement of the eclipses were 990 seconds later, on average, compared with the times when the Earth was at E_1 and Jupiter at J_1.

If R is the radius of Jupiter's orbit and r the radius of the Earth's orbit, light has to travel the distance J_1E_1, i.e. $(R - r)$, when the Earth is at E_1 and Jupiter at J_1. When the Earth is at E_2 and Jupiter at J_2, light has to travel a distance $(R - r) + r + r$, namely a distance $(R + r)$. If c is the velocity of light in km s^{-1}, it takes light $(R - r) \div c$ seconds to travel from J_1 to E_1 and $(R + r) \div c$ seconds to travel from J_2 to E_2. The difference between these two times is 990 seconds according to Cassini's and Rømer's measurements. This means that:

$$\frac{(R+r)}{c} - \frac{(R-r)}{c} = 990 \text{ seconds}$$

$$\therefore \frac{(R+r)-(R-r)}{c} = \frac{2r}{c} = 900 \text{ seconds}$$

$$\therefore c = \frac{2r}{990} = \frac{r}{495} \text{ km s}^{-1}$$

Here Rømer had an equation linking the velocity of light, c, with the average distance of the Earth from the Sun, r. These are the two fundamental dimensions in astronomy. If r is known, c can be calculated, and vice versa. The distance from the Earth to the Sun was, however, not well known in 1675. Rømer used a value of 143 million km. His value for the velocity of light was thus:

$$c = \frac{r}{495} = \frac{143\,200\,000}{495} = 289\,293 \text{ km s}^{-1}$$

R.S. McMillan and J.D. Kirzenberg repeated the experiment in 1971. Their value for the retardation in the time of the eclipses was 1008 seconds, and using the modern value for the distance from the Sun to the Earth, 149 600 000 km, their value for the velocity of light came to:

$$c = \frac{r}{504} = \frac{149\,600\,000}{504} = 296\,825 \text{ km s}^{-1}$$

This is in fairly good agreement with today's value of 299 792.458 km s^{-1} for the speed of light, which has been accepted by the International Astronomical Union at its conference which was held at Grenoble in 1976.

Using today's value of r, 149 600 000 km, Rømer would have obtained a value of 302 222 km s^{-1} for the speed of light.

The speed of light is tremendously great, but not infinite.

A prerequisite for astronomy was therefore to obtain a good value for the distance from the Earth to the Sun. This distance is called the astronomical unit, abbreviated AU.

Distance of the Moon

Because the Sun is so very far away, it subtends a barely perceptible angle of parallax at the Earth's distance, even if the points of observation are at two extremities of the Earth. The Moon is very much nearer, as Aristarchus had shown, and it will be easier to determine the Moon's distance from the Earth as a first step into the vastness of space. If a telescope is aimed at the Moon from the point E on the Earth, the angle between the zenith (the point vertically overhead) and the Moon's direction will be the

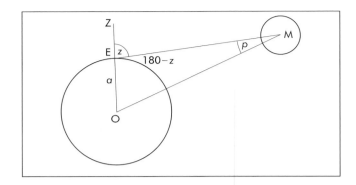

Figure 5.3 The Moon's parallax

angle ZEM. The correct angle between the zenith and the Moon's direction is the angle ZOM, where O is the Earth's centre (Figure 5.3).

Now, angle ZEM + angle OEM = 180 degrees, since ZEO is a straight line. The sum of the angles of a triangle = 180 degrees. Therefore angle (EOM + EMO) + angle OEM = 180 degrees. Therefore ZEM = angle (EOM + EMO). Angle EMO is the parallax of M, as subtended by the points E and O, respectively on the Earth's surface and at its centre.

The sine rule (Appendix 1, section 1.2) states that, in any triangle, the sine of an angle divided by the opposite side, is constant.

Thus $\quad\dfrac{\sin \text{EMO}}{\text{EO}} = \dfrac{\sin \text{OEM}}{\text{OM}}$

Therefore $\quad\dfrac{\sin p}{a} = \dfrac{\sin(180°-z)}{d}$

and $\quad \sin p = \dfrac{a}{d} \sin z, \quad$ because

$\sin(180° - z) = \sin z \quad$ (Appendix 1, section 1.7)

When the object observed is on the horizon of E

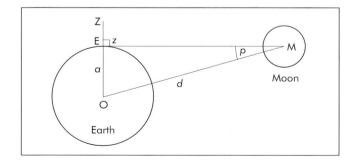

Figure 5.4 Horizontal parallax, angle EMO = P

Distance of the Moon

(Figure 5.4), the zenith angle ZEM = 90 degrees. The sine of 90 degrees is equal to 1. Therefore the horizontal parallax, P, is given by:

$$\sin P = \frac{a}{d}$$

Therefore $\sin p = \sin P \sin z$.

Because the angles p and P are always very small, the sine-values can be equated to the circular measure of the angles (Appendix 1, section 1.8). Therefore $p = P \sin z$, where P is the horizontal parallax.

In order to measure the Moon's parallax, and thus its distance, the zenith angles of the Moon must be measured simultaneously from two widely separated points on the Earth, preferably situated on the same longitude, i.e. angles z_1 and z_2 must be measured (Figure 5.5).

If the latitudes of points A and B, north and south of the equator, are known, figure OAMB can be solved, and the angles of parallax, p_1 and p_2, and the Moon's distance OM can be calculated.

In practice the Moon's equatorial horizontal parallax is arrived at by calculation, although the points of observation are not situated on the equator.

Simultaneously with determining the zenith angles z_1 and z_2 (Figure 5.5) the differences between the Moon's meridian zenith angle and those of a distant star X, having nearly the same right ascension and declination as the Moon, are measured, i.e. the angles MAX = a_1 and MBX = a_2 are measured. The star X is so far away that it has zero parallax.

At A the parallax of the Moon is the angle AMO = $P \sin z_1$ and the parallax at B is the angle BMO = $P \sin z_2$. But angle AMO + angle BMO = angle AMB.

$$\therefore \text{angle AMB} = P \sin z_1 + P \sin z_2$$
$$= P(\sin z_1 + \sin z_2)$$

Draw MX′ parallel to AX and BX:

$$\therefore \text{angle AMB} = \text{angle}\,(\text{AMX}' - \text{BMX}')$$
$$= a_1 - a_2$$
$$\therefore P(\sin z_1 + \sin z_2) = a_1 - a_2, \quad \text{so that}$$
$$P = \frac{a_1 - a_2}{\sin z_1 + \sin z_2}$$

To find the value of P, the Moon's horizontal parallax, we need to know the two zenith angles, z_1 and z_2, as well as angles a_1 and a_2. By means of this simple expedient, it is

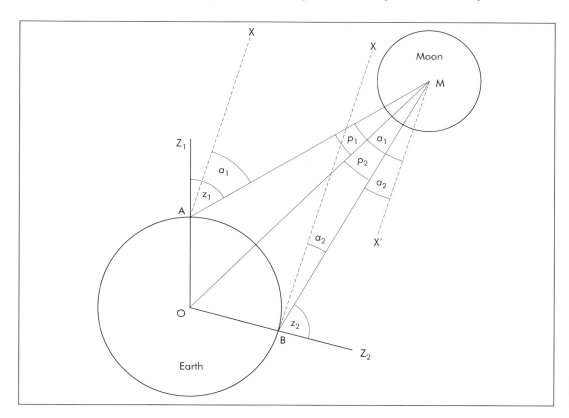

Figure 5.5
Measuring the Moon's equatorial horizontal parallax

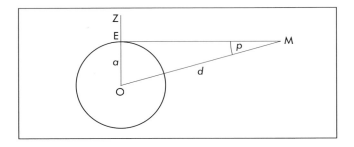

Figure 5.6 The Moon's distance from Earth

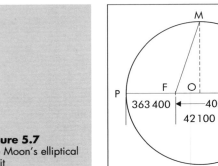

Figure 5.7 The Moon's elliptical orbit

not necessary to know the latitudes of observation points A and B. However, it is necessary to have a very accurate value of the Earth's equatorial radius. Today's accepted value is 6378.14 km. This is the value of a in Figure 5.6.

The most accurate value of the Moon's average horizontal equatorial parallax is $57'2.5''$. In arc seconds this is 3422.5. In circular measure (Appendix 1, section 1.8) this angle $= (3422.5)(3.14159) \div (180)(3600)$. This works out to be 0.016 593, which is the Moon's parallax in radians. Because the angle is so very small (less than 1 degree), the sine of the angle is equal to the angle itself, in radians. Therefore

$$P = 0.016\,593 = \sin P = \frac{a}{d} = \frac{6378.14}{d}$$

Thus $d = \dfrac{6378.14}{P} = \dfrac{6378.14}{0.016\,593} = 384\,400$ km

This is the Moon's average distance from the Earth, that is half the major axis, PA, of its elliptical orbit around the Earth (Figure 5.7). It is equal to the distance from the Earth, F to M, at the extremity of OM, the semi-minor axis. When the Moon is nearest to the Earth, at perigee P, its distance is 363 400 km on average: and when it is furthest, at apogee A, its distance is 405 500 km. The length of the major axis of the Moon's orbit is therefore $PF + FA = 768\,900$ km. The distance between the foci, $FF' = 405\,500 - 363\,400 = 42\,100$ km. The eccentricity of the Moon's orbit is equal to the distance between the foci divided by the major axis, namely $42\,100 \div 768\,900 = 0.0548$.

In its mean position BB′ (Figure 5.8) the Moon's diameter subtends an angle BOB′, equal to $31'05''$. At perigee, AA′, the angle $AOA' = 32'53''$. At apogee, CC′, the angle is $29'28''$. The distances from the Earth are inversely proportional to the angles subtended by the diameter of the Moon:

$$\frac{\text{Dist. OA}}{\text{Dist. OB}} = \frac{\text{angle BOB}'}{\text{angle AOA}'} = \frac{31'05''}{32'53''} = \frac{1865''}{1973''}$$

$$\therefore \text{Dist. OA} = \frac{1865}{1973} \times \text{Dist. OB}$$
$$= \frac{1865}{1973} \times 384\,400 = 363\,400 \text{ km}$$

At apogee:

$$\text{Dist. OC} = \frac{31'05''}{29'28''} \times \text{Dist. OB}$$
$$= \frac{1865}{1768} \times 384\,400 = 405\,500 \text{ km}$$

In the above calculations we have made use of the fact that the apparent size of an object is inversely proportional to its distance – the further the object, the smaller it appears to be; the smaller the distance, the larger it appears to be.

The angular diameter of the Moon can be used to

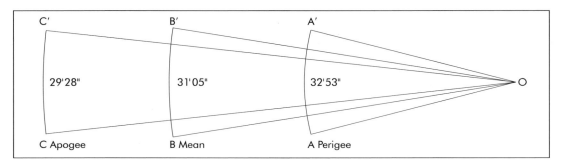

Figure 5.8 The Moon's apparent angular diameter

The Moon's Diameter

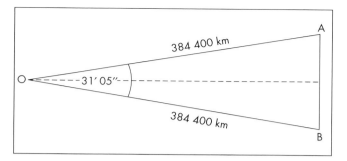

Figure 5.9 The Moon's diameter

calculate its actual diameter, since its distance is known. In Figure 5.9, the average circular angle subtended by the Moon's diameter, 31′05″, is represented by the angle AOB, where A and B are the extremities of the Moon's diameter. AB thus passes through the Moon's centre. OA and OB are each equal to 384 400 km. The angle AOB = 31′05″ = 1865 seconds.

This equals $\dfrac{1865(3.141\,59)}{180(3600)}$ radians

$= 0.009\,041\,77$ radians

$= \sin \text{AOB} = \dfrac{\text{AB}}{\text{OA}}$

∴ AB = OA(0.009 041 77)

$= 384\,400\,(0.009\,041\,77) = 3476$ km

Thus the Moon's diameter is 3476 km.

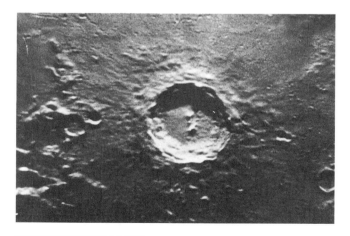

Figure 5.10 Crater Copernicus, 90 km wide

THE measurements of any object on the Moon's surface can be determined in the same way, because the Moon's distance is known. The crater Copernicus (Figure 5.10), for example, has an angular diameter of 48.5 arc seconds. To convert this angle to circular measure, multiply by $\pi\,(= 3.141\,59)$ and divide by 180(3600) (Appendix 1, section 1.8).

$48.5'' = \dfrac{48.5(3.141\,59)}{180(3600)} = 0.000\,235\,13$ radians

∴ $\dfrac{\text{Diameter of crater}}{\text{Distance of Moon}} = 0.000\,235\,13$

∴ Diameter of crater $= 0.000\,235\,13\,(384\,440)$

$= 90.4$ km

i.e. diameter of Copernicus crater = 90.4 km

To find the heights of the mountain peaks on the Moon, the angular measures of the shadows of the peaks have to be measured, that is the lengths such as AB in Figure 5.11, cast by the peak whose height is AC. The zenith angle of the incident rays of sunlight, CBZ, has also to be measured.

The shadow AB of a peak in the Apennine Mountains (Figure 5.12 *overleaf*) subtends an angle of 3.25 arc seconds when the zenith angle of the Sun's rays, angle CBZ, is 48 degrees.

$3.25'' = \dfrac{3.25\,(3.141\,59)}{180\,(3600)} = 0.000\,015\,76$ radians

∴ $\dfrac{\text{Length of shadow}}{\text{Distance of Moon}} = 0.000\,015\,76$ radians

∴ Length of shadow $= 0.000\,015\,76\,(384\,400)$

$= 6.056$ km $=$ AB

∴ $\tan 42° = 0.9 = \dfrac{\text{AC}}{\text{AB}}$ (Appendix 1, section 1.1)

∴ AC $= 0.9\,(\text{AB}) = 0.9\,(6.056)$ km

$= 5.450$ km $= 5450$ metres

$= 17\,880$ feet

Figure 5.11 Height of peak on Moon

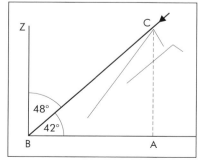

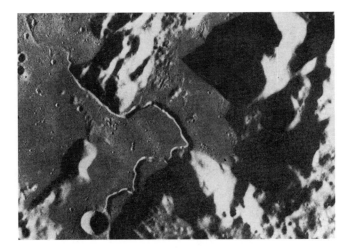

Figure 5.12 The Apennine Mountains on the Moon

Table 5.1 Corrections for refraction

Zenith angle	Correction	Zenith angle	Correction	Zenith angle	Correction
0°	0.0″	35°	40.8″	70°	2′ 38.8″
5°	5.1″	40°	48.9″	75°	3′ 34.3″
10°	10.3″	45°	58.2″	80°	5′ 19.8″
15°	15.6″	50°	1′ 9.3″	85°	9′ 47.8″
20°	21.2″	55°	1′ 23.4″	87°	14′ 28.1″
25°	27.2″	60°	1′ 40.6″	89°	24′ 21.2″
30°	33.6″	65°	2′ 04.3″	90°	33′ 46.3″

Because the angle of parallax of the Moon is so small (less than 1 degree) great care must be exercised while doing the measurements. The micrometer screw in the eyepiece must be very accurate and the right ascension and declination circles must be accurately adjusted. The clock drive of the telescope must be in perfect working order.

To obtain good results for the Moon's parallax, the difference between the angles a_1 and a_2 in Figure 5.5 must be as large as possible. The observing stations must therefore be as far apart as possible. That is the reason why the observatories selected were those of Greenwich and Cape Town. But Cape Town is not on the same longitude as Greenwich, being 18°22′ east. Corrections have therefore to be made for the small change in declination of the Moon during the lapse of time between meridian passage at Greenwich and Cape Town.

Because the angles being measured are so very small, account must be taken of refraction of light in the atmosphere. When light passes from a rarefied medium to a denser one, the rays are bent towards the normal (perpendicular to the surface) – see Figure 5.13.

The effect is that the light describes a curved path from A to B to O and is seen in the direction OS′ instead of OS, parallel to AB. Refraction therefore decreases the zenith angle from ZOS to ZOS′. The correction that has to be applied is proportional to the tangent of the zenith angle if the angle is fairly small. In Table 5.1 the corrections required for various zenith angles, are given.

On the horizon, where the refraction correction is about 34 minutes, the Sun and Moon will be seen just above the horizon when they are actually just below the horizon, because their diameters are about 32 minutes. Thus their rising times are accelerated and their setting times retarded, by about 2 minutes 8 seconds. Because refraction on the lower limbs of the Sun and Moon is about 4′42″ more than on the upper limbs, these bodies have a flattened shape when they are on the horizon. The amount of flattening is about 12% (Figure 5.14).

When measuring the Moon's parallax, corrections have to be made to compensate for the fact that the Earth is not a

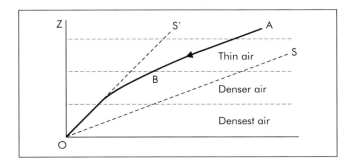

Figure 5.13 Refraction of light in the atmosphere

Figure 5.14 Flattening of the setting Sun as a result of refraction

perfect sphere. Also, a much more complicated formula is used, but the principle is the same as that described here. Today, radar is used to measure the Moon's distance.

The Parallax of the Sun

The Sun is much further away than the Moon and its angle of parallax cannot be measured directly because of the Sun's glare. The Sun's parallax therefore has to be determined by indirect methods by using a planet such as Mars when it is in opposition and therefore at its nearest to the Earth.

We have seen (page 30) that the Moon's parallax increases the zenith angle of the Moon. Similarly, when a celestial body is east of the prime meridian, its right ascension, relative to the stars, is increased by parallax; and when the body is west of the prime meridian, its right ascension is decreased by parallax (Figure 5.15).

In the figure, $C\gamma$ represents the direction 00 h 00 m, that is, the First Point of Aries, M and N the planet, when east and west of the meridian, and YZBA the sky.

When the planet is at M, east of the prime meridian, its right ascension is γB, with reference to C, the Earth's centre. From the point of observation, O, however, the planet is seen in the direction OMA and its right ascension is γA, which is greater than γB by an amount equal to BA. The arc BA is equal to the angle p, which is the parallax of M as seen from O, on the surface of the Earth.

When the planet is at N, west of the prime meridian, its right ascension is γY as seen from O, while its correct right ascension is γZ, as seen from C. γY is less than γZ by an amount YZ, which is the parallax of N as seen from O, and which is equal to the angle p.

The increase or decrease of the right ascension is proportional to the sine of the hour angle.

The procedure is therefore as follows: The right ascension of the planet is read off just after it rises in the east, and then read again just before setting in the west. The change in the planet's position is then due to:

(i) the parallax of the planet, and
(ii) the movement of the planet relative to the background of the stars.

If the readings are repeated the next night, the planet's own motion relative to the stars can be identified and subtracted from the movements (i) and (ii), to yield the alteration in right ascension due to parallax.

Account must be taken of refraction, which is considerably larger than the angle of parallax.

In Rømer's time, the parallax of Mars was taken as being 14 arc seconds.

Assuming Kepler's value of 1.524 as the ratio between the distance of Mars from the Sun and the distance of the Earth from the Sun, then the parallax of the Sun compared with the parallax of Mars will be the inverse, namely $1 \div 1.524$, because parallax is inversely proportional to the distance. Thus:

$$\frac{\text{Parallax of Sun}}{\text{Parallax of Mars}} = \frac{1}{1.524}$$

$$\therefore \text{Parallax of Sun} = \frac{1}{1.524} \times 14'' = 9.186''.$$

$$= \text{Angle ASO (Figure 5.16)}$$

$$\therefore \frac{\text{Earth's radius}}{\text{Sun's distance}} = \frac{OA}{OS} = \sin ASO$$

$$\therefore \text{Sun's distance} = \frac{OA}{\sin 9.186''} = \frac{6378}{0.000\,044\,53}$$

$$= 143\,200\,000 \text{ km}$$

This was the value which Rømer used.

When Mars is in opposition and, at the same time, at perihelion and the Earth also at aphelion (a fairly rare event), the parallax of Mars can be as much as 23.5″. Today's best value for the average parallax of Mars is 13.4″. Using this value the Sun's parallax $= 13.4'' \div 1.524 = 8.793''$. The sine of 8.793″ is 0.000 042 63. Thus the Sun's distance is equal to the Earth's radius divided by sin 8.793″, that is

$$6378 \div 0.000\,042\,63 = 149\,610\,000 \text{ km}$$

Later, when the Minor Planets were discovered, it was

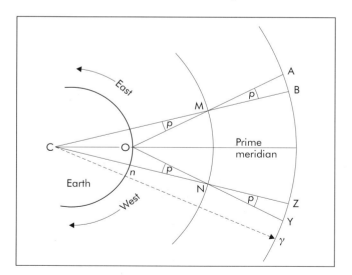

Figure 5.15 The effect of parallax on right ascension

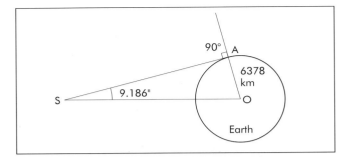

Figure 5.16 The parallax of the Sun

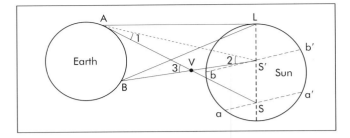

Figure 5.17 Sun's parallax by means of the transit of Venus

found that some of these approach the Earth closer than does Mars. In 1931, the minor planet Eros came to within 24 million km of the Earth, compared with Mars' closest approach of 56 million km.

An extensive programme was undertaken by many observatories to measure the parallax of Eros in 1931. The value of the solar parallax which this programme yielded was 8.790″, from which the distance of the Sun was calculated to be 149 670 000 km.

The most modern methods for determining the Sun's distance make use of radar. The radar waves are bounced off the planet and the time for the journey to and fro is measured. Today the velocity of light is known very accurately and therefore the distances of the planets and the Sun can be determined with great accuracy. The value of the Sun's distance accepted by the International Astronomical Union in August 1976 is 149 597 870 km. This value of the average distance of the Sun from the Earth is known as the Astronomical Unit. It forms the basis of all determinations of distances in the Solar System and it also serves as the base line on which the distances of the stars are calculated.

The equatorial horizontal parallax of the Sun which is derived from this value of the astronomical unit (AU) is 8.794″. It is worth noting that David Gill who was stationed at Cape Town Observatory from 1879 to 1906, and who was a leading astronomer in his time, measured the Sun's parallax from the Island of Ascension in 1877 by using the planet Mars. He obtained a value of 8.78″ for the average equatorial horizontal parallax of the Sun. This was a remarkable achievement.

It is of historical value to note that Edmond Halley, having seen the transit of Mercury across the face of the Sun in 1677, suggested that the transit of Venus would be a good opportunity to determine the parallax of the Sun. The next transit was due in 1761. The method suggested by Halley is illustrated in Figure 5.17.

The observer at A sees Venus as a small black dot moving across the face of the Sun from a to a′, and the observer at B sees it move across the Sun from b to b′. At an agreed moment of time, A measures the angle SAL between Venus as seen against the Sun and the Sun's limb at L. At the same time, B measures the angle S′BL between Venus as he sees it against the face of the Sun, and the limb of the Sun at L. This angle is equal to angle S′AL. Subtracting this angle from the angle measured by A gives angle SAS′. Call it angle 1. A and B also measure the directions of Venus, V, against the background of the stars by noting its positions some time before and after the transit. This gives angle 3, which is the parallax of Venus. Angle AS′B (= angle 2) is the Sun's parallax. It is equal to angle 3 minus angle 1.

Now, the Sun's parallax (angle 2) divided by the parallax of Venus (angle 3) is the inverse of 0.723, which is the relative distance of Venus from the Sun, according to Kepler's third law.

IN a modification of this method, suggested by L. Delisle, the two observers, instead of being north and south of each other, are to be as far east and west of each other as possible and their longitudes must be known accurately. The two observers note the times on their synchronised chronometers of first and last contact, as Venus moves across the face of the Sun. The set-up is illustrated in Figure 5.18, looking down on the north.

AV_1P = moment of first contact for A
AV_2Q = moment of last contact for A
BV_3P = moment of first contact for B
BV_4Q = moment of last contact for B.

From these time readings, it is easy to calculate the times taken by Venus to move from V_1 to V_3 and from V_2 to V_4.

Corrections have to be made for the proper motions of Venus during these time lapses.

Triangles APB and AQB can therefore be solved, because the relative distance of Venus from the Sun, 0.722 91, is

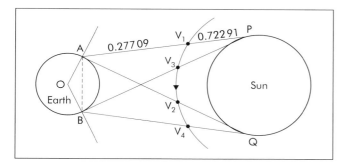

Figure 5.18 Delisle's method for measuring the Sun's parallax by the transit of Venus

known from Kepler's third law.

If the sidereal period of Venus is X, and that of the Earth, Y:

$$\text{then } \frac{X}{Y} = \frac{224.5 \text{ days}}{365.25 \text{ days}} = 0.614\,647\,5$$

The square of $0.614\,647\,5$ is $0.377\,791\,55$, which is equal to the cube of the relative distance of Venus from the Sun. The one-third power of $0.377\,791\,55$ is $0.722\,91$, as shown in the figure.

The distance V_1V_3 is known in circular measure and is equal to the angle APB. The sine of this angle = AB ÷ BP. From this the distance of the Earth from the Sun can be calculated.

The mathematics employed is more complicated, but the principle is the same.

By using the measurements made during the transits of Venus in 1761 and 1769, Simon Newcomb, in 1890, found a value of 8.79″ for the Sun's parallax.

The transits of Venus occur at intervals of 8, 122, 8 and 105 years. Five observed transits occurred in the years 1639, 1761, 1769, 1874 and 1882. The next will occur in 2004 and 2012.

On account of the inherent difficulties connected with observations directly on the Sun, these methods have fallen into disuse.

The Heliometer

An instrument which was very useful in determining the Sun's parallax was the heliometer, perfected by John Dolland in 1754. It consisted of a refracting telescope of which the objective lens was divided into two. The two half-lenses could each form an image independently, and could slide across the optical axis. The images of two objects, very close together, were, by sliding the half-lenses, brought together to form a single image. The angular separation of the two objects was read off on a micrometer. The angular measure of the diameter of a planet could then be read off. If the planet's distance was known, its actual diameter could be calculated. The heliometer has been replaced by photography, in which the separation of objects can be measured by means of a microscope.

The Aberration of Light

About the year 1726, James Bradley discovered that the zenith angles of stars showed annual deviations of about 20 arc seconds about their mean positions. This could not be ascribed to parallax, since all stars were affected, whether near or far. Nor could it be due to refraction, since stars overhead were no less affected than those near the horizon. The reason was that the speed of the Earth in its orbit was a measurable fraction of the velocity of light, which was not infinitely great, as Rømer had shown. This deviation of stars from their correct positions is called aberration. It can be compared to vertically falling raindrops, which appear to fall at a slant as soon as one moves forward (Figure 5.19).

Let C be the centre of the objective lens of a telescope, represented by AC, where A is the eyepiece. The true direction of a star, X, is XCA′. By the time the ray of light reaches the eyepiece, it has moved from A to A′ – a very small amount, yet measurable. If t is the time that it takes light to travel from C to A′ at a velocity c, then $CA' = ct$. In the same time, the Earth has moved from A to A′, in the direction AB, at a speed of v. Therefore $AA' = vt$.

If the Earth had not moved, the telescope, which points to the star, would be pointing in the direction A′C. Because of the Earth's speed, the telescope has to point in the

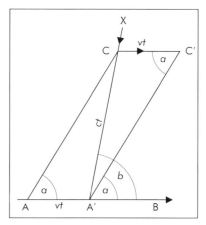

Figure 5.19 The aberration of light

direction A'C'.

CAA'C' is a parallelogram in which angle CC'A' = angle CAA'.

In triangle CA'C', according to the sine rule (Appendix 1, section 1.2):

$$\frac{\sin CA'C'}{CC'} = \frac{\sin CC'A'}{CA'}$$

But $CC' = AA' = vt$, and $CA' = ct$

$$\therefore \sin CA'C' = \frac{CC'}{CA'} \sin CC'A'$$

$$\therefore \sin(b - a) = \frac{vt}{ct} \sin CAA' = \frac{v}{c} \sin a$$

Because the angle $(b - a)$ is very small, we can use angular measure and say:

$$b - a = \frac{v}{c} \sin a$$

But $\frac{v}{c} = \frac{\text{Velocity of Earth}}{\text{Velocity of light}} = \text{constant}$

Therefore $(b - a) = K \sin a$, where a is the angle that the direction of the star makes with the direction of motion of the Earth. The Earth's direction of motion is along the tangent to the Ecliptic and it lies in the plane of the Ecliptic.

The greatest value that the angle of aberration $(b - a)$ can have is 20.5″, when $\sin(a)$ has its maximum value of 1, that is, when the angle CAB between the star's direction of motion and that of the Earth is approximately 90 degrees. As this angle becomes less than 90 degrees, the angle of aberration gets less than 20.5″.

Speed of the Earth in its Orbit

The average distance of the Earth from the Sun is 149 597 870 km. Using this value and considering the orbit to be a circle, the circumference of the orbit will be 2π times the radius. The time to cover this distance is 1 year. Therefore the speed of the Earth in its orbit is:

$$\frac{2(3.14159)(149597870)}{(365.25)(24)(3600)} = 29.785 \text{ km s}^{-1}$$

This is the value of v in the equation:

$$b - a = \frac{v}{c} \sin(a)$$

When $a = 90°$, $\sin(a) = 1$. Therefore

$$\frac{v}{c} = \frac{29.785}{c} = b - a = 20.5'' = \frac{(20.5)(3.14159)}{(180)(3600)}$$

$$\therefore c = 299\,688 \text{ km s}^{-1}$$

(20.5″ has to be converted into radians)

The value of c is very close to the most accurate value yet obtained, showing that the aberration constant is correct.

The first experimental determination of the velocity of light was made by H.L. Fizeau in 1849. He sent pulses of light between the teeth of a rapidly rotating cogwheel, to a reflecting mirror 6 km away. The speed of the rotating cogwheel was increased until a steady reflection was obtained. The time for the light to cover the to-and-fro journey was thus equal to the time for a cog to move through a distance equal to its own width. From this, Fizeau obtained a value of 315 000 km s^{-1} for the speed of light.

J.B.L. Foucault used a rotating mirror instead of a cogwheel.

A.A. Michelson and E.W. Morley controlled the speed of the rotating mirror by means of light reflected from a mirror on a tuning fork of frequency 256 vibrations per second. Sunlight falling on the rotating mirror was focused by a lens of focal length 45 metres on to a reflector 600 metres away. The deflection undergone by the sunlight was a measure of the amount of turning which the mirror had undergone. From this, the time for the to-and-fro journey of the ray of light could be found. Michelson and Morley obtained a value of 299 860 km s^{-1} for the velocity of light. This was very close to today's accepted value of 299 792.458 km s^{-1}.

Particles or Waves?

Although Newton considered light as a stream of particles, the wave theory championed by C. Huygens gained ground. Many experiments have shown that light undergoes interference, namely that the amplitudes of the waves can alternately increase or decrease. This can only be adequately explained if it is accepted that light is propagated in waves.

To explain the propagation of light waves through space, physicists supposed that space was filled with "something" which they called ether. This meant that the motion of the Earth through the ether should cause an ether wind, which would affect the velocity of light in various directions. Michelson and Morley adapted their experiment to measure this ether wind. Their result was negative, showing that the velocity of light is constant in all directions. This is a fundamental principle in Einstein's Theory of Relativity.

Fundamental Data

Velocity of light 299 792.458 km s^{-1}.
Astronomical Unit (AU) 149 597 870 km.
Sun's equatorial horizontal parallax 8.794″.
Aberration constant 20.5″.
Equatorial radius of Earth 6378.14 km.
Average distance of Moon 384 400 km.

Chapter 6
Surveying the Planets

Astronomy made rapid progress during the 17th, 18th and 19th centuries, thanks to the telescope. With the telescope it was possible to sight with precision and to determine the positions of stars and planets very accurately. The filar micrometer, in the eyepiece, made it possible to measure angular separations of very small angles.

1. Synodic Periods

The synodic period of a planet (or the Moon) is the time that elapses between two successive appearances of the body in the same relative position with regard to the Sun and the Earth. If, for example, the angle between Mars and the Sun, is 180 degrees, that is, Mars rises in the east when the Sun sets in the west (Mars then being in opposition), then Mars will be in opposition again after a lapse of 780 days. This is the synodic period of Mars. When Venus is directly in line between the Earth and the Sun, it is said to be at inferior conjunction. The time lapse before the next inferior conjunction, is 583.9 days. This is, therefore, the synodic period of Venus. The Moon's synodic period is 29 days 12 hours 44 minutes 2.9 seconds, or 29.530 589 days. Synodic periods can be measured from any position, relative to the Earth and Sun. (The data for all the planets appear in Tables 7.3 and 7.4 on pages 84 and 85.)

2. Sidereal Periods

The sidereal period is the lapse of time between two successive appearances of a planet in the same position, relative to the background of the stars, as viewed from the Sun. It is equal to the length of the planet's year, which is the time it takes the planet to complete one revolution round the Sun. Jupiter takes 11 years 314 days 2 hours 45 minutes and 36 seconds = 11.86 years to make one revolution around the Sun and to return to the same position against the background of the stars, as seen from the Sun. Compare this with Jupiter's synodic period, which is only 1 year 33.5 days – the time for it to appear in the same position, relative to the Earth and Sun.

3. Distances from the Sun

By applying Kepler's third law, the relative distance of a planet from the Sun, in terms of the Earth's distance, can be calculated. It is not necessary to know the Earth's actual distance in km, in order to find the relative distance, which is stated in Astronomical Units (AU). The relative distance of the furthest planet, Pluto, is 39.46 AU. The width of the Solar System is therefore 78.92 AU.

4. Distances in Kilometres

To find the actual distance of a planet from the Sun in km, the Earth's mean distance must be known. This is equal to one half of the major axis of the Earth's elliptical orbit around the Sun. The distance which has been accepted by the International Astronomical Union is 149 597 870 km. This is the value of the Astronomical Unit.

5. Eccentricities of Orbits

The actual distances of the planets from the Sun are continually changing, because their orbits, like that of the Earth, are ellipses. When a planet is at its closest to the Sun, at perihelion, it will be seen to move at its greatest speed against the background stars. This can be easily observed. When the planet is furthest from the Sun, at aphelion, it moves at its slowest. By noting planetary positions throughout the planet's sidereal period, the eccentricity of the orbit can be determined. The orbit of Venus is the least eccentric of all the planetary orbits. It is the nearest to a circle, the eccentricity being only 0.0068. The eccentricity is found by dividing the distance between the two foci of the ellipse by the major axis. (In Figure 5.7, FF' indicates the two foci of the Moon's orbit.) The mean distance of Venus from the Sun is 108 160 000 km. The length of the major axis is twice this, namely 216 320 000 km. Because the eccentricity is 0.0068, the distance between the foci is equal to $0.0068 \times 216\,320\,000$, and this works out to 1 471 000 km. Thus, although the orbit is almost a circle, the two foci are still about 1.5 million km apart.

In contrast to the eccentricity of Venus' orbit, that of Mercury's orbit is 0.206 and the foci are 23 854 800 km apart. At aphelion, Mercury is 70 827 400 km from the Sun and 44 972 600 km at perihelion. The ratio between these two distances, is 1575 : 1000. Because the intensities of light and heat are inversely proportional to the squares of the distances, the light and heat at Mercury's perihelion are $(1.575)^2 = 2.48$ times as great as at aphelion.

6. Inclinations of Orbital Planes

The planes in which the orbits of the planets lie are all very close to the plane of the Ecliptic, which is the plane in which the Earth revolves around the Sun. The inclinations of the planetary orbits to the Ecliptic can be measured very accurately because the stars serve as a frame of reference for noting the positions of the planets. The points where a planetary orbit crosses the Ecliptic are called the nodes. The point of cut where the planet moves from south to north of the Ecliptic is called the ascending node, and where the planet moves from north to south the descending node.

The inclinations of all the orbits are less than $3\frac{1}{2}$ degrees, except those of Mercury, 7 degrees, and Pluto, $17°19'$; that of the Moon is $5°09'$. The Solar System therefore apparently lies in a very flat disc.

The points where the celestial equator (which is a projection of the Earth's equator) cuts the Ecliptic are called the First Points of Aries and Libra, designated by the symbols γ and Ω. In the days of Hipparchus, ± 140 BC, the First Point of Aries was situated on the western boundary of the constellation Aries, the Ram. On account of precession, this point has moved westwards, right through Pisces, and is now entering Aquarius, the Water Bearer. All the planetary orbits are subject to precession. The amounts of precession agree with those calculated on the basis of Newton's law of gravitation, except that of Mercury. Einstein's theory of relativity, however, fully explains the precession of Mercury's orbit.

7. Inclinations of the Equators

The axes around which the planets rotate are not at right angles to the planes of their orbits, but deviate by various amounts. The nett result is that the planetary equators are inclined to the planes of the planetary orbits and to the Ecliptic. The bands in Jupiter's atmosphere are taken as parallel to the equator, which shows an inclination of $3°07'12''$ to the plane of Jupiter's orbit, which, in its turn, is inclined by $1°18'$ to the Ecliptic.

The ice caps on the poles of Mars facilitated the determination of the inclination of the axis and of the equator of Mars. Mars' equator has an inclination of $25°12'$ – a remarkable correspondence with that of the Earth, $23°27'$, and that of Saturn, $26°43'$, and Neptune, $29°33.6'$. The equatorial planes of the other planets have inclinations varying between $1°32'$ in the case of the Moon to $97°51.6'$ in the case of Uranus and $177°18'$ in the case of Venus. Because the last two angles are greater than 90 degrees, it means that the rotations of Uranus and Venus are retrograde, from east to west instead of west to east, as are those of all the other planets. Recently it was found that the rotation of Pluto is also retrograde, having an equatorial inclination of 118 degrees.

The inclination of the Earth's equator is the cause of the seasons, and a planet such as Mars must also experience seasons. The dark markings on Mars do undergo seasonal changes, as do the ice caps on the poles. During the summer of a given hemisphere it was noticed that the dark markings enlarge, and that they shrink in the winter. It was alluring to ascribe the seasonal changes on Mars to vegetation flourishing in summer and dying off in winter. Some astronomers, notably G. Schiaparelli and P. Lowell, were convinced that they saw straight-line markings on Mars. Schiaparelli called them "canali" and this gave rise to the notion that they must be canals, which must have been constructed by intelligent beings. To be visible from Earth, the canals had to be very wide. Lowell suggested that the straight lines are actually strips of vegetation astride the canals. The majority of astronomers could, however, not see any straight-line markings. At best, Mars is difficult to observe because it is rather small and far off. We know today that Mars is subject to global dust storms, especially when it is at perihelion. These dust storms serve to blur the visibility. The Earth's atmosphere is also not conducive to good seeing. Astronomers had hoped that it might, one day, in two or three hundred years' time, have become possible to station a telescope in orbit outside the Earth's atmosphere and to obtain better views of Mars from there!

8. Periods of Rotation

The dark blotches on Mars (Figure 6.1) made it possible to measure the time of one revolution. The time that elapsed between the taking of the two photographs was 76 minutes. In that time the planet rotated through 18.5 degrees. To turn through 360 degrees would take $76(360) \div 18.5 = 1478.9189$ minutes, which is equal to 24 hours 38.9 minutes. The best value found for the length of the Martian day is 24 hours 37 minutes 23 seconds. It is only slightly longer than the Earth day. The Martian day is called a "sol".

In the case of the planets that are too far away to show markings, changes in brightness were found to repeat. In the case of Mercury, no markings were visible. Mercury is always very close in direction to the Sun and the planet is visible only during the hours of dusk or early dawn. Because Mercury revolves so close to the Sun, astronomers thought that it must be highly probable that Mercury always keeps the same face turned towards the Sun, under the influence of the Sun's strong gravitational field. If that were so, the day on Mercury would be equal in length of duration to the sidereal period. It would be a parallel case to that of the Moon which always keeps the same face turned towards the Earth. These cases are termed synchronous rotation. One hemisphere of Mercury would thus be perpetually bathed in the merciless rays of the Sun, while the other hemisphere would be frozen in perpetual night. It was not before 1965 that it was discovered by means of the radio telescope that the period of rotation of Mercury is 58.6462 days in length. This is exactly two-thirds

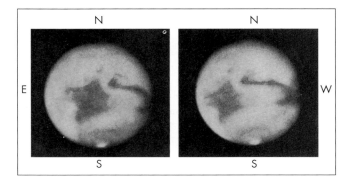

Figure 6.1 During the lapse of time between the two photographs, Mars has rotated eastwards (towards the left)

Figure 6.2 Venus, shrouded in clouds – photographed by Mariner 10 in 1973

of the length of the sidereal period; the periods of rotation and revolution of Mercury are thus in resonance.

The surface of Venus (Figure 6.2) shows no markings – it has an evenly bright appearance. Astronomers accepted that Venus is shrouded in white clouds and that its period of rotation would remain unknown until the clouds were penetrated. The Pioneer space probe, which was launched in 1978, revealed that the cloud layers rotate once in 4 days. The period of rotation, however, was found to be 243 days and the rotation is retrograde.

Jupiter (Figure 6.3, *overleaf*), accompanied by its four large satellites, Io, Europa, Ganymede and Callisto, which were discovered in 1610 by Galileo, shows irregular bands in its atmosphere; and 20 degrees south of the equator is the Great Red Spot, which has been visible on and off for the last three hundred years, sometimes red and then paler. Noting the positions of these markings gave Jupiter a period of rotation of 9 hours 55 minutes 30 seconds.

Saturn (Figure 6.4, *overleaf*), with its extensive system of rings, is one of the most beautiful sights in the Universe. The markings in the atmospheric bands of Saturn are much

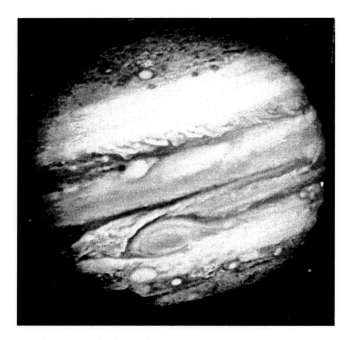

Figure 6.3 Jupiter, photographed by Voyager 1 in 1979

fainter than those of Jupiter. The sidereal period is 10 hours 39 minutes.

Saturn's Rings

The rings which revolve in the plane of the planet's equator give the planet an unique appearance. Its magnificence cannot be over-emphasised. Saturn's orbit is inclined by 2°29'22" to the Ecliptic. The equator is inclined by 26°43' to the plane of Saturn's orbit. The rings can thus show a tilt of as much as 29°12'22" to the plane of the Earth's orbit. The rings are most tilted when the planet is in the constellation Taurus, and again 14.73 years later, in Sagittarius.

Twice every 29.46 years (the sidereal period of Saturn) the Earth moves through the plane of the rings, when Saturn is in Leo and Aquarius, respectively. Then the rings become invisible – only a thin dark line can be seen on the equator. This shows that the rings are very thin, being about 10 km thick. At times when the rings became invisible, searches for small satellites close to Saturn were made.

The eminent astronomer E.E. Barnard measured the width of the rings in 1896. He found that they subtend an angle of 40" at the Earth. With a good knowledge of Saturn's distance of 1277 million km at opposition, he found the actual width to be 277 000 km. The system of rings is, however, entirely inside the Roche limit. In 1846, E.A. Roche calculated that a liquid satellite, orbiting within 2.44 radii of a planet's centre, would be broken up by the torque exercised by the planet's gravitational field, and spread out into a disc, revolving around the planet. Saturn's satellites are not liquid, but, at the beginning of their existence, they must have been rather loosely aggregated. Such a satellite could have been broken up and spread into a disc. It could possibly have been a comet that came too close to Saturn. Such a satellite must be above a certain size, because Saturn's gravitational forces on the near and far sides of the satellite had to be sufficiently different to exercise shear (Figure 6.5).

The force on the near side, A, must be sufficiently greater than the force on the far side, B, so that the tides would be

Figure 6.4 Drawing of Saturn by James E. Keeler in 1888, using the 91-cm refractor of Lick Observatory

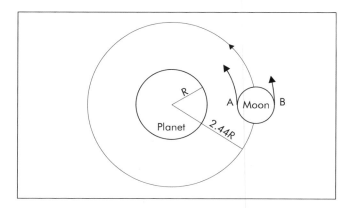

Figure 6.5 Torque exercised by the gravitational force of a planet

Saturn's Rings

Figure 6.6 Finely divided rings in Saturn's A ring

strong enough to bring disruption about. The particles at A are drawn away from the centre of the satellite, and they, in their turn, are drawn away from the particles at B. The orbital speed of the particles at A is also greater than that at B, indicated by a longer arrow at A than at B. The result is that the particles at A will shear away from those at B. Beyond 2.44 radii from the planet's centre, it is fairly safe.

If we multiply Saturn's radius of 60 000 km by 2.44, we get 146 400 km. The outer edge of Saturn's rings is $277\,000 \div 2 = 138\,500$ km from the planet's centre, which is well within the Roche limit of 146 400 km.

There are gaps in the rings, that is, regions where there are no, or very few particles to be found. One such gap is the Cassini gap between the bright A and B rings, and it is 4800 km wide. There is also the Encke gap, which is much narrower. These gaps have been caused by the gravitational forces exercised by the satellites Mimas and Enceladus. The period of revolution of Mimas is exactly twice that of any particles formerly in the Cassini gap. After every two revolutions of such particles, they found themselves in opposition to Mimas, at its closest. The gravitational force of Mimas thus set up a resonance that drew the particles away from the B ring towards the A ring.

However, the Voyager space probes have revealed a multiplicity of narrow gaps even in the brightest and therefore densest parts of the A and B rings, as is seen in Figure 6.6.

The radius of the inner edge of the rings was found by Barnard to be 71 000 km from Saturn's centre. This is only 11 000 km above the tops of the cloud layers which enshroud Saturn. Because particles nearer to Saturn move faster than those further away, there are constant collisions between the particles. There is thus a constant rain of particles falling on to Saturn. These particles are heated to incandescence by friction in Saturn's atmosphere of hydrogen – a constant fireworks' display. The particles being drawn towards Saturn constitute the Crepe ring. In time to come, Saturn will absorb most of the particles in the Crepe ring, but those in the outer rings will be firmly held in place by Saturn's inner satellites.

Confirmation that the rings consist of separate particles was supplied by J.E. Keeler in 1895. He examined the Doppler effect on the lines in the spectra of the inner and outer parts of the rings. By setting up the spectroscope so that the slit was in line across the rings, he found that the lines of the spectrum sloped: towards the blue at b and less so at a (see Figure 6.7 *overleaf*). That meant that the rings could not be solid – the part at b was approaching faster than that at a. On the other side of the planet, the slope of the spectral lines at e had a greater slope towards the red than the part at f. This meant that e was receding faster than f. The body of the planet itself showed lines sloping towards the blue at c and towards the red at d, that is c was approaching and d receding. The spectral lines of the planet showed greater slopes because the planet rotates faster than the rings. All in all, Spinrad and Giver (Figure 6.7) rendered a spectacular result.

If we apply Kepler's third law, we see that the particles nearest to Saturn revolve at a speed 1.4 times that of the furthest particles (Table 6.1).

Stated the other way round: the furthest particles move at 0.72 times the speed of the inner particles.

Table 6.1

	Inner particles	Outer particles
Radius	71 000 km	138 500 km
Relative radius = R	1	1.95
$R^3 = T^2$	1	7.4
Period of rotation = T	$\sqrt{1} = 1$	$\sqrt{7.4} = 2.72$
Circumference of orbit = $2\pi R$	6.28	12.25
Speed = $2\pi R \div T$	$6.28 \div 1$ = 6.28	$12.25 \div 2.72$ = 4.5
Relative speed	$6.28 \div 4.5$ = 1.4	$4.5 \div 4.5$ = 1
	$6.28 \div 6.28$ = 1	$4.5 \div 6.28$ = 0.72

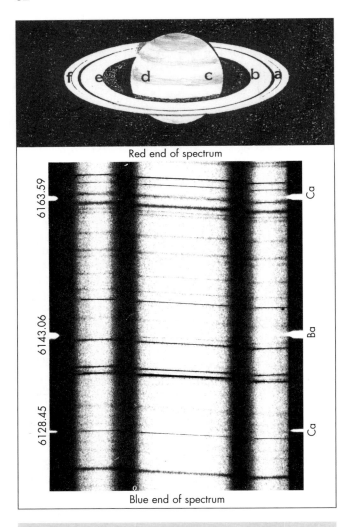

Figure 6.7 Doppler effect of Saturn's rings – demonstrated by H. Spinrad and L.P. Giver, using the high-dispersion spectrograph of Lick Observatory, 1964

Table 6.2		
	Angular diameter	
Planet	Furthest from Earth	Nearest to Earth
Mercury	4.9″	12.9″
Venus	9.9″	64.0″
Mars	3.8″	25.1″
Jupiter	30.5″	49.8″
Saturn	14.7″	20.5″
Uranus	3.5″	4.0″
Neptune	2.1″	2.3″
Pluto	0.1″	0.1″

9. Equatorial Diameters

In order to measure the diameter of a planet, the magnification offered by the telescope is essential. It is also necessary to collect enough light, so that the aperture of the objective lens or the primary mirror becomes important.

Friedrich Wilhelm Herschel, later famous and well known as William Herschel, constructed a 100-cm mirror of speculum, an alloy of copper and tin. William Parsons (Lord Rosse) constructed a mirror of diameter 180 cm in 1854. It was the largest telescope at that time.

To measure the diameter, the angular diameter must be determined. For this purpose a filar micrometer is essential. To convert the angular diameter into km the distance of the planet must be known. If Mars has an angular diameter of 18″, when it is 0.52 AU from the Earth, this angle equals $18 \div 3600 = 0.005$ degrees. The sine of this angle is equal to 0.000 087 27. When this is multiplied by 0.52 (149 597 870) we get 6789 km for the diameter of Mars. This is very close to the accepted value of 6787 km.

Table 6.2 contains the angular diameters of the planets when nearest to, and furthest from, the Earth.

When Venus is at its closest to the Earth, at inferior conjunction, it is a thin crescent and its maximum angular diameter is 64″. Venus is at its brightest midway between greatest elongation and inferior conjunction. On the other side of the Sun, at superior conjunction, the angular diameter of Venus is only 9.9″.

That Venus showed phases was discovered by Galileo (Figure 3.9, page 22). The aperture of his telescope was a mere 2 cm.

J.H. Schroter had succeeded, as long ago as 1790, in measuring the diameter of Venus by a method similar to measuring the diameter of the Moon. His value was 12 200 km. In 1833, F.W. Bessel found that Jupiter's diameter subtends an angle of 37.6″ at the Sun. That gave Jupiter a diameter of 142 000 km, which is 11 times that of the Earth. Saturn is a good second largest with a diameter of 122 800 km, as found by J. Bradley in 1719. In 1835, Bessel obtained a value of 112 600 km. Today's accepted values are 142 796 km for Jupiter and 120 000 km for Saturn. In 1859, U.J.J. Leverrier obtained a value of 4935 km for the diameter of Mercury. Today's accepted value is 4878 km.

10. Relative Diameters

Dividing the diameter of a planet by that of the Earth gives the planet's relative diameter, the Earth's diameter being taken as 1. The relative diameters of the planets are then as follows:

Jupiter 11.19, Saturn 9.407, Uranus 3.98, Neptune 3.888, Venus 0.949, Mars 0.532, Mercury 0.382, Moon 0.2725 and Pluto 0.179.

11. Polar Diameters

The Earth's diameter through its poles is a bit less than that through its equator. The difference is 43 km, namely 0.3371%. The equatorial diameter, 12 756 km, is 1.003 382 4 times the polar diameter of 12 713 km.

The Moon, Mercury and Venus have no flattening at the poles. The greatest amount of flattening is in the case of Saturn, namely 10.7625%. Saturn's surface is invisible because of the clouds in its atmosphere, and this value is that of the cloud deck. We can assume, however, that the clouds are near to the surface and parallel to it. Jupiter's flattening amounts to 6.482%, Uranus 2.4%, Neptune 2.59% and Mars 0.5157%. Pluto appears to have no flattening.

12. Volumes of the Planets

When the diameter and the degree of flattening are known, the volume can be calculated. Jupiter has the greatest volume, 1318 times that of the Earth; then comes Saturn with 745 times the Earth's volume. The Sun's volume is 1 330 000 times that of the Earth.

13. Masses of the Sun and the Planets

As knowledge of the precise distance of the Earth from the Sun improved, it became possible to calculate the relative masses of the Sun and the planets in comparison with that of the Earth.

By means of experiments with the pendulum it was found that the Earth causes a body to fall through a distance of 490.333 cm in the first second after it is released. The Earth's force of gravity acts as if all the Earth's mass is concentrated at the centre of the Earth. At the equator, the Earth's surface is 6378 km from the centre. The Sun's distance is

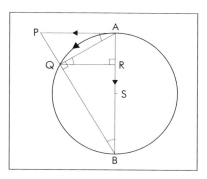

Figure 6.8 Distance through which the Earth falls in one second

$$149\,600\,000 \div 6378 = 23\,455.6$$

times the Earth's radius.

The force of attraction of the Earth at a distance of 23 455.6 times its radius is $1 \div (23\,455.6)^2$ times the Earth's force at its surface, because the force is inversely proportional to the square of the distance.

At this distance, equal to the average radius of the Earth's orbit around the Sun, the Earth's force will cause a body to fall, in the first second, through a distance of

$$490.333 \div (23\,455.6)^2 = 8.9125 \times 10^{-12} \text{ km}$$

The distance through which the Sun causes the Earth to fall in 1 second can be found by reference to Figure 6.8.

If there is no force acting on the Earth at A, drawing it in the direction AS, towards the Sun, the Earth would move in a straight line to P, where AP is a tangent to the circle. But the Sun causes the Earth to move, in one second, from A to Q, that is, the Earth falls through a distance PQ.

The length of the chord AQ (or arc AQ) is equal to the circumference of the Earth's orbit, divided by the number of seconds in one year.

$$\text{Thus, arc AQ} = \frac{2\pi(149\,600\,000)}{365.25\,(24)\,(3600)}$$

$$= 29.785\,653 \text{ km}$$

The arc AQ can be considered equal to the chord AQ, and PQ can be taken to be equal to AR without any loss of accuracy, because PQ is very small and, instead of doing the calculation for 1 second, we could do it for one-thousandth of a second.

Because AP and RQ are at right angles to AB, AP and RQ are parallel. Therefore angle PAQ = angle AQR.

Since AB is the diameter of the circle, angle AQB is a right angle.

Triangles AQR and ABQ have angle QAR common and angle ARQ = angle AQB (both = 90 degrees). Thus triangles AQR and AQB are similar (Appendix 1, section 1.4). The ratios between sides opposite equal angles are therefore equal. (Use angle AQR = angle ABQ and angle ARQ = angle AQB.) Therefore

$$\frac{AR \text{ (opp. ang. AQR)}}{AQ \text{ (opp. ang ABQ)}} =$$

$$\frac{AQ \text{ (opp. ang. ARQ)}}{AB \text{ (opp. ang. AQB)}}$$

Therefore $AR = \frac{AQ^2}{AB}$

$$\therefore PQ = AR = \frac{AQ^2}{AB} = \frac{(29.785\,653)^2}{2(149\,600\,000)} \text{ km}$$

$$= 0.000\,002\,965\,190\,9 \text{ km}$$

$$= 29.651\,909 \times 10^{-7} \text{ km}$$

$$= 2.965\,190\,9 \text{ mm!}$$

This is the distance that the Earth falls towards the Sun in 1 second, when it is at its average distance of 23 455.6 Earth radii from the Sun.

Now: Mass of Sun ÷ Mass of Earth = distance which the Sun makes a body fall in 1 second, when 23 455.6 Earth radii from the Sun, ÷ distance which the Earth makes a body fall in 1 second when it is 23 455.6 Earth radii from the Earth. That is

$$\frac{\text{Mass of Sun}}{\text{Mass of Earth}} = \frac{29.651\,909 \times 10^{-7}}{8.912\,5 \times 10^{-12}}$$

$$= 3.327\,002\,4 \times 10^5$$

$$= 332\,700$$

The mass of the Sun is therefore 332 700 times the mass of the Earth!

On page 18, it was mentioned that the mass of the Earth is 5.9742×10^{24} kg. The mass of the Sun is therefore 332 700 times greater, namely 1.988×10^{30} kg.

Mass of Planet by means of Satellite

The mass of a planet which has a satellite can be calculated by equating the planet's gravity to the centrifugal force of the satellite in a stable orbit around the planet.

According to Newton's law of gravitation, the attractive force which the planet exercises on the satellite, equals

$$F_p = \frac{GMm}{r^2}$$

where M is the mass of the planet, m the mass of the satellite, r the distance from the centre of the planet and G the universal constant of gravitation, which equals

$$6.672 \times 10^{-11} \text{ N m}^2 \text{ kg}^{-2}$$

The centrifugal force of the satellite is

$$F_c = \frac{mv^2}{r}$$

where v is the velocity of the satellite in its orbit in m s^{-1} (page 30).

Therefore

$$F_p = \frac{GMm}{r^2} = F_c = \frac{mv^2}{r}$$

$$\therefore M = \frac{rv^2}{G} \quad (m \text{ cancels})$$

In the case of Jupiter's satellite, Io, $r = 421.6 \times 10^6$ metres. The circumference of Io's orbit is therefore

$$2\pi r = 2(3.141\,59)(421.6) \times 10^6 \text{ metres}$$

Io covers this distance in 1.769 138 days, which is 152 853.5 seconds. Therefore Io's velocity is

$$\frac{2(3.141\,59)\,421.6 \times 10^6}{152\,853.5}$$

$$= 0.017\,33 \times 10^6 = 17.33 \times 10^3 \text{ m s}^{-1}$$

M, the mass of the planet, is given by

$$M = \frac{rv^2}{G} = \frac{421.6 \times 10^6 (17.33 \times 10^3)^2}{6.672 \times 10^{-11}}$$

$$= \frac{126\,618 \times 10^{12}}{6.672 \times 10^{11}} = 1897.8 \times 10^{24} \text{ kg}$$

Today's accepted value for the mass of Jupiter is 1898×10^{24} kg. How does the mass of Jupiter compare with that of the Earth?

$$\frac{\text{Mass of Jupiter}}{\text{Mass of Earth}} = \frac{1898.8 \times 10^{24}}{5.9742 \times 10^{24}}$$

$$= 317.8 \text{ Earth masses}$$

The mass of the Sun is therefore $332\,700 \div 317.8 = 1047$ times the mass of Jupiter.

In 1687, Newton calculated the mass of the Sun to be 1067 times that of Jupiter. F.W. Bessel found a value of 1047 in 1841 by taking all four of Jupiter's satellites into consideration. His value agrees exactly with the modern value.

In 1802, P.S. Laplace found, by using measurements of G.D. Cassini, that the mass of the Sun is 3359 times that of Saturn. Today's accepted value is 3496.

To determine the masses of the planets Mercury and Venus, which have no satellites, the perturbations they cause on the orbits of minor planets and comets have to be measured.

The total mass of all the large planets, taken together, is 0.001 342 56 that of the Sun, that is, 0.134 256%! The Sun is thus 745 times as massive as all the planets.

14. Densities

The densities are found by dividing the masses by the volumes (Table 6.3).

It is remarkable that the giant planets have low densities; Saturn could actually float on water! Their nuclei must therefore consist of small solid cores and the overlying layers must be very extensive. The fact that their satellites consist largely of ice indicates that they must also have an abundance of ice and/or water. They have deep atmospheres and their dimensions are those of the cloud decks which enshroud them.

The densities in the table are average values. The seismic waves which radiate from the epicentres of earthquakes show that the interior of the Earth is not homogeneous. The primary, or p-waves, are longitudinal waves – they oscillate in the line of motion. They are reflected by the Earth's nucleus, where they undergo refraction. This alters the speed of the waves, showing that the core of the Earth is denser than the overlying layers.

A model that works out at the correct value of the density allocates a core, consisting largely of iron and nickel, to the Earth's nucleus. The iron and nickel must be in the liquid state because the temperature in the core is very high. On account of the very high pressure reigning there, the density of the iron/nickel is higher than normal. The Earth's magnetism has its origin in this liquid core.

Table 6.3

Planet	Density (g cm^{-3})
Mercury	5.43
Venus	5.24
Earth	5.515
The Moon	3.34
Mars	3.94
Jupiter	1.33
Saturn	0.704
Uranus	1.24
Neptune	1.58
Pluto	1.1

The secondary s-waves are transverse waves that oscillate across the line of motion, and they are reflected by an interruption in the semi-liquid magma of the mantle, at a depth of 2900 km.

15. Gravities

The gravity exercised by a planet depends upon its mass and the radius: the force of gravity is proportional to the mass of the planet and inversely proportional to the square of the radius of the planet. The mass of Jupiter is 317.8 times that of the Earth and its radius 11.19 times the Earth's radius. The force of gravity on Jupiter's surface, compared with that on the Earth's surface, is therefore

$$317.8 \div (11.19)^2 = 2.538$$

times the gravity on the Earth's surface.

16. Velocities of Escape

Closely connected to the surface gravity of a planet, is its velocity of escape, that is, the velocity that must be attained to break away from the planet's gravity and move off into space, without falling back to the planet. The velocity of escape is equal to:

$$\sqrt{2\,G\,M \div R} \text{ cm s}^{-1}$$

G is the universal constant of gravitation, $= 6.672 \times 10^{-11}$ N m^2 kg^{-2}; M is the mass of the planet and R its radius. In the case of the Earth, $M = 5.9742 \times 10^{24}$ kg and $R = 6378 \times 10^5$ cm.

Escape velocity from the Earth works out at 11.2 km s^{-1}.

Any spacecraft wanting to depart from the Earth must attain a velocity of 11.2 km s^{-1}; if not, it will fall back to the Earth again. This velocity is only 1/267 of 1% of the velocity of light.

THE escape velocity from the Moon's surface is 2.375 km s^{-1}. The Moon's speed in its orbit around the Earth, not counting its orbit around the Sun, is 1.023 km s^{-1}. At the Moon's distance from the Earth, the escape velocity from the Earth's gravitational field is 1.44 km s^{-1}. The Moon's speed is thus only 29% less than that required for the Moon to break away from the Earth.

We can calculate the strength of the Earth's grip on the Moon, by comparing it to the Sun's grip on the Moon, by applying Newton's law of gravitation.

The relation between the Sun's force on the Moon, compared with the Earth's force, is given by:

$$\frac{G M_s m}{R^2} = \frac{G M_e m}{r^2}$$

where M_s is the mass of the Sun; M_e the mass of the Earth; R the distance Sun to Moon (= distance Sun to Earth); r the distance Earth to Moon; and m the mass of the Moon. Therefore:

$$\frac{\text{Sun's force}}{\text{Earth's force}} = \frac{M_s}{R^2} \div \frac{M_e}{r^2} = \frac{M_s}{M_e} \times \frac{r^2}{R^2}$$
$$= \frac{332\,700}{1} \times \frac{384\,400^2}{149\,600\,000^2}$$
$$= 2.1966$$

The Sun's force of attraction on the Moon is thus more than twice as strong as that of the Earth. This means that the Moon is not really a satellite of the Earth, but a planet, which incidentally finds itself in the neighbourhood of the Earth.

If the Moon is a planet in its own right, its orbit around the Sun must always be concave towards the Sun.

In Figure 6.9, E_1 and E_2 indicate the positions of the Earth, separated by a time lapse of one day. The angle $E_1 S E_2$ will then be 1 degree, because, seen from the Sun, the Earth moves through $360 \div 365.25 = 0.9856$ degree, and this is almost 1 degree. The Moon moves through 360 degrees in 27.321 661 days, that is, through 13.176 degrees in 1 day. We can call this 13 degrees and it equals angle $KE_2 M_2$.

The Moon's positions M_1 and M_2 correspond to the Earth's positions E_1 and E_2. The broken curve represents the Moon's orbit.

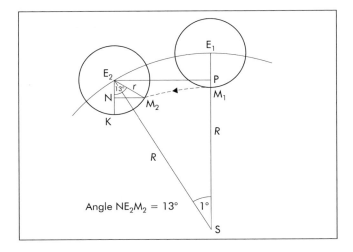

Figure 6.9 The Moon's orbit is always concave towards the Sun

After the lapse of 1 day, the Earth has moved a distance $E_1 P$ nearer to the Sun; while the Moon has moved KN further from the Sun. The question is whether $E_1 P$ is greater than or smaller than KN.

$(SP = SE_2 \cos 1°, \text{ and } E_2 N = E_2 M_2 \cos 13°)$

$$\therefore E_1 P = E_1 S - SP = R - R \cos 1°$$
$$\text{and } KN = E_2 K - E_2 N = r - r \cos 13°$$
$$\text{Now } E_1 P - KN = R(1 - \cos 1°) - r(1 - \cos 13°)$$

But $R \div r = 149\,600\,000 \div 384\,400$ so that $R = 389r$

$$\therefore E_1 P - KN = 389r(1 - 0.9998) - r(1 - 0.9744)$$
$$= r[389(0.0002) - (0.0256)]$$
$$= 0.0522r$$

The difference $E_1 P - KN = 0.0522r$ is a positive amount, because r itself is positive. In the cases of all the satellites of the other planets, this amount is negative. $E_1 P$ is thus greater than KN. In the one-day interval, the Moon has thus moved nearer to the Sun: That means that the Moon's orbit is concave towards the Sun. This therefore serves as a mathematical proof that the Moon is a planet of the Sun.

When we compare the Moon to the other satellites in the Solar System, we see that:

1. The Moon's distance from the Earth is disproportionately large.
2. The Moon's mass relative to the Earth is much greater than the masses of other moons relative to their planets.
3. The Moon's orbit is inclined to the Earth's equator by $23°27' \pm 5°09'$, while all the other true satellites revolve in the plane of their planet's equator.
4. The Moon's orbit is inclined to the Ecliptic by only $5°09'$, which is less than the inclinations of the orbits of Mercury and Pluto.
5. The Moon's orbit is always concave towards the Sun, as are the orbits of the other planets.
6. The Sun's gravitational force on the Moon is greater than that of the Earth.

The Earth–Moon system can therefore be considered as a double planet. Pluto and its moon, Charon, seem to be in the same category.

17. Velocities in Orbits

Since 149 600 000 km is the average distance of the Earth from the Sun, we can consider the Earth's orbit to be a circle with radius 149 600 000 km. While going once around its orbit, the Earth covers a distance $= 2\pi R = 2(3.141\,59) \times$

Velocities in Orbits

149 600 000 km. If this is divided by the number of seconds in a year (365.25 × 24 × 3600), the result will be the velocity, V, of the Earth in its orbit, that is

$$V = \frac{2\pi \,(3.141\,59)\,149\,600\,000}{(365.25)\,(24)\,(3600)}$$

$$= 29.786 \text{ km s}^{-1}$$

According to Kepler's third law, $T^2 = R^3$, where T is in years and R in astronomical units.

$$\therefore T = R^{3/2}, \text{ or inversely } R = T^{2/3}$$

The velocity V of a planet in its orbit is proportional to $(R \div T)$. In the form of an equation, $V = K(R \div T)$, where K is the velocity of the Earth in its orbit, namely 29.786 km s^{-1}.

Because $T = R^{3/2}$, the above equation becomes

$$V = K(R \div R^{3/2})$$

that is $V = K(1 \div R^{\frac{1}{2}})$, or $V = K(1 \div \sqrt{R})$

The velocity is therefore inversely proportional to the square root of the radius of the orbit.

The graph of V (the velocity) against $\sqrt{R}$ (the square root of the average distance from the Sun) is a hyperbola (Figure 6.10).

The graph of V against the inverse of the square root of R, namely $1 \div \sqrt{R}$, is a straight line through the origin.

Table 6.4 Velocities and distances of the planets

Planet	Period = T	$T^{\frac{1}{3}} = \sqrt{R}$	Velocity $V = \frac{29.786}{\sqrt{R}}$	$1/\sqrt{R}$
Mercury	0.241	0.6223	47.86	1.607
Venus	0.615	0.8504	35.03	1.176
Earth	1.000	1.000	29.786	1.000
Mars	1.881	1.2344	24.13	0.810
Jupiter	11.86	2.2805	13.06	0.4385
Saturn	29.46	3.0885	9.644	0.324
Uranus	84.08	4.3809	6.799	0.228
Neptune	164.8	5.4826	5.433	0.1824
Pluto	248.4	6.286	4.738	0.1591

Table 6.4 contains the data from which the graphs in Figures 6.10 and 6.11 have been drawn.

The velocities are average values over the whole orbit. Mercury, for example, has a velocity of 38.7 km s^{-1}, while its velocity at perihelion is 56.6 km s^{-1}.

The velocity of a planet in its orbit can be measured directly, by measuring its change of position in arc seconds. Corrections have to be made for the angle which the plane

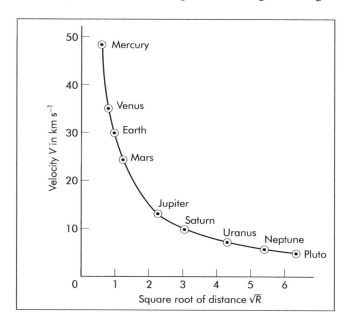

Figure 6.10 Graph of planet's velocity against $\sqrt{R}$

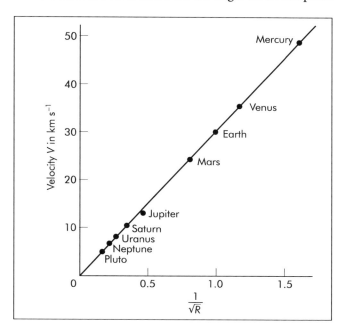

Figure 6.11 Graph of planet's velocity against $\frac{1}{\sqrt{R}}$

of the orbit makes with the line of sight, and for the Earth's motion in the interim.

At greatest elongation, the velocity of an inner planet can be determined by means of the Doppler effect on the spectral lines.

18. Reflective Powers

If the size and the distance of a planet from the Sun are known, the amount of light it receives from the Sun can be calculated. The ratio between the amount of light a planet reflects and the amount it receives is known as the reflective power or albedo.

The Moon reflects 12% of the light it receives from the Sun, therefore its albedo is 0.12.

During new Moon, when the Moon is a thin crescent, the dark portion of the Moon can be vaguely seen. This is known as earthshine and is caused by light reflected from the Earth on to the dark portion of the Moon and then back again to the Earth. From the intensity of this doubly reflected light, it has been calculated that the albedo of the Earth is 0.367. Venus, shrouded in white clouds, has a very high albedo of 0.65. The albedo of Mercury is only 0.106, indicating that its surface must be fairly dark. Astronomers had expected the surface of Mercury to be like that of the Moon, and that it would be found to be covered with craters.

19. Temperatures

The temperature of a distant object can be measured by making use of a thermopile, which consists of a series of thermocouples (Figure 6.12).

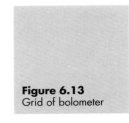

Figure 6.13 Grid of bolometer

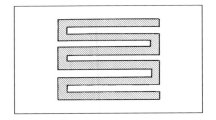

A thermocouple consists of two thin rods: one a brittle metal such as bismuth, and the other a metal such as antimony. The rods are electrically insulated by mica, but their ends are soldered or welded together. When heat falls on the one welded end, while the other end is cold, an electric current arises and flows from the one rod to the other, as shown by the arrows A and B in Figure 6.12. The current can be amplified so that its strength can be read off on a sensitive galvanometer. The deflection of the galvanometer needle is proportional to the current strength, which, in its turn, is proportional to the temperature difference between the ends of the thermocouple. In this way the temperature of the distant object can be derived.

Another instrument that can be used to measure radiation is the bolometer. It makes use of the sensitivity of the electrical conductivity, arising from temperature changes in a platinum grid (Figure 6.13).

The platinum grid is 1/500 mm thick and is covered with a layer of carbon. As the temperature of the grid rises, its electrical resistance changes and these changes can be detected by means of an arrangement such as a Wheatstone bridge, in which the resistance of the bolometer grid, R_1, is compared with that of a standard resistance, R_2 (Figure 6.14). The contact, X, of the galvanometer, G, can slide along a stretched conductor AB, until there is no deflection in the galvanometer when contact is made at X. The ratio of the resistance R_1 to that of R_2 is then equal to the length AX to the length XB.

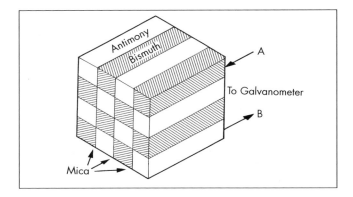

Figure 6.12 Thermopile of eight couples

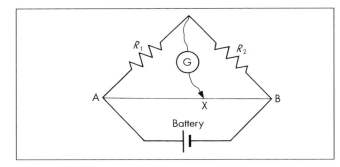

Figure 6.14 The Wheatstone bridge

Table 6.5

Time on Mars	Temperature
6 a.m.	−93°C
8 a.m.	−40
10 a.m.	+5
12 m.d.	+27
2 p.m.	+15
4 p.m.	−29

A temperature difference as small as 1/10 000°C can be registered by this means. Determinations of the temperature of the Moon have shown that a point on the surface reaches a temperature of 137°C (410 K) after seven days of continuous sunshine. After another seven days, at sunset, the temperature falls to −10°C (263 K).

The Moon always keeps the same face towards the Earth. Since it takes 29 days to complete one orbit around the Earth and to return to the same position, relative to the Sun, a point on the Moon's surface has 14.5 days of sunshine, followed by 14.5 days of continuous darkness, during which the temperature drops to −169°C (104 K).

The light and heat received by a planet from the Sun is, in part, reflected, but at longer wavelengths such as the infra-red. Measurements of these long-wave reflected rays have not been very accurate and reliable results were obtained only when space probes visited the planets.

However, in 1954 W.M. Sinton and James Strong, using the 5-metre Hale telescope on Mount Palomar, obtained the results given in Table 6.5 for temperatures on the surface of Mars, at its equator. At a latitude of 20 degrees, the mid-day temperature was 10°C, and at 60 degrees latitude it was −51°C. Mars is thus too cold for water to exist in liquid form or as a vapour. Any water on Mars must be in the form of ice, most of it as permafrost.

According to Sinton and Strong, the darker areas are slightly warmer. They must thus be higher-lying and covered with rough material.

Spectroscopic tests on the atmosphere of Mars could not detect any water vapour.

20. Moons of the Planets

The first planetary moons to be discovered were the four large moons of Jupiter. Forty five years later, in 1655, C. Huygens discovered Titan, the largest moon of Saturn. G.D. Cassini discovered Iapetus in 1671, Rhea in 1672 and Tethys and Dione in 1684. Thus Saturn took the lead from Jupiter. The last moon to be discovered visually was Amalthea, closer to Jupiter than Io, by E.E. Barnard in 1892. The small moonlets were discovered photographically or by space probes.

21. Planetary Atmospheres

The rim of Mars has a fuzzy appearance, not sharply defined like that of the Moon. When Mars occults a star by moving in front of it, there is a gradual decrease in the brightness of the star. At reappearance, the light of the star brightens gradually, not suddenly as in the case of the Moon. In the spectrum of a star, taken during disappearance or reappearance, dark lines appear. These dark lines are due to absorption by gas. These phenomena revealed that Mars has an atmosphere. The spectrum also revealed that Mars contains carbon dioxide (CO_2) in its atmosphere. The faintness of the lines showed that the atmosphere is very rarefied. No signs of oxygen were detected, nor any signs of water vapour. The ice caps at the Martian poles had to be ascribed to frozen carbon dioxide with an admixture of water ice. The temperature on Mars is low enough for carbon dioxide to freeze. In summer the frozen carbon dioxide sublimates into the gaseous state.

A direct spectrum of a planet also shows dark lines. This is due to sunlight being absorbed as it passes down into the atmosphere and back again after reflection from the surface. The wavelengths of these lines reveal the nature of the gases in the atmosphere.

Mercury has no dark lines in its spectrum, which is simply that of reflected sunlight. Occultations of stars by Mercury are also instantaneous, showing that Mercury has no appreciable atmosphere. It may have traces of hydrogen and helium.

96% of the atmosphere of Venus consists of carbon dioxide. This gas has a hothouse effect, so that the temperature on Venus is about 460°C.

The spectra of the giant planets showed methane (CH_4) and ammonia (NH_3) in their atmospheres. The boiling point of methane is −126°C, which is just about the same as Jupiter's measured temperature of −120°C. The methane is definitely in the gaseous state. Ammonia freezes at −42°C, which is far higher than Jupiter's atmospheric temperature. The ammonia must therefore be in the state of fine crystals. Jupiter's atmosphere consists largely of hydrogen, with an admixture of helium. This also holds for the other giant planets. Because they are so massive, they were able to hold on to the hydrogen and helium of the primeval nebula from which the Sun and its planets condensed. The escape

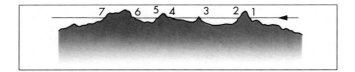

Figure 6.15 Grazing occultation of a star

velocities of the massive planets are high enough to prevent the hydrogen from escaping into space.

The colours in Jupiter's atmosphere, and to a lesser degree in the atmosphere of Saturn, were difficult to explain until those planets were visited by space probes. Besides the Great Red Spot, Jupiter also has white spots, brown stripes and vortices, showing that there is great turmoil. By means of the Doppler effect, it has been found that winds parallel to Jupiter's equator blow at speeds of 400 km h^{-1}.

That the Moon has no atmosphere is revealed by the fact that the light of a star is cut off instantly when the star is occulted by the Moon. At reappearance, the light appears instantly. If the Moon had an atmosphere, these changes in the starlight would be gradual.

When a star is in line with the northern or southern rims of the Moon, a grazing occultation takes place (Figure 6.15).

The straight line in the diagram indicates the apparent path of the star, as the Moon moves eastwards (from left to right). The star appears to move from right to left. At point 1, the starlight vanishes suddenly and then reappears equally suddenly at point 2. At point 3, an instantaneous vanishing and reappearance takes place. Between points 4 and 5, extinction takes place for a few moments. Between points 6 and 7 the extinction is of longer duration. Here there is a broader ridge, while at point 3 there is a sharp peak.

Amateur astronomers render good work by monitoring grazing occultations. The observers spread out in a north–south direction. At one end of the line, the Moon will be seen to miss the star, and no occultation will take place. Further along the line, one or more disappearances of the star's light will occur. The times of these events are carefully noted. At the other end of the line of observers, only one disappearance will take place, followed by reappearance after a considerable time.

Based on a knowledge of the latitude of the line of observers, a profile of the particular rim of the Moon can be drawn.

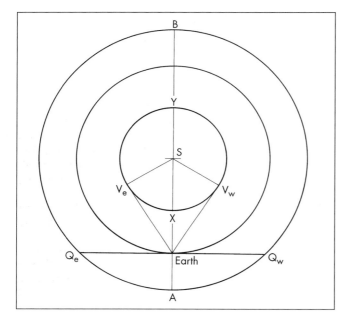

Figure 6.16 Planetary positions

A	Opposition
B	Conjunction
Q_e	Eastern quadrature
Q_w	Western quadrature
X	Inferior conjunction
Y	Superior conjunction
V_e	Greatest eastern elongation
V_w	Greatest western elongation

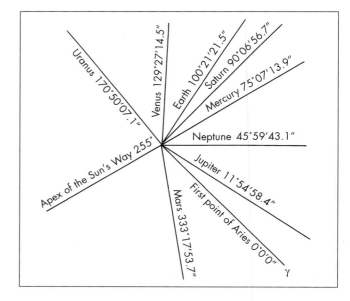

Figure 6.17 Orientations of the perihelions of the planetary orbits

Chapter 7

New Planets, Comets and Meteors

1 Outermost Planets

Uranus

The first addition to the five planets that were known since antiquity was made by William Herschel (Figure 7.1). On 13 March 1781 he noticed a bluish-green spot in the constellation Gemini. At first, he thought it was a comet, but on subsequent nights he saw that the motion of the body was not like that of a comet. It had to be a planet!

Every 24 hours its position changed by 42.2″. How long would it take, at this speed, to go right around the sky and return to its original position? It would take $360(3600) \div 42.2 = 30\,710.9$ days, which equals 84.08 years! This is more than twice the 29.46 years that it takes Saturn to complete one revolution around the Sun. And what would be its distance from the Sun? Kepler's third law states that the cube of the average distance, R^3, is proportional to the square of the period, T^2, where R and T are multiples of the Earth's average distance from the Sun and its period of one year.

$$\text{Thus } R^3 = T^2$$
$$\text{i.e. } R^3 = (84.08)^2$$
$$= 7069.45$$
$$\therefore R = \sqrt[3]{7069.45}$$
$$= 19.19$$

The average distance from the Sun of the new planet was thus 19.19 AU which equals 19.19 (149 600 000) km. This works out to 2871×10^6 km, namely 2 871 000 000 km. The estimated size of the Solar System was doubled by this discovery and the first idea of the vastness of interplanetary space took root in men's minds.

In due course the new planet was named Uranus. Its diameter was found to be four times that of the Earth and its mass 14.5 times.

In 1787 Herschel discovered two moons revolving around Uranus: Titania at 440 000 km and Oberon at 584 000 km from the centre of Uranus, respectively. W. Lassell discovered another two: Ariel at 191 000 km and Umbriël at 266 000 km from the centre of Uranus. A fifth satellite, Miranda, was discovered as late as 1948 by G.P. Kuiper at a distance of 129 000 km from Uranus. The five satellites (Figure 7.2, *overleaf*) were found to revolve in a plane tilted by 97°53′ to the plane of the planet's orbit, a plane inclined by 46′23″ to the Ecliptic. This is also the angle of inclination of the planet's equator. Because this angle is greater than 90 degrees, Uranus spins on its axis in a retrograde direction, from east to west. The five satellites revolve in the plane of the equator of Uranus and their motion is also retrograde. Because the axis is

Figure 7.1
William Herschel

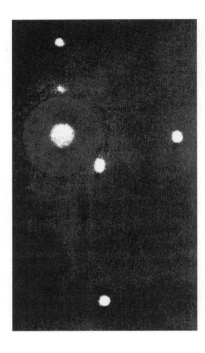

Figure 7.2 Uranus and its five moons – photo by courtesy of W.M. Sinton

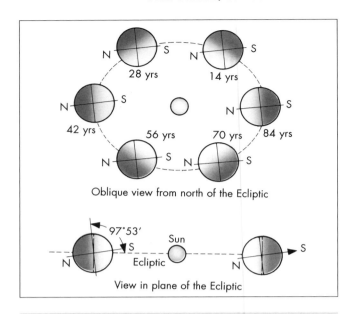

Figure 7.3 Polar axis of Uranus inclined by 97°53′

inclined to the Ecliptic by an angle very nearly a right angle, it means that the poles of the planet are alternately pointed towards the Sun, as is illustrated in Figure 7.3. At the right-hand side of the diagram the north pole has continual sunshine for 28 years, while the south pole has perpetual night. During this time the equator has the Sun on the horizon for 28 years. For the next 14 years, the mid-latitudes have alternating day and night, while the Sun lingers on the horizon at the poles. Then the south pole gets its turn to have 28 years of sunshine as we see on the left-hand side of the figure.

The period of rotation about its axis is about 16 hours. Points near the equator have the Sun attaining higher elevations for 21 years, during which the length of daylight decreases from 16 to about 8 hours. Thereafter, for another 21 years the reverse takes place.

The intensity of sunshine on Uranus is very much less than on Earth, because it is inversely proportional to the square of the distance, that is, $1 \div (19.19)^2 = 0.0027$ or 1/370. The apparent diameter of the Sun as seen from Earth is about 30′. From Uranus the apparent diameter of the Sun is $30 \div 19.19$, namely 1.56′ – a soup plate compared with a marble.

Very little could be learned about Uranus because of its bland appearance – an even blue-green or aquamarine with no distinguishable whorls or stripes in the atmosphere. It was difficult to determine the length of the Uranian day.

In 1977 calculations showed that Uranus would occult a star, SAO–158687, on 10 March. The United States National Aeronautics and Space Administration (NASA) launched a flying observatory, the Kuiper Airborne Observatory, from Perth in Australia. It flew over the Indian Ocean to a height of 12 500 metres. Fifty one minutes before the calculated time of the occultation, the photometer on board the "KAO" showed a momentary diminution of the light of the star. Could it be a moonlet? Nearly four minutes later there was another instantaneous loss of light from the star. Then followed another three extinctions and only after that, the occultation of the star took place. After the occultation, five extinctions, all of them momentary, took place. Four were symmetrical in time with the last four before occultation; but the fifth was one minute early.

The fact that the momentary extinctions occurred again after the occultation by the body of the planet proved that they could not have been caused by small moonlets, but had to be due to very narrow rings. Figure 7.4 shows the depths and distribution in time of the five occultations before and the five after, the main occultation. The widths of the rings could not be more than 10 km. The rise in the background light intensity on the right-hand side of the photometer tracing was due to the onset of dawn.

The occultation was timed to be visible from South Africa a few hours later. Observations made at the South African Observatory at Cape Town confirmed the occurrence of five extinctions before and after the main occultation.

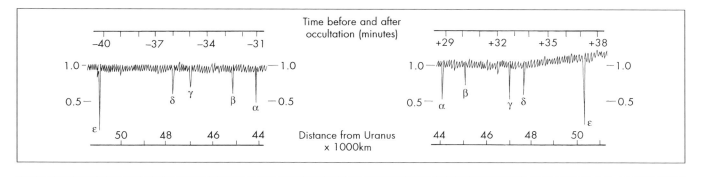

Figure 7.4 Occultations of SAO-158687 by five of Uranus' rings

So Uranus became known as the second planet to be possessed of rings. Further occultations in later years showed that Uranus had another four rings, bringing the total to nine.

The Titius–Bode Rule

Nine years before the discovery of Uranus, Johann Bode called attention to a rule which Johann Titius had added to a book by Karl Bonnet. The rule, known today as the Titius–Bode rule, starts with the figure 0, then adds 3 and subsequently doubles the previous figure. To each of the figures in this series, 4 is added and then divided by 10 (Table 7.1).

Bode pointed out that the figures in the last column agree fairly well with the relative distances of the planets from the Sun, the Earth being the third planet at a distance of 1.0 AU. There does appear to be a gap at 2.8, and no known planets fit the distances 19.6 and 38.8.

When Uranus was discovered, its distance from the Sun of 19.19 AU was fairly close to the value of 19.6 for the eighth planet in the Titius–Bode rule (Table 7.2).

2 Minor Planets

By the end of the 18th century, interest in the Titius–Bode rule had become very real. Fired by Herschel's discovery of Uranus, astronomers searched avidly for the "missing planets", especially for one that would be at a distance of 2.8 AU from the Sun. Great was the excitement when G. Piazzi announced his discovery on 1 January 1801 of a planet whose orbit was inclined to the Ecliptic by $10°36'28''$, but whose period of 4.61 years meant that its average distance from the Sun had to be 2.77 AU, which is very close to the 2.8 of the Titius–Bode rule. The planet was named Ceres.

For full measure, H.W.M. Olbers discovered another

Table 7.1

Series	Add 4	Divide by 10
0	4	0.4
3	7	0.7
6	10	1.0
12	16	1.6
24	28	2.8
48	52	5.2
96	100	10.0
192	196	19.6
384	388	38.8

Table 7.2

	Distances of planets	
Planet	Titius–Bode	Actual
Mercury	0.4	0.39
Venus	0.7	0.725
Earth	1.0	1.0
Mars	1.6	1.524
?	2.8	?
Jupiter	5.2	5.2
Saturn	10.0	9.54
Uranus	19.6	19.19

planet on 28 March 1802. Its orbit was steeply inclined to the Ecliptic, making an angle of 34°43'9". Its average distance from the Sun was also 2.77 AU. It was named Pallas.

Then, on 1 September 1804, K.L. Harding discovered a third planet, much closer to the Ecliptic, having an inclination of 13°1'9" to the Ecliptic. Its average distance from the Sun, was found to be 2.67 AU and it was named Juno.

The cup overflowed when Olbers discovered his second planet on 22 March 1807. The plane of its orbit was found to be much closer to the Ecliptic, being inclined by only 7°7'55.7". Its average distance from the Sun turned out to be 2.36 AU, rather far from 2.8 AU. It was named Vesta.

By 1850, another nine planets were discovered; a further 21 by 1855; and another 24 by 1860! Today the total is far in excess of 4000. The orbits of at least 3000 have been tested by observations made during at least three oppositions.

These planets are very small and they are correctly known as "Minor Planets" – formerly they were called asteroids.

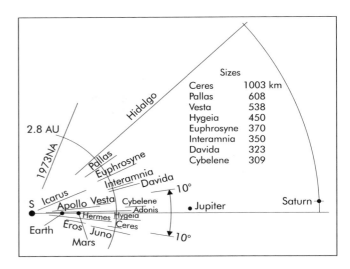

Figure 7.5 Inclinations of the orbits of some of the Minor Planets

Sizes of the Minor Planets

The first discovered, Ceres, is also the largest, having a diameter of 1003 km. Pallas is the second largest, being 608 km in diameter. Next comes Vesta at 538 km. Hygeia has a diameter of 450 km.

There are altogether 33 Minor Planets larger than 200 km in diameter, and 41 between 100 and 200 km.

The best method of measuring the diameters of such small bodies is to measure the time lapses when they occult stars. The duration of the occultation is the time that it takes the Minor Planet to move a distance equal to its diameter. The distance of the planetoid from the Earth must be known in order to calculate its diameter. Amateur astronomers can render valuable work in this respect. Measurement of the duration of an occultation of a star by the Minor Planet, Eunomia, by M.D. Overbeek of Edenvale, Transvaal, South Africa, enabled Gordon Taylor of Greenwich Observatory to calculate the diameter of Eunomia. He obtained a value of 261 km. The previous value was 206 km.

The calculated times of occultation by Minor Planets, as well as their locations on Earth, are published regularly, for example in *Sky and Telescope* magazine.

Inclinations of Orbits

The minor planet whose orbit has the greatest inclination to the Ecliptic, is No. 1973NA, with an inclination of 67 degrees. Next is No. 944, Hidalgo, at 47°30'. Its diameter has been measured as being 15 km. Cincinnati, No. 1373, follows at 38°54', and then Pallas at 34°43'09". There are 12 whose orbits are inclined by 20 to 26 degrees; 45 with inclinations between 10 degrees and 20 degrees. At least 1250 have inclinations of less than 10 degrees (Figure 7.5).

Brightnesses

The largest Minor Planets are spherical in shape. The smaller ones are probably irregular in shape, because their brightnesses vary as they spin and present varying areas to the light of the Sun.

In 1931 W. Finsen and J. van den Bos, using the 66-cm refractor of the Johannesburg Observatory, with a magnification of 720 times, noticed that the Minor Planet Eros is cigar-shaped, when it made its closest approach to the Earth, coming to within 23 million km.

The brightest of all the Minor Planets is Vesta, magnitude 6.5 – only just too faint to be seen with the naked eye. Ceres, although larger, has a magnitude of 7.4 and is invisible to the naked eye. Next in the brightness scale is Iris, magnitude 7.8; Pallas, 8.0; Hebe and Eunomia, 8.5; and Juno, 8.7. The rest are fainter, of magnitudes 10 to 15 and higher. (The higher the magnitude, the fainter the object is – see Chapter 10.)

Eccentricities of Orbits

The average of the eccentricities of the eight largest Minor Planets is 0.144 – noticeably eccentric but not very much so. The difference between the aphelion and perihelion distances of Ceres is 0.39 AU; and for Euphrosyne (No. 31), 1.14 AU. In Figure 7.5 the lines drawn next to the names of the Minor Planets indicate the differences between the aphelion and perihelion distances.

Hidalgo, No. 944, has a very large eccentricity of 0.657. Its distance from the Sun varies between 2.0 and 9.61 AU. The projection of its orbit comes close to the distance of Mars from the Sun and reaches to slightly beyond the distance of Saturn. The only times that it can come close to Jupiter and Mars are when it crosses the plane of the Ecliptic. Hidalgo was discovered by W.H.W. Baade in 1920.

Distribution of the Minor Planets

There are about forty Minor Planets whose orbits lie totally or partly within that of Mars. The rest have orbits lying between the orbits of Mars and Jupiter. The histogram of the distribution of 2000 of the Minor Planets (Figure 7.6) shows that the large majority have distances from the Sun of 2.15 to 3.35 AU.

The zones of avoidance, known as the Kirkwood gaps (after D. Kirkwood) are due to the gravitational forces exercised by Jupiter. The 3:1 zone, for example, contains very few Minor Planets because after every three orbits around the Sun, the planets in that zone find themselves in opposition to Jupiter, that is at their closest to Jupiter. The continued repetition of Jupiter's force of attraction would have had a resonance effect on the Minor Planets and would have drawn them out of their orbits lying within those zones. The 5:2 zone is completely devoid of Minor Planets. It is the zone in which five revolutions of the Minor Planets coincided with two revolutions of Jupiter.

The Greek letter mu (μ) indicates the mean of the distances from the Sun of all the Minor Planets, namely 2.75 AU.

The total mass of all the planetoids has been computed at 1/3000th of the mass of the Earth. These bodies may be representative of the planetesimals from which the planets of the Solar System formed by a process of accretion.

The Earth Grazers

There are at least twelve Minor Planets whose orbits come not only within that of Mars, but also within the Earth's orbit. They are the Earthgrazers or Apollo-asteroids, named after the Minor Planet, Apollo, that approached the Earth to within 10 480 000 km in 1932. Icarus, whose orbit has an eccentricity of 0.83, came to within 6 million km of the Earth in 1967. In 1936 Adonis came to within 2.25 million km; in 1976 the Minor Planet 1976UA came within 1.2 million km. On 30 October 1937, Hermes came as close as 800 000 km, but it was never seen again. The perihelions of the orbits of Apollo and Adonis are 0.647 and 0.441 AU respectively; they are thus within the orbit of Venus. Icarus comes within the orbit of Mercury and rounds the Sun at 0.187 AU at its closest.

Maybe it was an asteroid that collided with the Earth 65 million years ago, a collision which led to the extinction of the dinosaurs and many other species. The "safe" spaceship, Earth, is apparently subject to such disasters. At many places on Earth a thin layer of iridium has been found, the age of which is apparently 65 million years. The iridium could have been formed from the fall-out of the

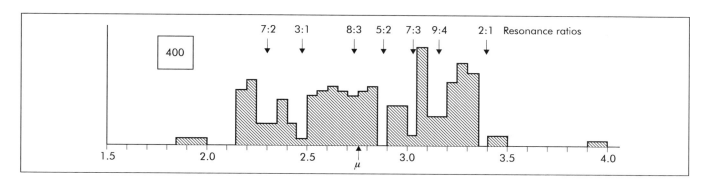

Figure 7.6 Histogram of distribution of distances from the Sun of 2000 Minor Planets. Mean distance μ = 2.75 AU

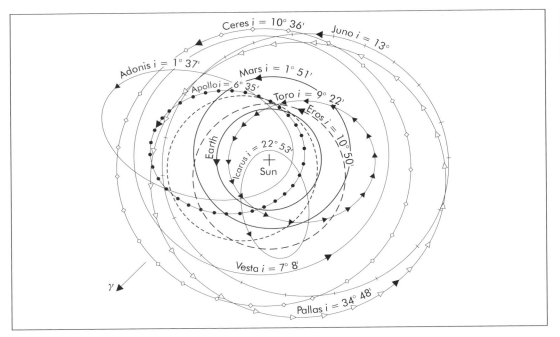

Figure 7.7
Orbits of Ceres, Pallas, Juno and Vesta; and five Earthgrazers, projected on to the plane of the Ecliptic

cloud of gas and dust which formed over the site of the collision and which must have enshrouded the Earth in a cloud, impenetrable by sunlight.

The Earthgrazers (Figure 7.7) are very small, being less than 30 km in size. If one of them struck the Earth, it would probably not wipe out all life forms, but the loss of life and the damage done would be tremendous. The best avoiding action that could be taken would be to launch a rocket to go and nudge the asteroid into a safe orbit around the Earth. Such a rocket would constantly have to be in readiness, for launching at a moment's notice.

Figure 7.8 The Minor Planet Ida

Another unique Minor Planet is No. 2060, Chiron, discovered in 1977 by C. Kowal, by means of the Schmidt camera at Mount Palomar. The orbit of Chiron lies almost entirely between the orbits of Saturn and Uranus. Its perihelion distance is 8.5 AU and its aphelion 18.5 AU. The inclination of the orbit to the plane of the Ecliptic is only 6.9 degrees.

There are probably many asteroidal bodies revolving in orbits between those of the giant planets. Besides Chiron, nine other bodies have been found in orbits in resonance with Neptune's orbit.

One of the first photographs of a Minor Planet is that taken by the probe Galileo of Ida, while Galileo was on its way to Jupiter (Figure 7.8).

On 28 August 1993, Galileo passed within 2400 km of the 52 km long Minor Planet. Its surface is covered in craters, showing how prevalent smaller bodies like meteorites have been.

Discovering Minor Planets

Many Minor Planets have been discovered by taking photographs on successive nights of the same parts of the sky and then comparing the photographs by means of a blink comparator (Figure 7.9).

The two identical negative plates, P_1 and P_2, are placed

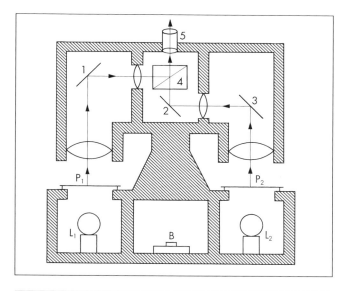

Figure 7.9 The blink comparator

above the lamps L_1 and L_2. The positions of the plates are adjusted so that they form only one image when viewed through the eyepiece, 5. The two lamps are then alternately switched on and off, by means of the pushbutton B, or by means of an electric motor. The light from the lamps passes through the ground glass plates below the negatives. The negatives are thus evenly illuminated and the light rays then follow the paths indicated by the arrows, to the eyepiece. If there happens to be an object, such as a planet or a comet, on the plates, it would have changed position in the interval between the taking of the two photographs, while the stars occupy identical positions. The planet or comet will then be seen to jump to and fro as the negatives are alternately illuminated, as shown in Figure 7.10. The arrows indicate the object that has moved.

The moving body can then be monitored to determine whether it is a planet or a comet.

The Trojan Asteroids

There are about forty asteroids caught in the gravitational field of Jupiter. They are known as the Trojans. Their perihelions vary between 4.44 and 5.08 AU, and their aphelions between 5.37 and 5.98 AU. Their mean distance from the Sun is 5.2 AU, the same as that of Jupiter. Half of them revolve in a clump, 60 degrees east of Jupiter, and half 60 degrees west of Jupiter.

The points 60 degrees east and west of Jupiter are the 4th and 5th Lagrange points. These are points of stable gravity where bodies undergo no gravitational perturbations. Four of the Trojans are between 90 and 180 km in diameter.

Perturbations of Uranus

While the excitement about the discovery of the planetoids ran high at the start of the nineteenth century, it was found that there was something wrong about Uranus. It was usually ahead of its calculated position, as if it had undergone an extra push in its orbit. Could it be that there was something wrong with Newton's law of gravity?

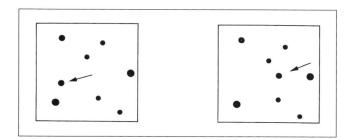

Figure 7.10 Blink comparator photographs. Planetoid has changed position

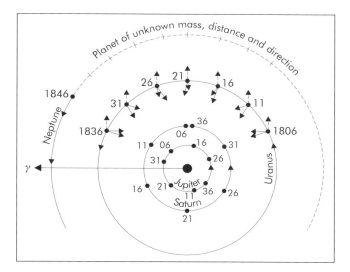

Figure 7.11 Perturbations of Uranus by Jupiter, Saturn and unknown planet, 1806–1836

Or could there be an unknown planet in the vicinity, causing the perturbations? After 1821, it was found that Uranus started lagging behind its calculated positions. If there was an unknown planet in the vicinity, its position must have been such that it was overtaken by Uranus round about 1821 and that it had to be further from the Sun than Uranus.

Figure 7.11 (*previous page*) shows the positions of Jupiter, Saturn and Uranus during the years 1806, 1811 and so on up to 1836, indicated by 06, 11, ... 36. The solid arrows radiating from Uranus show the gravitational attractions of Jupiter and Saturn on Uranus. The broken arrows show the supposed attractions by a planet of unknown mass, direction and distance. From 1816 to 1821, Jupiter and Saturn tended to draw Uranus nearer to the Sun, but this acceleration was not sufficient to explain the speeding-up which Uranus underwent.

Adams, Leverrier and Neptune

In 1843 the 24-year-old J.C. Adams made a mathematical analysis of the perturbations of Jupiter and Saturn on Uranus. He correctly attributed the excess of the perturbations to a planet beyond Uranus. He set about calculating where that planet was likely to be to exercise those perturbations. In 1845 he sent his calculations to G.B. Airy at Greenwich, indicating where the planet was likely to be found. Airy dragged his heels.

In the meanwhile, the 34-year-old U.J.J. Leverrier had done a similar analysis and calculation. On 15 September, he sent his prognostications to J.G. Galle at Berlin. Galle and H. D'Arrest set to work at once, and immediately, on 23 September 1846, found the missing planet, within 1 degree of the position indicated by Leverrier! That is twice the apparent width of the full Moon. The position calculated by Adams was out by $2\frac{1}{2}$ degrees. The new planet received the name of Neptune.

This mathematical achievement created a great impression and emphasised the accuracy and validity of Newton's law of gravitation. To be able to calculate the position of a planet of unknown mass, distance and direction was something unheard of.

A computer program, worked out by A.S. Hilton, Director of the Computer section of the Astronomical Society of Southern Africa, which made use of the longitudes and latitudes of Uranus and Neptune, showed that the difference in longitude between Uranus and Neptune was, on average, 0.4 degree, and the difference in latitude, 1.3 degrees during the whole year of 1821. The two planets were thus within the same field of any ordinary telescope during that year.

The astronomers who studied Uranus failed to notice Neptune as a planet. Some had marked it as just another faint star. Because photography did not exist and there was no blink comparator, and because the planet moves so slowly, it was not noticed as a planet.

Neptune

The sidereal period of Neptune was found to be 164.8 years; that gave it an average distance from the Sun of 30.07 AU. According to the Titius–Bode rule, its distance should be $\{(2 \times 192) + 4\} \div 10 = 38.8$ AU. The "rule" therefore does not hold in Neptune's case, and it must be assumed that agreements in the cases of the other planets are mere coincidences. Neptune's mass was found to be 17.3 times that of the Earth and its diameter (50 000 km) is 3.9 times the Earth's diameter.

In the year of Neptune's discovery, W. Lassell discovered that it had a large moon, 3000 km in diameter, revolving at a distance of 354 000 km from the planet's centre. It was called Triton.

More than 100 years elapsed before G.P. Kuiper discovered a second moon in 1949. It was called Nereid and found to have retrograde revolution in a very eccentric orbit, of eccentricity 0.749, and having an inclination to the plane of Neptune's equator of 27 degrees. Neptune's equator is inclined to the plane of the Ecliptic by $28°48'$. The plane of Nereid's orbit can thus come within 2 degrees of the Ecliptic. Nereid is probably a captured planetoid.

New Moons

Besides the moons already known to revolve around Saturn, William Herschel discovered another two in 1789: Mimas at 185 700 km, and Enceladus at 238 200 km from Saturn. In 1848 Hyperion was discovered by G.P. Bond, in an orbit at a distance of 1 484 000 km from Saturn. E.C. Pickering discovered the moon Phoebe in 1898. It is the most remote of Saturn's moons, at a distance of 12 950 000 km. It has a very eccentric orbit, inclined by 150 degrees to Saturn's equator. It therefore has retrograde revolution in a plane which is only $3°16'$ away from the Ecliptic. It is most likely a captured planetoid. In 1966 Janus, the moon nearest to Saturn, was discovered by A. Dollfus. Its distance was found to be 151 450 km from Saturn. At that stage, Saturn was credited with ten moons, but Jupiter had taken the lead.

In 1892 E.E. Barnard discovered Amalthea, the moon nearest to Jupiter, at a mean distance of 181 500 km and thus nearer than the four Galilean moons. Amalthea was the last satellite to be discovered visually, without the aid of photography. Jupiter's five nearest moons revolve in orbits that coincide with the plane of Jupiter's equator. They are thus true satellites.

Ten times further than Callisto, four satellites were found, revolving at distances between 11 110 000 km and 11 743 000 km from Jupiter. Their orbits are inclined to the plane of Jupiter's equator by about the same angle of 27 degrees. The eccentricities of their orbits vary between 0.1 and 0.2. They could possibly be captured asteroids or captured comet nuclei.

Even further, at distances between 21 200 000 km and 23 700 000 km from Jupiter, there are another four satellites. The inclinations of their orbits lie between 147 degrees and 163 degrees and the eccentricities vary between 0.17 and 0.4. They have retrograde revolution and must be captured asteroids. These eight satellites are very small, between 20 and 100 km in size, and they are irregular in shape. They were all discovered by means of photography.

Photography

David Gill, who was stationed at the Cape Town observatory in South Africa, was one of the first astronomers who applied photography to astronomy. His first photograph was that of the great comet of 1882.

Photography has the advantage over the human eye that the photographic plate can store the photons of light, whereas the eye is only momentarily sensitive.

Because the photographic plate can store light, it makes objects visible that cannot be seen by eye.

The most usual method of using photography in astronomy is to attach the camera to the eyepiece. The light rays entering the camera are parallel and therefore the camera has to be set at infinity. To take a time exposure, the mechanism of the telescope's clock drive must keep pace accurately with the diurnal westward motion of the sky. To achieve this end, the object being photographed can be kept constantly on the crosswires in the finderscope. The longer the time exposure, the more detail can be captured and the fainter the objects that can be seen. Very faint objects are today being registered by means of the charge coupled device.

Another advantage of photography is that the camera lens serves to increase the focal length of the telescope. If the aperture of the primary mirror is, say, 150 mm and the focal length is 1800 mm, the telescope is rated as f/12. Attaching a 50-mm camera increases the focal length to 3300 mm, and the telescope becomes an f/22. The camera therefore yields an extra degree of magnification.

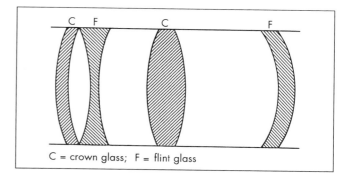

Figure 7.12 Lenses of the Papadopoulos astrographic camera

Photography is of inestimable value in making spectrograms, because the light can be collected over a period of time. The spectrograms can then be studied at leisure.

Many amateur astronomers have joined the teams which monitor certain parts of the sky by repeatedly photographing allotted portions of the sky, so that any sudden change, such as the explosion of a star, can be captured in the act. This happened with supernova 1987A.

The photographic plate is, however, not equally sensitive to all the different wavelengths of light, and the sensitivity is not on a level with that of human eyesight. On photographic plates "blue" stars appear brighter than equally bright "red" stars. The brightnesses seen by the eye through a telescope are therefore not the same as on the photographic plates of star atlases.

To overcome this difficulty, Christos Papadopoulos of Johannesburg had a special camera built by the firm Zeiss. His combination of crown and flint glass completely eliminated chromatic aberration (Figure 7.12). The aperture was 125 mm and the focal length 625 mm, yielding a ratio of f/5. In front of the field lens, Papadopoulos placed a green filter of thickness 3 mm. With this system, he succeeded in obtaining photographs that depicted the stars in their true relative brightnesses, irrespective of their colour, and the stars had the same brightnesses as seen by the eye. Figure 7.13 shows the Southern Cross, taken in the usual manner on the left, and as taken by Papadopoulos on the right. To the eye, the stars Alpha, Beta and Gamma appear equally bright. In the usual photograph, the stars Alpha and Beta, which are "blue" stars, appear much brighter than Gamma, which is a "red" star. The filter and lens system employed by Papadopoulos depict the brightnesses in agreement with that seen by the human eye.

Papadopoulos covered the whole sky in 456 plates, each 330 mm square. Each photograph covered 15 degrees and of that the central 11 degrees was printed, with overlaps of

Figure 7.13
Constellation Crux – Southern Cross
a Usual photo
b Papadopoulos photograph

1 degree. The whole undertaking took 12 years. For this monumental achievement, Papadopoulos was awarded the David Gill medal, which is awarded only for meritorious astronomical work.

Pluto

C.W. Tombaugh searched through thousands of plates looking for the unknown planet beyond Neptune. In 1930 he studied the two photographs, shown in Figure 7.14, in the blink comparator. One of the faintest points of light was seen to jump to and fro as the plates were alternately illuminated. This was the long-searched-for planet! It received the name Pluto.

The sidereal period of 248.4 years gave it an average distance from the Sun of 39.46 AU – fairly close to the Titius–Bode 38.8 for the eighth planet! But Pluto is the ninth planet. The orbit is inclined by 17 degrees to the Ecliptic, and it has a high eccentricity of 0.248. At perihelion its distance from the Sun is less than that of Neptune. Pluto reached its perihelion on 5 September 1985. From 21 January 1979 to 14 March 1999, Pluto will be nearer to the Sun than Neptune.

On 2 July 1978, J. Christy, using the 1.54-metre reflector of the US Naval Observatory and with the aid of a charge coupled device, produced a photograph of Pluto that showed a hump, which must have been due to a satellite of the planet (Figure 7.15). This moon was given the name Charon. Its greatest distance from Pluto is 19 640 km. At

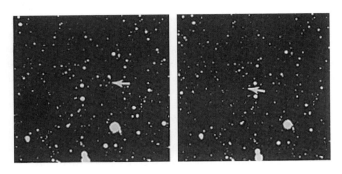

Figure 7.14 Plates on which C.W. Tombaugh discovered Pluto in 1930

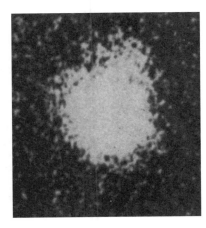

Figure 7.15 Charon forms hump on Pluto's image

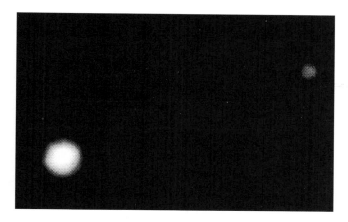

Figure 7.16 Complete resolution of Pluto and Charon by the repaired Hubble Space Telescope

Pluto's distance, this separation equals only 0.8 arc seconds in angular measure; that is why the telescope could not resolve the two bodies. The first resolution of Pluto and Charon was achieved in 1990 by the Hubble Space Telescope. After being repaired by the astronauts, Hubble completely resolved Pluto and Charon (Figure 7.16).

Previously it had been noticed that the brightness of Pluto was variable, with a period of 6.4 days. Now the period of revolution of Charon was found to be 6.4 days. Charon's orbit is almost vertical to the plane of the Ecliptic. From 1985 to 1990, Charon alternately passed in front of Pluto and suffered eclipse by Pluto; hence the variability of the light.

This phenomenon will be repeated only after 124 years, when Pluto is on the other side of its orbit, at aphelion.

Charon's period is 6 days 9 hours 17.5 minutes. From this, it is possible to calculate the masses of Pluto and Charon. Pluto's mass turns out to be 1/400 of that of the Earth. The diameters of Pluto and Charon work out to be 2284 and 1192 km respectively. Charon is thus, in proportion to its primary, the largest satellite in the Solar System. Pluto and Charon can also be considered to be a double planet.

A model of the cross-section of Pluto, by W. McKinnon and S. Mueller, shows that Pluto has a fairly large core of dehydrated rock, of radius 812 km. On top of that, there is a layer of water ice, 210 to 320 km thick. The crust, only 10 km thick, consists of ices of methane, carbon dioxide and carbon monoxide. Sunlight converts these ices into dark-coloured hydrocarbons. Pluto's albedo of 0.4 is due to fine crystals of methane in the atmosphere.

Scale Model of the Solar System

From the data of Tables 7.3 and 7.4 on pages 84 and 85, a scale model of the bodies in the Solar System can be made, by dividing the sizes by 2×10^9 (Figure 7.17, *overleaf*). To make a scale model of the distances, we shall have to divide by 40×10^9. The Sun is then only 3.48 cm in size, slightly larger than a pingpong ball. We could conveniently place the bodies on a rugby field. If we place the Sun, as a pingpong ball, on the northern dead ball line, Mercury is a speck of dust 1/5 mm in size, 1.45 metres from the Sun. Venus, 0.3 mm in size, is 2.7 metres and the Earth, 1/3 mm, 3.74 metres from the Sun. Mars, 1/6 mm, is 5.7 metres from the Sun. The giant planet, Jupiter, is on this scale 3.57 mm in size and 19.5 metres from the Sun, between the northern goal posts. Moving southwards, Saturn, 3 mm in size, and its rings, 6.9 mm wide, is 35.7 metres south of the Sun, two-thirds of the way to the 22-metre line. Uranus, 1.3 mm in size, is 71.75 metres from the Sun, just about on the centre line of the rugby field. Neptune, 1.25 mm in size is 112.5 metres from the Sun, 10 metres short of the southern 22-metre line. Pluto, just 0.075 mm, lies 148 metres from the Sun, beyond the southern dead ball line, among the lower seats.

How far would the nearest star, Alpha Centauri, be on this scale? Let us wait for the answer until we have found out how to measure the distances of the stars!

3 Ephemeral Visitors

Comets

When a bright comet appeared in the skies of Europe in 1066, William Duke of Normandy interpreted it as a sign that he should take up arms and invade England. The English King, Harold, in his turn, saw the comet as presaging the fall of his Kingdom. And he was dead right! At the battle of Hastings, he was struck in the eye by an arrow and he died. William established himself in England and brought about a radical change in the history of England. The English language underwent a far-reaching change.

Comets as Members of the Solar System

Tycho Brahe was able to prove that comets are not atmospheric phenomena, as Aristotle had said, but moved beyond the atmosphere, and further than the Moon.

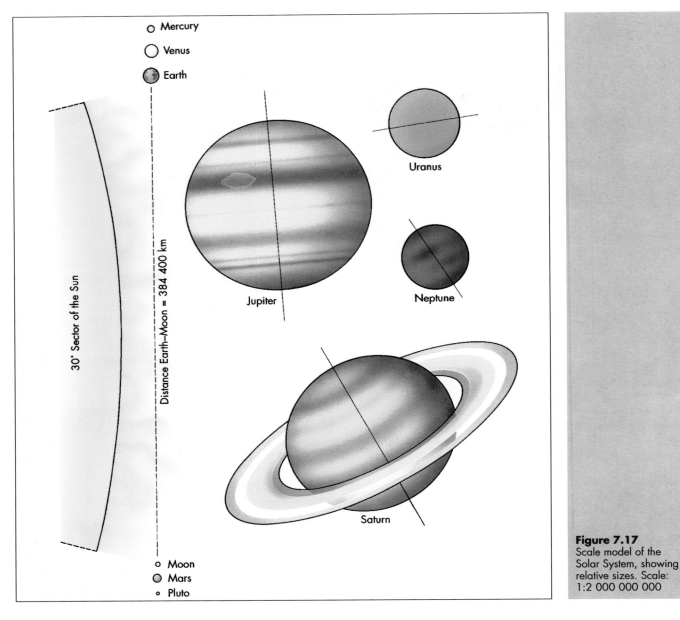

Figure 7.17
Scale model of the Solar System, showing relative sizes. Scale: 1:2 000 000 000

That comets belonged to the Solar System and were subject to the law of gravitation was proved when the bright comet which made its apparition in 1682 appeared again in 1759, as Edmond Halley had predicted. Halley noticed that the orbit of the comet of 1682 seemed to coincide with those of the comets of 1607 and 1532. By applying Newton's law of gravitation, Halley found that the comet will make its next return to the precincts of the Earth and Sun in 1758. That meant that it is periodic in nature, with a period of 75–76 years. The intervals between its appearances from 1531 up to 1758 were 76, 75 and 76 years. The comet was seen again on Christmas night 1758 and it passed through perihelion on 12 March 1759. Halley's prediction caused a sensation and it showed that the law of gravitation was of universal application.

It was fitting that the comet was named after Halley. Incidentally, the comet which appeared in 1066 was also Halley's comet. After 1759, the comet returned to the Sun

Figure 7.18 Halley's comet, photograph by South African Observatory, 7 May 1910

again in 1835, 1910 and 1986. In 1910 the comet's appearance was most spectacular (Figure 7.18), because the Earth was on the same side of the Sun as the comet's perihelion, which it reached on 20 April 1910. At that time, the tail stretched right across the sky. It was a truly magnificent sight and on 18 and 19 May, the Earth passed through the gas and dust of the tail, but nobody was any the worse off for it.

Halley's comet became visible to the naked eye again in November 1985, when it was seen in the evenings just after sunset. But it had already been spotted in October 1982, by means of a charge coupled device attached to the 5-metre Hale telescope on Mount Palomar.

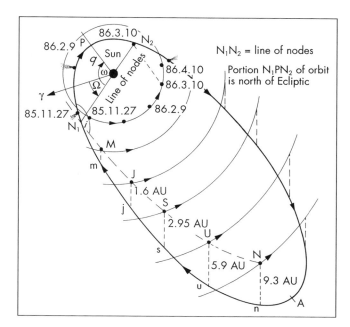

Figure 7.19 (Not to scale) Orbit of Halley's Comet, A–P–A. N_1–M–J–S–U–N are in the plane of the Ecliptic

On 27 November, the comet came to within 0.62 AU, while approaching the Sun, after it had passed from south to north of the Ecliptic, at its ascending node, 18 days previously (Figure 7.19). Then the Earth and the comet separated again. The comet sank lower and lower in the west, until it got lost in the glare of the Sun. It passed through perihelion on 9 February 1986, when it was 0.587 AU from the Sun. Unfortunately, the Earth was on the other side of the Sun, 1.55 AU distant. At that time the comet was inside the orbit of Venus, whose average distance from the Sun is 0.723 AU. On 10 March, Halley's comet passed through its descending node (from north to south of the plane of the Ecliptic); its distance from the Earth was then 0.98 AU. Being west of the Sun, it was visible before dawn. Halley reached its closest approach to the Earth on 10 April 1986, when it was 0.42 AU from the Earth and 1.33 AU from the Sun. The tail had become noticeably shorter and was pointing away from the Earth.

The orbits of comets are very eccentric (usually about 0.75). Figure 7.19 shows the orbit of Halley's comet, A–P–A, in relation to the orbits of the planets. The planetary orbits lie very close to the plane of the Ecliptic, N_1–M–J–S–U–N. The plane of the orbit of Halley's comet makes an angle of $i = 162°24'$ with the plane of the Ecliptic. Because this angle is greater than 90 degrees, it means that the comet has retrograde revolution (clockwise) as seen from the north, while the planets all revolve counter-clockwise.

The shortest distances between the comet and the outer planets are indicated in the figure. When Jupiter is at J and Halley's comet is at j, the perturbation of Jupiter is at its greatest. Taking the Sun's mass as 1047 times that of Jupiter, and its distance as 5.2 AU, the effect of Jupiter's gravitational pull compared with that of the Sun, is

$$\frac{1}{1047} \times \frac{(5.2)^2}{(1.6)^2} = 1\%$$

The effects of the other giant planets are between 0.3% and 0.54% that of the Sun. The orbit of Halley's comet is thus very stable, but it does gyrate, so that the ascending node, N_1, moves $2\frac{1}{2}$ degrees eastwards per revolution. After 144 revolutions (10 900 years) the orbit will have gyrated through 360 degrees. The other elements of the orbit also change gradually but regularly. In the last 1000 years the inclination to the Ecliptic has changed by only 1%.

The Sun always occupies one focus of the elliptical orbit of a comet (or planet). The distance between the foci divided by the major axis of the ellipse (distance from perihelion to aphelion) is equal to the eccentricity, which in Halley's case equals 0.967. Since the perihelion distance is 0.587 AU, the distance between the foci equals the length of the major axis minus 2 (0.587). This distance divided by the length of the major axis gives the eccentricity:

Table 7.3 Planetary data: the terrestrial planets

	Mercury	Venus	Earth	Moon	Mars
1. Synodic period (days)	115.88	583.9	—	29.530589	780
2. Sidereal period (days)	87.97	224.7	365.25	27.321661	687
(years)	0.241	0.615	1	0.0748	1.881
3. Distance from Sun (Earth = 1)	0.387	0.723	1	±1	1.523
4. $\times 10^6$ km	57.90	108.16	149.6	149.2 – 150	227.8
5. Eccentricity of orbit	0.206	0.0068	0.017	0.0549	0.093
6. Inclination of orbit	7°	3°23'40"	0°	5°09'	1°51'
7. Inclination of equator	0°	177.3°	23°27'	1°32'	25°12'
8. Period of rotation	58.6462 d	243.01 d	24 h	27.321661 d	24.623 h
9. Equatorial diameter (km)	4878	12104	12756	3476	6787
10. Diameter (Earth = 1)	0.3824	0.949	1	0.2725	0.532
11. Polar diameter (km)	4878	12104	12713	3476	6752
12. Volume (Earth = 1)	0.056	0,858	1	0.0203	0.150
13. Mass (Earth = 1)	0.05527	0.815	1	0.0123	0.1074
14. Density ($g\,cm^{-3}$)	5.43	5.24	5.515	3.34	3.94
15. Gravity (Earth = 1)	0.377	0.905	1	0.166	0.38
16. Escape velocity ($km\,s^{-1}$)	4.25	10.36	11.2	2.375	5.02
17. Velocity in orbit ($km\,s^{-1}$)	47.86	35.03	29.786	30.8	24.13
18. Albedo	0.106	0.65	0.367	0.12	0.16
19. Temperature: minimum	−85°C	445°C	−88°C	−169°C	−125°C
maximum	430°C	482°C	58°C	137°C	26°C
20. Number of moons	0	0	0	0	2
21. Atmosphere	Traces of He, H_2, A, Ne	96% CO_2, 3% N_2, A, HCl, CO, $HFSO_2$,	78% N_2, 21% O_2, A, CO_2, H_2O	Traces of H_2, He, A, Ne	95% CO_2, 2.7% N_2, A, O_2, H_2

Planetary Data

Table 7.4 Planetary data: the outer planets

	Jupiter	Saturn	Uranus	Neptune	Pluto
1. Synodic period	1 y 33.5 d	1 y 12.75 d	1 y 4.331 d	1 y 2.2 d	1 y 1.45 d
2. Sidereal period (years)	11.86	29.46	84.08	164.8	248.4
3. Distance from Sun (Earth = 1)	5.200	9.54	19.19	30.06	39.46
4. $\times 10^6$ km	777.9	1427	2871	4497	5909
5. Eccentricity of orbit	0.048	0.056	0.047	0.0086	0.248
6. Inclination of orbit	1°18′	2°29′22″	0°46′	1°46′30″	17°19′
7. Inclination of equator	3°7′12″	26°43′	97°53′	29°33.6′	118°
8. Period of rotation	9.9249 h	10.65 h	17.24 h	16.1408 h	6.387 d
9. Equatorial diameter (km)	142 796	120 000	50 800	49 600	2284
10. Diameter (Earth = 1)	11.19	9.407	3.98	3.888	0.179
11. Polar diameter (km)	133 540	107 085	49 576	48 315	2284
12. Volume (Earth = 1)	1318	745	64.49	60.05	0.0125
13. Mass (Earth = 1)	317.8	95.16	14.5	17.2	0.0025
14. Density (g cm^{-3})	1.33	0.704	1.24	1.58	1.1
15. Gravity (Earth = 1)	2.538	1.075	0.914	1.14	0.078
16. Escape velocity (km s^{-1})	59.56	35.55	21.33	23.75	1.32
17. Velocity in orbit (km s^{-1})	13.06	9.644	6.799	5.433	4.738
18. Albedo	0.44	0.47	0.35	0.35	0.4
19. Temperature	−120°C	−150°C	−200°C	−215°C	−230°C
20. Number of moons					
(pre space age)	13	10	5	2	1
(after space age)	16	17	15	8	1
21. Atmosphere	90% H_2, 10% He, NH_3, CH_4, C_2H_2, CO, C_2H_6, PH_3, H_2O	94% H_2, 6% He, NH_3, CH_4, C_2H_2, CO, C_2H_6, H_2O	80% H_2, 15% He, NH_3, CH_4, C_2H_2, C_2H_6, H_2O	85% H_2, 15% He, NH_3, CH_4, C_2H_2, C_2H_6, H_2S	CH_4, NH_3

$$\frac{\text{Major axis} - 2(0.587)}{\text{Major axis}} = \epsilon = 0.967$$

$$\therefore \text{Major axis} - 2(0.587) = 0.967 \times \text{Major axis}$$

$$\therefore 0.033 \times \text{Major axis} = 1.174$$

$$\therefore \text{Major axis} = 35.575 \text{ AU}$$

The aphelion distance of Halley's comet is therefore 35.575 − 0.587 = 35 AU (nearly). So the comet makes its turn half-way between the orbits of Neptune and Pluto. At its widest the orbit is 9.1 AU.

The elements of the orbit of Halley's comet are as follows:

Inclination of orbit = i = 162°24′
Longitude of ascending node = Ω = 58°15′
Longitude of perihelion = ω = 112°
Perihelion distance = q = 0.587 AU
Eccentricity of orbit = ϵ 0.967
Time of perihelion passage = T = 1986-02-09

When a comet is near to the Sun, it is difficult to say whether its orbit is an ellipse, parabola or hyperbola (Figure 7.20). In the case of a parabola, the eccentricity is equal to 1, and the comet will not return to the Sun. Orbital measurements are needed when the comet is further away.

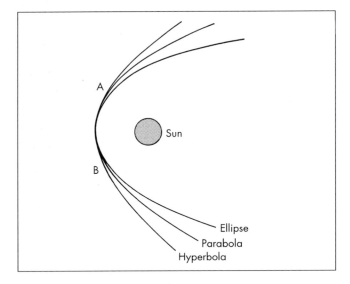

Figure 7.20 Difficult to distinguish between ellipse, parabola and hyperbola when comet is between A and B

Inclinations of Cometary Orbits

The inclinations to the Ecliptic of the orbits of comets are rather large. Out of a group of 46 well-known comets, there are 16 with inclinations between 5 and 10 degrees; 20 between 10 and 15 degrees; and 10 between 15 and 20 degrees. The inclinations could have been even larger in the past, as is shown by non-periodic comets. The giant planets could have drawn the periodic comets nearer to the plane of the Ecliptic.

Discovering Comets

When a comet is far from the Sun, it is a faint smudge. Observations, night after night, will reveal that the smudge is moving against the background stars. Measurements of the right ascension and declination, shortly after discovery, are of paramount importance to calculate the orbit. When an astronomer discovers a comet, he must hasten to send a telegram to the Central Bureau of Astronomical Telegrams, Cambridge, Massachusetts, USA. The abbreviated address is as follows:

TWX 710 320 6842, Astrogram, Cam.

Details of the coded figures that have to be used are given on page 152 of the book *Halley's Comet*, by Donald Tattersfield, published by Blackwell, 108 Cowley Rd, Oxford, OX4 1JF, UK.

The Life of a Comet

As a comet approaches the Sun, a coma develops around its nucleus. The coma consists of hydrogen and can stretch over millions of kilometres. As it gets nearer to the Sun, the tail begins to develop and it gets longer and longer until perihelion is reached. Then, as the comet moves away from the Sun, the tail gets shorter, until it ceases to exist. The coma and tail consist of gases and dust driven out of the nucleus by the heat of the Sun. They are visible by the sunlight which they reflect. The tail points away from the Sun.

A spectroscopic survey made by the South African Observatory, Sutherland, on 15 March 1986, of Halley's comet shows that the dust in the tail consists of finely-divided silicon, and magnesium, in proportions which are similar to those in ages-old meteorites, and to those in the Sun.

Figure 7.21 The nucleus of Halley's Comet, showing gas and dust streaming out. Copyright (1986) Max-Planck-Institut für Aeronomie, Lindau-Harz, Germany by courtesy Dr. H.U.Keller.

The Russian probe, *Vega I*, also found carbon dust and showed that the gases in the tail consist of hydrogen, oxygen and nitrogen, as well as organic molecules. The gas and dust particles were found to come chiefly from four separate active spots in the nucleus – the same as appeared in the 1910 apparition of the comet. Figure 7.21 was taken by the probe *Giotto*, which was launched by the European Space Agency and which passed through the coma of the comet. The team of scientists responsible for the fly-by of *Giotto* was led by Dr H. Uwe Keller of the Max Planck Institute for Astronomy. This was the first photograph ever taken of the nucleus of a comet. The gas and dust escaping from the nucleus are illuminated by sunlight.

Dirty Snowballs

F.L. Whipple proposed the theory that comets consist of a mixture of ices and dust particles, like a dirty snowball, and that they are no more than 50 km in size. Spectroscopic analysis had shown that the coma contains ions such as hydrogen H^+, hydroxyl OH^-, carbon dioxide CO_2^+, carbon monoxide CO^+, nitrogen N_2^+, and the CH^+ radical. The hydrogen and hydroxyl ions are derived from water H_2O. The other ions derive from carbon dioxide, carbon monoxide, ammonia (NH_3) and methane (CH_4), as well as cyanogen (CN) and hydrocyanic acid (HCN). They are all frozen, and dust particles are occluded in the ices.

When a comet nears the Sun, these ices evaporate. The jets of gas so formed blow the dust particles out of the nucleus. The jets of gas and dust reflect sunlight. Some sunlight is absorbed and radiated as fluorescence.

Because the tails of comets always point away from the Sun, there must be some force, coming from the Sun, which drives the gas and dust away. In 1950, L.F. Biermann suggested that an electrically charged "wind" must radiate from the Sun to bring about this effect. Satellites which were launched during the 1960s confirmed the existence of such a "wind", and found that it consists of protons and electrons, respectively positively and negatively charged, that have their origin in the Sun. Almost 90% of the Sun consists of hydrogen.

This solar wind, pressing on the comet, causes the coma and tail to assume a stream-lined shape, pointing away from the Sun. The tail can be made up of two parts: (1) consisting of gaseous ions, pointing directly away from the Sun; and (2) consisting of fine dust particles, which sometimes assume a curved shape, as for example in Comet West. This scimitar shape must have been the cause of the dread with which men viewed comets through the ages. The curved tail usually appears near perihelion, when the comet is moving at right angles to the direction in which the tail points. The dust particles further away from the comet move more slowly and lag behind. Donati's Comet of 1858 was a good example of a comet with a curved tail.

The Sun's magnetic field wraps itself around the coma and tail. The magnetic field becomes concentrated, so that whirls originate in the tail.

All moving bodies spin, and the spin of the nucleus of a comet gives the issuing gases a twist. If the comet is moving in a direct orbit and spins in a direct direction, the reaction of the issuing gas streams will propel the nucleus forward and speed it up (Figure 7.22). At its next appearance the comet will be early. If the comet spins in a retrograde direction, while moving in a direct orbit, the issuing gases will tend to hold the nucleus back so that the comet will be retarded, causing it to arrive late at its next appearance.

Encke's Comet, for example, is speeded up by $2\frac{1}{2}$ hours at each return to the Sun. Halley's Comet is retarded by 4.1 days between two appearances. Out of 20 periodic comets, 9 are speeded up, 9 are retarded and 2 show no difference. These two must therefore be very slow spinners.

When a comet moves away from the Sun, it receives less and less heat, so that the generation of gas decreases. Less particles of dust are then freed and the tail becomes shorter, until it ceases to exist. The frozen nucleus then goes on its

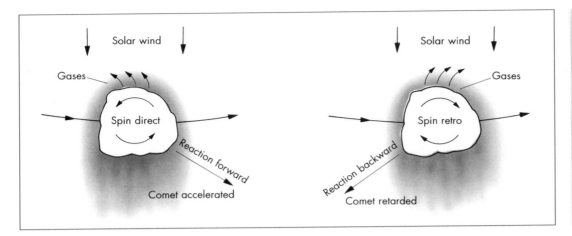

Figure 7.22
Gas jets from the nucleus either speed up or retard the comet

course towards its aphelion, before turning back towards the Sun, perhaps after many years. Comets in parabolic and hyperbolic orbits do not return to the Sun.

With each passage by the Sun, a comet loses a lot of material, so that a comet cannot last forever. There must therefore be a source of cometary material from which comets are replenished.

Oort's Cloud

J.H. Oort formulated the theory that there exists a sphere of cometary material, enshrouding the Sun at a distance of 40 000 to 100 000 astronomical units. This is known as the Oort Cloud. The material in the cloud consists of the remains of the primeval nebula out of which the Sun and planets condensed. At a distance of 100 000 AU, the material is subject to perturbations by passing stars. From time to time these disturbed clumps of material start tumbling towards the Sun, reaching its environs after, perhaps, a million years.

As criticism of this theory, one must bear in mind that the Sun and the stars in its neighbourhood describe more or less parallel orbits around the centre of the Milky Way. Their distances apart therefore do not change noticeably, so that there can be no talk of passing stars. Later, we shall see that the fastest moving star, Barnard's star, will not reach the precincts of the Sun before 9600 years, and then it will still be further away than the present nearest star.

It is more likely that the cometary material actually has its origin in the Sun, being constantly replenished by the solar wind. The substances out of which comets consist, hydrogen, carbon, oxygen, nitrogen and metals such as iron, silicon and magnesium, are all found in the solar atmosphere and are blown out together with the solar wind.

Far beyond the furthest planets, the solar wind will come to rest, relative to the Sun, and all the matter will be in the form of ices. Variations in the strength of the solar wind will cause turbulence in the material, causing some of it to be dislodged, even if ever so slightly, but enough to send the perturbed clumps hurtling towards the Sun, at first imperceptibly, but speeding up with the lapse of time.

It is also possible that interstellar dust and gas particles may perturb the cometary material and set it moving in clumps towards the Sun.

Possibly, the newly discovered discs of dust which radiate in the infra-red around the stars Vega and Beta Pictoris are samples of cometary material around those stars.

Perturbation of Cometary Orbits

When cometary nuclei come into the region of the giant planets, they will undergo perturbations. The nuclei will be drawn closer to the plane of the Ecliptic. The periods will be altered and the orbits themselves may be altered from direct to retrograde, and vice versa. Comets in parabolic or hyperbolic orbits may have their orbits changed to ellipses and thereby become captives of the Sun. In Figure 7.23a a comet in a parabolic direct orbit is deflected by Jupiter into a retrograde elliptical orbit, and in Figure 7.23b a retrograde parabolic orbit becomes a direct elliptical orbit. It is also possible for a comet to be thrown out of the Solar System, as shown in Figure 7.24.

There are no less than 16 periodic comets that have aphelion distances of 5 to 5.4 AU, approximately Jupiter's average distance from the Sun. These comets are known as the Jupiter Family of Comets.

In 1886, Brookes' Comet whizzed around Jupiter, between the four large moons. They were not perturbed

Comets

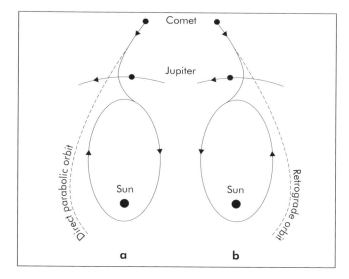

Figure 7.23 The capture of a comet. **a** Direct orbit becomes retrograde **b** Retrograde orbit becomes direct

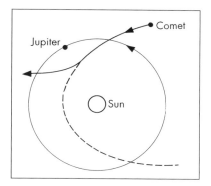

Figure 7.24 Comet thrown out of the Solar System by Jupiter

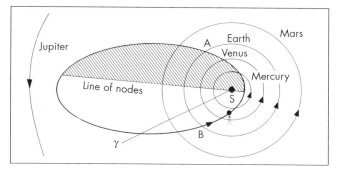

Figure 7.25 Orbit of Encke's Comet

at all, but the comet's period was altered from 29 years to 7 years! Calculations of the mass of the comet, based on these perturbations, showed that the mass is about 1/200 000th that of Jupiter's moon, Io.

The comet with the shortest period is Encke's Comet, discovered by J.F. Encke. Its period is 3.3 years and decreases by 3 days per century. Since its discovery, the comet has visited the Sun more than 50 times. Its perihelion distance of 0.34 AU brings it within the orbit of Mercury. Its aphelion lies at 4.09 AU and the orbit is inclined to the Ecliptic by 12 degrees. The line of nodes almost coincides with the major axis. The eccentricity is 0.85 and the motion is direct (Figure 7.25).

The figure shows the orbit as crossing the Earth's orbit at points A and B, but actually the comet is far north and south of the Ecliptic at those points, being 29 and 25 million km away respectively. The shaded portion of Encke's orbit lies south of the ecliptic.

A beautiful and bright comet was discovered in 1969 by Jack Bennett of Pretoria, South Africa. It was comet 1969i

Figure 7.26 Bennett's Comet over the Alps. Photograph by C. Nicollier

Figure 7.27 Jack Bennett

Figure 7.28 Dust tail of Comet West

Figure 7.29 Comet Shoemaker–Levy-9, taken by Hubble Space Telescope's wide field camera. Courtesy of H.A. Weaver and T.E. Smith, Space Telescope Science Institute

and its period is about 1700 years (Figures 7.26 and 7.27).

A very interesting comet of modern times is No. 1975n, discovered by R. West. Its curved tail was densely filled with dust. It was found that the nucleus had divided into four fragments (Figure 7.28).

The same fate befell the comet discovered in 1826 by W. Biela. Its period was found to be 6.6 years, with perihelion at 0.86 AU and aphelion at 6.19 AU.

Biela's Comet returned in 1832, but not in 1839. When it returned in 1846, the nucleus had split into two pieces, 282 000 km apart. At the return of 1852, the two fragments were 2.3 million km apart. And the comet was never seen again.

On 27 November 1872, there was a brilliant display of meteors (shooting stars) with radiant (the point from which they seem to radiate) near the star Gamma Andromedae, at the exact spot where the Earth's orbit crosses that of Biela's Comet. The display was probably caused by fine particles, left over from the comet, burning up in the atmosphere, showing that the nucleus must have totally disintegrated.

Besides Jupiter being able to alter the orbits of comets and throw them out of the Solar System, it also has the ability to capture a comet, as it did with the comet Shoemaker–Levy-9. This comet was discovered by Eugene and Carolyn Shoemaker, assisted by David Levy and Philippe Bendjoya on 23 March 1993. It had a very strange appearance. A photograph taken by the Hubble Space Telescope revealed the comet as broken up into 22 fragments, strung out like a string of pearls (Figure 7.29).

Figure 7.30 Impacts of 20 July 1994, taken in the infra-red at 2.3 microns. Copyright (1986) Max Planck-Institut für Aeronomie, Lindau-Harz, Germany by courtesy Dr. H.U. Keller.

The disruption of the nucleus must have taken place when the comet passed through perijove (nearest point in its orbit to Jupiter) on 8 July 1992, and the nucleus must have been within the Roche limit. The orbit around Jupiter had a period of two years, but the next close approach in July 1994 would see the fragments crash into Jupiter's atmosphere. And that is how it happened. The first fragment was calculated to crash on 16 July 1994 at 20 h Universal Time, and the last on 22 July 1994 at 8 h 19 m. The points of collision were to be located a few degrees beyond Jupiter's eastern limb, so that the collision would not be visible in optical telescopes. After about five hours, the sites of collision would have rotated into view from Earth. Each collision made a black spot, somewhat larger than the Earth. Infra-red telescopes were able to capture the heat glow of the collisions, as is seen in Figure 7.30, where three of the explosions are visible at the wavelength of 2.3 microns in the infra-red.

The satellite IRAS (Infra-red Astronomical Satellite), while scanning the skies for infra-red radiation, also discovered several comets, among others a comet which was simultaneously discovered by Araki and Alcock.

Meteors

Besides the Bielid meteor shower, visible in the early morning hours, between 15 and 20 November, in the constellation of Andromeda, there are several other meteor swarms connected to the orbits of comets. The Perseids are visible in the evenings from 25 July to 12 August in the constellation Perseus. From 16 to 21 October, a swarm is visible in the evenings in the constellation of Orion, which coincides with the orbit of Halley's Comet. Another swarm of particles, blown out from the nucleus of Halley's Comet, is visible in the evening between 1 and 6 May, in the constellation Aquarius. The Leonids, in Leo, appear in the orbit of the comet Temple–Tuttle, from 15 to 17 November. They are the longer streaks in Figure 7.31.

Meteor Showers

The best time to spot meteors is the early hours, 02 h to 04 h, because the part of the Earth's atmosphere moving forward in the Earth's path from midnight to dawn, indicated by M–D in Figure 7.32, is then overtaking the meteors.

The speeds at which meteors enter the Earth's atmosphere are greater, ± 70 km s^{-1}, in the early hours, because the Earth's velocity in its orbit has then to be added to the speeds of the meteors. Meteors have to catch up on the Earth when they enter the evening atmosphere, S–M, and their speeds are then only ± 40 km s^{-1}.

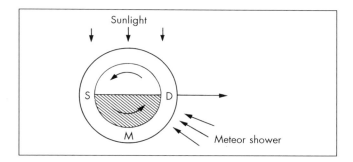

Figure 7.32 Best time for spotting meteors

Figure 7.31 The Leonids, 15 to 17 November in Leo

Table 7.5 Best meteor showers

Shower	Maximum	Right Ascension h m	Decl.
Quarantids	3 Jan.	15 28	+50°
Lyrids	21 Apr.	18 08	+32
Perseids	12 Aug.	03 04	+58
Orionids	21 Oct.	06 24	+15
Taurids	8 Nov.	03 44	+14
Leonids	17 Nov.	10 08	+22
Geminids	14 Dec.	07 28	+32

Observations (Table 7.5) have strengthened the supposition that meteors derive from comets, and that they largely consist of small grains that have been driven out of the nuclei of comets by the escaping gases. When the Earth moves through a region through which a comet has passed, it picks up these fine grains which then burn up in the atmosphere, on account of the friction generated by their swift motion.

If a meteor is large enough, it will form a spectacular fire ball with a trailing tail. The speeds of such bolides have been measured at 40 to 70 $km\,s^{-1}$. Despite the great amount of friction, some bolides may not burn up before striking the surface of the Earth. Samples of these meteorites are to be found in many museums.

The Gibeon Meteorites

In the Transvaal Museum, Pretoria, South Africa, there are two pieces of the Gibeon meteorites, of which 52 pieces were found near Gibeon in Namibia. These two pieces weigh 410 and 200 kg respectively. The South African Museum at Cape Town has one piece which weighs 650 kg. The rest of the Gibeon meteorites are housed at the Windhoek Museum. Figure 7.33, showing these fragments at Windhoek, was donated by Dr C.N. Williams of Bedfordview, Johannesburg.

These rocky meteorites are called chondrites. They consist largely of enstatite and chrysolite.

Other large meteorites are the Rateldraai meteorite, found near Kenhardt, weighing 550 kg, and the Humansdorp meteorite, weighing 600 kg.

Figure 7.33 Dr C.N. Williams and the Gibeon meteorites

Figure 7.34 Dr C.N. Williams and the Hoba Meteorite, Grootfontein, Namibia

The Hoba Meteorite

The largest meteorite known lies where it landed, near Grootfontein in Namibia. It is known as the Hoba meteorite and is a more or less rectangular block of iron and nickel, 3.8 metres by 2 metres by 1 metre, and of mass at least 60 tons. It consists of 82.4% iron, 16.4% nickel, 0.76% cobalt and traces of carbon, sulphur, chromium, zinc, gallium, germanium and iridium (Figure 7.34).

It is difficult to comprehend how such massive blocks of iron–nickel and other silicaceous rocks could have blown out of the nuclei of comets by jets of gas. One should seek elsewhere for the origin of these meteorites. Such clumps of iron–nickel could easily have been formed when the matter out of which the Sun and planets condensed was originally cast out by an exploding star – a supernova. In such an explosion all manner of atomic transformations take place. Iron is the end-product of the fusion of lighter atomic nuclei in the nucleus of a star, before it explodes. Under the tremendous pressure prevailing in the matter of an exploding star, any iron–nickel formed would have been compressed into very dense clumps.

Rocky meteorites such as the Gibeon meteorites could have had their origin in similar circumstances or they could have been cast out of the Moon when meteors and asteroids crashed on to it. Meteorites found in Antarctica contain anorthosite, a soda-lime feldspar, which occurs on the Moon and could easily have been cast out with the formation of a crater. Tectites which have been found in

Figure 7.35 Crater Tycho, showing rays of material which was cast out

Australia, and elsewhere, are flattish, rounded marbles of button-like appearance that show signs of having melted and solidified twice. They could have been cast out when the crater Tycho was formed by an asteroid which crashed on to the Moon. From this crater, which is near to the Moon's south pole, rays of material which was cast out radiate in several directions. Some rays are 2700 km long (Figure 7.35). The explosion of the crash would have sent material hurtling into space. The heat generated would have turned everything to gas. On cooling, droplets would be formed and, if these had velocities in excess of the Moon's escape velocity of 2.375 $km\,s^{-1}$, they would have left the Moon's gravitational sphere. After many circuits of the Sun, these droplets, by now frozen, could have entered the Earth's gravitational field and plummeted into Earth's atmosphere, to melt again, before crashing and solidifying for the second time. All tectites show signs of having solidified twice.

Craters on the Earth

Unlike the Moon, Mercury and other bodies in the Solar System, the Earth has no craters dating from the original accretion of matter from the primeval nebula from which the Sun and planets condensed. The reason is because the water on the Earth's surface has eroded all traces of the original craters. The craters that do exist on the Earth were formed in recent times. None of them are near a million years old.

Meteor Crater

Meteor Crater, or Barringer Crater, in Arizona, USA, is 1200 metres wide and 174 metres deep, and the ridges of the rim rise 52 metres above the surrounding plain (Figure 7.36). It was formed about 30 000 years ago by a meteorite whose mass has been estimated to have been 100 000 tons and of size 100 metres. The heat of the collision would have vaporised everything near and a mushroom cloud of dust would have risen skywards. The outward pressure waves would have forced the rock layers outwards, thus forming the raised rim.

There are also other signs of craters on Earth. About 70 are known, of sizes ranging from 30 to 140 metres. Most of them have been discovered by satellites.

In Ontario, Canada, there is the Sudbury Structure, which is the remains of a crater 59 by 27 km in size. Radioactive dating has fixed the age at 1.8 milliard (10^9) years – one of the oldest craters known to exist on Earth.

Algeria boasts two craters: the Amguid Crater, 450 metres wide and 30 metres deep; and the Talemzane Crater, 1750 metres wide and 70 metres deep. The latter is situated 120 km from Laghouat.

Figure 7.36 Meteor Crater, Arizona, USA

Figure 7.37 Crater Serra da Congalha, Brazil

In Alaska, situated at 151°23′ west and 66°7′ north, is the Sithylemenkat Crater, 12.4 km wide and 500 metres deep. It was traced on photographs taken by the Landsat satellite.

The aviator G. Winter noticed that the Serra da Congalha in northern Brazil, 46°52′ west and 8°05′ south, is actually a crater. It is 3 km wide and 350 metres deep, and encircled by a second ring-shaped range of hills, 12 km in diameter (Figure 7.37). The photograph was taken by G. Winter from a height of 10 km.

In southern Germany, there is a group of dents which have been identified as the remains of craters, 1 to 3.5 km in size, which were formed 14.8 million years ago. They are situated near Nordlingen.

If a large city is hit by a meteorite as large as the one that formed the crater in Arizona, the loss of life would be frightful and the damage to property incalculable. Because three-quarters of the Earth's surface is covered by seas and oceans, we can assume that, on average, three out of every four meteorites will crash into the sea. P. Piper and L.F. Jausa wrote about a crater, 45 km in diameter, situated 200 km south-east of Nova Scotia. It is under 113 metres of water and the crater itself is 2800 metres deep. It has a peak in the centre (*Nature*, 18 January 1987).

Besides the sea, there are also very large uninhabited parts of the Earth, such as the Sahara and Antarctica, where meteors can crash with relative safety. Figure 7.38 shows a meteorite, weighing 8 kg, which was found in Antarctica. The constitution of the rock agrees very well with that of certain rocks on Mars, and it is possible that it had its origin on the Red Planet.

Tunguska

The Earth can, of course, also be struck by the nucleus of a comet. At 7 h 20 m on 30 June 1908, there appeared a blinding fire ball, much brighter than the Sun, in the cloudless sky over the Tunguska Valley in Siberia (Figure 7.39). Then followed a thunderous explosion, which was heard as far

Figure 7.38 Visitor from Mars

Figure 7.39 Destruction of the forests at Tunguska, 1908

away as 1000 km. The Earth shuddered and, in nearby villages, people were thrown off their feet. Several herds of reindeer were scorched to death. Luckily, it was a very sparsely populated area.

If it was a meteor that struck the Earth, remains of nickel–iron or chondritic rocks would have been found. But nothing of this sort was ever found in the area. There is also no sign of a crater. However, trees as far as 30 km from the centre of the explosion were flattened, over an area of 2150 km². The trees lay pointing radially outwards from the point of collision.

If the cause was not a meteor, then it must have been the nucleus of a comet. In that case, there would be no remains because the nucleus of a comet consists of fine particles, bound in ices. The dust particles would vaporise and then congeal to form microscopic globules of glassy appearance. Tiny spherules of this nature have been found there.

If the nucleus of a comet enters the Earth's atmosphere, the friction between it and the air would raise the temperature to incandescence. The nucleus of the comet would tend to vaporise, but the surrounding air would act as an immovable barrier, so that the pressure would rise – a case of an irresistible force meeting an immovable object! The temperature would have risen so high that the incandescent fire ball would be as bright as the Sun. And that is what was seen. Then an unimaginable explosion would take place. According to calculations, this must have happened at a height of 6 to 8.5 km above the surface. The outgoing pressure blast would have flattened the trees and the heat would have done the scorching. Figure 7.40 shows

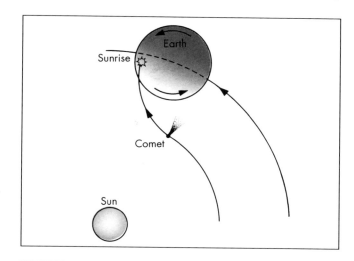

Figure 7.40 Possible orbit of comet which collided with the Earth in 1908

the possible orbit of a nucleus of a comet, which could have struck from the south-east, in the early daylight hours.

Nobody was aware of the approaching catastrophe. Nobody had monitored the comet. According to the figure, it was in line with the Sun and therefore not visible.

Calculations indicate that the force of the explosion was equal to that of a 12-megaton nuclear explosion. "Spaceship Earth" lives dangerously!

Chapter 8
Exploring the Solar System

1 The Advent of the Rocket

The First Artificial Satellite

The year 1957 was International Geophysical Year, during which scientists of many nations co-operated in research on the Earth and its environs. On 4 October 1957, the world stood amazed when it was announced that the first artificial satellite, Sputnik, had been launched and that it was revolving in a stable orbit with perigee (nearest point to Earth) of 228 km and apogee (furthest from Earth) of 947 km.

When the rocket reached its elliptical orbit around the Earth, the nose cone of the rocket opened and set the satellite free, so that its antennae could deploy. The satellite itself had a diameter of 58 cm and a mass of 83.6 kg.

The plane of the elliptical orbit was inclined by 65 degrees to the Earth's equator. Thus the satellite could be seen from all points on Earth, at various times. It was a bright "star", moving fairly rapidly among the background stars. Its radio signals were clearly audible. Besides its radio transmitter and receiver, the satellite had instruments on board to measure the temperature and density of the atmosphere and also the concentration of electrons.

The Principle of the Rocket

The principle on which the rocket works is Newton's Third Law: to every action there is an equal but opposite reaction. The fuel (paraffin) and oxidising agent (liquid oxygen) are pumped to the ignition chamber in the top of the exhaust where ignition is initiated by means of an electric spark. The gases freed by the combustion exercise great pressure in all directions, but because they escape from the open end of the exhaust, there is a nett resultant pressure against the head of the combustion chamber, thus propelling the rocket upwards (Figure 8.1). If a balloon is inflated and

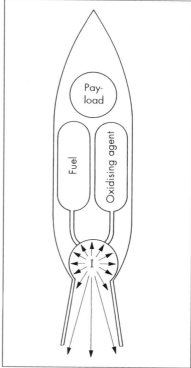

Figure 8.1
Principle of the rocket;
I = ignition chamber

then suddenly released, the compressed air in the balloon escapes through the opening and thrusts the balloon aimlessly forward. The jet of ejected gases propels the rocket in the same way. Because the rocket is streamlined, it moves in a straight line. A rocket moves faster in a vacuum than in air, because in a vacuum it experiences no wind resistance.

The thrust of the rocket motor is equal to the product of the mass of gas cast off per second and the speed of the escaping gases. Therefore, it is preferable that the speed of the jet of gas be as high as possible. The thrust developed by the Sputnik booster was 200 000 kg. It was sufficient to give the rocket a speed of 28 000 km h^{-1} that is 7.7 km s^{-1}. At this speed a body will not fall back to the Earth, provided it is outside the atmosphere where there is no air resistance. The rocket will then be in orbit. The orbit will be an ellipse, and in the limiting but rare case, a circle.

The instruments aboard Sputnik I functioned for 21 days. After 96 days and 1400 revolutions around the Earth, it fell back into the atmosphere, where it burned out. This showed that at the height of Sputnik's perigee, 228 km, there is enough air to slow down the rocket.

Sputnik I heralded the start of the Space Age – man's first step to break the shackles of his earth-bondage. Man would now be free to develop a cosmic consciousness. To astronomers, it was the dawn of the long-awaited age of space-exploration.

Soon, in 1958, there followed the launching of the Explorer satellites. Explorer I was launched on 31 January 1958 by means of a Jupiter-C booster. The satellite was very small, weighing only 14 kg, but nevertheless it required a rocket of length 35 metres, and a thrust of 35 375 kg to place it in orbit, having a perigee of 360 km and apogee of 2532 km. While moving so far away from the Earth, Explorer was able to explore the electrically charged belts, called the Van Allen belts. In those belts electrically charged particles (ions) are trapped in the magnetic field of the Earth.

The Rocket Breaks Away from Earth's Gravitational Field

Luna I was the first space probe to break away from Earth's gravity and enter the gravitational field of another celestial body, the Moon. It was launched on 2 January 1959 by means of a booster which had a thrust of 263 000 kg; its mass was 361 kg. It passed by the Moon at a distance of 5955 km and went into an orbit around the Sun. To overcome Earth's gravity, the rocket had to reach a speed of 11.2 km s^{-1}, or 40 320 km h^{-1}.

Luna III took photographs of the far-side of the Moon, the side which man had never seen. The photographs were not very clear, but they did reveal a surface more cratered than the side facing the Earth. There are no large plains, or maria, on the far side of the Moon.

The first man in space followed soon. On 12 April 1961 Yuri Gagarin, in the capsule Vostok, made a complete circuit of the Earth and made a safe landing after reaching an apogee of 324 km. The mass of Vostok and its contents was 4725 kg, which required a massive booster, having a thrust of 509 840 kg, as well as a second stage having an extra 90 260 kg thrust. The trip took 89 minutes.

Scouting the Moon

Several other series of satellites were soon launched: Rangers, Lunar Orbiters, Surveyors, Pioneers, Mariners, Vikings and Voyagers, as well as Russian probes such as Venera to Venus.

The Rangers were launched to home in on various parts of the Moon, photographing the areas, before crashing on the surface. The boosters used were the Atlas–Agena-D, of mass 118 tons. The thrust of the five engines totalled 166 460 kg. Its diameter was 3 metres and length 24 metres. When the burn was completed, its speed was 28 166 km h^{-1}. On top of the Atlas, the Agena second stage, 7.7 metres in length, delivered an additional thrust of 7250 kg and this enabled the rocket to reach the escape velocity of 11.2 km s^{-1}.

The Rangers took a total of 17 000 photographs of 2 million km^2 of the Moons surface, for transmission to Earth by radio.

The same boosters also launched the Lunar Orbiters – their purpose was to find landing sites. The photographs were used to compile the first Lunar Atlas, including the far side of the Moon. Figure 8.2 shows the inside of the crater Copernicus. The central peaks, just below the centre of the photograph, rise to almost 1.5 km.

The Lunar Orbiters were fitted with four solar panels to provide electricity for the various instruments. The scanning system signalled the scenes, one pixel at a time, to the Earth. An important finding of the Orbiters was that the gravity of the Moon is not the same all over: it is stronger in the maria than in the highlands. There appear to be concentrations of mass in the maria; these are called mascons. While orbiting the Moon, the Lunar Orbiters experienced perturbations of their orbits when passing over the maria. W.L. Sjogren and P.M. Muller of the Jet Propulsion Laboratory calculated that there must be discs of dense material, 200 km wide and 15 km thick, near the surfaces of the maria. These discs of dense material are the

The Advent of the Rocket

Figure 8.2 Lunar Orbiter photograph of the inside of crater Copernicus

results of collisions with asteroids or meteors; or what is more likely, loosely compacted clumps of dense material, which came from the primeval nebula from which the Sun, planets and their satellites condensed by the process of accretion. The plains of the maria were formed when these collisions liberated so much heat that the material on the surface melted and flowed as lava, covering many of the formerly formed craters. The craters we do see in the maria are much more recent in origin. Figure 8.3 shows a region near the crater Marius, in Oceanus Procellarum, where the lava flows congealed into ridges and escarpments.

Figure 8.3 Lava flow in Oceanus Procellarum

There are few craters to be found in the vast Mare Imbrium, showing that the maria were formed late in the accretion process, after most of the material that went to form the Moon had already compacted on the Moon's surface.

Soft Landings on the Moon

The Surveyor space probes succeeded in making soft landings on the Moon, because they were equipped with retro-rockets, which slowed them down in their descent to the Moon's surface. They found the Moon's surface firm enough to take the weight of a space vehicle. The surface is therefore not a sea of fine dust in which everything would sink away, as some had thought.

The Atlas–Centaur booster which launched the Surveyors was 45 metres long and 3 metres in diameter. The total thrust developed was 209 000 kg. The retro-rockets had a thrust of 4536 kg. Surveyor I made the first fully controlled soft landing in the Sea of Storms (Oceanus Procellarum) on 1 June 1966 after a flight of duration $63\frac{1}{2}$ hours. In the next six weeks, Surveyor took 11 150 photographs in all directions from the horizon up to close by, and signalled them to Earth.

Manned Orbital Flights

The Mercury capsules were designed to carry one man in orbital flight; they were launched by the Atlas booster. John Glenn completed three orbits of the Earth, each taking $88\frac{1}{2}$ minutes. The perigee was 161 km and the apogee 261 km above the surface of the Earth. The highest speed attained was 7.84 km s^{-1}. The acceleration at launch was 7.7g, that is 7.7 times the Earth's gravity, and the same force was experienced on the return to Earth. The friction caused by the atmosphere raised the temperature of the air against which the broad end of the capsule thrust to 5260°C, which made the air under the capsule glow orange in colour. It was almost as hot as the Sun's surface! To protect the capsule against this high temperature, the base of the conical capsule was covered with resin and fibreglass. The resin boiled away and evaporated, thus exercising a cooling effect. Because fibreglass is a bad conductor of heat, the temperature of the floor of the capsule did not rise above 500°C. The conical shape of the capsule ensured very little friction against the sloping sides. Insulating material kept the temperature of the inside of the capsule within bearable limits. Besides the flight by John Glenn, three other astronauts completed flights in Mercury.

The next step was to launch the astronauts in pairs. For this purpose, the larger Gemini capsule was used. It weighed 3700 kg, compared with 1800 kg of Mercury. The greater thrust needed was provided by the Titan-2, which delivered a thrust of 240 360 kg. The rocket was 33 metres in length and stood as high as a ten-storey building! Twelve flights were carried out in Gemini capsules. During nine of the flights, docking manoeuvres were practised and four space walks were undertaken. For the space walks, the astronauts had to wear space suits with life support systems on their backs.

The Apollo Missions

The Apollo capsules were designed to land the first men on the Moon. The capsule carried three astronauts. Apollo-VIII made the first flight around the Moon: it circled the Moon 10 times at a height of 110 km above the lunar surface. During the last circuit, the thrust engine was switched on when the capsule was on the far side of the Moon. In this way the speed of the capsule was increased and, as it rounded the Moon, it pulled away from the Moon and sped on its way to the Earth, where it splashed down in the Pacific Ocean, 11 seconds before its pre-destined time!

Apollo-11

The 6 cubic metres of space inside the Apollo-11 capsule provided sufficient space for the three astronauts, Neil Armstrong, Edwin Aldrin and Michael Collins. The booster rocket, the Saturn-V which was designed by Wernher von Braun, was launched from Cape Kennedy at 09h 32 m, on 16 July 1969. The Saturn-V (Figure 8.4) was the most successful of all boosters and was a magnificent piece of engineering ingenuity. Its size was overwhelming. The diameter of the first two stages was 10 metres. Together with the third stage, the height was 85.6 metres. The Apollo space vehicle consisting of the Lunar Excursion Module in its housing, the service module, the command module (containing the astronauts) and the escape tower, had a length of 25 metres. The total height was thus 110.6 metres – as high as a 36-storey building! The total mass was 2 913 000 kg, or 2913 tons!

The first two stages each had five engines. The first stage contained 802 420 litres of paraffin as fuel and 1 300 900 litres of liquid oxygen as oxidising agent. Paraffin is a compound of carbon and hydrogen, having the general formula:

$$C_nH_{2n+2}$$

In pentane, $n = 5$. The reaction between pentane and oxygen, is as follows:

$$C_5H_{12} + 8O_2 \rightarrow 5CO_2 + 6H_2O$$

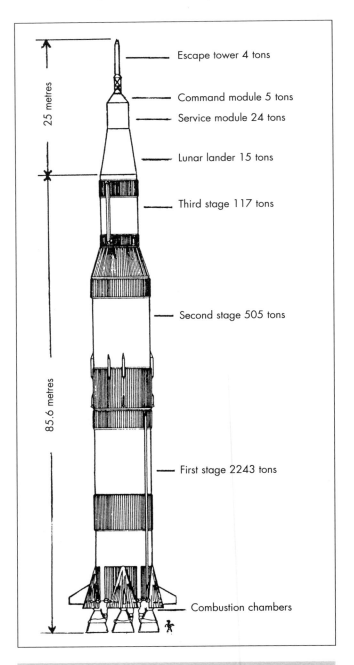

Figure 8.4 The Saturn-V rocket and the Apollo-11 space vehicle

The reaction is accompanied by the liberation of 92 million joules per kilogram of paraffin which is burned. The temperature attained in the combustion chambers is 3400°C. The gases produced, carbon dioxide and steam, leave the combustion chamber as a jet of tremendous velocity. The total thrust delivered by the first stage is 3 425 000 kg. This is 17% more than the total mass, which therefore gets lifted into the sky. The paraffin and liquid oxygen are pumped into the combustion chambers at a rate of 12 800 litres per second. The diameters of the five exhausts from which the jets emerge are each 4.6 metres. To combat the tremendous heat of the escaping flaming gases, 40 000 litres of water are sprayed into the concrete pit into which the flaming gases jet. If this were not done, the launching tower and the rocket itself would melt from the heat! The products of combustion form towering clouds to the sides of the launching pad and rise more than 300 metres into the sky. All in all, a most spectacular sight!

The author had the privilege of attending the launch of Apollo-11 and can testify that it is a never-to-be-forgotten sight. As a safety precaution, the spectators were kept 5.6 km from the launching platform. Because of this separation, the deafening roar of the five engines could not be heard until 16 seconds after the moment of ignition – it sounded like the simultaneous firing of thousands upon thousands of machine guns – and the Earth shuddered!

By the time that the sound was audible, the rocket had risen to the top of the launching tower, 130 metres high. Some seconds later, the pressure wave of the expanding gases reached the spectators and the wind beat the tufts of grass flat on the ground. Figure 8.5 is of the night launch of Apollo-17.

The first stage fired for only 2 minutes 40 seconds and by that time the rocket had reached a height of 61 km and a speed of 8530 km h^{-1}. When the burn of the first stage was finished, the bolts holding the first stage broke and the empty shell of the first stage fell into the ocean, 660 km down range. Simultaneously, the firing of the second stage commenced and the spectators could not help cheering.

The second stage contained 1 033 300 litres of liquid hydrogen, as fuel, cooled to −253°C, and 340 270 litres of liquid oxygen, cooled to −172°C. This stage developed a thrust of 526 165 kg, while it burned for 6 minutes. The reaction between the liquid hydrogen and liquid oxygen, which were pumped to the combustion chambers under very high pressure, was as follows:

$$2H_2 + O_2 \rightarrow 2H_2O$$

This reaction yields 120 million J kg^{-1} of hydrogen burned.

As the rocket then sped away, it seemed, on account of perspective, as if it were dropping down to the horizon. After six minutes, the speed had reached 24 625 km h^{-1} and the height 183.5 km. Then the second stage broke loose and fell back to Earth, 4200 km from the launching pad.

The total mass was now reduced to 5.6% of what it was at launch. After 9 minutes of flight, the third stage fired for $2\frac{1}{2}$ minutes. The speed was now 28 800 km h^{-1} (8 km s^{-1}). This was sufficient to maintain orbital flight. In the winking of the eye, the computer controlling the flight had calculated the orbit as having a perigee of 190.6 km and an apogee of 192 km. This was very close to being a circle. The third stage, 6.57 metres in diameter and 10.5 metres long, and of mass 117 tons, had a single engine. The 243 000 litres of liquid hydrogen and 77 590 litres of liquid oxygen which it carried were able to yield a thrust of 104 325 kg.

Off to the Moon

Apollo then completed one revolution around the Earth and when three-quarters of the second orbit had been completed, after the flight had been in progress for 2 hours 44 minutes 16.2 seconds, the third stage fired again for 5 minutes 47 seconds. This increased the speed by 3.182 km s^{-1}. The speed was then 11.182 km s^{-1}, which

Figure 8.5 Night launch of Apollo-17

was adequate for the space vehicle to break away from its orbit around the Earth and speed on to the Moon.

After 3 hours 17 minutes, Apollo separated from the third stage, when it was 9000 km from Earth. The space vehicle then consisted of the landing module, the service module and the command module. The vernier jets were fired to separate the service module and the command module from the third stage. They turned right around and approached the third stage. The housing around the Lunar Landing Module opened and the conical end of the command module docked with the Lunar Landing Module and drew it away from the third stage. Columbia, now consisting of the Lunar Landing Module, the command module (containing the three astronauts) and the service module, then turned around again and the engine of the service module was fired for 3.4 seconds, so as to head towards a point 333 km from the limb of the Moon. The used-up third stage then went on its way, to pass around the Moon, missing it by 4334 km and then going into orbit around the Sun.

The Columbia then made use of its vernier jets to give it a rolling motion, so that all sides could receive heat from the Sun, while they coasted towards the Moon.

After the flight had been in progress for 13 hours 30 minutes, the astronauts turned in for sleep. During their sleep, all systems were monitored and controlled from Earth.

At 26 hours 45 minutes 58 seconds, the second course adjustment was made, aiming for a point 116 km above the lunar surface. After 38 hours, it was time for sleep again – a 12 hour long sleep.

While coasting towards the Moon, the velocity of Columbia was continually decreasing, but it still was high enough to reach the Moon's gravitational field. One can easily calculate how far on the route from the Earth to the Moon the point is situated where the gravities of the Earth and Moon are equal (Figure 8.6).

Suppose the two gravities are equal at C. According to Newton's law of gravitation:

$$\frac{G \times \text{mass of Earth}}{AC^2} = \frac{G \times \text{mass of Moon}}{BC^2}$$

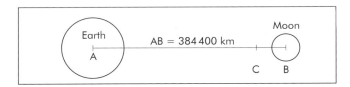

Figure 8.6 Point where gravities of Moon and Earth are equal

The mass of the Earth is 81 times that of the Moon. Therefore:

$$\frac{\text{mass of Earth}}{\text{mass of Moon}} = \frac{AC^2}{BC^2} = \frac{81}{1} \quad \therefore \frac{AC}{BC} = \frac{9}{1}$$

But $BC = (384\,400 - AC)$

$$\therefore \frac{AC}{(384\,400 - AC)} = \frac{9}{1}$$

$$\therefore AC = 9(384\,400) - 9AC$$

$$\therefore 10\,AC = 3\,459\,600$$

$$\therefore AC = 345\,960 \text{ km}$$

When this point was reached, after 61 hours 40 minutes of flight, the speed was 3281 km h^{-1}. Then the space vehicle started accelerating towards the Moon. It attained an orbit with pericynthion (point nearest the Moon) of 113.5 km and apocynthion (point furthest from the Moon) of 312.6 km above the surface of the Moon. The aim had been 113.2 by 312.2 km! Once in this orbit around the Moon, the astronauts retired to sleep.

When 75 hours 49 minutes 50 seconds had elapsed, the retro-rockets of the service module were fired for 6 minutes 2 seconds when Columbia was on the far side of the Moon, so as to reduce speed and make the orbit more circular.

The fuel used by the service module was dimethyl-hydrazine ($CH_3CH_3N_2H_2$) and the oxidising agent was nitrogen peroxide.

After 80 hours 11 minutes 36 seconds, when Columbia was on the far side of the Moon, the retro-rockets were fired again for 17 seconds. This changed the orbit to 99.6 by 121.7 km above the Moon's surface.

Now Armstrong and Aldrin crawled through the tunnel, connecting the command module to the landing vehicle, so as to test all the systems.

Landing on the Moon

When 100 hours 12 minutes had elapsed, during the twelfth orbit around the Moon, the landing vehicle, with Armstrong and Aldrin on board, was disconnected from the command module. After another 1 hour 24 minutes 41 seconds, the descent to the surface of the Moon commenced. During the descent, the vernier rockets on the landing module were fired to turn the module right around so that its landing legs pointed downwards. The landing engine was then fired for 13 minutes to decrease the speed of descent. It had a thrust of 4477 kg. In $5\frac{1}{2}$ minutes the speed was down to 2200 km h^{-1}; after $6\frac{1}{2}$ minutes, down to 1540 km h^{-1}; after $8\frac{1}{2}$ minutes, the speed was down to 500

The Lunar Landing

km h^{-1}; the height above the surface was 2.3 km, and the distance from the point of leaving lunar orbit was 472 km. Attached to the legs of the landing vehicle were antenna-like feelers, which, when they touched the surface, would automatically shut off the retro-motor.

102 hours 45 minutes 40 seconds into the flight, a soft landing, at 1 m s^{-1}, was made in Mare Tranquillitatis. At that time the retro-motor had fuel left for another 10 seconds of firing. At Cape Kennedy the time was 22 hours 56 minutes on 20 July 1969.

The two astronauts on board then prepared themselves to leave the landing vehicle, dressed in their space suits with the life support systems on their backs. Neil Armstrong was the first to set foot on the lunar surface, saying: "This is one small step for a man, but a giant leap for mankind". Then Aldrin followed.

They collected 21.75 kg of rock samples to bring to Earth and set out a few scientific experiments: a laser reflector to reflect laser rays from the Earth; a seismometer to monitor moonquakes and the collisions of meteors; aluminium foil to collect solar wind particles; and a radio transmitter to signal readings to Earth.

Ascent from the Moon

Armstrong and Aldrin spent 2 hours 40 minutes on the lunar surface and then re-entered the landing vehicle. The lunar module was so designed that the lower half could serve as a launching platform, from which the upper half could take off as is shown in Figure 8.7. The upper half had a thrust of 1588 kg and could lift the two astronauts back into lunar orbit, where Michael Collins had been orbiting the Moon in the command module, attached to the service module.

The time of departure from the Moon was calculated so that the meeting of the two craft could take place with facility. Docking of the two craft took place after 3 hours 41 minutes. Then Armstrong and Aldrin crawled back into the command module. The top half of the lunar landing vehicle was then uncoupled, its motor was turned on and it was sent off to crash on to the Moon's surface. The first task that the seismometer executed was to monitor the moonquake and transmit the seismogram to Earth. A remarkable finding was that the registering needle quivered for four hours. This showed that the upper layers of the Moon are

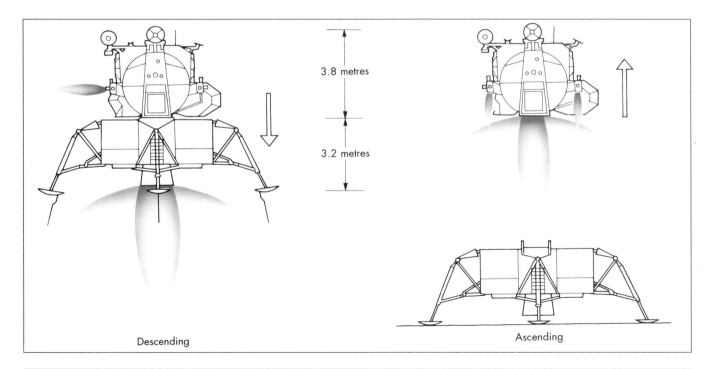

Figure 8.7 The Lunar Landing Vehicle

not very firmly compacted, which is why echoing of the moonquake took place.

When Apollo, now consisting only of the command module and the service module, reached the far side of the Moon, 135 hours into the flight, its engine was switched on for $2\frac{1}{2}$ minutes. This extra thrust enabled the vehicle to gain sufficient velocity to leave Moon orbit and go on its course back to Earth.

After 150 hours 29 minutes 54 seconds, a course adjustment was made by firing the motor of the service module for 10.8 seconds.

Now the space vehicle was speeding on its way back to Earth. After 194 hours 49 minutes 19 seconds, the service module was disconnected from the command module, its motor was switched on automatically and it was sent to a fiery death in the Earth's atmosphere.

Landing on the Earth

The command module, containing the three astronauts, was now alone. Its mass was slightly less than 0.2% of the mass that was launched from the Earth! Its speed had increased to 40 000 $km\,h^{-1}$ and it now had to be aimed to enter the Earth's atmosphere at the correct angle. If the angle was too small, it would bounce off into space, just like a small flat stone bounces off a body of still water, if it is thrown fast enough. The astronauts and the Command Module would then be lost forever! If the angle was too steep, the friction with the air could make the vehicle evaporate! As it was, friction ionised the air rushing past the vehicle, causing a break in radio communication.

When radio contact was re-established, Columbia had already lost so much speed that it could deploy its parachutes and make a comparatively soft "landing" in the Pacific Ocean. The total duration of the trip was 195 hours 18 minutes 35 seconds – 30 seconds late by schedule!

Man's first visit to an extra-terrestrial body had been a resounding success! It also proved that the astrometric measurements of the past had been correct.

Figure 8.8 gives an overview of the whole trip.

Later Apollo missions (Table 8.1) were more extensive and longer times were spent on the surface of the Moon. Apollos 15, 16 and 17 each had a battery-driven car, the Rover, which enabled them to cover longer distances (Figure 8.9).

While Apollo-11 was orbiting the Moon, it took many photographs of the lunar surface. Figure 8.10 shows part of the far side (left) and part of the near side (right).

On the way back to Earth, Apollo-17 took the photograph of the Earth shown in Figure 8.11. It is a magnificent sight, showing most of the continent of Africa and Arabia. Between the snow-white clouds, the blue of the atmosphere is most striking. North of the green belt astride the equator, the brown of the sands of the Sahara desert can be seen. The South Pole is completely covered in clouds and ice. Several cyclones, each with its tell-tale cold front, are sweeping past South Africa, where the south coast is receiving a good fall of rain. The Earth is unique among the planets in having temperature conditions such that water can exist in its three states, solid, liquid and gas, or vapour, thus producing weather and climate, which are essential for life to exist.

Research on the Moon

Each of the Apollo missions set up an array of scientific experiments on the Moon. Besides the laser reflectors and seismometers, the following sets of experiments were also set up:

1. measuring the composition, energy and speed of positive ions, and detecting any traces of atmosphere that may be ejected from volcanoes;
2. measuring the energies of protons and electrons of the solar wind;
3. measuring the temperature gradient of the soil below the surface.

Each group of experiments had a power supply, consisting of a radio-isotope, thermo-electric generator, which also provided the energy for the radio receivers and transmitters. The generators consisted of rods of beryllium which contain 3.8 g plutonium, enclosed in graphite wrappings. The plutonium generated 1480 watts of heat, which was converted into 63.5 watts of electricity. Part of the heat generated was used to keep the apparatus warm during the 14-day-long lunar night.

One of the most important instruments used by the astronauts on the Moon was the ultra-violet camera and spectrograph. The camera was sensitive to ultra-violet light of wavelengths shorter than 1600 Å; these wavelengths did not penetrate the atmosphere but were received on the Moon's surface. The camera was an f/1.0 Schmidt telescope with an aperture of 75 mm. The incident rays went through the correction plate and were reflected by a spherical mirror on to the image receptor, which was covered with potassium bromide. The ultra-violet light set electrons free from the receptor and they were focused by a cylindrical magnetic field to pass through a hole in the middle of the primary mirror, where they impinged on a sensitive film. This electronographic camera was 10 to 20 times faster than an optical camera. Visible light had no effect on the photo-

To the Moon and Back

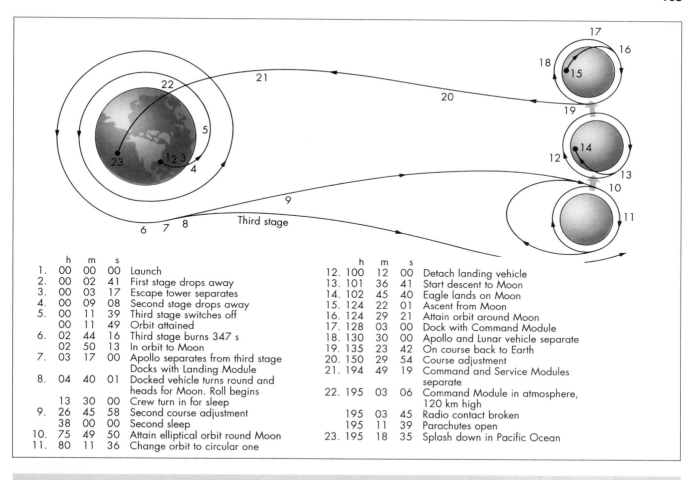

	h	m	s			h	m	s	
1.	00	00	00	Launch	12.	100	12	00	Detach landing vehicle
2.	00	02	41	First stage drops away	13.	101	36	41	Start descent to Moon
3.	00	03	17	Escape tower separates	14.	102	45	40	Eagle lands on Moon
4.	00	09	08	Second stage drops away	15.	124	22	01	Ascent from Moon
5.	00	11	39	Third stage switches off	16.	124	29	21	Attain orbit around Moon
	00	11	49	Orbit attained	17.	128	03	00	Dock with Command Module
6.	02	44	16	Third stage burns 347 s	18.	130	30	00	Apollo and Lunar vehicle separate
	02	50	13	In orbit to Moon	19.	135	23	42	On course back to Earth
7.	03	17	00	Apollo separates from third stage	20.	150	29	54	Course adjustment
				Docks with Landing Module	21.	194	49	19	Command and Service Modules separate
8.	04	40	01	Docked vehicle turns round and	22.	195	03	06	Command Module in atmosphere,
				heads for Moon. Roll begins					120 km high
	13	30	00	Crew turn in for sleep		195	03	45	Radio contact broken
9.	26	45	58	Second course adjustment		195	11	39	Parachutes open
	38	00	00	Second sleep	23.	195	18	35	Splash down in Pacific Ocean
10.	75	49	50	Attain elliptical orbit round Moon					
11.	80	11	36	Change orbit to circular one					

Figure 8.8 The course of Apollo-11 to the Moon and back

Table 8.1 Apollo missions

No.	Time on Moon h	min	Exploration time h	min.	Mass of samples (kg)
11	21	36	2	40	21.75
12	31	31	7	45	34
14	32	30	9	17	43
15	66	54	19	08	77
16	71	14	20	15	94
17	74	59	22	04	110

Figure 8.9 Geologist H. Schmitt at large boulder with battery-driven car at left

Figure 8.10 The Moon's far side (left) and near side (right)

Figure 8.11 The Earth as seen from space

cathode. When the camera was turned towards the Earth, the sensitive film registered only the ultra-violet glow which enveloped the Earth's atmosphere. This glow stretched to a distance of 25 000 km beyond the atmosphere. The glow was generated by hydrogen ions in the solar wind, of Lyman alpha and beta wavelengths, namely 1215 and 1026 Å.

When the telescope was turned through 90 degrees, it registered the spectrum formed by a reflecting grating. This spectrograph can measure the radial velocity of a distant object, as high as 3700 km s^{-1}. It makes use of the fact that the redshifts can be detected when the spectrum lines move from the short ultra-violet, right into the red end of the spectrum.

Both Apollo 15 and 16 launched a satellite from the service module into an orbit around the Moon, 63 to 86 km above the surface. The satellites were used to monitor the Moon's reaction to the solar wind and to measure the Moon's gravity, so that more could be learned about the mascons in the maria.

Lunar Conferences

Each year, during January, commencing in 1970, the American National Aeronautics and Space Administration (NASA) have held a conference to discuss findings regarding the Moon. In 1972, for example, 200 papers were read and discussed.

The rock samples of Apollo 11 and 12 show great chemical differences. These rocks contain no water and are deficient in volatile elements, such as sodium, potassium and rubidium. There is also a paucity of lead. However, the rocks are richer in uranium and thorium. These rocks have radioactive ages between 3.1 and 3.9 thousand million years (10^9), while the soil of the maria is 4.5 thousand million years old. This value agrees with the age of the Solar System. All the samples show multiple origins, that is, they underwent several meltings: solidifying, fracturing and remelting and resolidifying. This was caused by the continued bombardment by meteors and other material, plunging to the surface. Small globules, or spherules, are to be found all over. They formed from tiny particles, melting and solidifying again. Calculations show that a pressure of 500 000 earth-atmospheres would have been great enough to cause vaporisation of the rocks at the points of collision.

The lunar material is of two main kinds: maria type and highland type. In some places the highland type underlies the maria type. This came about when the molten maria type flowed over the highland type and submerged it. The maria type became liquid when large meteors or asteroids crashed on to the surface. There are also places where dykes (material pushed up from below) were formed.

Because the maria have comparatively few craters, it means that they were formed after the greatest concentration of plummeting material had occured, during the accretion process whereby the Moon and the other planets, were formed. Where an asteroid struck, great heat was developed, so much so, that the whole area was flooded with lava from the point of collision. Mare Orientale (Figure 8.12), just beyond the Moon's eastern limb, 15 degrees south of the equator, is a beautiful example of a collision between the Moon's surface and a large meteor, asteroid or other clump of matter. In the centre is the plain, consisting of solidified lava, 430 km in diameter. Around it are three circular ranges of material that was thrown out by the explosion of the collision. The widest range has a diameter of 1200 km. Beyond the widest range there are loose clumps of material that were thrown out.

There is a similar formation on Mercury, the Caloris Basin, and two on Jupiter's moon, Callisto, namely Valhalla and Asgard. Mare Imbrium, on the Moon, is much larger than Mare Orientale. Peaks in the ranges ringing Mare Imbrium rise to a height of several thousand metres.

The rock samples of the several Apollo missions have differences. Many samples are rich in potassium and trace elements. The ratios of the amounts of calcium, aluminium, rare elements, uranium and thorium are five times those in meteorites and the Sun.

By means of the laser reflectors, differences as small as 15 cm in the positions of the stations on the Moon were detected. The fine librations of the Moon, that is, the oscillations in longitude and latitude, indicate that the nucleus of the Moon is not liquid. If there is a liquid core, it must be very small. This explains why the Moon's magnetic field is so very weak – 1/10 000th that of the Earth – because the lack of a liquid core makes it impossible for the dynamo effect to develop, which is necessary for a magnetic field to be established. Individual rocks, however, do show stronger magnetism. This must have been induced during the collisions.

The long durations of the oscillation times of moonquakes indicate that the upper layers of the Moon must be heterogeneous. At a depth of 20 km, the speed of seismic waves increases by 7 km s^{-1}. The upper layers are definitely basaltic, but lower down there is more eclogite and anorthosite. At 60 km depth, there is a further increase in the speed of seismic waves, of 9 km s^{-1}. Below this level, the rocks are most probably olivine, of density 3.2 g cm^{-3}.

The measurements of temperature, made by the temperature gradient meters, indicate that the inner material of the Moon is in continuous motion and that heat is transferred by convection, that is, the heat is transferred by the motion of the particles themselves as they move from place to place. On the surface the day temperature reaches 93°C, and at night the temperature drops to −155°C.

The surface of the Moon is covered by a layer of dust, called the regolith. In the maria it is 4 to 5 metres deep, but on the highlands it is 8 to 10 metres thick. Apollo-17 found places in the Taurus–Littrow area where the regolith is 30 metres thick. The regolith was formed by the bombardment over the millennia by tiny meteoritic particles. The particles themselves also form part of the regolith. This bombardment has had a churning action on the Moon's surface.

Cross-section of the Moon

The seismometer readings indicate that the Moon's crust is thicker than that of the Earth. The thickness varies between 32 and 68 km. The thicker crust means that the temperature of the nucleus must be lower than that of the Earth. Below the crust, the upper mantle has a thickness of 230 km and the lower mantle 670 km. The layer below the lower mantle consists of weak material; it is called the asthenosphere and is 480 km thick. The radius of the nucleus is 290 km and its temperature is 1500°C, which is much lower than that of the Earth (Figure 8.13).

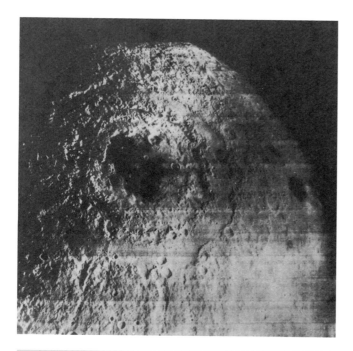

Figure 8.12 Lunar Orbiter photograph of Mare Orientale

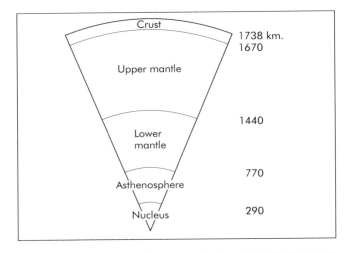

Figure 8.13 Sector of the Moon

It can be expected that the smaller body will be colder than the larger Earth. The crust on the far side of the Moon is thicker than on the near side. There are no large maria on the far side.

In the asthenosphere, the material which is a shell around the nucleus is not able to withstand deformations by outside forces, whether pressure from the overlying layers, or gravitational forces.

The crash of the third stage of Apollo-11 on the Moon's surface caused the needle of the seismograph to vibrate for four hours. The material in the lunar crust echoed and re-echoed.

Radioactive dating of the rock and soil samples gave ages between 3.1 and 3.9 thousand million years for the fine soil of the maria, but 4.5 to 4.7 milliard years for the highland material. The collisions by the large lumps of matter occurred towards the end of the accretion process, whereby the Sun and the planets were formed. At that stage the material of the primeval nebula had collected into planetesimals and it must have been some of these that formed the maria. Mare Orientale is the youngest of the maria.

The Lunar Craters

Meteoritic material is comparatively scarce on the Moon, so that the craters could not have been formed by ordinary meteors. Most of the craters vary in size between 10 and 100 km. On the near side of the Moon, there are 61 craters more than 100 km in diameter; 163 between 51 and 100 km; and 306 between 11 and 50 km in diameter. The size which occurs most frequently is 48 km.

The floors of most of the craters are lower than the levels outside the crater walls. The surface material must therefore have been dislodged and thrown out, together with the material of the falling lumps. The outward thrust of the surface material formed the inner crater walls. Figure 8.14 is a good example of this: many craters have peaks in the centre. These were formed by counter pressure on to the centre. The floor of crater Walter, which is the lowest of the three large craters in Figure 8.14 is one km lower than the surrounding surface.

Craters such as Tycho and Copernicus have rays radiating outwards. These rays consist of lighter-coloured material which was thrown out by the force of the collision. The material may contain much anorthosite. The rays stretch over hundreds of kilometres. Craters such as these

Figure 8.14 From top to bottom: Craters Purbach, Regiomontanus and Walter. Large crater at left is Deslandres

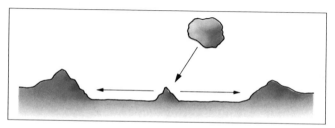

Figure 8.15 Crater forming

were most likely formed by very dense clumps, probably iron–nickel, or by asteroids that had compacted into dense masses. Smaller meteorites have formed smaller craters on top of the older craters.

The walls of these are usually conical, meeting at a point at the bottom. They do not have flat floors. Many of these can be seen in the figure.

Most of the craters do not have rays such as Tycho, showing that most of the craters must have been formed by loosely compacted clumps of matter, which was the prevailing condition at the height of the accretion process.

Even if a clump of material fell at a slant, it would still form a round crater, because the explosive forces act radially outwards from the point of impact (Figure 8.15). There are some craters which have wrinkles outside the crater walls. These form across the line of descent and are on opposite sides of the crater. The forward rim of the crater is then higher than the backward rim, showing that the material was pushed forward.

Many craters have floors of solidified lava, showing that melting took place at the time of impact. It is also possible that reaction pressure forced dikes up from below the surface, thus forming the central peaks. It will require inspection by man to obtain certainty about this.

The sizes of the lumps of material that formed craters of various sizes can be calculated (Figure 8.16).

In the figure, the floor of the crater is taken as being $\frac{1}{2}$ km lower than outside. The cross-sectional area of the rims of a crater 48 km in diameter is approximately 8 km². The amount of material in the whole rim is then given by:

$$2\pi r(8) = 2(3.14159)(24)(8) = 1206 \text{ km}^3$$

where the radius of the crater $(r) = 24$.

The volume of lunar material gouged out from the floor is

$$\pi r^2 h = (3.14159)(24)^2(\tfrac{1}{2}) = 904 \text{ km}^3$$

The volume of matter that was contributed by the in-falling lump is then equal to

$$1206 - 904 = 302 \text{ km}^3$$

If such a lump were more or less spherical in shape, its volume can be calculated as follows, where r is the radius of the lump:

$$\text{Volume} = \frac{4}{3}\pi r^3 = 302$$

$$\therefore r = \sqrt[3]{\frac{302 \times 3}{4\pi}} = \sqrt[3]{72.9} = 4.16 \text{ km}$$

Thus, the diameter of a lump of material that formed a crater of diameter 48 km is approximately 2(4.16) = 8 km.

Crater-forming went on apace for the first 1500 million years of the life of the Solar System. At the end of that period, the Moon was more or less evenly covered with craters, as the far side shows. After that, collisions with asteroids occurred, thus forming the maria where the lava, formed by the heat of the collisions, flooded the surrounding areas. It is therefore not strange that there are concentrations of mass in the maria, because that is where the heavy material of the asteroids is lodged.

The median of the formation time of the highlands was about 4.6 milliard years ago; and that of the maria, 3.5 thousand million years ago. Because there are comparatively few craters in the maria, it shows that the accretion process was just about completed 3.5 thousand million years ago.

IN section 2 of this chapter, we shall see that the accretion process took place throughout the Solar System. There can therefore be no doubt about the veracity of the theory that the bodies of the Solar System compacted together by means of the process of accretion.

The rocket thus made it possible for space probes to examine the planets at close range, and became the instrument whereby scouting of the planets could take place.

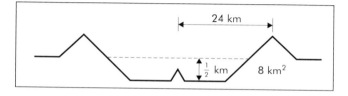

Figure 8.16 Amount of material in the rim of a crater, 48 km in diameter

2 Scouting the Planets

Mercury

It was always difficult to observe Mercury, because of its proximity in direction to the Sun. The space probe Mariner-10, however, lifted the veil and transmitted thousands of photographs to Earth.

Mariner-10 was equipped with two TV-cameras, each having a field of $\frac{1}{2}$ degree (as wide as the width of the full Moon). Each image was divided into 700 scanning lines of 832 pixels (photo-elements) each. Mariner-10 also had a sensor for electrically charged particles; two ultra-violet spectrometers; magnetometers; an infra-red radiometer; antennae for high and low gain capacities; as well as the solar panels for the generation of electricity to activate the instruments and the radio reception and transmission.

The probe was launched on 3 November 1973. The first thing it did was to focus on the Moon from a distance of 112 000 km in order to calibrate its instruments.

On 5 February 1974, Mariner passed by Venus (Figure 8.17), and took 3000 photographs, commencing when it was 5760 km from Venus – one photo every 42 seconds. Venus then gave Mariner a slingshot acceleration towards Mercury, where it arrived on 29 March 1974 (Figure 8.17).

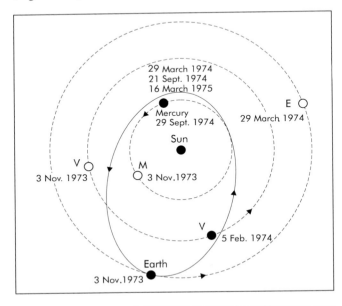

Figure 8.17 The orbit of Mariner-10 past Venus, on its way to Mercury

Six days earlier, it had started photographing Mercury, when it was still 5.4 million km away. When Mariner was $3\frac{1}{2}$ hours from Mercury, the resolution of the photographs was 3 km. At its closest to Mercury, the resolution was down to 500 to 100 metres.

The excellent TV-photos revealed a landscape richly covered in craters. Mercury's surface appearance is closest to that of the Moon of all bodies in the Solar System. Many of the craters overlap. The accretion process therefore had proceeded in the same way as in the case of the Moon. All of the younger craters are smaller, showing that the larger clumps of material had previously plummeted down to form the planet.

Besides the craters, there are escarpments as high as $1\frac{1}{2}$ km. There are many ridges and ranges, circular basins and plains. Mercury is very similar to the Moon's highlands and the far side of the Moon. The density of the craters is about the same as that on the Moon. So too is the ratio of depth-to-diameter of the craters. Mercury has many folded ranges, where lava solidified after being formed during the collisions.

The colour of the surface is generally dark. Here and there, lighter coloured craters are found, where the material which has been cast out is younger than the rest of the surface. The brightest crater has been named after G.P. Kuiper, the famous planetologist, who was a member of the Mariner-10 management team. Kuiper passed away 47 days after the launch of Mariner-10.

There is a very large circular basin, named the Caloris Basin, which has concentric ranges stretching as far as 1350 km. Figure 8.18 shows a portion of the basin, on the left-hand side of the photograph. Caloris basin is reminiscent of Mare Orientale on the Moon, and Valhalla and Asgard on Jupiter's moon, Callisto. Figure 8.18 is representative of most of the surface of Mercury.

Mercury also has rift valleys, where the lower layers have subsided. One of these valleys is 10 km wide and 100 km long. There are very few straight cracks, but there are escarpments, 100 km long.

No sign of an atmosphere could be found. Recently, small deposits of ice have been found at one of the poles.

Extensive flooding by silicaceous lava indicates that the bombardment must have generated great heat. There are no traces of water and therefore the craters have undergone no weathering, except for that due to micrometeorites, as on the Moon. The solar wind has caused some erosion.

Mariner-10 found Mercury's density to be 5.45 g cm^{-3}. This agrees very well with the previously determined value of 5.5 g cm^{-3}. Mercury must have a nucleus of dense materials, like the Earth.

After Mariner-10 passed by Mercury, its orbit carried it back to the Earth's orbit, and then, on 29 September 1974, it

Mercury

Figure 8.18 The Caloris Basin on Mercury

flew past Mercury again, almost at exactly the same spot in its orbit. In the 176 days which had elapsed, Mercury had revolved twice around the Sun.

The magnetometers registered the magnetic field of Mercury as being between 0.011 to 0.023 that of the Earth.

After another two of Mercury's years, Mariner-10 passed it again on 16 March 1975, thus making its third visit.

The Rotation Period of Mercury

Before Mariner-10, it was never possible to make out any markings on the surface of Mercury. Astronomers had assumed that its rotation period around its axis and its period of revolution around the Sun, 88 days, must be equal, because Mercury is so near to the Sun and it would therefore always keep the same face turned to the Sun, just as the Moon always keeps the same face turned to the Earth.

R.B. Dyce, G.H. Pettingill and I.I. Shapiro published an article in the *Astronomical Journal*, Vol. 72 (1967) p. 351, in which they set out their findings with regard to Mercury's period of rotation, as revealed by their radar investigation in 1965, made with the radio telescope at Arecibo, Puerto Rico.

They had sent short pulses, of 1 to 5 ten-thousandths of a second, at a frequency of 430 megahertz (4.3×10^8 Hz) to Mercury and monitored the reflected waves. Their premise was that the radar waves reflected from points near to Mercury's equator, and thus on the vertical joining the Earth to Mercury, would have a shorter distance to travel than radar waves reflected from points near to Mercury's limb. The idea is illustrated in Figure 8.19.

Radar waves transmitted from the Earth and reflected from points such as A and B, near to the limb of the planet, have to travel a distance of 2d further than waves reflected from the point O, near to the equator.

If the spin of the planet is direct, the point B will be approaching the Earth and the frequency of the reflected wave will be increased from f to $(f + \Delta f)$.

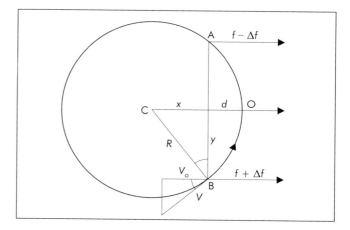

Figure 8.19 Effect of planet's rotation on the frequency of reflected radar waves

The point A will be receding from the Earth and the frequency of the reflected wave will be decreased from f to $(f - \Delta f)$. These differences in frequency can be measured very easily. The time delays were measured in microseconds (10^{-6} seconds), and caused a broadening of the spectrum lines (Figure 8.20).

The reflected radar waves were received from points A and B, 210 microseconds ($=\Delta t$) after the reception from O. The calculation proceeds as follows:

1. During the time delay Δt (= 210 microseconds), the radar waves travel a distance of $2d$, at the speed of light (= c). Therefore, d is given by:
$$d = \tfrac{1}{2} \times c \times \Delta t = \tfrac{1}{2}(3 \times 10^8) \times 210 \times 10^{-6}$$
$$\therefore d = 0.0315 \times 10^6 \text{ metres}$$

2. If R is the radius of Mercury (= 2.42×10^6 metres), the values of x and y in Figure 8.19 can be calculated as follows:
$$x = R - d = 2.42 \times 10^6 - 0.0315 \times 10^6$$
$$\therefore x = 2.3885 \times 10^6 \text{ metres}$$

According to Pythagoras' Theorem (Appendix 1, section 1.5):
$$y^2 = R^2 - x^2$$
$$\therefore y^2 = (2.42 \times 10^6)^2 - (2.3885 \times 10^6)^2$$
$$= 5.8564 \times 10^{12} - 5.7049 \times 10^{12}$$
$$= 0.1515 \times 10^{12}$$
$$\therefore y = \sqrt{0.1515 \times 10^{12}} = 0.3892 \times 10^6 \text{ metres.}$$

3. The speed of rotation, V, of the planet, can now be calculated. At the point B, the speed of rotation is represented by the length V. This speed has a component V_o along the line of sight to Earth.
The two triangles in the diagram are similar (Appendix 1, section 1.4), therefore:
$$V \div V_o = R \div y$$

To find the value of V_o we must make use of the Doppler effect, in terms of the small change in frequency Δf, that is
$$V_o \div c = \Delta f \div f$$

where the measured frequency $f = 4.3 \times 10^8$ Hz. The small change in frequency, Δf, is read off from the spectrum (Figure 8.20). The change in frequency is read off at each maximum of the signal strength, namely 1.28

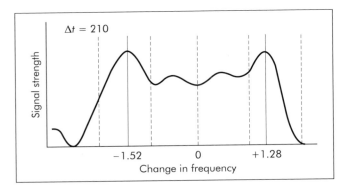

Figure 8.20 Part of radar spectrum

and −1.52. The average of these two deviations is 1.4. (The minus sign is ignored, because we need only the magnitude of the deviation, not its direction.) Δf is half of 1.4 = 0.7.

Therefore $\dfrac{V_o}{c} = \dfrac{\Delta f}{f}$

$$\therefore V_o = \dfrac{c \times \Delta f}{f} = \dfrac{3 \times 10^8 \times 0.7}{4.3 \times 10^8} = 0.488 \text{ m s}^{-1}$$

Now, $\dfrac{V}{V_o} = \dfrac{R}{y}, \therefore V = \dfrac{V_o R}{y}$

$$\therefore V = \dfrac{0.488 (2.42 \times 10^6)}{0.3892 \times 10^6} = 3.03 \text{ m s}^{-1}$$

that is, the angular velocity $V = 3.03$ m s^{-1}
The circumference of Mercury is:
$$2\pi R = 2(3.141\,59)2.42 \times 10^6$$
$$= 1.52 \times 10^7 \text{ metres}$$

To go once around the planet, B takes
$$1.52 \times 10^7 \div 3.03 = 0.5017 \times 10^7 \text{ seconds}$$

There are 86 400 seconds (= 8.64×10^4) in one day of 24 hours. Therefore Mercury's period of rotation is $0.5017 \times 10^7 \div 8.64 \times 10^4 = 58.07$ days.

THIS remarkable achievement by R.B. Dyce, G.H. Pettingill and I.I. Shapiro was the first indication that Mercury is not locked in synchronous rotation with the Sun – that its period of rotation is not equal to its period of revolution; but rather is in resonance, its day being exactly two-thirds of the length of Mercury's year.

What did Mariner-10 find from its vantage point close in to Mercury? By applying the Doppler effect on the frequency of reflected waves, it found that the length of

Venus

In October 1975 two Venera space probes made soft landings on Venus. They transmitted findings for about an hour each, before they were overcome by the very high ambient temperature, which they found to be 485°C! The pressure of the atmosphere of Venus is 90 times that of the Earth's atmosphere, and consists of 96.6% of carbon dioxide.

Pioneer Venus-1 and Venus-2 followed in 1978 and also liberated probes that made soft landings, having made use of the very dense atmosphere to slow their descent. They also functioned for about an hour each.

Venus is completely shrouded in clouds (Figure 6.2, page 59). The cloud layer is between 63 and 67 km above the surface at the top of the clouds; the densest part is at a height of 49 to 52 km, and the lowest layer 33 km above the surface. The fine drops constituting the clouds consist of sulphuric acid! Below the clouds, there is a layer of fine sulphur crystals. The water vapour content is less than 1/6000th that of the Earth. The atmosphere disperses the sunlight, so that visibility is poor.

The atmospheres of Venus, Earth and Mars are given in Table 8.2.

The cloud decks of Venus rotate around the planet in four days, while the planet has a retrograde rotation of 243 days. The equator is inclined at 178 degrees to the Ecliptic and the orbit is inclined at $3°23'39.8''$.

The sidereal day of Venus is 243 earth days, while the sidereal period, the time taken to orbit the Sun, is 225 days. How long is the day on Venus?

In 243 days, Venus moves through $\frac{243}{225} \times 360 = 388.8$ degrees (Figure 8.21).

When Venus is at position A, the arrow points to a star in the zenith. The direction of the setting Sun is at right angles to the star. After Venus has moved through 30 degrees in its orbit, from A to B, a point on the equator would have moved through

$$\frac{30}{388.8} \times 360 = 27.7° \ (= \text{angle GBF})$$

While the arrow has moved westwards from G to F, through 27.7 degrees, the terminator (the dividing line between dark and light) has moved eastwards from G to E, through 30 degrees. The direction in which the arrow points has therefore moved through 30 + 27.7 = 57.7 degrees away from the direction of the Sun (= angle EBF).

At the point D, where the planet has moved through 93.46 degrees in its orbit from A, the arrow has moved through an angle of

$$\frac{93.46}{388.8} \times 360 = 86.54°$$

Relative to the Sun, the arrow has turned through 93.46 + 86.54 = 180 degrees, so that the whole night has passed and the Sun is just rising in the west. The length of the night was thus (93.46 ÷ 360) 225 = 58.4 earth days. Day and night together last for 2(58.4) = 116.8 earth days.

The Venusian year thus consists of 225 ÷ 116.8 = 1.926 36 Venus days.

The rotation period of Venus and the fact that it has retrograde spin were proved in 1965 by means of radar. The Pioneer Venus probes corroborated these findings.

The Pioneers also made a relief map of Venus, using the fact that radar waves reflected from high areas return more quickly than from lower lying areas. About 60% of the surface lies within 500 metres of the modal radius (the most general radius); 5% is lower than 2 km; 20% consists of low-lying plains; 70% consists of undulating plains; and 10% consists of highlands.

The largest highland is Ishtar Terra, between 50 and 75 degrees north latitude. It stretches 120 degrees east–west: 2700 km south to north and 5000 km east to west. The folded mountains of Ishtar Terra are very steep. The highest mountain, Maxwell Montes, rises 12 km above the modal plane. (Mount Everest is 8.8 km above sea level.) The photograph shown in Figure 8.22, made by Venera 15 and 16, shows the plain Lakshmi Planum and the Akna Mountains. Also visible is the caldera of the volcano Colette, $1\frac{1}{2}$ km high and 180 km in length. Its lava flows reach 300 km in length. The highest range rises to 4 km above the ambient plains.

Another highland, Aphrodite Terra, astride the equator, is 10 000 km long and 3000 km north to south.

Venus has several funnel-shaped features and rounded structures, with humps or domes. It would appear that Venus undergoes tectonic activity and shows signs of vulcanism. Craters are also to be found.

Table 8.2 Atmospheres of planets

Gas	Venus	Earth	Mars
carbon dioxide	96.6%	0.03%	95.0%
nitrogen	3.2	78.1	2.7
oxygen	minimal	20.9	0.2
argon	minimal	0.9	1.6
water vapour	minimal	1.0	minimal

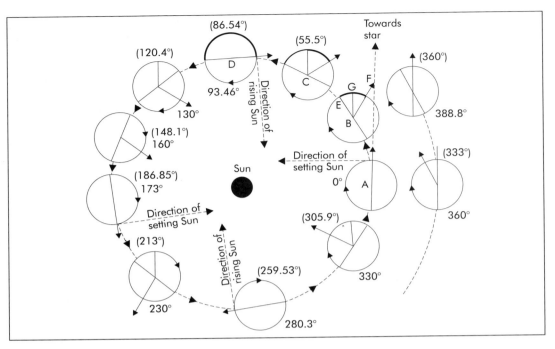

Figure 8.21 Movement of a point on Venus' equator, during one rotation of the planet on its axis

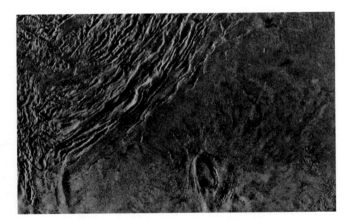

Figure 8.22 Akna Mountains and Lakshmi Planum on Venus

The magnetometers found no magnetic field. The solar wind therefore penetrates to the upper layers of the atmosphere, where there is an ionised layer.

The space probe Magellan reached Venus in 1990 and completed an extensive survey of the planet.

The Earth

Many volumes have been written on the surveys of the Earth made by satellites during the past 35 years. Figure 8.23 shows the South Western Cape Province, South Africa. It was made by the Landsat satellite, in blue, green and infrared light. Vegetation appears red and brown, and built-up areas blue and green. From such photographs, geologists can determine the mineralogy of the area.

The greatest difference between the Earth and the other terrestrial planets, Mercury, Venus, Mars and the Moon, is the water and ice which the Earth possesses. The Earth's craters have long since been erased by water erosion. Only recently formed craters remain on Earth.

The bottoms of the oceans consist of basaltic rocks, the so-called SIMA, or silicon–magnesium rocks; while the continents contain lighter rocks, the SIAL, or silicon–aluminium rocks. The continental rocks are older than the rocks of the floors of the oceans. This is because molten rock continually wells up between the tectonic plates under the oceans. This forces the plates apart and mountain ranges congeal on the floors of the oceans. However, many parts of the continents are overlain with solidified lava, which is younger than the continental rocks.

The Earth

Figure 8.23 Landsat photograph of the Cape of Good Hope and the South Western Cape Province, South Africa

A very interesting discovery regarding the Earth made by satellites was that of the Van Allen radiation belts, which envelop the Earth. One belt is situated three-quarters of an earth radius, and the other $3\frac{1}{2}$ radii above the surface (Figure 8.24). The inner radiation belt is about half an earth radius thick, and the outer belt one radius thick. The belts have trapped electrified particles (ions) between the magnetic lines of force. The particles spiral to and fro between the magnetic poles. Incoming radiations, which may be inimical to life, are trapped in the Van Allen belts, named after J.A. Van Allen who designed the magnetometers of the first Explorer satellites, which discovered the radiation belts.

When the solar wind meets up with the Earth's magnetic field, the wind is deflected and it wraps itself around the bow shock, where the solar wind and the magnetic field meet. On the side of the Earth, away from the Sun, the bow shock trails off into space (see Fig 9.22, page 150).

Above the Earth's magnetic poles, the electrified particles have free access to the Earth's surface, because the lines of the magnetic field are concentrated downwards towards the surface. The electrified particles interact with the molecules of the upper air, causing a glow from time to time. These glows are the Aurora Borealis or northern lights, and Aurora Australis, or southern lights. Figure 8.25 of the

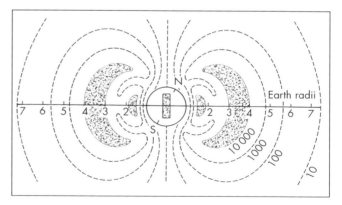

Figure 8.24 The Earth's magnetic field and the Van Allen radiation belts (speckled)

Aurora Borealis was taken from Fairbanks, Alaska, by Akira Fujii. The aurorae occur just after outbursts of solar flares.

The nature of the Earth's magnetic field leads to the concept that the Earth can be considered to have a bar magnet near its centre, in the liquid core, and that it is inclined to the axis of spin.

Figure 8.25 Aurora seen from Fairbanks, Alaska. Photograph by Akira Fujii

Figure 8.26 Mariner-9's course correction to enter Mars' orbit

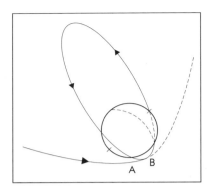

The northern pole is situated north of Canada, below the point 75 degrees north, 101 degrees west. The southern pole is below the point 66°30′ south, 139°30′ east, under the coast of Antarctica.

The strength of the magnetic field of a planet is expressed as the equivalent dipole magnetic moment. It takes into consideration the strength of the magnetic field, the angle that it makes with the axis of spin and the displacement away from the geometric centre of the planet. The dipole moment divided by the third power of the planet's radius gives the value of the magnetic field on the surface at the equator. The strength of the field varies with longitude. According to this system, the Earth's average magnetic field strength is 0.308 gauss. One gauss is subdivided into 100 000 gammas. 10^9 gamma equals 1 tesla. The field strengths in Figure 8.24 are in gammas.

Mars

The first space probe to visit Mars was Mariner-4, which was launched on 28 November 1964. It sailed past Mars on 14 July 1965 at a distance of 9800 km. Mariner-4 was the first probe to visit a planet. It took 21 photographs of Mars: the first close-up of the surface of another planet revealed an arid, desert-like expanse, peppered with craters. This was in direct contradiction of a world criss-crossed with canals, as P. Lowell and G.V. Schiaparelli had painted.

In 1969 Mariner-6 and -7 followed. They took thousands of photographs, and there was no sign of "canals". Some dark areas were found not to be dark at all. The ice caps at the poles were found to consist mainly of carbon dioxide and the density of the atmosphere was found to be only 1/100th that of the Earth, and 95% of what there is consists of carbon dioxide. This agreed with spectroscopic findings made during the years.

Mariner-9, launched on 30 May 1971 (Figure 8.26), turned on its retro-rockets at the right moment when it reached point A (Figure 8.26). The backward thrust of 136 kg acted for 15 minutes 15.6 seconds. Thereby it lost enough speed to be captured by the gravitational field of Mars, instead of sailing off into space along the broken line, past B. At that moment, the journey had taken 167 days and covered 459 million km from Earth.

The orbit around Mars of Mariner-9 had an apomartian (furthest point) of 17 300 km and a perimartian (nearest point) of 1385 km. Later, these distances were altered to 17 100 and 1650 km, and a period of 11.98 hours. (These data are by courtesy of Bruce C. Murray, *Scientific American*, January 1973.)

Mariner-9 was the first space probe to launch a satellite around another planet. Mariner-9 functioned well for a year and returned thousands of photographs.

When Mariner-9 first reached Mars, the planet was at perihelion (nearest point in its orbit to the Sun), and as had happened so often in the past, the planet was shrouded in a global dust storm, so that Mariner couldn't see a thing! The Jet Propulsion Laboratory, Pasadena, sent signals to Mariner, to direct its cameras on the two satellites of Mars, Phobos and Deimos (Figure 8.27).

Both moons are very small: Phobos is 25 km by 20 km, and Deimos 12 km by 10 km. They are both covered with craters and irregular in shape. Phobos orbits Mars in 0.31 marsdays. It therefore appears to rise in the west and set in the east, after being in the sky for $4\frac{1}{2}$ hours. In that time, it goes through half of its phases. Deimos orbits in 1.23 marsdays. It therefore stays above the horizon of any particular point for $2\frac{1}{2}$ days and goes through its phases twice.

Figure 8.27 Mars' moon Phobos

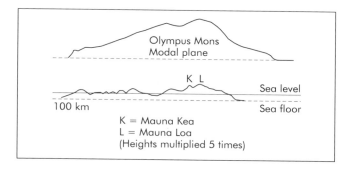

Figure 8.29 Comparison between Olympus Mons and Mauna Kea and Mauna Loa, Hawaii

At perihelion Mars receives 45% more heat from the Sun than at aphelion. This is the cause of the dust storms.

The first objects to become visible as the dust gradually settled were the great volcanoes: the three on Tharsis ridge, Arsia Mons, Pavonis Mons and Ascraeus Mons, and largest of all, Olympus Mons (Figure 8.28). In the figure, Olympus Mons is encircled by clouds of carbon dioxide crystals.

Formerly Olympus Mons was known as Nix Olympica (snows of Olympus) because it was visible through earthbound telescopes as a bright spot.

Figure 8.29 gives some idea of the size of Olympus Mons, when compared with the largest volcanoes on Earth, Mauna Loa and Mauna Kea in the Hawaiian Islands.

The caldera at the peak is 65 km in diameter. The mountain is 25 km high and it stands on a base 500 km in diameter. Mauna Loa, by contrast, rises 4.2 km above sea level, and 10 km above the floor of the ocean. Mons Olympus is the largest volcano known in the Solar System.

To the east of the three volcanoes of Tharsis ridge, the greatest rift valley in the Solar System, Valles Marineris, stretches over a distance of 4000 km between 5 degrees and 10 degrees south of Mars' equator. Its average width is 100 km and its depth is 6 km in places (Figure 8.30).

Figure 8.28 Olympus Mons towers above clouds of carbon dioxide crystals

Figure 8.30 The rift valley Valles Marineris shows subsidence of the opposite wall

Figure 8.31 Dune field in crater Proctor

Figure 8.32 Crater Yuty

In the figure one can see how the material on the opposite escarpment has subsided. It appears to consist of dry, sandy material.

The dry sand in the crater Proctor has been blown by the wind into a dune field, 150 km across (Figure 8.31).

Most of the craters on Mars are partially blown over by sand. An exception is crater Yuty, 20 km in diameter and having a massive central peak (Figure 8.32).

Although no trace of liquid water is to be found on Mars, there are indications that torrential floods swept over the surface in the past (Figure 8.33). The water flowed from left to right and seeped into the sandy soil where it froze.

The rift valleys were not eroded by water, but were caused by subsidences.

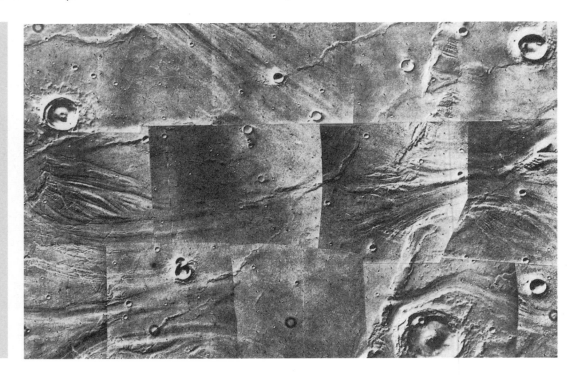

Figure 8.33 Tracks of torrential floods on Mars

Life on Mars?

The water which streamed over the surface of Mars must have come from the volcanoes, during their eruptions. The very size of Olympus Mons testifies that it must have been active over a great length of time. Ninety five per cent of the gases belched forth by volcanoes consist of steam. The steam would have rapidly condensed in Mars' rarefied and very cold atmosphere, causing repeated floods. The seething masses of water would have eroded wide channels and even broken through crater walls, before suddenly vanishing into the soil, where they froze. The volcanoes do not appear to be active at present.

VIKING 1 and 2, launched in 1975 and 1976, were destined to do research on the surface of Mars to find out if there is, or ever was, any life in the Martian soil.

Each of the Vikings launched a landing vehicle on to the surface of Mars. Retro-rockets and parachutes enabled them to make soft landings. Each of the craft had a scoop at the end of a long boom, which could stretch out in order to collect soil. This was then retracted into the craft and placed in a compartment where it could undergo tests (Figure 8.34).

Three tests were conducted:

1. One sample was heated for five days in a compartment containing an atmosphere similar to that of Mars, but having carbon dioxide tagged with radioactive carbon (carbon-14). The compartment was irradiated by artificial sunshine to promote photosynthesis. After an incubation period lasting 11 days, the compartment was heated to 626°C to decompose any organic material. The atmosphere was then flushed out by means of a stream of helium, to a detector which would detect any carbon-14. If any biological material had been present, it would have absorbed the carbon dioxide containing carbon-14 through the process of photosynthesis. The detector would then have been able to detect the carbon-14.

   The sample was then heated to 700°C in order to convert any organic matter into carbon dioxide. Then it was tested again for radioactive carbon-14. Two peaks in the amount of carbon-14 were detected, but they were not strong enough to say with certainty that they derived from organic or biological material.

2. In the second experiment, the sample of soil was put in a compartment containing "plant food". The atmosphere was monitored to determine whether gases containing carbon-14 were liberated. When the first few drops of plant food were added, the radioactivity underwent a sudden, large increase. But after one marsday (sol) the level of radioactivity subsided. After a further seven days, more plant food was added. Again, carbon-14 was detected, but the level fell off again. These drops in the level of radioactivity could not be reconciled with the existence of biological material.

3. In the third experiment, the sample was placed in a porous pot, above a container of plant food. After one week, the container of plant food was slowly raised so that the food could penetrate through the pores of the pot containing the sample. Suddenly there was a strong reaction. Oxygen was liberated but it was 15 times more than could be expected from any biological material. This could have been a chemical reaction. Carbon dioxide and nitrogen were also set free but after a time they fell off.

Figure 8.34 Viking scoops up soil from the Martian surface

All in all, it could not be said with certainty that the experiments had proved the existence of biological material, nor could the existence of biological material be written off as non-existent.

What can be said with certainty is that it is astonishing that such experiments could be carried out on the surface of Mars and be controlled from Earth, which at its closest is 56 million km away. Telemetry had become a wonder science.

The age-old dream that there may be life on Mars has to be forgotten. In the Solar System, we are alone!

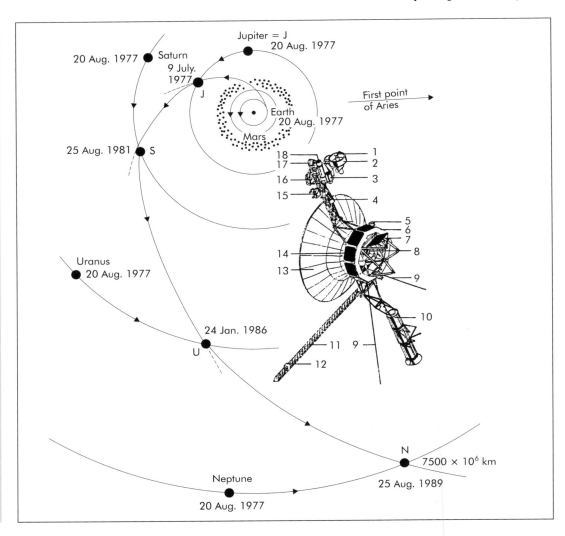

Figure 8.35
The Grand Tour of the Planets – Voyager-2's trajectory

Jupiter

Pioneer-10, launched in March 1972, reached Jupiter in December 1973 and sailed past at a distance of 131 000 km. This was the first probe to visit Jupiter.

Jupiter then gave Pioneer a slingshot acceleration and it sped off into space. In 1987 it crossed the orbit of Pluto and moved off in the direction of Leo.

Pioneer-11, launched in April 1973, came to within 46 000 km of Jupiter in December 1974. Jupiter accelerated it in such a manner that it headed towards Saturn. On 1 September 1979, it passed within 3500 km of Saturn's outer ring and then sped off in the direction of the constellation Aquarius. Both Pioneers are to be monitored until they leave the domain of the solar wind.

The Pioneers revealed that Jupiter has a very strong magnetic field which stretches for 108 Jupiter radii. The strength of the field is 12 times that of the Earth.

The American National Aeronautics and Space Administration (NASA) reached the peak of their space exploration with the launching of the two Voyager probes.

Voyager-1 was launched on 5 September 1977 and Voyager-2 on 26 August 1977. Voyager-1 was destined to visit Jupiter and Saturn, and Voyager-2 all four giant planets, Jupiter, Saturn, Uranus and Neptune. The alignment of the four planets in 1977 was such that they could be visited one after the other in a single trajectory (Figure 8.35). A similar alignment occurs only every 176 years. By making use of the accelerations which each planet could impart to the probes, the trip to Neptune could be accomplished in 12 years, whereas a direct trip to Neptune would require 30 years.

Voyager-2 would cover a total distance of 7500 million km by the time it reached Neptune.

Although Voyager-1 was launched 16 days after Voyager-2, it was aimed at reaching Jupiter on 5 March 1979, while Voyager-2 was to reach Jupiter on 9 July 1979.

The Voyager space probes were electronic wonders, as an examination of their abilities will show. They each contained:

(1) Ultra-violet spectrometer;
(2) Infra-red spectrometer and radiometer;
(3) Photopolarimeter, to measure the polarisation of light;
(4) Low-energy particle scanner;
(5) Attitude control and propulsion unit;
(6) Compartment with electronic systems;
(7) Instrument calibration panel and radiator;
(8) Fuel tank for propulsion;
(9) Long antennae;
(10) Thermo-electric radio-isotope generator, to supply 30-volt direct current;
(11) Extendable boom;
(12) Magnetometer to measure magnetic fields;
(13) High-gain parabolic antenna, 3.66 metres in diameter;
(14) Radiators;
(15) Cosmic ray monitor;
(16) Interplanetary plasma scanner;
(17) Wide-angle TV camera;
(18) TV camera with telephoto lens.

In order to keep the parabolic antenna continually directed towards the Earth, use was made of the Sun and the bright star Canopus as guides. The craft had twelve attitude control jets, to keep it stable. Thermo-electric energy had to be used, because solar panels, as in the Mariners, would not be able to produce enough electricity, the Sun being too weak at that distance, namely $1 \div (5.2)^2 = 1/27$ of the intensity of sunlight at the distance of the Earth. At the distance of Neptune, the sunlight is only $1 \div (30.06)^2 = 1/903$ of the intensity at the Earth.

The two 10-metre extendable booms kept the nuclear energy generator far enough away from the instruments.

The radio receiver and transmitter kept contact with the Earth. The radar measured distances.

The magnetometers were able to measure the field strengths in three dimensions and also monitored the solar wind.

Voyager was able to analyse gases and fine particles, as well as the gases in Jupiter's atmosphere at various depths. It measured the pressures at various depths and the energy spectra of electrons in the radiation belts around Jupiter.

The video cameras revealed worlds that could not be seen through the most powerful telescopes.

The first pictures were taken when Jupiter was still 75 million km distant. Voyager-1 returned 18 000 videos. This was only a small fraction of its total scientific measurements.

The bow shock, where Jupiter's magnetic field encounters the solar wind, was found to be 6 million km from Jupiter (= 85 Jupiter radii). Here the velocity of the solar wind suddenly decreases from 417 to 110 km s^{-1}, and the temperature increases ten-fold. The closest that the bow shock came to Jupiter was 3.3 million km, or 47 Jupiter radii. The magnetic tail trails off far beyond Saturn's orbit.

When the videos of Jupiter were made into a film, Jupiter's rotation and the motions of the gas streams, parallel to the equator, could be clearly seen. Figure 8.36 shows Jupiter, with Io (left) and Europa (right) in the foreground. In these photographs, by courtesy of NASA and the Jet Propulsion Laboratory, Pasadena, the colours are exaggerated so as to differentiate various substances.

The film showed that the gases in the Great Red Spot spiral counter-clockwise. Since the spot is situated 20 degrees south of the equator, the Great Red Spot is an anticyclone, in which the gases spiral downwards under high pressure (Figure 8.37).

The Great Red Spot is therefore not an area of rising gases, as had previously been supposed. The average dimensions of the Great Red Spot are 26 000 km by 14 000 km. It can thus easily swallow two spheres, each the size of the Earth. The white spot, next to the Great Red Spot, consists of fine crystals of ammonia.

Figure 8.36 Jupiter, Io and Europa

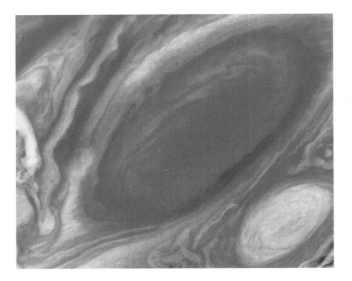

Figure 8.37 Jupiter's Great Red Spot

The colours of the gases in the atmosphere vary from white, through red to black. At times the Great Red Spot is much paler, even white. The white spots are most likely ammonia and its compounds, such as ammonium hydro-sulphide (NH_4SH). Some white spots could be water ice. The yellow and orange gases are situated lower down in the atmosphere than the white spots. Spectroscopic analysis has revealed the presence of gases such as methane (CH_4), ethane (C_2H_6), acetylene (C_2H_2), phosphine (PH_3), carbon monoxide (CO) and germanium tetra-hydride (GeH_4).

Ninety per cent of Jupiter's atmosphere, however, consists of hydrogen, and helium constitutes about 4.5%.

Jupiter's strong gravitational field gathers in fine particles of sulphur, which are spewed out by Io's volcanoes. These sulphur particles, as well as crystals of sulphur dioxide, probably contribute to the colours of the gases in the atmosphere.

One hundred km above the clouds, the temperature is 150 K. At 35 km above the clouds, the temperature is down to 100 K. In the clouds the temperature starts rising and at the lowest levels it reaches 250 K. On account of the increasing pressure, the temperature rises with greater depth.

The trajectories of Voyager 1 and 2 are shown in Figures 8.38 and 8.39. The launching times and the routes followed were chosen so that all the large moons could be viewed from close by.

The closest approach to the four moons is given in Table 8.3.

The Voyagers also discovered three small moonlets: Thebe, 1979 J1, 110 by 90 km, revolves in 16 h 11 m 17 s at a height of 151 000 km above the cloud tops; Adrastea, 1979 J2, much smaller, egg-shaped, 25 by 20 by 15 km, revolves in 7 h 9 m 30 s at a height of 57 000 km; and Metis, 1979 J3, 40 km in size, revolves in 7 h 4 m 29 s at a height of 56 200 km. These discoveries brought Jupiter's satellite tally to 16 (Table 8.4).

The gaseous belts in Jupiter's atmosphere move from west to east at various speeds. On the equator, the speed is 360 km h^{-1}. At 7 degrees north and south, the speed is 576 km h^{-1}. In these belts, differently coloured ovals often occur. At 18 degrees north, the speed is 140 km h^{-1} but in the opposite direction, east to west, and at 18 degrees south, the speed is 270 km h^{-1}. At the boundaries between the different belts, great turbulence takes place.

Jupiter's rotation period can be measured by taking the time that it takes any distinctive swirl in the gases to make one rotation. At 9 degrees from the equator, the period turns out to be 9 h 50 m 30.003 s. The belts further north and south take about 5 minutes longer. An average value of 9 h 55 m 40.632 s has been accepted for the mid-latitudes (40–50 degrees). Nearer the poles the rotation period is longer.

Jupiter's radio emissions, however, have a period of 9 h 55 m 29.710 s. This has a close connection with the position in its orbit of Io. Since the magnetic field is integrally connected with the body of the planet, this period is taken as Jupiter's sidereal period of rotation, irrespective of the periods of rotation of the gas belts.

The magnetic field is complex and is inclined by 10 degrees to the axis of rotation.

At a distance from Jupiter, the magnetic field is bipolar, but the north and south poles are the reverse to those of the Earth. At the top of the cloud decks, the strength of the magnetic field is 4.2 gauss, that is, 13.6 times that of the Earth.

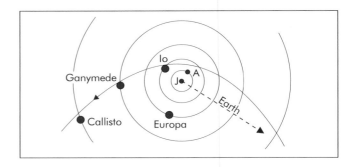

Figure 8.38 Trajectory of Voyager-1 among Jupiter's moons

Jupiter's Moons

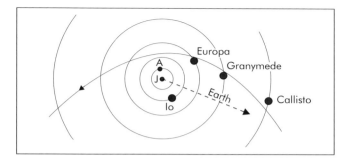

Figure 8.39 Trajectory of Voyager-2 among Jupiter's moons

Table 8.3 Closest approach of Voyagers to Jupiter's moons

Moon	Voyager – 1	Voyager – 2
Amalthea	425 000 km	558 270 km
Io	22 000	1 130 000
Europa	732 000	204 030
Ganymede	130 000	59 530
Callisto	130 000	212 510

The smallest of these moons, Metis and Adrastea, revolve inside the orbit of Amalthea, 127 600 and 128 400 km from Jupiter's centre.

The diameters of the four Galilean satellites were somewhat refined by the Voyagers. The diameters of Io and Europa have been increased by 7 and 12 km to 3630 and 3138 km, respectively. The diameters of Ganymede and Callisto have been decreased by 14 and 20 km, and are now known to be 5262 and 4800 km, respectively. It speaks volumes for the accuracy of the measurements made throughout the years that the values were so close to the actual values.

Jupiter can be considered as having four sets of satellites:

1. the inside four tinies, of irregular shapes;
2. the four Galilean satellites, discovered independently of each other by Galileo and Marius von Gunzenhausen in 1610;
3. the group of four whose orbits are inclined by 24 to 29 degrees and which revolve at distances from Jupiter

Table 8.4 Jupiter's four families of satellites

Satellite	Dist. from Jup. centre (×1000 km)	Diameter (km)	Period of revolution (days)	Inclination of orbit to equator	Discoverer
Metis (1979 J3)	127.6	40	0.294 78		S.P. Synnott, 1980
Adrastea (1979 J2)	128.4	25 × 20 × 15	0.298 26		E. Danielson and D. Jewitt, 1980
Amalthea	181.3	270 × 166 × 150	0.498 18	0.4°	E.E. Barnard, 1892
Thebe (1979 J1)	222.4	110 × 90	0.674 5	0.8°	S.P. Synnott, 1980
Io	421.6	3630	1.769 1	0.04°	Galileo, Marius, 1610
Europa	670.9	3138	3.551 2	0.47°	Galileo, Marius, 1610
Ganymede	1 070	5262	7.154 6	0.21°	Galileo, Marius, 1610
Callisto	1 883	4800	16.689 0	0.51°	Galileo, Marius, 1610
Leda	11 094	16	238.72	26.07°	C. Kowal, 1974
Himalia	11 480	186	250.566 2	27.63°	C.D. Perrine, 1904
Lysithea	11 720	36	259.22	29.02°	S.B. Nicholson, 1938
Elara	11 737	76	259.652 8	24.77°	C.D. Perrine, 1905
Ananke	21 200	30	631 retro	147°	S.B. Nicholson, 1951
Carme	22 600	40	692 retro	164°	S.B. Nicholson, 1938
Pasiphae	23 500	50	735 retro	145°	P.J. Melotte, 1908
Sinope	23 700	36	758 retro	153°	S.B. Nicholson, 1914

between 11 094 000 and 11 737 000 km; and
4. the most distant group of four, at distances between 21 200 000 km and 23 700 000 km, and with orbits inclined to Jupiter's equator by 145 to 164 degrees. Because these angles are greater than 90 degrees, these satellites have retrograde revolutions.

THE videos made of the four Galilean satellites were most astonishing.

Io

Everybody was amazed to see that Io is covered with sulphur and crystallised sulphur dioxide. The colour of the surface varies between white and yellow, through red to black. Sulphur can melt and solidify at various temperatures, and the temperature at which it solidifies determines whether the solid will be bright yellow, orange, red or black. The black spots on Io are the caldera of the volcanoes that spew sulphur vapour and sulphur dioxide (Figure 8.40).

When the first videos of Io were received, Miss Linda Morabito decided, by way of overtime work, to project the image of Io again, but to blank out the bright glow of Io itself, to see if there were any small satellites in the background. She saw an umbrella-shaped spout on the rim of Io, formed by one of the volcanoes. The spouts reach heights of 100 to 280 km.

The layers below Io's surface must therefore be hot enough to make sulphur boil (448°C). The heat derives from the convective movement caused by Jupiter's gravitational force and by centrifugal forces, both of which are stronger on the side of Io which faces Jupiter than on the side turned away from Jupiter. The four Galilean satellites have rotation periods equal to their periods of revolution, so that they each keep the same face turned towards Jupiter.

The four large moons revolve within the magnetosphere of Jupiter. In the neighbourhood of Io, particles of sulphur, sodium, potassium and magnesium, as well as electrified particles, formed by the bombardment of cosmic rays from Jupiter, form a sort of torus, almost coinciding with Io's orbit. Between Jupiter and Io a strong electric current of 5 million amperes under a potential of 400 000 volts, flows at right angles to the plane of Io's orbit. This electric current also contributes to the heating of Io's interior.

The absence of craters on Io is due to its surface being continually renewed by outpourings from the volcanoes.

The adjustment to Io's diameter means that its density is 3.53 g cm^{-3}. Figure 8.41 is a model of Io's cross-section which works out to the correct density.

The model envisages a rocky core of radius 660 km, a mantle of liquid silicates, 1135 km thick, and a crust, 20 km thick, consisting mainly of sulphur and sulphur dioxide. This layer is in constant convective motion.

Europa

The surface of Europa (Figure 8.42) is in marked contrast to that of Io. The surface is an extensive plain of ice, with long,

Figure 8.40 Io, covered in sulphur and sulphur dioxide. Volcano on rim spouts sulphur and sulphur dioxide

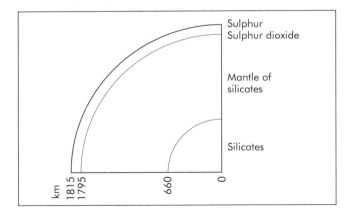

Figure 8.41 Cross-section of Io

Jupiter's Moons

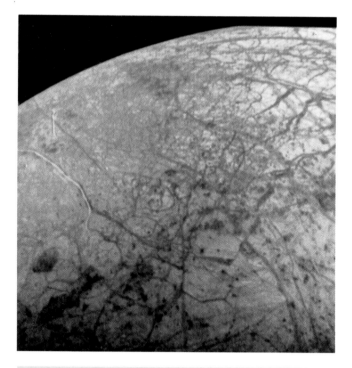

Figure 8.42 Europa's ice-covered surface

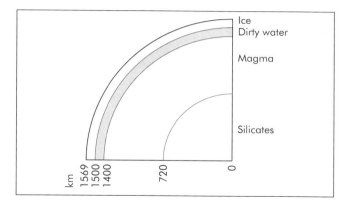

Figure 8.43 Cross-section of Europa

linear, winding cracks, filled in with darker material. Some of the cracks are 40 km wide and 1000 km long. There are also narrow ridges, 10 km wide and 100 km in length. These linear features are called linea. A ridge is called a flexus, and a spot a macula.

Only three craters, 18 to 25 km wide, have been found. Because ice tends to flow, craters in the ice are soon levelled off. The surface therefore remains young.

The density of Europa works out at 3.03 $g\,cm^{-3}$. The model of cross-section of Europa, which works out at this density, is depicted in Figure 8.43.

The rocky, silicate core of Europa has a radius of 720 km. Above it is a layer of liquid magma, 680 km thick. Above that is a layer of muddy water, 100 km thick, and it is topped by the ice layer, 70 km thick.

Ganymede

Ganymede retains it pre-eminence as the largest satellite in the Solar System. It is $1\frac{1}{2}$ times as large as the Moon and 384 km larger than Mercury. Ganymede's diameter is 5262 km.

With very large telescopes, it is just possible to see a dark patch on Ganymede. It has now been called Galileo Regio, and is 3200 km wide and covers one-third of the hemisphere which is turned away from Jupiter (Figure 8.44).

Ganymede has many craters, but they are not as densely spread as those of the Moon and Mercury. The crust must therefore be younger. The craters occur on top of folded terrain, looking as if a giant currycomb had scraped the surface. There are broad, long trenches, called sulci, containing ice. Many craters have white crests, consisting of ice, which formed after the collisions. Some craters have

Figure 8.44 Ganymede

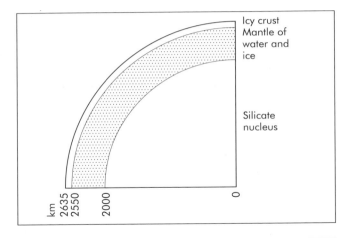

Figure 8.45 Cross-section of Ganymede

Figure 8.46 Asgard on Callisto

white streaks, radiating outwards. These streaks consist largely of ice.

Figure 8.45, a cross-section of Ganymede, shows an icy crust, about 85 km thick, over a layer of water and ice, 550 km thick. Ganymede's density works out at 1.93 g cm^{-3}. In order that this value be correct, Ganymede must have a very large core of radius, at least 2000 km, consisting of silicate rocks. Convective movements in the mantle of water and ice are the cause of the folded terrain. Movements of the crust would not allow craters to remain intact for very long.

Callisto

Callisto is very similar to Ganymede: its surface is somewhat darker and it has many more craters, most of them topped with ice. There are two huge circular ridge systems, Valhalla and Asgard. They each have concentric ridges, spreading over a distance of 600 km. These must have been formed by a collision with a very large meteor or small asteroid. The heat of the collision would have melted the ice of the surface and, as the water waves moved outwards, the water would have frozen to form permanent patterns. Valhalla and Asgard (Figure 8.46) are reminiscent of Mare Orientale on the Moon and Caloris Basin on Mercury, although in the latter two cases, much material was cast out.

Callisto's density of 1.6 g cm^{-3} indicates an ice crust 300 km thick over a mantle of water and soft ice, 1000 km thick and a silicate core of radius 1200 km (Figure 8.47).

The folding on the surface is not as marked as that of Ganymede.

Amalthea

Amalthea was discovered by E.E. Barnard in 1892 – the last moon to be discovered visually, without photography as an aid. For over 80 years it was known as the innermost moon

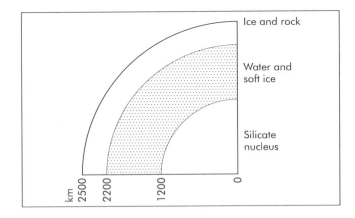

Figure 8.47 Cross-section of Callisto

of Jupiter. In 1979, three small moons were discovered by the Voyager space probes, of which two, Adrastea and Metis, revolve within the orbit of Amalthea.

Amalthea is reddish, of irregular shape and measures 270 by 166 by 150 km. The Voyager photos show two craters on Amalthea along with two hillocks. Voyager found that Amalthea is hotter than can be expected from the amount of heat it receives from the Sun. This may be due to the electric currents induced by the magnetic field of Jupiter.

Jupiter's Far-Away Moons

There are two sets of four satellites which revolve much further from Jupiter than Callisto. The four, whose names end in the letter "a" – Leda, Himalia, Lysithea and Elara – revolve at a mean distance of 11 500 000 km from Jupiter, at least six times Callisto's distance. Their orbits are inclined by about 27 degrees to Jupiter's equator. Their mean period of revolution is about 254 days, or 15 times that of Callisto. They could be captured asteroids.

The names of the second group of four end in the letter "e" – Ananke, Carme, Pasiphae and Sinope. Their average distances from Jupiter vary between 21.2 and 23.7 million km, about twice as far as the first group. Their orbits are inclined to Jupiter's equator by angles between 145 and 164 degrees. Therefore they are in retrograde revolution. Their orbits are also very eccentric, that of Pasiphae being $\varepsilon = 0.378$.

If J and J_1 are the two foci of the orbit of Pasiphae, then the distances x are equal to $0.378(23\,500\,000) = 8\,883\,000$ km, where $23\,500\,000$ is Pasiphae's average distance from Jupiter.

Therefore Pasiphae's distance from Jupiter varies between 14 617 000 and 32 383 000 km. If Pasiphae's greatest distance from Jupiter (taking it in line with the Sun) is projected on to the plane of the Ecliptic, its effective distance from Jupiter becomes 26 526 600 km, and 751 373 400 km from the Sun. The Sun's force of attraction on Pasiphae, compared with Jupiter's force, is:

$$\frac{\text{Sun's mass}}{\text{Jupiter's mass}} \times \frac{(26\,526\,600)^2}{(751\,373\,400)^2}$$
$$= 1047 \times 0.001\,246\,38 = 1.3$$

Thus, when Pasiphae is furthest from Jupiter and nearest to the Sun, the Sun's force is 1.3 times that of Jupiter, or 30%

stronger. The orbits of the four outer moons therefore undergo large perturbations on account of the Sun's gravitation. However, for the largest portions of their orbits, these moons are firmly in the grip of Jupiter. Their orbits and distances from Jupiter are continually changing. They must be considered to be captured asteroids.

The sizes of the outer moons are worked out according to their brightnesses and are only approximately correct.

Jupiter's Ring

When Voyager passed round the far side of Jupiter, it noticed that Jupiter has a faint ring. The ring became visible because of the dispersion of light from the infinitesimally small particles comprising the ring. The average size of the particles is 8 micrometres (= 8 thousandths of a mm). The brightest, and outermost, part of the ring is 126 380 ± 140 km from the centre of Jupiter. The radius of the inner edge of the ring is 125 580 km. The particles are, however, spread throughout the space between the ring and Jupiter itself; they form a haze over the atmosphere. Carbon monoxide was unexpectedly discovered there. The ring is orange coloured, probably on account of sulphur particles, originating in Io's volcanoes.

When Voyager moved into Jupiter's shadow, it noticed lightning discharges in Jupiter's atmosphere.

The satellites Metis and Adrastea revolve next to the outer edge of the ring and act as shepherding moons, keeping the ring particles from dissipating.

The ring and these two satellites revolve within the Roche limit, which is 174 200 km from Jupiter's centre. The two

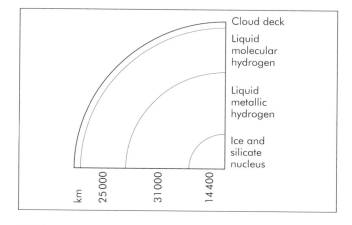

Figure 8.48 Cross-section of Jupiter

moonlets do not get broken up, because they are very small and are solid rocks.

A model of the composition of Jupiter (Figure 8.48) indicates that there is a sudden change 1000 km below the surface, where the hydrogen becomes liquid. At this boundary, the temperature reaches 2000 K and the pressure 5600 times the Earth's atmospheric pressure. The layer of liquid hydrogen is about 25 000 km thick, and contains 4.5% helium. As the depth in this layer of molecular hydrogen increases, both the pressure and the temperature increase; 25 000 km down, the pressure reaches 3 million Earth atmospheres and the temperature 11 000 K. Now the hydrogen undergoes another change and becomes metallic hydrogen which is a good conductor of electricity. This layer is about 31 000 km thick. Below that, Jupiter has a comparatively small core of radius of 14 400 km, consisting of ice and silicate rocks. At its centre, the pressure in Jupiter rises to 100 million times the pressure of the Earth's atmosphere and the temperature rises to 30 000 K. This is the source of the heat which enables Jupiter to radiate twice as much energy as it receives from the Sun.

Slingshot Acceleration by Jupiter

As the Voyagers neared Jupiter, their velocities increased, owing to the gravitational attraction of Jupiter. Their trajectories were controlled so that they would not plunge into Jupiter, but make full use of Jupiter's gravity to gain speed as they passed around the planet. This slingshot acceleration enabled the Voyagers to speed off to Saturn. Voyager-1 reached its nearest point to Jupiter, of 350 000 km, on 5 March 1979, and Voyager-2 reached a point 714 000 km from Jupiter on 9 July 1979. Voyager-1 reached Saturn on 12 November 1980, and Voyager-2 passed Saturn on 26 August 1981. Without the aid of the acceleration imparted by Jupiter, the trip, launched with the same thrust, would have taken nine years to reach Saturn, instead of the 3 years 69 days of Voyager-1 and the 4 years of Voyager-2.

Saturn

The first space probe to visit Saturn was Pioneer-11. It passed by Saturn, at a distance of 3500 km outside the rings, on 1 September 1979. Its spectroscope showed that the rings largely consist of small lumps of ice, as had been expected. Pioneer revealed details of the rings, which had not been expected. Also 3000 km outside the outer A ring, Pioneer found an unknown ring, 50 km wide. This is the F ring or Pioneer ring. Figure 8.49 gives some idea of the fine divisions between the rings. It also reveals the phenomenon of "spokes" that rotate as spokes and not as would be expected from Kepler's third law. This is because these "spokes" consist of magnetically trapped particles, which rotate with the magnetic field.

Figure 8.49 Fine divisions in Saturn's rings

Saturn's magnetic field is not as strong as that of Jupiter, but stronger than the Earth's field. Saturn's magnetic axis coincides with the axis of rotation.

In 1966, A. Dollfus discovered what was then Saturn's tenth moon, Janus, 14 000 km outside the A ring, having a period of 17.975 hours. In 1976, S. Larson and J.W. Fountain discovered a satellite, apparently in the same orbit as Janus – moon 1980 S3 – now known as Epimetheus.

Pioneer-11 discovered a further four satellites.

Both Pioneer-11 and Voyager-1 flew past the "underside" or unlit side of the rings. They were thus able to analyse dispersed sunlight and this proved that the rings consist of small particles of ice. There are also larger, rough pieces in the rings. It is quite remarkable to see how much material there is in the Cassini gap.

Voyager-1 took the photograph of Saturn shown in Figure 8.50 when it was 18 million km away.

By the time Voyager-2 had reached Saturn, it had covered 2 228 957 024 km from the Earth, at an average speed of 17.68 km s^{-1}.

Saturn's Satellites

As in the case of Jupiter's satellites, Voyager again revealed astonishing sights of Saturn's satellites.

The densities that have been worked out are only slightly more than 1 g cm^{-3}; this means that they consist largely of ice. Titan, the largest moon, has a density of almost

Saturn's Moons

Figure 8.50 Saturn

Figure 8.51 The ice-moon, Mimas

2 g cm^{-3}. It must therefore have a sizeable rocky core. The density of Saturn itself is 0.7 g cm^{-3}; it is lighter than water. Besides ice, Saturn must have a very extensive layer of hydrogen and helium. See Table 8.5 for other details.

Mimas

Mimas, whose diameter is 390 ± 10 km, has one very large crater, 130 km in diameter, with rims 9 km above the floor. (Figure 8.51). The crater has a larger central peak. The rest of the surface is covered with small craters, which have a smooth appearance, showing that the surface is largely ice, because ice tends to flow with the passage of time. Its density of 1.2 g cm^{-3} is further proof that it consists largely of ice. There are a few rifts.

Enceladus

Enceladus (Figure 8.52) has an even whiter appearance than Mimas. It is pure ice and has an albedo of 1, making it a perfect reflector of light. It has many flat craters and there are ridges where ice has welled up, in some cases deforming the craters. The ridges are also flattened, owing to the flowing of the ice. The diameter is 500 km, within 20 km either way, and the density works out to 1.1 g cm^{-3}.

Figure 8.52 Enceladus

Table 8.5 Saturn's satellite system

Satellite	1980	Dist. from Sat. centre (km)	Diameter (km)	Period of revolution (days)	Density (g cm^{-3})	Orbital inclination	Discoverer
Atlas	S 28	137 670	40 × 20	0.6019		0.3°	Voyagers 1980, 81
Prometheus	S 27	139 353	140 × 100 × 80	0.6130		0.0°	Voyagers 1980, 81
Pandora	S 26	141 700	110 × 90 × 70	0.6285		0.0°	Voyagers 1980, 81
Epimetheus	S 3	151 422	140 × 120 × 100	0.6945		0.3°	J.W. Fountain and S. Larson, 1976
Janus	S 1	151 472	220 × 200 × 160	0.6945		0.1°	A. Dollfus, 1966
Mimas		185 520	392	0.9424	1.2	1.5°	W. Herschel, 1789
Enceladus		238 020	500	1.3702	1.1	0.0°	W. Herschel, 1789
Tethys		294 660	1060	1.8878	1.0	1.9°	G.D. Cassini, 1684
Telesto	S 25	294 660	34 × 28 × 26	1.8878			Voyagers 1980, 81
Calypso	S 13	294 660	34 × 22 × 22	1.8878			B.A. Smith, 1980
Dione		377 400	1120	2.7369	1.4	0.0°	G.D. Cassini, 1684
Helene (= Dione-B)	S 6	377 400	36 × 32 × 30	2.7369		0.0°	J. Lecacheux and P. Laques, 1980
Rhea		527 040	1530	4.5175	1.3	0.4°	G.D. Cassini, 1672
Titan		1 221 830	5150	15.9454	1.9	0.3°	C. Huygens, 1655
Hyperion		1 481 100	410 × 260 × 220	21.2766		0.4°	G.P. Bond, 1848
Iapetus		3 561 300	1460	79.3302	1.2	14.7°	G.D. Cassini, 1671

Tethys

Tethys is considerably larger, having a diameter of 1060 km. It has a large rift valley, 1000 km long and 100 km wide, with a depth of 2 to 3 km. There are many craters and Voyager-2 found a very large astrobleme where a huge chunk of ice must have crashed. The floor of this crater has been raised back to the ambient level of the surface. The ridges are flattened. The density of Tethys works out to 1.0 g cm^{-3}. Thus it must be pure ice, although its colour is somewhat dark.

Dione

Dione's diameter is 1120 km, and it has many flattened craters. The largest crater is nearly 100 km in diameter. The floor and central peak have been raised. There are many small craters overlying the older larger craters. The density of 1.4 g cm^{-3} indicates that Dione must have a sizeable rocky core. Its reflectivity is also very good (Figure 8.53).

Figure 8.53 Dione

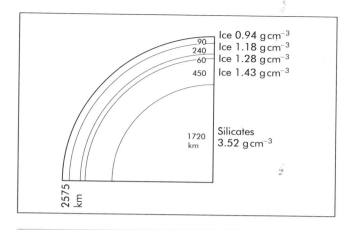

Figure 8.54 Cross-section of Titan

Titan

The largest of Saturn's moons, Titan, was discovered by C. Huygens in 1655. Its diameter of 5150 km is 1674 km greater than that of the Moon and 272 km greater than Mercury's diameter. Earth-based observations showed that Titan has an atmosphere $1\frac{1}{2}$ times the density of the Earth's. Voyager flew to within 5000 km of Titan and found that its atmosphere consists largely of nitrogen, with traces of methane, hydrocyanic acid (HCN) and hydrocarbons. Voyager carried out the analysis by transmitting radio waves through the atmosphere. By this means the pressure and temperature at various levels could be determined. The average surface temperature is 92 K, that is, −181 degrees below the freezing point of water. It is possible that liquid nitrogen falls on Titan as rain.

Titan is the densest of Saturn's satellites, the density being 1.9 $g\,cm^{-3}$. The model of the composition of Titan, made by C. Alexander and R. Reynolds, gives Titan a rocky core of radius 1720 km (Figure 8.54).

Hyperion

Hyperion is irregular in shape, 410 by 260 by 220 km and revolves at a distance of 1 481 100 km from Saturn. It has many craters, a rocky structure and is darker than the other moons.

Iapetus

At more than twice the distance from Saturn of Hyperion, Iapetus revolves 3 561 300 km from Saturn. It has been noticed that its light is variable, and that one side must be darker than the other. Voyager corroborated this. Iapetus is in synchronous rotation, keeping the same face turned to Saturn. Its preceding hemisphere is darker than the other, having an albedo of 4%, so that it is almost black. The plane of the orbit of Iapetus is inclined by 14.7 degrees to the plane of Saturn's equator, while the fifteen satellites within its orbit revolve in the plane of Saturn's equator.

Phoebe

Almost four times further, Phoebe revolves at 12 952 000 km from Saturn. The orbit is inclined by 159 degrees to the plane of Saturn's equator, and it therefore has retrograde revolution. The eccentricity of the orbit is $\varepsilon = 0.163$. Phoebe is probably a captured asteroid. With a low albedo of 5%, it is not one of the ice-moons.

Saturn's Tiny Moons

These are all irregular chunks of rock, of densities about 2 $g\,cm^{-3}$. The nearest to Saturn is Atlas (or 1980 S28). It revolves close the outermost ring, at a distance of 137 670 km. This is only 890 km beyond the outer edge of the A ring. This moon is responsible for the sharp demarcation of the A ring.

Next is Prometheus (1980 S27) just inside the F ring, 139 353 km from Saturn. Just outside the F ring, at 141 700 km, another moonlet, Pandora, together with Prometheus, carry out the function of shepherding moons. They cause the F ring to have braided form.

Further from Saturn are two moons: Epimetheus (1980 S3) and Janus (1980 S1). Janus was discovered from Earth by A. Dollfus in 1966. The difference between the radii of the orbits of these two moons is only 50 km, which is less than their own sizes (Table 8.5). Their orbital speeds are almost identical. As the inner of the two overtakes the outer, it gets speeded up by the outer moon, so that it moves outwards, finding itself further from Saturn than the moon, which then occupies the inner position. These outer changes of position take place at intervals of four years.

Between 170 000 and 171 000 km there is a faint ring, the G ring, which was discovered by Voyager, and beyond that there is an even fainter ring, 300 000 km wide.

The following tiny moons are the pair Telesto and Calypso. They are almost equal in size (Table 8.5) and revolve in the orbit of Tethys. One is 60 degrees ahead of Tethys and the other 60 degrees behind, that is, they occupy the 4th and 5th Lagrange points of stable gravity.

The last of the small moons is Helene (1980 S6), also known as Dione-B, because it revolves in Dione's orbit, but on the other side of Saturn.

Saturn therefore keeps the lead with a total of 17 satellites. Jupiter has 16.

Saturn's Clouds

Saturn's clouds do not show the marked differences in colour, nor the turbulence of Jupiter's clouds. There are eddies, but on a smaller scale. The greatest velocity of the cloud streams is on the equator, where velocities of 1440 km h^{-1} prevail. Away from the equator, the velocities of the gas streams decrease.

The length of Saturn's day is 10 h 39 m. It has been determined, as in the case of Jupiter, by the periodicity of the radio emissions, these depending on the rotation of the magnetic field.

The following gases have been traced in Saturn's atmosphere: ammonia (NH_3), water vapour (H_2O), sulphuretted hydrogen (H_2S) and ammonium hydrosulphide (NH_4SH), as well as the extensive envelope of hydrogen, with an admixture of helium.

The composition of the planet agrees very well with that of Jupiter. Below the atmosphere of hydrogen and helium, there is a layer of liquid hydrogen; then a layer of metallic hydrogen, on a core consisting of ice and silicate rocks.

When the Voyagers passed around Jupiter and Saturn, they moved in planes coinciding with the planes in which the satellites revolve. This was not to be the case at the next planet, Uranus.

Uranus

Saturn's slingshot acceleration sped Voyager-2 on its way to Uranus, a distance greater than the distance it had travelled from the Earth to Saturn (Figure 8.35, page 120). After having covered 2230 million km, it now had 2570 million km to go to reach Uranus. This part of the journey took 4 years 152 days, and Voyager-2 reached Uranus on 24 January 1986.

The plane of Uranus' equator is tilted to the Ecliptic by 97°53′. The five known moons revolve in the plane of the equator. Voyager-2 could therefore not move in the plane of the orbits. They were at right angles to Voyager's course, somewhat like a target at which shots are fired, but Voyager's velocity was much greater than that of any bullet, 16 km s^{-1}.

During December 1985, as Voyager was closing in on Uranus, it found that none of the five moons was exactly on its calculated position. The fault lay with Uranus' mass. The accepted value was apparently one-quarter per cent too little, an amount more or less equal to the mass of Mercury. A course correction had to be made by firing the propulsion rockets for 14 minutes, on 24 December 1985.

A radio signal from Earth to the region of Uranus takes $2\frac{1}{2}$ hours to cover the distance. Voyager therefore had to be programmed in advance, so as to be able to direct its instruments in turn on Uranus and the five moons. Voyager's first target was Titania, which it had to scan 2 hours 51 minutes before zero hour (Figure 8.55). Titania was then at a distance of 370 000 km; then followed Ariel, 126 000 km away at 1 hour 38 minutes before zero. Miranda, 29 000 km distant, was scanned at 0 hours 55 minutes before zero. Uranus came into view at zero hour. Then followed Oberon, 469 000 km away at 1 hour 47 minutes after zero, and finally Umbriël, 323 000 km away at 2 hours 53 minutes after zero. Voyager-2 thus had only 5 hours 44 minutes at its disposal to do the whole job. And what a magnificent job it did! Because the time available for the transmission of the impulses was so short, Voyager was programmed to transmit only the differences between succeeding pixels. In this way, many more pixels could be transmitted. When Voyager reached its closest point to

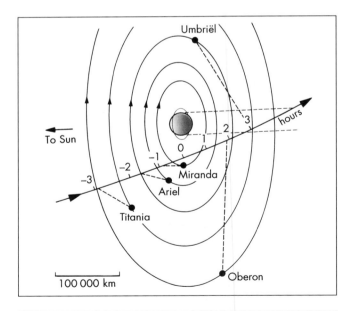

Figure 8.55 Course followed by Voyager-2 through Uranus' system

Uranus' Moons

Figure 8.56
Oberon

Uranus, it was only 100 km off course! It was on course for its rendezvous with Neptune in August 1989.

In 1977 when Uranus occulted the star SAO 158687, spectroscopic analysis of the starlight which passed through Uranus' atmosphere revealed that Uranus has hydrogen and methane (CH_4) in its atmosphere. Methane absorbs red light very well, with the result that Uranus appears to be very blue. Voyager could discern only a few white stripes of ammonia crystals. Helium comprises 12 to 15% and there are also traces of acetylene (C_2H_2) in the atmosphere. The top layer of the atmosphere over the south pole, which was directly under the Sun, had a temperature of 750 K. The north pole, which was in shadow, had a temperature of 1000 K. The north pole was so warm because the molecules cool down very slowly on account of the rarefied nature of the upper layers. With greater depth, the temperature falls quickly – a temperature as low as 51 K was recorded. Because of the 98 degree tilt of the axis of spin of Uranus, there is very little difference between the temperatures at the equator and the poles.

Uranus' magnetic field, which is tilted to the axis of spin by 60 degrees, is stronger than the Earth's magnetism, and is displaced by some 8000 km towards the north pole. This means that Uranus' magnet is not situated in the rocky core, but in the liquid layer, which is 9600 km thick, surrounding the core. The magnetic field has a period of rotation of 17.24 hours, which is a bit slower than the period of rotation of the cloud layers. The effect of this is that the winds in the clouds always blow from west to east.

Uranus' nine rings showed up very well in the images returned by Voyager. There is also a very faint tenth ring. All the rings are very narrow and dark in colour. There is also a broad band of finely divided material, 3000 km wide, and 37 000 km from Uranus' centre. When Voyager moved into Uranus' shadow, the rings showed up as 30 fine rings.

Voyager also monitored the occultation of the star Beta Persei by Uranus and obtained sharply defined extinctions by the rings.

The albedos of Uranus' five moons are very low. To get good images, time-exposures had to be taken. This required the scanning platforms, housing the video cameras, to make compensatory movements to prevent the images being smeared, owing to Voyager's high velocity.

OBERON (Figure 8.56), the furthest of the five moons, has many craters with dark floors where dirty water has frozen. The craters also have rays where water which was cast out, froze.

TITANIA, second from the outside, has a very dark surface, with many shallow craters, some of which have ice-covered rims. There are long, deep rifts. The whole surface is apparently covered with soot (Figure 8.57).

UMBRIËL is also very dark and is crater-strewn. Its spectrum shows the presence of ice. The surface appears to be old (Figure 8.58).

Figure 8.57
Titania

Figure 8.58
Umbriël

Figure 8.59 Ariel

Table 8.6 Uranus' satellite system

Satellite	1986	Distance from Uranus' centre (km)	Diameter (km)	Period of revolution (days)
Cordelia	U7	49 770	50	0.335 034
Ophelia	U8	53 800	50	0.376 409
Bianca	U9	59 170	50	0.434 578
Cressida	U3	61 780	60	0.463 569
Desdemona	U6	62 680	60	0.473 650
Juliet	U2	64 350	80	0.493 066
Portia	U1	66 090	80	0.513 196
Rosalind	U4	69 940	60	0.558 060
Belinda	U5	75 260	50	0.623 500
Puck	U10	86 010	170	0.761 833
Miranda		129 390	480	1.413 479
Ariel		191 020	1158	2.520 379
Umbriël		266 300	1172	4.144 177
Titania		435 910	1580	8.705 872
Oberon		583 520	1524	13.463 239

ARIEL has many deep rifts, shallow craters and folded terrain. There are blotches consisting of frozen water or methane (Figure 8.59).

MIRANDA, the closest to Uranus, shows signs of tremendous deformation. Besides the many shallow craters, there is a large chevron pattern and a huge folded area, reminiscent of the Circus Maximus of Rome. Miranda has escarpments 20 km in height, and most of the surface appears to consist of ice. It seems as if a huge ice chunk has collided with Miranda and fused with it (Figure 8.60).

The orbits of the five large satellites are almost perfect circles. The greatest eccentricity is 0.005. The orbit of Miranda is inclined to the plane of Uranus' equator by 4.2 degrees, but the other four moons have inclinations less than 0.4 degree. The system is very regular and must have been so since the time of its formation. The great deformations on Miranda may bear testimony to the great crash which caused its orbit to tilt by 4.2 degrees from the plane of the equator.

THE surprises with regard to the Uranian system were still to come. Voyager-2 discovered no less than ten small moons – the largest 170 km in size, and the smallest between 80 and 50 km. They are all irregular and very dark. Uranus' system of moons is set out in Table 8.6.

After flying through the Uranian system, Voyager's course correction jets were fired on 14 February 1986, for $2\frac{1}{2}$ minutes. That was all that was required to set its course to Neptune.

Neptune

With 4800 million km completed, Voyager-2 still had 2700 million km to go to reach Neptune, for an appointment on 25 August 1989. The goal was a distance of 4900 km above Neptune's north pole.

Voyager's first finding regarding Neptune was that its magnetic field is weaker than that of Uranus and inclined by 47 degrees to the spin axis of Neptune. The magnetic field rotates once every 16.11 hours. This has to be Neptune's period of rotation.

It had been suspected that Neptune has at least three partial rings. Voyager found two well-demarcated partial rings and two very faint, scattered rings. The rings revolve

Figure 8.60 Miranda

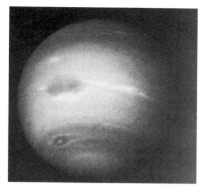

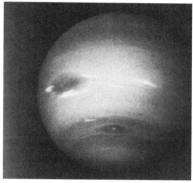

Figure 8.61
Neptune's Great Black Spot and Small Black Spot. Photos taken 17.6 hours apart

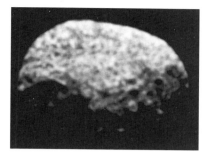

Figure 8.62
Proteus, Neptune's second largest moon

Triton

Voyager came to within 40 000 km of Triton and revealed a spectacular world. Triton's orbit is inclined to Neptune's equator by 160 degrees. Neptune's equator is inclined to the Ecliptic by 29°33.6'. Thus, the angle between the Sun's direction and Triton's polar axis can be as little as 52 degrees. Because Triton is in synchronous rotation around Neptune, the poles of Triton are exposed to sunlight for long periods. The frozen methane and nitrogen evaporate and blow over to the pole which is in darkness. There the gases freeze. This exposition was given by A.M. Thorpe (*Sky and Telescope Magazine*, May 1989). And this is exactly what Voyager found. The pole, seen by

at distances of 41 900, 53 200, 56 000 and 62 900 from Neptune's centre.

Neptune is fairly evenly blue in appearance, because the methane in its atmosphere absorbs red light. Twenty two degrees south of the equator, there is a large, dark spot which rotates counter-clockwise, like Jupiter's Great Red Spot. The black spot rotates around Neptune in 18.3 hours, which is a bit longer than Neptune's rotation period. Further south at latitude 54 degrees, there is a small black spot, with white clouds of frozen methane in the middle. There are also streaks of frozen methane in the atmosphere (Figure 8.61).

The two magnificent photographs of Neptune show how the small spot has overtaken the large spot, after one rotation (left to right). The small spot laps the large spot every five days.

Speeds in the cloud layers vary from 70 to 1170 km h^{-1}.

Neptune's diameter has now been corrected from 49 600 to 49 528 km ± 20 km, and the polar diameter from 48 315 to 48 680 ± 30 km.

Voyager-2 discovered six moons, of size 60 to 415 km. The largest of these, named Proteus (Figure 8.62), is therefore larger than Nereid, the furthermost of Neptune's known moons.

Voyager then sailed on to inspect Neptune's large moon, Triton.

Figure 8.63 Triton

Voyager, had extensive ice and frost deposits, looking somewhat like the skin of a cantaloup. There are many plumes of gases blowing towards the other pole (Figure 8.63).

Triton's surface has great plains and folded terrain. The shallow craters are ice tipped. The crust of frozen methane and nitrogen is calculated to be 175 km thick. Below that, there is a layer of semi-liquid ammonia, methane and water, 150 km thick, and a silicate core, of radius 1025 km. Triton's diameter is thus 2700 km.

Nereid was 4.7 million km from *Voyager*. Its half-moon phase had a greyish appearance. Its albedo is 12%.

AND now, *Voyager*s are on their way beyond the confines of the Solar System: *Voyager*-2 at an angle of 48 degrees southwards away from the Ecliptic. It will be monitored for another 25 years, submitting data on the solar wind, until it crosses the heliopause, where the solar wind ceases in interstellar space. Farewell *Voyager*!

Voyager-2 was certainly, by far the best "astronomer" of all time. It will be difficult to replace it. Telemetry has produced wondrous results (see Tables 8.7 and 8.8).

Table 8.7 Neptune's satellites

Satellite	1989	Dist. from Nep. centre	Diameter (km)	Period of revolution
Naiad	N6	48 000 km	60	0.30 days
Thalassa	N5	50 000	80	0.31
Despina	N3	52 500	150	0.33
Galatea	N4	62 000	160	0.43
Larissa	N2	73 600	200	0.55
Proteus	N1	117 600	415	1.12
Triton		354 290	2720	5.88
Nereid		5 510 660	300	360.2 retro

Table 8.8 Neptune's rings

Name	Dist. from Nep. cent.	Width (km)	Description
1989 N3R	41 900 km	1700	40–70% dust
1989 N2R	53 200	15	40–70% dust
1989 N4R	56 000	5800	Wide sheet, Little dust
1989 N1R	62 900	50	3 dusty arches

Chapter 9
The Sun

Situated about 150 million km from the Earth, the Sun occupies the centre of its planetary system. The mean distance of the Sun has been computed – by sending radar waves to certain planets and monitoring the reflected waves – as 149 597 870 km.

All the planets, their satellites, the comets and meteors are captives in the Sun's gravitational field. The planets revolve in orbits around the Sun. Mercury is closest to the Sun at a mean distance of 57.9 million km, and Pluto is the furthest, being 5909 million km distant (on average).

The Sun appears to move through 360 degrees annually against the background of the stars, which seems to be immovable. The path that the Sun traces against the background of the stars is called the Ecliptic. The Ecliptic is therefore the plane in which the Earth revolves around the Sun.

The apparent motion of the Sun is due to the fact that the Earth revolves around it. At night, the starry sky is that part of the expanse which is opposite in direction from the Sun. At midnight, the prime meridian is 180 degrees removed from the Sun (Figure 9.1).

During summer in the northern hemisphere, the Sun is north of the celestial equator, which is the projection of the Earth's equator against the sky. The Earth is then on the left-hand side of Figure 9.1. During the northern winter, the Sun is south of the celestial equator and the Earth is on the right-hand side of Figure 9.1. In between these two positions, the Sun crosses the equator: from north to south, on 23 September, and again from south to north, on 21 March. These two dates are known as the equinoxes, when day and night are equal in length over the whole world.

When the Sun crosses the equator, it is moving at its fastest in the north–south directions. After crossing the equator, the north–south component of the Sun's motion progressively reduces, until it comes to a stop on 21 December and 21 June. These dates are known as the solstices. On the solstices, the Sun's motion, north–south, comes to a stop and its direction of motion changes. From the solstice on 21 December, the days in the northern hemisphere become progressively longer, and shorter in the southern hemisphere. Around 21 March, the change is at its most rapid. The lengthening of the days in the northern hemisphere and shortening in the southern hemisphere then slow down until the longest and shortest days, respectively, are reached on 21 June. From 21 June to 21 December, the reverse takes place.

This phenomenon occurs because the Earth's axis of

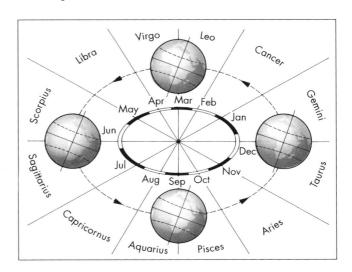

Figure 9.1 The Earth's motion around the Sun; the seasons and the constellations of the Ecliptic

Table 9.1

Dates of the month	20	21	22	23	
Equinox, March	11	11			times
Solstice, June		18	4		times
Equinox, September			3	19	times
Solstice, December		6	16		times

spin is inclined to the vertical to the Ecliptic, by an angle of 23°27′. The Earth's axis always points in the same direction. Actually, the axis describes a circle against the sky, having a radius of 23°27′, in 25 765 years (see Figure 2.8, page 17). This does not, however, affect the year-to-year state of affairs.

The inclination of the Earth's axis is the cause of the seasons (see Figure 2.7, page 16). On 21 or 22 December, the Sun is vertically above the Tropic of Capricorn, which is 23°27′ south of the equator. It is then mid-winter in the northern hemisphere and mid-summer in the southern hemisphere. Seen from the Earth, the Sun is then in the constellations Scorpius/Sagittarius. At night, the constellations Taurus and Gemini, on the Ecliptic, are visible. Orion is also very prominent. At this time of year, 2000 years ago, the Sun was in Capricorn.

On 21 March, the Sun is vertically above the equator and, from the Earth, the Sun is seen in the constellations Aquarius/Pisces. It is now spring in the northern hemisphere and autumn in the southern hemisphere. At night, the stars of Leo and Virgo are prominent on the Ecliptic.

On 21 or 22 June, the Sun is vertically above the Tropic of Cancer. It is summer in the north and winter in the south. The Sun is in the direction of the constellations Taurus/Gemini. At night, the stars of Scorpius/Sagittarius are prominent on the Ecliptic.

On 23 September, the Sun is again vertically above the equator, in the direction of Leo/Virgo. At night, the stars of Aquarius and Pisces are prominent. It is autumn in the northern hemisphere and spring in the southern hemisphere.

Because the year does not consist of a whole number of days (365.242 20 days, taken as 365.25), and because an extra day is added to February every fourth year, the dates of the equinoxes and solstices vary somewhat between 20 and 21 March; 21 and 22 June; 22 and 23 September; and 21 and 22 December. The number of times the equinoxes and solstices have fallen on the different dates is given in Table 9.1, as it applied to the period 1970 to 1991. Usually these dates are referred to as 21 March, 21 June, 23 September and 21 December.

Eclipses

While the Earth completes one revolution around the Sun in one year, the Moon revolves 13.38 times around the Earth. If the orbit of the Moon was in the same plane as that of the Earth's orbit around the Sun, the Moon would come into line with the Sun at every new Moon and a total eclipse of the Sun would take place; and at each full Moon, the Moon would move into the Earth's shadow and a lunar eclipse would take place.

The Moon's orbit is, however, inclined to the Ecliptic by 5°09′. The Moon's orbit cuts the Ecliptic twice during each revolution around the Earth, at the nodes. An eclipse of the Sun can only occur if the Moon is at, or very close to, a node, when it is new Moon. The nodes have retrograde movement along the Moon's orbit and complete one revolution in 18.6 years.

On the left of Figure 9.2, the Moon is indicated as being at M_1 and M_2. The line of nodes is not in line with the Sun. The Moon's shadow therefore does not fall on the Earth, nor does the Moon move into the Earth's shadow. Therefore no eclipses can occur.

On the right-hand side of Figure 9.2, the line of nodes passes through the positions M_3 and M_4 of the Moon. When

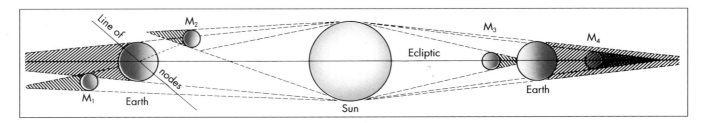

Figure 9.2 (*Left*) Nodes of Moon's orbit not in line with Sun and Earth. No eclipse possible. (*Right*) Nodes of Moon's orbit in line with Sun and Earth. Eclipses possible

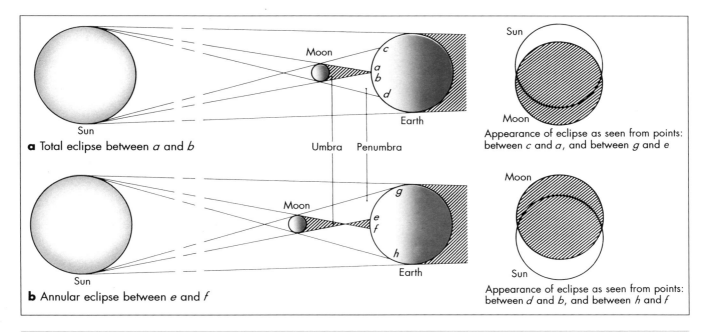

Figure 9.3 Solar eclipses: **a** Total; **b** Annular

the Moon is at M_3 (new Moon), it is in line with the Sun, so the Moon's shadow will fall on the Earth. In the darkest part of the Moon's shadow (the umbra), a total eclipse of the Sun will be visible. The umbra moves over a narrow strip of Earth and it is only in that strip that the eclipse will be total. To the sides of the strip, the eclipse will be partial.

When the Moon is at M_4, it moves through the Earth's shadow and a lunar eclipse is visible. The Earth's umbra is very much broader than the Moon. Thus a lunar eclipse is visible over a whole hemisphere and it lasts some hours, as compared with the few minutes' duration of a total solar eclipse.

Because of the eccentricity of the Moon's orbit (0.0549 ± 0.011), the angular diameter of the Moon varies between 29′28″ at apogee (furthest from the Earth) to 32′53″ at perigee (nearest to the Earth). The eccentricity of the Earth's orbit is 0.017, so that the angular diameter of the Sun varies from 31′28″ at aphelion to 32′ 33″ at perihelion. When the Moon is at apogee, its angular diameter of 29′28″ cannot completely cover the disc of the Sun, even when the Earth is at aphelion (when the Sun's angular diameter is 31′28″). When this condition prevails, the Sun, even at the height of eclipse, will still be visible as a thin ring around the black disc of the Moon. This is called an annular eclipse and is illustrated in Figure 9.3b.

Between points e and f in the figure, the highest degree of eclipse will be the annular form. Points on Earth, represented by g to e and h to f, will experience only a partial eclipse.

When the Moon's angular diameter is equal to, or greater than, that of the Sun, points on Earth, as between a and b in Figure 9.3a, will experience a total eclipse. Points between c and a and between d and b will experience a partial eclipse.

If the Sun is not sufficiently near to a node of the Moon's orbit (Figure 9.4), no place on Earth will experience a total eclipse of the Sun.

Places that experience a partial eclipse of the Sun are situated in the partial shadow of the Moon, that is the

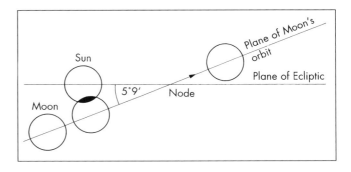

Figure 9.4 Moon's centre not on Ecliptic

penumbra.

When the Moon is eclipsed, it does not become altogether black because the sunlight is refracted by the Earth's atmosphere, so that some light does reach the Moon's surface. On account of the dust particles in the atmosphere, the shorter wavelengths of the sunlight are dispersed and the Moon takes on a bronze colour.

The maximum number of eclipses that can take place in any one year is 7 and, in that case, 4 or 5 of the 7 will be solar eclipses. The minimum number in any one year is 2, both of which will be solar eclipses.

In any particular region of the Earth, total eclipses of the Sun are very rare. Since 1874, for example, South Africa has experienced only one total solar eclipse, on 1 October 1940, and the next will be on 4 December 2002, visible from the northern part of the Kruger Game Reserve.

The Babylonians discovered that eclipses repeat every 18 years 11 days – 6585.32 days to be exact. This period is called the Saros. A Saros consists of 223 lunations (the Moon's synodic month being 29.530 589 days in length). The 0.32 of a day means that the next succeeding eclipse will be located 120 degrees west in longitude. Australia experienced a total eclipse of the Sun on 21 September 1922. If we add 6585.32 days, we come to the date of 1 October 1940, when a total eclipse was visible in South Africa. If another Saros period is added, we come to the date 12 October 1958, when a total eclipse was visible in South America. The next was 23 October 1976 and then 3 November 1994. Adding another Saros to this date brings us to 14 November 2012.

THE author had the opportunity of viewing the total solar eclipse of 1 October 1940, visible from Cape Province, South Africa (Figure 9.5). The umbra of the Moon's shadow moved from west to east over Calvinia, Murraysburg, Cradock and Cathcart. The width of the umbra was 190 km. On a col between two hills west of Murraysburg, there was a view extending 50 km to the west. At 14 h 59 m, the first perceptible bite was taken out of the Sun's disc. Systematically, the Moon covered more and more of the Sun's disc, until totality began at 16 h 8 m 47 s. As the moment came closer, the excitement of the spectators grew. It became appreciably darker and cooler. In the west, the Moon's conical shadow was visible as it hung in the atmosphere. Light and dark bands swept over the ground as the last points of light on the Sun's limb vanished. These points of light, on the limb of the Sun, are called Baily's Beads, named after F. Baily, who, it is reported, saw them first in 1836. Now the Sun was a jet-black disc and spread all around it was the most beautiful pearly, translucent glow of the Sun's atmosphere against the black of the sky. The sight was breath-taking. Known as the corona, it spreads more than

Figure 9.5 Total eclipse of the Sun, viewed from Murraysburg, 1 October 1940

two solar diameters all round the Sun. Fine observation revealed the thin reddish-yellow ring of the chromosphere encircling the black disc of the Moon.

A few of the brighter stars became visible, although some light filtered through from the north and south. The light was about equal to that of full Moon.

After the 3 minutes 38 seconds of totality, one's eyes had become partly adapted to the twilight, and then suddenly, with a blinding flash, the first ray of sunlight burst forth from the lower limb of the Sun. The corona vanished. The thin crescent of sunlight gradually waxed and, an hour later, it was all over.

One should never look directly at the Sun during an eclipse. If a smoke-darkened glass is used, it must be so black that one's hand is invisible directly before one's eyes. Strict precautions have to be taken in using a telescope to view the Sun. A 75 or 100 mm objective lens is ample.

When viewing a faint object, a star diagonal can be used to reflect the image at right angles to the optical axis. The diagonal face of the prism is silvered, and all the light is reflected into the eyepiece (Figure 9.6).

If however, the telescope is used to view the Sun, a Herschel wedge must be used (Figure 9.7). Here the prism is not silvered, and it is turned around so that most of the light goes right through and only a minute amount is reflected from the diagonal face to the eyepiece. The eyepiece must also have a black filter. Number 12 welding goggles are good for this purpose. A mylar filter in front of the objective provides good protection for the optics. It is best to project the image of the Sun on to a sheet of stout, white cardboard in a box to keep out extraneous light.

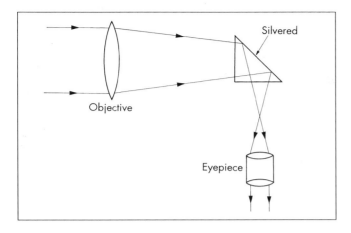

Figure 9.6 Star diagonal for viewing faint objects

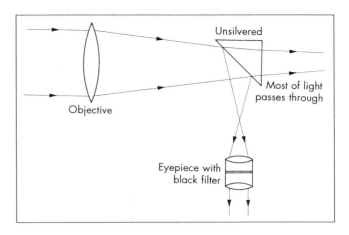

Figure 9.7 Herschel wedge for viewing the Sun

The Sun's Photosphere

The luminous surface of the Sun is called the photosphere. Just about all the visible light of the Sun is radiated from here. The first thing one notices is that the brightness of the photosphere drops off near the limb of the Sun. This shows that the Sun is a gas, of which the temperature decreases with greater distance from the Sun's centre. The photosphere is partly transparent, down to a depth of about 400 km. At the middle of the Sun's disc, most of the light comes from the hotter, lower layers of the photosphere. Near the Sun's limb, where the line of vision is tangential to the surface, most of the light comes from the upper, cooler layers of the photosphere, which therefore appear darker. This effect is very marked in blue and violet light. In the wavelengths of radio, ultra-violet and X-rays, the opposite holds, because these wavelengths originate in the Sun's atmosphere, above the photosphere, where the temperature is higher than in the photosphere. The opacity of the gases in the lower layers of the photosphere is due to the higher temperature that prevails there. Negative hydrogen ions, that is hydrogen atoms that have an extra electron, are the greatest single cause of the continuous absorption of ordinary light waves. However, many frequencies do penetrate the photosphere. The concentration of hydrogen ions decreases very rapidly with height in the photosphere. Measurements of density indicate that the pressure in the lower layers of the photosphere is about 10% that of the Earth's atmospheric pressure at sea level. In the upper layers of the photosphere, the pressure is only 0.01% of that value. Nevertheless, the photosphere has a sharply defined surface, showing that a sudden and marked drop in pressure occurs there.

The Sun's Spectrum

Most of the data about the Sun have been obtained through the spectroscope. The Sun has a continuous spectrum, crossed by dark lines (Figure 9.8, *overleaf*). The dark lines are produced by the absorption of certain wavelengths of sunlight by atoms, the most plentiful of which are listed in Table 9.2. Hydrogen is the most abundant of all the atoms in the Sun's atmosphere, and helium is a fairly good second.

Table 9.2 Atoms in Sun's atmosphere

Element	% mass
Hydrogen, H	73.46
Helium, He	24.85
Oxygen, O	0.77
Carbon, C	0.29
Iron, Fe	0.16
Neon, Ne	0.12
Nitrogen, N	0.09
Silicon, Si	0.07
Magnesium, Mg	0.05
Sulphur, S	0.04
Rest	0.10

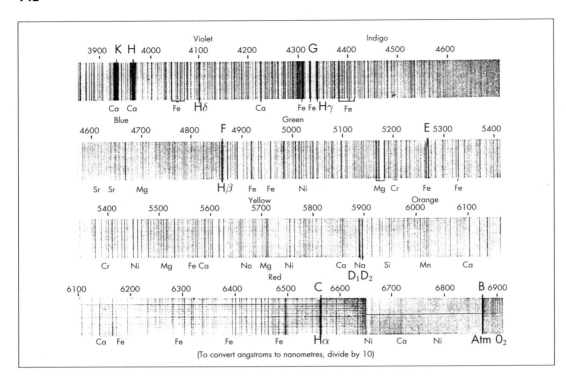

Figure 9.8
The Sun's spectrum

The spectra of stars show that hydrogen and helium comprise most of the matter of which stars are built. This is the surest sign that stars are bodies like the Sun. Put the other way round, the Sun is nothing but a star.

The dark lines in the Sun's spectrum become stronger with height above the photosphere, because the layer of absorbing atoms becomes progressively greater.

The Sun's spectrum has many dark lines due to metals – especially iron. Two lines of ionised calcium, the K and H lines, of wavelengths 393.3 nanometres (3933 Å) and 396.7 nm, are very prominent. Metallic lines also appear in the spectra of stars. The K and H lines of ionised calcium also appear in the spectra of the most distant objects, such as galaxies that are thousands of millions of light years distant. These lines are used to determine the distances of objects that are very far away.

In the yellow part of the spectrum, the D_1 and D_2 lines of sodium, at 589.6 and 589.0 nm, are very prominent. In the red part of the spectrum, the B line of oxygen at 687.0 nm is not due to the Sun, but is formed by oxygen in the Earth's atmosphere. It is one of the telluric lines.

At least 70 of the elements which occur on Earth have been identified in the Sun's spectrum.

In 1868, J. Janssen and N. Lockyer independently discovered that prominences, seen on the limb of the Sun, radiate a certain definite wavelength of light, at 656.3 nm. They discovered this by setting the slit of the spectroscope tangentially to the limb of the Sun. By widening the slit, they were able to obtain the shapes of the prominences. This wavelength is that of the hydrogen alpha line, one of the Balmer series. In Figure 9.8, four of the hydrogen lines of the Balmer series are visible: $H\alpha$ 656.3 nm; $H\beta$ 486.1 nm; $H\gamma$ 434.0 nm; and $H\delta$ 410.2 nm.

In Chapter 4, page 41, it was mentioned that N. Lockyer discovered helium in the solar spectrum, 28 years before it was discovered on Earth.

The Spectroheliograph

In 1892 G.E. Hale and H. Deslandres, also independently, developed an instrument whereby an image of the Sun can be obtained in a single wavelength, for example, the hydrogen alpha wavelength 656.3 nm. Hale and Deslandres' instrument is the spectroheliograph (Figure 9.9).

In the spectroheliograph, the image of the Sun passes through a narrow slit in a receptor plate. The narrow beam of light is then rendered parallel by means of a collimator lens. Next the light passes through a series of high-dispersion prisms which cause the wavelengths to diverge. To gain further divergence, a diverging lens is used. In this

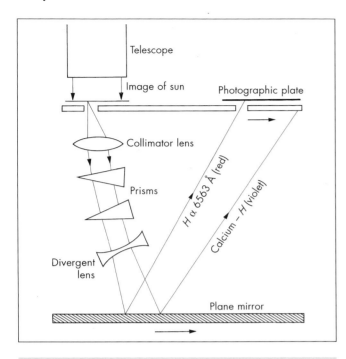

Figure 9.9 The principle of the spectroheliograph

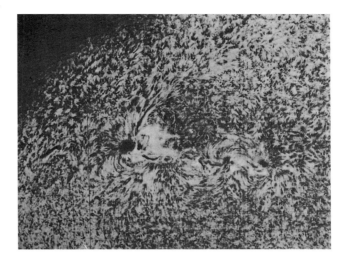

Figure 9.10 The Sun in hydrogen alpha light

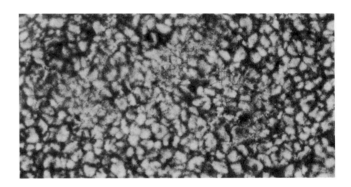

Figure 9.11 Granular appearance of the photosphere

way, the different colours, and therefore the different wavelengths of the narrow slit of sunlight, are separated. After reflection from the plane mirror, the hydrogen alpha wavelength, 656.3 nm, is selected and this ray returns to a second slit in the receptor plate, where it passes through to the photographic plate. This plate is moved in unison with the movement of the scanning slit, so that an image of the Sun is recorded in only one wavelength. The position of the second slit determines which wavelength is selected. Figure 9.10, photographed by the Hale Observatories, shows the Sun in the light of hydrogen of wavelength 656.3 nm. If the K and H wavelengths of calcium are selected, the image of the Sun will be in the violet part of the spectrum. If the second slit is moved further to the right, an image in the ultra-violet will be obtained. If the second slit is moved to the left, an image in the infra-red can be obtained.

Today, filters are being made which let through only one wavelength, thus eliminating the need for the spectroheliograph.

Photographs indicate that the surface of the Sun has a "granular" appearance (Figure 9.11). The average size of a granule is about 2 arc seconds. At the Sun's distance, this means 1000 km. The granular appearance is due to boiling up of hot gases from the lower layers, by convection. At the surface the hot gases move radially outwards and then sink at the edges of the granule, after cooling down. By applying the Doppler effect, it has been found that the hot gases rise at 0.5 km s^{-1}. Because the descending gases at the edges of the granules are cooler, the edges appear darker.

The duration of a granule is about 8 minutes. Groups of 100 or more form super-granules that last 12 to 24 hours. Between the granules there are pointed tongues of flame, the spicules, that shoot up to heights of 10 000 km.

Sunspots

Most prominent in the photosphere are the sunspots, which

Figure 9.12 Sunspots and plage

appear from time to time as black spots (Figure 9.12). They were first seen by J. Fabricius, C. Shreiner and Galileo, round about 1610. However, there are references to sunspots in Chinese annals dating back 2150 years. Sunspots must therefore be a phenomenon which persists.

From 1826 to 1843, S.H. Schwabe made a thorough study of sunspots. He found that they increase and decrease in number and size in a period of 10 years. R. Wolf found that the cycle has a period of 11 years.

The sizes of sunspots vary from scarcely visible, when they are called pores, to 40 000 km in diameter, that is, 1% of the Sun's surface area. A group of sunspots has been known to stretch over 100 000 km. The largest spot known was that of April 1947, which covered 17 900 million km^2; its diameter was 75 000 km.

When a sunspot cycle commences, the spots appear 30 to 40 degrees north and south of the Sun's equator but, as the cycle progresses, the spots appear nearer and nearer to the equator.

By watching the sunspots day by day, it is seen that they move from west to east across the Sun's disc, showing that the Sun rotates from west to east (Figure 9.13).

At the Sun's equator, one rotation is completed in 26.8 days; at 30 degrees north and south, in 28.2 days; at 60 degrees, in 30.8 days; and at 75 degrees, in 31.8 days. This shows that the Sun's surface is a gas. The average period of rotation is taken as 25.38 days. As seen from the Earth, the period of rotation appears to be 27.275 days. This is so because the Earth moves eastwards in its orbit and the sunspots must catch up on the Earth.

If it is assumed that the neutral line between the spots of the two hemispheres marks the Sun's equator, then it appears that the Sun's equator is inclined to the Ecliptic by 7.25 degrees. On 6 June and 6 December the line of nodes (the line joining the points of intersection of the plane of the Sun's equator and the plane of the Ecliptic) points towards the Earth. On those dates, the sunspots move in straight lines from west to east across the Sun's disc. Between these dates, the spots move in curves which depend on the Earth's distance from the line of nodes.

The period of rotation of the Sun can be determined by beaming radar waves to the Sun. The Doppler effect shows that the waves reflected from the western limb, which approaches the Earth, have increased frequency, while those from the eastern limb, which recedes from the Earth, have lower frequency. The radar finding gives the Sun a period of rotation of 26 days at the equator and 37 days near the poles. The sunspots thus rotate 4–5% faster than the Sun does. A point on the Sun's equator moves through 12.4 to 15.4 degrees per day.

In 1769, A. Wilson noticed that a sunspot seemed to be

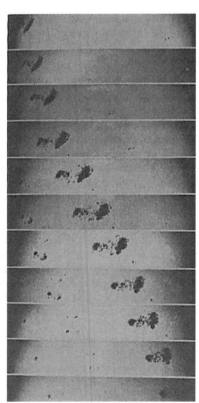

Figure 9.13 Sunspots show rotation of Sun

hollow when it neared the Sun's limb, somewhat like a saucer. This is known as the Wilson effect. It is apparently an illusion. The darkest part of a sunspot, the umbra, is more transparent than the photosphere, so that one can see deeper into the umbra and this gives it the appearance of being hollow. One can see deeper into the umbra since the gases are less dense there, because of the explosive occurrence in a sunspot.

Around the black umbra of a sunspot, there is a semi-shadow (the penumbra), but it is still darker than the photosphere. The penumbra is usually 2.5 times the width of the umbra. In the penumbra there are light and dark stripes, or filaments, radiating from the umbra. Spectroheliograms show that the filaments and fibrils have the same pattern as is obtained when iron filings are sprinkled over a bar magnet. Moving electrons generate a magnetic field, which restricts the movement of the electrons in lines and curves, the lines of force. The solar material gets caught in the lines of force and thus shows striated patterns.

The umbra and penumbra must be cooler than the photosphere, and that is why they appear dark against the background of the luminous photosphere. Measurements made by means of the pyrheliometer show that the temperature of the umbra is 4240 K; that of the penumbra 5680 K; and that of the photosphere 6050 K. Sunspots are therefore dark by contrast with the bright photosphere.

The radiation from the umbra is 30% and that from the penumbra 70% of that from an equal area of photosphere.

In 1908, J. Evershed discovered, by applying the Doppler effect, that the gases in a sunspot flow as shown in Figure 9.14, where U represents the umbra and P the penumbra: the gases rise at the outer fringes of the penumbra and descend in the centre of the umbra.

Sunspots usually occur in pairs, although individual spots are known to have occurred. In 1896, P. Zeeman discovered that the spectrum lines of a source of light are broadened and even split when a strong magnetic field is applied. This is known as the Zeeman effect. H.A. Lorentz proved that the effect of the magnetic field on electrons causes the Zeeman effect.

The two members of a sunspot pair have opposite magnetic polarities: if the leading member is positive, the following member is negative, bearing similarity to the Earth's north and south magnetic poles. In the other hemisphere of the Sun, the poles are reversed. In the umbra of a sunspot, the strength of the vertical component of the magnetic field is between 2000 and 4000 gauss (0.2 to 0.4 tesla). By comparison, the Earth's magnetic field has a strength of 0.3 gauss.

The Sunspot Cycle

Since records have been kept, it has been found that the sunspot cycle varies between 7 and 17 years. Interwoven with this period is a period of 80 years for repetitions of the greatest maximums. E. Maunder found that during the 70 years from 1645 to 1715 there were just about no sunspots. This period is known as the Maunder minimum.

At the start of a cycle, the first spots appear at latitudes of 30 to 40 degrees, north and south of the Sun's equator. As the cycle progresses, the spots appear nearer to the equator – at maximum they are 15 degrees from the equator. As the maximum wanes, the spots get still nearer to the equator and, when minimum is reached, they lie between latitudes 7 and 5 degrees. When the last spots occur, the first spots of the next cycle make their appearance.

If the locations of the spots are plotted on a graph, a characteristic butterfly pattern is obtained (Figure 9.15). The distribution in time and latitude was formulated as a law by G.F.W. Spörer.

In order to quantify the number of spots, the Wolf relative sunspot number (or Zurich number) is used, given

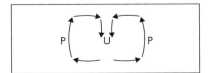

Figure 9.14
Gas flow in a sunspot

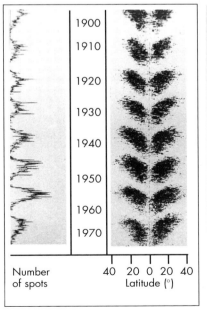

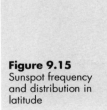

Figure 9.15
Sunspot frequency and distribution in latitude

by: $R = K(f + 10\ G)$, where f is the number of separate spots and G is the number of groups. K is a personal factor, depending on visual acuity. If, for example, there are three groups of spots, consisting of 3, 6 and 10 separate spots, then $f = 19$ and $G = 3$.

Therefore $R = K(19 + 10 \times 3) = 49$, if $K = 1$. At another time, there may be four groups of 2, 2, 3 and 4 each, so that $f = 11$. Then $R = K(11 + 10 \times 4) = 51$, if $K = 1$. (K will be greater than 1 if the individual's sight is weaker than normal.)

These two examples, consisting of 19 and 11 separate spots, nevertheless have almost the same Wolf relative number. A more accurate method would be to measure the total area which the spots comprise, although it would take up more time.

Amateur astronomers, especially in countries with good sunshine, can make valuable contributions by monitoring sunspots on a daily basis, and only a small telescope is necessary.

At the end of the 11-year cycle, the magnetic polarity reverses. Thus 22 years elapse before the original pattern is repeated. The reversal of polarity actually commences a year or so after maximum. The average length of time for maximum to be attained is 4.6 years; and the time to reach minimum 6.7 years.

The total duration of a group of sunspots is somewhat less than two weeks, although spots lasted more than 100 days in 1943.

Sunspot Theory

H.W. Babcock and R. Leighton put forward the theory that the Sun has a weak general magnetic field which is "frozen"

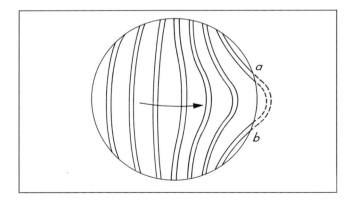

Figure 9.16 Sun's magnetic tubes of force

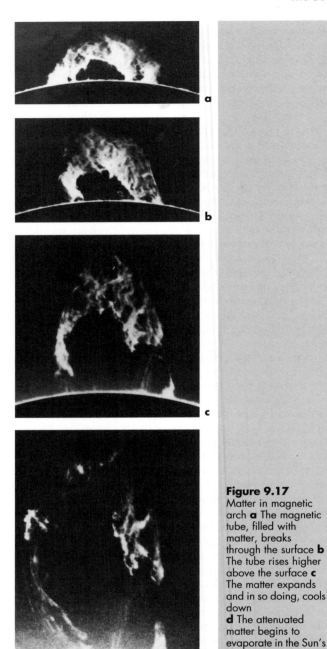

Figure 9.17
Matter in magnetic arch **a** The magnetic tube, filled with matter, breaks through the surface **b** The tube rises higher above the surface **c** The matter expands and in so doing, cools down
d The attenuated matter begins to evaporate in the Sun's atmosphere. On the surface the dark spots

into the matter of the Sun. If the Sun did not spin on its axis, the magnetic lines of force (or tubes of force) would run parallel from one pole to the other. On account of the Sun's differential rotation, the tubes of force undergo

The Chromosphere

Figure 9.18 Quiescent prominence

Figure 9.19 Magnetic arches

The Chromosphere

Just above the photosphere, there is a layer of rarefied gas, about 10 000 km thick. This is called the chromosphere; it is visible at the start and finish of a total eclipse of the Sun, when it is seen to have a slightly red tinge. That is why it is called the chromosphere, meaning sphere of colour.

If the slit of a spectroscope is set tangentially to the limb of the Sun, it will be noticed, during a total eclipse, that the chromosphere is not homogeneous. Sharp tongues of flame, the spicules, are to be seen; they occur at spots where a magnetic tube has made a single rupture of the Sun's surface. They have widths of 1000 km and lengths of 10 000 km. They also have high temperatures, between 10 000 K and 20 000 K. These jets of flame shoot out at speeds of 20 to 30 km s^{-1} and last for 5 to 10 minutes. The spicules are very bright near the Sun's limb, but against the bright disc of the Sun they appear dark. Calculation shows that there are about 500 000 spicules active at any one time, over the whole surface of the Sun. They are very prevalent between super-granules.

The chromosphere exhibits a network, corresponding to the granular appearance of the photosphere. There are also emission bubbles with dark rims, which occur up to the top of the chromosphere.

The orbiting space laboratory, Skylab, found that the chromosphere is rather darker above the Sun's poles and that there are voids in the corona, above the chromosphere. The solar wind, consisting of electrified parties, streams out from these voids.

At the poles of the Sun, the spicules are twice the length, hotter, at 50 000 K, and last 40 to 50 minutes.

The Sun's magnetic tubes of force are very concentrated above the poles. In these regions the spicules preponderate.

The north and south magnetic polarities at the Sun's poles have a strength of 5 to 6 gauss. The Sun does not, however, possess a simple, dipole magnetism, but a complex arrangement of poles.

In monochromatic light, the chromosphere displays filaments, 1000 to 2000 km thick, and which spread over distances as great as 10 000 km. These filaments are a property of the Sun's magnetism.

The chromosphere has prominences, flares and plages. The latter two are very bright areas of flame in the neighbourhood of sunspots.

The spectrum of the chromosphere shows many emission lines. As many as 3500 have been counted. The lines are faint because the chromosphere is very transparent, and thus a poor absorber and radiator of energy. Many of the spectral lines coincide with lines from the photosphere, showing that the same substances occur in both the photosphere and the chromosphere. Lines of

stretching near the equator (Figure 9.16). The first concentration of tubes of force occurs at latitude 40 degrees, where the magnetic field then becomes concentrated. The tubes bulge towards the surface, where they eventually break through, forming an arch in the photosphere, of matter being hurled upwards. This is then a sunspot. The expansion causes the matter to cool down, so that the umbra appears dark against the bright background of the photosphere. The pair of spots represented by *a* (in the northern hemisphere) and *b* (in the southern hemisphere) are each the location of an arch of material thrown up above the photosphere. Such an arch is vividly portrayed in the series of photographs of Figure 9.17. After reaching a certain height above the photosphere, the matter falls back to the surface, forming the characteristic prominences. The arches thrown up contain filaments which consist of matter constrained by the magnetism. These filaments have the form of iron filings strewn over a bar magnet. Matter dropping back to the Sun's surface is shown in Figure 9.18 of a quiescent prominence. Figure 9.19 shows tightly constricted magnetic arches.

As the sunspot cycle progresses, the concentration of magnetic tubes increases, and more and more spots appear.

The following spots of each pair tend to move nearer to the poles. Afterwards some magnetic tubes break below the surface, thus bringing about a reversal of the polarity.

Babcock and Leighton's theory does not explain how the Sun's magnetism originates, except by ascribing it to the magneto effect of fast-moving electrons. Moving electrons constitute an electric current, and an electric current always has a concomitant magnetic field.

ionised helium and some metals that are not present in the spectrum of the photosphere show that the temperature of the chromosphere is much higher than that of the photosphere. The upper layers of the chromosphere show spectral lines of singly ionised helium, of wavelength 30.4 nm. These indicate a temperature of 80 000 K. There are also lines of triply ionised oxygen, indicating a temperature of 100 000 K. The temperature increases with height in the chromosphere at a rate of 200 K km^{-1}, and it even reaches 500 000 K, while the density decreases dramatically to one ten-thousand-millionth of the density of the Earth's atmosphere at sea level. It could almost be called a vacuum.

Prominences

Closely connected to the photosphere and the chromosphere is the phenomenon of prominences. The first successful photograph of a total eclipse, showing prominences, was taken by Berkowski in 1851.

Prominences originate in sunspots, owing to the bursting out of magnetic tubes. They sometimes have beautiful shapes and usually reach heights of about 40 000 km. Figure 9.20 shows a prominence that leapt to a height of 380 000 km above the Sun's surface.

Prominences can have a breadth of 200 000 km and thicknesses of 8000 km. The temperatures of prominences reach 30 000 K, as compared with the 6050 K of the photosphere. The high temperatures are generated by the effect of the magnetic field on the particles in the prominences. The matter in the prominences is 100 times denser than the surrounding matter, because the magnetism squeezes the matter. During total eclipses of the Sun, prominences appear as white, because they are so hot. Space probes that photograph prominences in ultra-violet light and in X-rays see them as shadows.

Compared with the bright disc of the Sun, the prominences have a dark appearance, but when viewed on the Sun's limb they appear bright. The reason is that prominences largely consist of neutral, unionised hydrogen, which radiates on the wavelength of 656.3 nm of the hydrogen alpha light.

There are also eruptive prominences which eject matter great distances from the Sun, up to 60 000 km and more at speeds of 200 km s^{-1}. These prominences vanish rapidly into space. Other prominences have been known to last for a year.

The shapes of prominences vary from appearances such as "rain, funnels, loops, arches, tree trunks, hedges, hanging umbrellas to jet flames".

Figure 9.20 Huge prominence, 380 000 km high

Arch-shaped prominences link areas of opposite magnetic polarity. In these arches, matter flows downwards towards both magnetic poles. Eruptive prominences originate in flares that eject dangerous radiation at all wavelengths. They are most easily observed in the frequencies of hydrogen and calcium K lines.

Hydrogen alpha frequency shows that flares originate in bright points in plages. *Plage* is a French word meaning sea sand. A plage looks like bright, white sea sand. Plages always occur in the neighbourhood of sunspots, being on both sides of the neutral line linking the spots. The temperature in the centre of a flare can reach one thousand million (10^9) K. The magnitude of a flare can be enormous, thousands of millions of tons of material being hurled into space at speeds of 2000 km s^{-1} and ballooning in the chromosphere and corona. The oscillations engendered radiate radio waves. Flares always originate in complex magnetic fields. One flare can produce one-hundred-thousandth of the total energy output of the Sun. Measurements have shown that they can produce a flood of energy of 10^{21} joules per second, enough to boil away

4470 million tons of freezing water.

The Corona

The corona can stretch over 20 solar radii. Previously, the corona could be observed only during total eclipses of the Sun. The invention of the coronagraph in 1930 by B.F. Lyot made it possible to observe the corona at any time. The coronagraph contains a series of black, ring-shaped shields, which absorb scattered light. In front of the field lens of the eyepiece there is a black, conically shaped disc, which prevents any light from the photosphere from reaching the eyepiece. There is also a black occulting disc and a filter. Only the light of the corona can pass through to the eyepiece and the photographic plate.

In the lower layers of the corona are arch-shaped condensations, five to ten times denser than the corona. This condensed matter moves downwards towards the sunspots because of the constricting effect of the magnetic tubes of force. The spectrum of this part of the corona is continuous. A little higher up, the dark Fraunhofer lines begin to be visible and, higher up, emission lines appear.

At a height of 500 km above the photosphere, the temperature is 4200 K, but it rises to a million degrees in the corona, which stretches into space, beyond the chromosphere. Space probes have measured temperatures of $1\frac{1}{2}$ million degrees Kelvin. One may wonder how it is possible that the corona can have a higher temperature than the photosphere, considering that it is further from the source of heat, and if this is not in conflict with the Second Law of Thermodynamics. This law states that energy can only flow from high level (high temperature) to low level (low temperature). How can energy then flow from the photosphere at 6050 K to the corona at a temperature of $1\frac{1}{2}$ million degrees? The answer is that the energy level does not only depend on temperature. The total energy is greater in the photosphere than in the corona because of the higher density of matter in the photosphere, namely a million times. The Sun's atmosphere (chromosphere and corona) are very transparent to optical and infra-red radiation. The temperature is not derived from these radiations, but is the result of kinetic energy, that is energy of motion, of the electrically charged particles in the corona. These particles move at relativistic speeds, that is speeds close to the speed of light, and this is the source of their energy.

Sound waves of low frequency (like rumbling thunder) that originate in the photosphere and chromosphere as a result of the sudden drop in density in the corona, accelerate and become supersonic shock waves, causing highly energetic collisions between the particles in the corona, and this causes the temperature to rise to great heights.

Magnetohydrodynamic waves of the Sun's magnetic field also cause shock waves, which contribute to the rise in the temperature of the corona.

In 1869, very noticeable green lines were seen in the spectrum of the upper layers of the corona, which did not correspond to green lines of any known element on Earth. The supposition was made that, as in the case of helium, an unknown element, coronium, must be responsible for the green lines. In 1940 it was shown that these green lines are due to highly ionised atoms of ordinary metals, under very low pressures but at very high temperatures. These spectral lines are known as forbidden lines because they represent electron transitions (leaps) which are contrary to the rules of selection of quantum transitions. The spectral line of iron at 530.286 nm in the spectrum of the corona is due to an iron atom which has lost no fewer than 13 of its electrons. This ionised iron atom is designated as FeXIV, because neutral iron is written as FeI. The existence in the corona of highly ionised nickel and cadmium shows that that portion of the corona must be at a temperature of one thousand million degrees Kelvin.

The abundances of the elements in the corona correspond to those of the photosphere but their degrees of ionisation are different.

The shape of the corona changes as the sunspot cycle progresses. At minimum of the sunspot cycle, the corona is symmetrical and has long plumes above the Sun's equator. At maximum, the plumes are spread over the whole solar surface.

In 1980 the orbiting Skylab discovered certain transient phenomena in the corona: gigantic loops which were hurled into space. They usually originate in flares and do not fall back to the Sun; their velocities are between 200 and 900 $km\,s^{-1}$. These masses of ejecta are very prevalent during sunspot maxima. One of these ejecta was measured by Skylab to have a mass of 5 thousand million tons and a velocity of 470 $km\,s^{-1}$. Its total energy was 10 times that of the flare in which it originated; this is ascribed to the magnetic fields which accelerate the ejecta. The whole corona is brought into commotion by such rapidly moving masses of gas.

Skylab also discovered voids in the corona. They appear as dark blotches on ultra-violet and X-ray photographs, which can only be taken from outside the Earth's atmosphere. These voids occur frequently above the Sun's poles. The electrified particles which stream out through these voids constitute the solar wind. As long ago as 1900, O. Lodge drew attention to the possible existence of highly ionised atoms, stripped of their electrons. In 1958, E.N. Parker, while studying the tails of comets, discovered that the corona expands while it loses mass; he called this the solar wind.

The Solar Wind

The high temperature of the corona and the pressure of radiation provide the power behind the solar wind. The pointed tongues of flame (the spicules) augment the outward forces by casting matter into the corona. The particles hurled out from the Sun take 4 to 5 days to reach the Earth.

The space probe Mariner-2 found, on its way to Venus, that the solar wind consists of more or less equal numbers of electrons and protons, that is, hydrogen atoms which have been stripped of their electrons. The solar wind also contains small amounts of heavier atoms such as helium, which make up 4 to 5% of the total.

Near to the Earth, these particles of the solar wind have temperatures as high as 10 000 K, measured according to their kinetic energies (energies of motion).

Apart from the solar wind, the Sun loses about 4 million tons of mass per second, on account of the nuclear reactions in its core. The Sun has been doing this for about 5 thousand million years and it can go on at the present rate of mass loss for another 30 million million (3×10^{13}) years. This is 5000 times longer than the Sun can go on existing in its present state.

Interplanetary probes have found that the Sun's magnetic field spirals outwards into space (Figure 9.21).

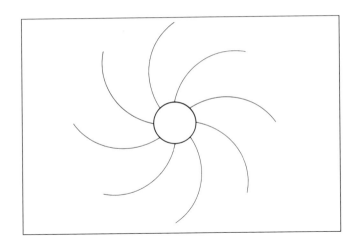

Figure 9.21 The Sun's magnetic field in space

These spiral "spokes" rotate with the Sun's rotation in a counter-clockwise direction as seen from the north. At the distance of the Earth, the magnetic lines of force make an angle of 45 degrees with the line joining the Earth and Sun.

On the Sun's side of the Earth, the Earth's magnetic field stretches as far as 8 to 10 Earth radii, but on the side away from the Sun, the magnetic field forms a long tail

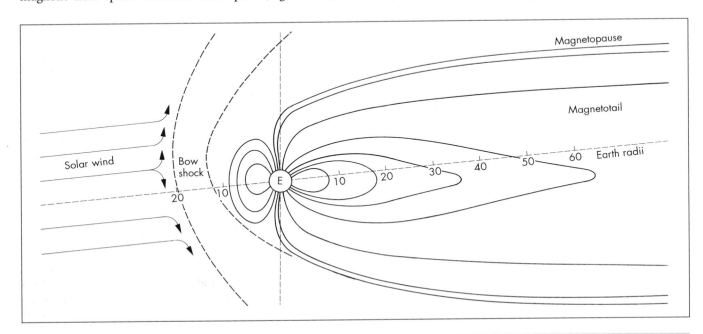

Figure 9.22 Effect of the solar wind on Earth's magnetosphere

The Sun's Radiation Energy

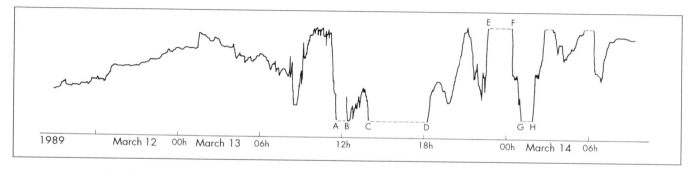

Figure 9.23 Magnetic storm on 13 March 1989 at Edenvale, Transvaal

(Figure 9.22), sweeping far beyond the Moon's orbit. It has been traced up to a distance of 1000 Earth radii.

The limit of the Earth's magnetosphere is known as the magnetopause. The Sun's lines of force fold themselves around the magnetosphere. Near to the Earth, the solar wind moves at eight times the velocity of sound and the electrified particles pile up against the magnetopause, forming the bowshock. The kinetic energy which the particles lose here is transformed into heat, so that the temperature of the plasma rises to one million degrees at the bowshock. Between the bowshock and the magnetopause, there is a region, known as the magnetosheath, containing distorted lines of force.

The particles emitted by a solar flare pile up against the magnetosphere and thereby increase the strength of the Earth's magnetic field. Figure 9.23 shows the magnetogram obtained by M.D. Overbeek, whose magnetometer is situated in Edenvale, Transvaal, South Africa. On 13 March 1989 there was a sudden and tremendous increase in the strength of the magnetic field, so much so that the needle of the magnetometer was deflected right off the scale at the points between A-B, C-D, E-F and G-H. The 50-mm displacement of the needle was due to an increase of 1000 gammas. The released magnetic energy accelerates the electrified particles, thus causing disturbances in radio communications. This is known as a magnetic storm. During the storm of 13 March 1989, the Aurora Australis (Southern Lights) was seen in Mid-Natal at a latitude of 29 degrees, which is no less than 61 degrees north of the south pole. This had never happened before in living memory.

The electrified particles of the solar wind become trapped in the magnetic lines of force and give rise to the Van Allen radiation belts (see Figure 8.24, page 115). Certain very energetic protons and electrons penetrate the Van Allen belts and they, together with ultra-violet rays and X-rays, ionise the upper layers of the Earth's atmosphere. The layer of air at which this happens is called the ionosphere. It is a good reflector of radio waves of wavelengths between 10 and 100 metres. This is detrimental to astronomical research on radio sources in space. The large flares produced on the Sun's surface therefore disturb radio reception. During maxima of the sunspot cycles, these disturbances can occur 10 times a year. During sunspot minima these disturbances do not occur.

Between 15 and 50 km above the Earth's surface, the ultra-violet rays in sunlight (210–310 nm in wavelength) dissociate oxygen molecules (O_2). The free oxygen atoms tend to recombine in threes to form ozone (O_3). The ozone absorbs ultra-violet rays, which are injurious to life, not only causing sunburn but also skin cancer. Certain solar protons ionise nitrogen atoms. They readily combine with the free oxygen atoms to form nitric oxide (NO). The nitric oxide tends to break down the ozone, which is therefore in a delicate state of balance.

Energetic solar particles which penetrate within 10 to 20 degrees of the magnetic poles cause oxygen to glow by absorbing the wavelength 557.7 nm, so that a greenish glow appears. This is known as the Northern and Southern Lights, or Aurora Borealis and Aurora Australis, which often appear during sunspot maxima.

The Sun therefore exercises a continuous influence on the Earth and the other planets.

The Sun's Radiation Energy

There is a theoretical concept known as "black body", which is able to absorb all the radiation falling on it, and which re-radiates the energy, without loss. Such a body is therefore in thermal equilibrium with its environment. The amount of energy which a black body radiates depends solely on its temperature. Each temperature has a definite wavelength of maximum emission. W. Wien derived a law

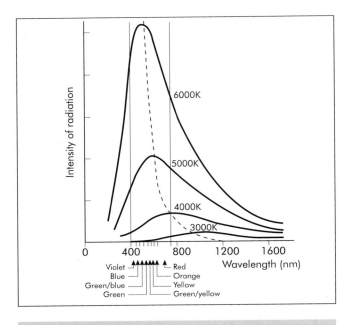

Figure 9.24 Graph of intensity of radiation against wavelength

which states that the wavelength at which maximum energy is radiated is inversely proportional to the absolute temperature of the radiating body, that is, the higher the temperature of the body, the shorter the wavelength of maximum emission of energy. The maxima are indicated in Figure 9.24 by the broken curve.

If, for example, a piece of metal is heated, it starts to glow dull red. As its temperature rises, the glow becomes orange, then yellow and eventually white and blue-white. At room temperatures, most energy is radiated in the infra-red at wavelengths of 10 microns (10 000 nm).

A black body at the temperature of the Sun radiates maximally on a wavelength of about 550 nm, in the yellow part of the spectrum. When peak emission of a black body takes place in the violet portion of the spectrum, the wavelength of maximum emission is about 400 nm.

The Sun and stars can be considered to approximate black bodies. The effective temperature of the Sun (or a star) is the temperature which an equally large black body must have to be able to radiate the same total amount of radiation as the Sun (or star). By this assumption the Sun's temperature works out at 5780 K, which has a peak of emission in the middle of the visible spectrum. The value is in good agreement with the measured temperature of 6050 K.

The greatest proportion of the radiation at a given wavelength originates in that particular layer of the Sun, or its atmosphere, in which the opacity to that wavelength ceases. Nearer to the core, that radiation is absorbed and, higher up, the Sun's atmosphere becomes transparent to that wavelength. The Sun's visible light originates in the photosphere; the ultra-violet radiation comes from the chromosphere and short radio waves, from just above the photosphere. Longer radio waves come from the lower layers of the corona.

When the Sun is normal, that is inactive, its energy flux fits that of a black body at 6000 K. At ultra-violet and X-ray frequencies, the Sun's radiation curve deviates from that of a black body. The strongest ultra-violet rays have a wavelength of 121.6 nm.

The Sun is a weak radiator of radio waves. At times its radio radiation gets stronger owing to electrons being accelerated by magnetic lines of force, that is, by synchrotron radiation, which has been produced on Earth by means of the synchrotron which accelerates particles to relativistic speeds.

The Sun's infra-red radiation comes from the photosphere and the lower layers of the chromosphere. The radiation curve fits that of black bodies at temperatures between 4000 K and 6000 K.

The Solar Constant

The solar constant is the total energy received from the Sun per second per square metre, outside the Earth's atmosphere, when the Earth is at its mean distance from the Sun. It is equal to 1370 watts per squre metre. 99% of the Sun's radiation takes place between wavelengths of 300 nm in the ultra-violet and 6000 nm in the infra-red. The remaining 1% is mainly in the ultra-violet between 300 nm and 120 nm.

The solar constant is constant to within 1% although it does vary over long periods; variations of 0.1% over a few days having been measured. Two decreases of 0.2% have been found to coincide with the appearance of large sunspots. A change of 0.5% to 1%, over a period of a century, could have a considerable effect on the Earth's climate, possibly leading to an ice age. A protracted decrease of 2.5% in the solar constant could plunge the Earth into a permanent ice age.

Standard Model of the Sun

The mass of the Sun, 1.9891×10^{30} kg (332 700 times the mass of the Earth) is not uniformly distributed in the Sun. The overlying layers exercise pressure which increases with

The Sun's Energy

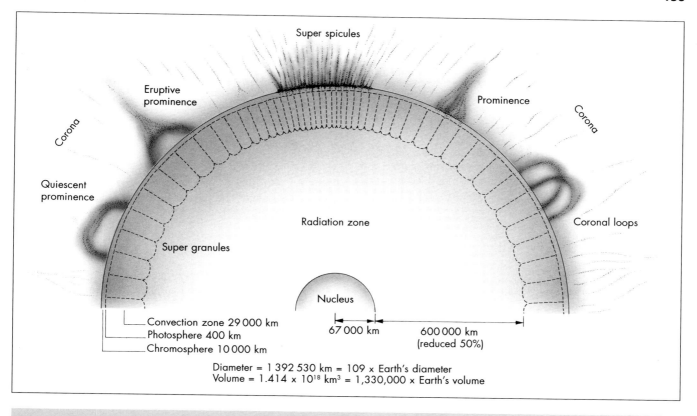

Figure 9.25 Standard model of the Sun

depth below the surface. At the Sun's centre the density is 98 g cm^{-3} (= 8.67 times the density of lead). The pressure reaches a value of 2×10^{11} Earth atmospheres, or 2×10^{16} N m^{-2}. This incredible pressure raises the temperature of the centre to 15 million degrees Kelvin.

The nucleus of the Sun, within 67 000 km of the centre (= one-tenth of the Sun's radius), contains half of the Sun's total mass. Here, all the atoms are totally ionised, that is, stripped of their electrons. This mixture of atomic nuclei and electrons is called a plasma. The atomic nuclei and electrons move about freely.

Above the nucleus of the Sun, there is a layer, 600 000 km thick, consisting mainly of hydrogen with an admixture of 10 to 20% of helium and 1% of other atoms, according to mass. This huge layer is called the radiation zone.

Above the radiation zone is a layer, 28 to 30 thousand km thick, in which the matter is constantly "boiling". It is called the convection zone (Figure 9.25).

Above this layer is the photosphere, 400 km thick; it is the visible surface of the Sun.

Above the photosphere is the chromosphere, about 10 000 km thick.

Beyond the chromosphere is the corona, which can reach to a distance of 20 solar radii.

The Sun's Energy

The great pressure and high temperature in the Sun's centre are sufficient to make the protons fuse together to form heavier atoms. The force of repulsion between the positive electric charges of the protons is overcome by a positive charge leaving the proton and coming free as a positive electron (positron). The proton which has lost a positive charge has thus become a neutral particle – a neutron. Together with a proton, the neutron forms a new nucleus, consisting of 2 mass units – a deuteron (or deuterium). A massless particle is shot out – a neutrino; it has no electric charge and apparently no mass, but does possess spin, and does not easily interact with matter.

A proton is a nucleus of hydrogen having mass of 1 and electric charge of 1. It is written as $_1^1H$. A deuteron consists of 1 proton and 1 neutron. Thus its mass is 2 but its electric charge is 1. It is written as $_1^2H$. The proton–proton reaction can thus be represented by

$$_1^1H + _1^1H \rightarrow _1^2H + \text{positron} + \text{neutrino}$$

(A positron can be written as e^+, an electron as e^- and a neutrino as ν.)

Although neutrinos have energies of 0.42 million electron volts (= 0.42 MeV), they seldom interact with matter and are difficult to trace. A positron and an electron will annihilate each other and form gamma rays of high frequency.

Two protons can also fuse together:

$$_1^1H + e^- + _1^1H \rightarrow _1^2H + \nu \quad (1.44 \text{ MeV})$$

The liberated neutrino carries away 1.44 MeV of energy.

The deuterons thus formed fuse with protons to form helium nuclei of mass 3 and electric charge 2. Energy is liberated as gamma rays, γ:

$$_1^2H + _1^1H \rightarrow _2^3H + \gamma$$

Two of these helium nuclei of mass 3 are then forced to fuse together by the great pressure to form a helium nucleus of mass 4 mass units. Two protons are set free:

$$_2^3He + _2^3H \rightarrow _2^4He + _1^1H + _1^1H$$

The process is now repeated, using the liberated protons. To arrive at this reaction six protons were consumed. Four of them have been bound in the helium nucleus and two set free. This reaction is responsible for the production of 99% of the Sun's energy.

Hans Bethe and Carl von Weizsäcker proposed this series of reactions in 1938, before the first nuclear bomb had proved that enormous amounts of energy are liberated by the fusion reaction.

These reactions take place in a billionth of a second. The energy liberated can be calculated as follows. The mass of the four protons which are consumed is 4.032 52 atomic mass units (amu). The mass of the helium $_2^4He$ that is formed is 4.003 89 amu. There is thus a nett loss of 0.028 63 amu, that is 0.7%. This mass lost is converted into energy, as Albert Einstein set out in his Theory of Relativity in 1905. He proved that mass and energy are equivalent: mass is a form of energy, and vice versa. The relationship between mass and energy is given by Einstein's famous equation:

$$E = mc^2$$

In this equation E is the energy in ergs, m is the mass destroyed and c is the velocity of light in cm s^{-1}.

When 4.03252 grams of solar material are converted into helium, the amount of energy liberated is given by:

$$E = mc^2$$
$$= 0.028\,63 (2.997\,92 \times 10^{10})^2$$
$$= 0.2573 \times 10^{20} \text{ ergs}$$
$$= 0.2573 \times 10^{13} \text{ joules}$$
$$= 0.2573 \times 10^{13} \div 4.18 \text{ calories}$$
$$= 0.615\,55 \times 10^{12} \text{ calories}$$

It takes 636 calories of heat to convert 1 gram of water at 0°C into steam. The mass of ice-cold water which can be converted into steam by $0.615\,55 \times 10^{12}$ calories is equal to:

$$0.615\,55 \times 10^{12} \div 636$$

or 9.6785×10^8 grams

The volume of this mass of water is 9.6785×10^8 cm^3. This volume is, in round figures, the volume of a dam of water 35 metres in diameter and 1 metre deep. This volume of ice-cold water will be converted into steam by the energy liberated from 4 grams of solar material which is converted from protons to helium, and then the dam will be empty!

The Sun loses 4.5 million tons of matter per second as a result of its nuclear reactions which convert matter into energy. It can go on doing this for 5 or 6 thousand million years before showing much change.

There are other nuclear reactions in the Sun and stars whereby heavier nuclei are formed and energy liberated.

Helium-3 can combine with helium-4 to form beryllium-7, whose charge is 4:

$$_2^3H + _2^4He \rightarrow _4^7Be + \gamma$$

The beryllium then fuses with an electron. The negative charge of the electron will neutralise the positive charge of one proton, converting it to a neutron. Because the positive charge has been reduced from 4 to 3, the nucleus has become a nucleus of lithium-7 with charge 3. A neutrino is set free:

$$_4^7Be + e^- \rightarrow _3^7Li + \nu$$

The next step is that the lithium-7 nucleus fuses with a proton. The nucleus splits into two helium-4 nuclei, each having 2 protons and 2 neutrons:

$$_3^7Li + _1^1H \rightarrow _2^4He + _2^4He$$

Instead of fusing with an electron, the beryllium-7 nucleus can fuse with a proton. The nucleus, containing five protons, is a nucleus of boron. Gamma rays are set free:

$$_4^7Be + _1^1H \rightarrow _5^8B + \gamma$$

The boron-8 nucleus is unstable and it ejects a positron, thus losing a positive charge. The nucleus now contains four protons, although the mass is still 8 amu because a proton has changed into a neutron. It is then a nucleus of beryllium-8. A neutrino (energy 14.04 MeV) is freed:

$$^{8}_{5}B \rightarrow \,^{8}_{4}Be + e^+ + \nu$$

The beryllium-8 is unstable and it readily splits into two helium-4 nuclei:

$$^{8}_{4}Be \rightarrow \,^{4}_{2}He + \,^{4}_{2}He.$$

In all these reactions the end result is helium-4 or $^{4}_{2}He$, a very stable nucleus, which is the ash of the fusion of protons.

The gamma rays produced by the annihilation of a positron and an electron have only 1/52 of the energy of the proton fusion reaction, in which only 1/140 of the matter is annihilated. If all the mass of the protons were annihilated, the energy liberated would be 140 times greater. The gamma rays are the main source of the radiation which the Sun pours out. The frequency of these rays is about 10^{25}.

The Sun in Balance

The gamma rays produced in the core of the Sun are absorbed by atoms in the radiation zone and re-emitted, but each time at a lower frequency and longer wavelength. The mean free path of the gamma rays is about 1 cm. It takes a very long time before these photons reach the photosphere – 10 000 to 100 000 years! By the time the photons reach the photosphere, their frequencies have been reduced from 10^{25} to 10^{14}, namely to visible light of wavelengths between 400 and 700 nm.

The outward radiation pressure, together with the gas pressure, oppose the gravitational force which tends to crush the Sun (or a star) to a point. In this way a balance is maintained between the outward and inward forces. Calculations show that the Sun has been in a state of balance for about 5 thousand million years. Radioactive dating shows that the moonrocks are 4.6 thousand million years old. The planets must, therefore, have been formed by accretion, simultaneously with the Sun, or just after.

The original composition of the Sun was probably 78% hydrogen, 20% helium and 2% carbon, nitrogen and oxygen. Owing to nuclear reactions, the hydrogen in the Sun's core now constitutes only 36%, and the remainder is helium ash. Outside the core, the constitution is approximately the same as originally.

Neutrinos

Neutrinos travel at the speed of light and can easily penetrate thousands of kilometres of lead. Apparatus which has been set up deep down in mines, in order to shield it from other radiations, contains large amounts of perchloroethylene (C_2Cl_4; detergent). The chlorine atoms of mass 37 amu in the detergent react with neutrinos, the chlorine becoming argon-37. The argon-37 is radioactive and can be separated from the liquid. By means of the mass spectrograph the argon atoms can be counted; it has been found that only one-third of the number of neutrinos which were expected has reacted with the chlorine-37 atoms in the detergent.

At present there is no explanation for this anomaly. Astrophysicists are grappling with the problem in order to refine our knowledge of the nuclear reactions which take place in the Sun and stars.

Chapter 10

Stars and Double Stars

1 Single Stars

The stars do not appear to be of equal brightness. About 125 BC, Hipparchus made the first classification of stars according to their brightnesses. His predecessor, Aristarchus (310–230 BC), had proved that the Sun, Moon and planets are at varying distances from the Earth, and that they do not move equally far from the Earth against a crystal sphere, as Aristotle had stated.

The differences in brightnesses of the stars can be ascribed to two factors: intrinsic differences in the stars themselves, and different distances from the Earth.

We have already seen that the Earth's radius subtends an angle of 8.79″ at the distance of the Sun. The Earth's radius, and even its diameter, are too small to subtend measurable angles at the distances of the stars. From extremities of the Earth, the directions to any star are parallel.

The diameter of the Earth's orbit around the Sun gives a base line of 2 × 149 597 870 divided by 12 756 = 23 455 times longer than the Earth's diameter.

As a result of the Earth's annual revolution around the Sun, the nearer stars ought to describe small ellipses against the expanse of the background stars, ellipses that are mirror-images of the Earth's elliptical orbit (Figure 10.1).

As the Earth moves counter-clockwise, from *a* to *b*, to *c*, to *d*, a nearby star ought to show movement from *1* to *2*, to *3*, to *4* (Figure 10.1). The naked eye cannot perceive any such movement. Tycho Brahe based his opposition to Copernicus' system on this fact. The refinement brought about by using a filar micrometer in the eyepiece of the telescope made it possible to determine the positions of stars accurately to a fraction of an arc second . The stars did show parallax on an annual basis. The idea of parallax can be illustrated as follows: stretch the arm out and hold the

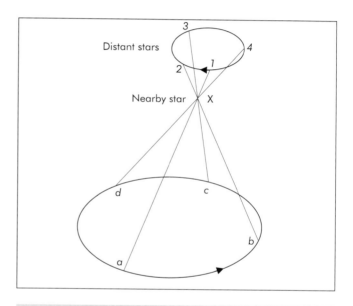

Figure 10.1 The nearer stars show parallax

index finger erect; then look at the finger with the left and right eyes alternately. The finger will be seen to jump to and fro from right to left against the background. The base line between the eyes does subtend an angle at arm's length. The annual shift in position of a nearby star is used to determine its distance. The principle is illustrated in Figure 10.2 (*overleaf*).

When the Earth is at A in its orbit, the difference in direction (angle *1*) between a bright, nearby star, X, and a faint, distant star, C, is measured. Six months later, when the Earth is at B, the difference in direction is again

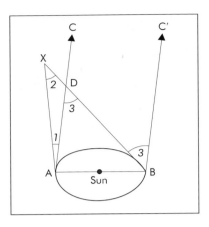

Figure 10.2 Determining the distance of a nearby star X

heliocentric direction SX. AS is the Earth's mean distance from the Sun, 149 597 870 km = r. If SX is the distance of the star from the Sun = d, and angle SAX is the angular distance of the star from the Sun's direction, then

$$\frac{\sin AXS}{AS} = \frac{\sin SAX}{SX} \quad \text{(Appendix 1, section 1.2)}$$

i.e $\quad \dfrac{\sin p}{r} = \dfrac{\sin SAX}{d}$

Thus $\sin p = \dfrac{r}{d} \sin SAX$ \hfill (1)

The greatest value of the sine of an angle is 1 when the angle is 90 degrees. When angle SAX = 90 degrees (as in the figure)

$$\sin SAX = 1 \text{ and } \sin p = r/d$$

The heliocentric annual parallax is indicated as:

$$\sin P = r/d$$

Thus $\sin p = \sin P \sin SAX$ \hfill [from (1)]

Since the angles P and p are very small (less than 1 arc second), their sine-values are equal to their angular measures (see Appendix 1, section 1.8).

$$\text{Thus } p = P \sin SAX$$

If the circular measures of P and p, in arc seconds, are represented by P'' and p'' respectively, then

$$p'' = \frac{180 \times 60 \times 60}{3.141\,59} \times \frac{r}{d} = 206\,265 \frac{r}{d} \quad (2)$$

$$\text{and } p'' = P'' \sin SAX$$

$$\text{so that } P'' = p'' \div \sin SAX$$

measured (angle 3). Because the distant star is so far away, it has zero parallax. Thus the lines AC and BC′ are parallel. Therefore angle ADB is equal to angle C′BD = 3. The exterior angle ADB of triangle AXD is equal to the sum of the interior, opposite angles DAX plus AXD, that is angle *1* plus angle *2* = angle *3*. Therefore angle *2* = angle *3* minus angle *1*.

Angle *2* is the angle which the diameter of the Earth's orbit subtends at the star X. The greater this angle, the nearer the star is; the smaller, the further away the star is. Angle *2* is the parallax.

The proper motion of star X can be eliminated by repeating the measurements after one year has elapsed, when the Earth is at point A again. Corrections have to be applied for refraction (see page 52) and aberration (see page 55). These two effects are each much greater than the angle of parallax.

Because the Earth's distance from the Sun varies from 147 055 000 to 152 141 000 km, the Earth's mean distance of 149 597 870 km is used as a base line. This is one-half of the major axis of the Earth's elliptical orbit around the Sun (Figure 10.3).

The annual parallax of the star X is the angle AXS = p = angle between the geocentric direction AX of star X and its

In order to determine the distance of a star, the values of p'' and angle SAX have to be measured. When the value of P'' has thus been found, the distance of the star may be calculated from equation (2):

$$\text{Distance of star} = d = 206\,265 \frac{r}{p''}$$

where r is the mean distance of the Earth from the Sun, namely 149 597 000 km.

That the determination of the parallax of a star is very difficult may be seen by the fact that the first parallax of a star to be determined was that of 61 Cygni, 0.3″ – three-tenths of one arc second! This parallax was determined by F.W. Bessel (Figure 10.4) in 1838. Bessel selected this star because it has a fairly large proper motion of 5.22″ per year and must therefore be close to the Solar System. Larger proper motions have since been detected, for example, Kapteyn's star 8.7″ and the record holder, Barnard's star 10.29″. The latter is rightly known as the runaway star. The

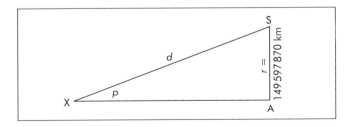

Figure 10.3 Heliocentric annual parallax

Figure 10.4 F.W. Bessel

Figure 10.5 R.T.A. Innes

co-ordinates of 61 Cygni are 21 h 06 m 24 s right ascension; +38°41.3′ declination, in the constellation of Cygnus.

The parallax of the nearest star, Alpha Centauri, was determined during 1831–1833 by T. Henderson at Cape Town Observatory, but his calculations were completed only in 1839. Alpha Centauri has a proper motion of 3.68″ and its heliocentric annual parallax is 0.76″. This means that no star has a parallax as large as 1 arc second!

The distance of Alpha Centauri from the Sun, d, is given by:

$$d = \frac{206\,265\,r}{p''}$$

i.e. $$d = \frac{206\,265\,r}{0.76} = 271\,400\,r$$

where r is the mean distance of the Earth from the Sun (= 1 AU). Alpha Centauri is therefore 271 400 times further than the Sun. The furthest planet, Pluto, is 40 AU from the Sun. The Solar System thus occupies 80 AU, on all sides. If we add another 40 AU all round, we can say that the Solar System needs 160 AU of space. If Alpha Centauri has a similar system of similar size, the space between the two systems will be 271 400 − 2(160) = 271 080 AU. Into this space 271 080 ÷ 160 = 1694 similar planetary systems could be fitted without causing overcrowding!

In km, the distance of Alpha Centauri is 271 400 × 149 597 870, which equals 40.6 × 10^{12} km, namely 40.6 million million km.

According to the scale model on page 81, the distance of Alpha Centauri from the Sun will be 40.6 × 10^{12} ÷ 40 × 10^9 = 1000 km (the distance from London to Marseilles or Washington to Quebec)!

To represent the distances of the stars more realistically, we could take as unit the distance which light travels in one year at a speed of 299 792.458 km s^{-1} = (299 792.458)(3600)(24)(365.25) km. This works out at 9.46 × 10^{12} km – a distance known as the light year. It therefore takes light 40.6 × 10^{12} divided by 9.46 × 10^{12} = 4.3 years to cover the distance from Alpha Centauri to the Earth. The distance of Alpha Centauri is thus 4.3 light years. The light year is a distance – in the same way as a distance of 20 km was referred to in the olden days as two hours on horseback.

Alpha Centauri is a double star and it also has a third member, 1°51′ to the south, with right ascension 0 h 9.9 m to the west of Alpha Centauri A and B. This third member was discovered in 1912 by R.T.A. Innes (Figure 10.5) of Johannesburg Observatory, South Africa. Its parallax is 0.762″ and its distance 4.28 light years. It is therefore the nearest individual star and is known as Proxima Centauri.

Proxima Centauri is 13 000 times fainter than the Sun. It is a flare star which, from time to time, suddenly brightens by 10% and then, just as rapidly, fades to its normal brightness.

The Parsec

The unit of distance used by astronomers is the parsec. The parsec is the distance at which the semi-major axis of the Earth's orbit (= 1 AU) subtends an angle of one second of arc (Figure 10.6, *overleaf*).

In Figure 10.6, A represents the Earth, S the Sun and X a hypothetical star. The angle AXS = $p = 1''$ is subtended by AS (= 149 597 870 km).

$$1'' = \frac{1 \times 3.141\,59}{180 \times 3600} \text{ radians} \quad \text{(see Appendix 1, section 1.8)}$$

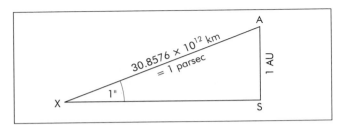

Figure 10.6 The parsec

$$1'' = 0.000\,004\,848 \text{ radians} = \sin 1'' = \frac{AS}{AX}$$

$$\therefore AX = AS \div 0.000\,004\,848$$

$$= \frac{149\,597\,870}{0.000\,004\,848} = 30.8576 \times 10^{12} \text{ km}$$

$$= \frac{30.8576 \times 10^{12}}{9.46 \times 10^{12}} = 3.26 \text{ light years}$$

The parsec equals 3.26 light years, which equals 206 265 AU.

The most refined measurement made by means of the trigonometrical method is an angle of parallax of one-hundredth of an arc second as heliocentric annual parallax, and measured on a photographic plate by means of a special microscope.

A parallax of 0.01 arc second indicates a distance of 100 times one parsec, namely 326 light years. The number of stars within the range of 300 light years is about 5000. Photographs, however, show that there are countless millions of stars. They must all therefore be further than 300 light years. How much further? At the beginning of the 20th century, the only method of estimating distances of stars greater than 300 light years was by means of their spectral types. Not until the 1920s was a new method found for determining stellar distances beyond the 300 light year limit. It was based on the work of Henrietta Leavitt, as we shall see in Chapter 12.

If the parallax of a star is known, its distance is simply the inverse of parallax. The parallax of Alpha Centauri is 0.76''; therefore its distance is $1 \div 0.76 = 1.316$ parsec. To convert this to light years, simply multiply by 3.26, giving 4.289 light years, or 4.29 correct to the second decimal place.

Table 10.1 contains the data for the nearest 25 stars. Actually, there are 32 stars, since seven of them are double stars. The intrinsic brightnesses of the stars, are given under the heading, Absolute Magnitude = M. Only three of the stars are intrinsically brighter than the Sun, namely Sirius A, Procyon and Alpha Centauri A. The Sun's absolute magnitude is 4.85. Alpha Centauri B, Tau Ceti and Epsilon Eridani are slightly fainter than the Sun. Twenty two are much fainter, being of spectral type dM, indicating that they are reddish in colour. They are known as the red dwarfs.

The Brightness of Stars

According to the scale of brightness, which was introduced by Hipparchus in 125 BC, a star of magnitude 1 is about 100 times brighter than a star of magnitude 6, which is the faintest that the naked eye can see. A greater magnitude number thus indicates less brightness, that is, the brightness is inversely proportional to the magnitude.

N.R. Pogson suggested that a difference of 5 (that is 6 − 1) in magnitude be set exactly equal to a ratio of 100 times in brightness. A brightness difference of 1 magnitude is thus equal to:

$$\sqrt[5]{100} = 2.512 \text{ (rounded off to 3 decimal places)}$$

A star of magnitude 1 is thus 2.512 times brighter than a star of magnitude 2, which in its turn is 2.512 times brighter than a star of magnitude 3, and so on. A star of magnitude 1 is thus $(2.512)^2$ times brighter than a star of magnitude 3 and $(2.512)^3$ times brighter than a star of magnitude 4, ... and $(2.512)^5$ times brighter than a star of magnitude 6.

If the magnitude of a star is a, then its brightness is $1 \div (2.512)^a$.

If star A has magnitude a and star B has magnitude b, their brightnesses can be compared as follows:

$$\frac{\text{Brightness of A}}{\text{Brightness of B}} = \frac{\frac{1}{2.512^a}}{\frac{1}{2.512^b}} = \frac{2.512^b}{2.512^a}$$

$$= 2.512^{(b-a)}$$

Similarly, the brightness of B compared with that of A is equal to $(2.512)^{(a-b)}$.

The value of 2.512 may seem rather clumsy, but the human eye sees ratios of brightness as differences. For example, when we have to divide a by b, we subtract the logarithm of b from that of a (Appendix 1, section 1.10). The logarithm of 2.512 is 0.4, that is, 2.512 equals 10 raised to the power of 0.4. With some practice, the eye can easily interpret a difference of 0.4 as a difference in brightness of 1 magnitude. The ratio of the brightness of a magnitude 1 star, compared with that of a magnitude 2 star, is therefore given by:

$$\frac{\frac{1}{2.512^1}}{\frac{1}{2.512^2}} = \frac{2.512^2}{2.512^1} = 2.512^1 = 10^{0.4}$$

Table 10.1 THE NEAREST STARS

Name of star	Abbr.	Co-ordinates 2000.0 Right ascension (RA)	Declination (Dec)	Parallax π''	Distance Parsec $\frac{1}{\pi}$	Light years $\frac{3.26}{\pi}$	Proper motion (arc seconds)	Magnitude Visual m	Absolute M	Spectral type
1. Proxima Centauri	α^3 Cen	14 29 43	−62 40.8	0.762	1.31	4.28	3.7″	10.7	15.1	dM 5e
2. Alpha Centauri A	α^1 Cen	14 39 37	−60 50.0	0.760	1.32	4.29	3.68	0.0	4.4	G2
Alpha Centauri B	α^2 Cen							1.4	5.7	K 1
3. Barnard's star	M 15040	17 57 49	+04 41.6	0.545	1.83	5.98	10.3	9.5	13.4	dM 5
4. Wolf 359	W 359	10 56 29	+07 00.9	0.421	2.38	7.74	4.84	13.5	16.8	dM 6e
5. Lalande 211851	L 211851	11 03 20	+35 58.2	0.398	2.51	8.19	4.78	7.49	10.5	dM 2
6. Sirius A	α^1CMa	06 45 09	−16 43.0	0.377	2.65	8.65	1.32	−1.46	1.4	A1
Sirius B	α^2CMa							8.7	11.4	DA
7. UV Ceti A	UV Cet A	01 39 01	−17 57.0	0.365	2.74	8.93	3.35	12.4	15.2	dM 6e
UV Ceti B	UV Cet B							12.9	15.9	dM 6e
8. Ross 154	R 154	18 49 50	−23 50.1	0.351	2.85	9.29	0.67	10.6	13.3	dM 5e
9. Ross 248	R 248	23 41 55	+44 10.5	0.316	3.16	10.3	1.58	12.2	14.7	dM 6e
10. Epsilon Eridani	ϵ Eri	03 32 23	−09 29.8	0.303	3.30	10.8	0.97	3.73	6.1	K2
11. Ross 128	R 128	11 47 45	+00 48.3	0.298	3.36	10.9	1.40	11.1	13.5	dM 5
12. 61 Cygni A	61 Cyg A	21 06 23	+38 41.6	0.293	3.41	11.1	5.22	5.22	7.6	K5
61 Cygni B	61 Cyg B	21 06 24	+38 41.1					6.04	8.4	K7
13. Luyten 789-6	L 789-6	22 35 42	−15 36.0	0.292	3.42	11.2	3.27	12.2	14.5	dM 6
14. Procyon	αCMi	07 39 18	+05 13.5	0.288	3.47	11.3	1.25	0.38	2.6	F5 IV
15. Epsilon Indi	ϵ Ind	22 02 29	−56 50.0	0.285	3.51	11.4	4.69	4.67	7.0	K5
16. Struve 2398 A	Σ2398	18 42 45	+59 37.9	0.280	3.57	11.6	2.29	8.9	11.2	dM 4
Struve 2398 B								9.7	12.0	dM 5
17. Groombridge 34 A	GRB 34	00 18 23	+44 01.4	0.278	3.60	11.7	2.91	8.1	10.3	dM 2e
Groombridge 34 B								10.8	13.1	dM 4e
18. Tau Ceti	τCet	01 43 32	−15 59.9	0.275	3.64	11.9	1.92	3.5	5.7	G8
19. Lacaille 9352	Lac 9352	23 05 52	−35 51.2	0.273	3.66	11.9	6.87	7.3	9.6	dM 2
20. BD +5° 1668		07 27 24	+05 13.5	0.263	3.80	12.4	3.73	9.8	12.0	dM 4
21. Lacaille 8760	Lac 8760	21 17 15	−38 52.1	0.255	3.92	12.8	3.46	6.68	8.8	dM 1
22. Kapteyn's star	HD 33793	05 11 40	−45 01.1	0.251	3.98	13.0	8.79	8.8	10.8	dM 0
23. Ross 614	R 614	06 29 24	−02 48.8	0.251	3.98	13.0	0.97	11.1	13.1	dM 4e
24. Krüger 60 A	Kr 60	22 28 00	+57 41.8	0.249	4.02	13.1	0.87	9.8	11.8	dM 4
Krüger 60 B								11.4	13.4	dM 5e
25. BD-12° 4523	W 1061	16 30 18	−12 39.7	0.243	4.12	13.4	1.18	10.1	12.0	dM 5e

Figure 10.7 Constellation Crux

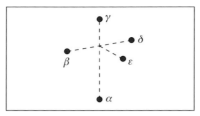

The brightness of a magnitude 3 star compared with that of a magnitude 5 star is given by $(2.512)^{(5-3)} = 2.512^2 = 6.31$.

The brightness of Alpha Crucis, of magnitude 0.87, compared with that of Beta Crucis, magnitude 1.25, is

$$2.512^{(1.25-0.87)} = 2.512^{0.38}$$
$$= 1.42$$

Alpha Crucis is therefore 1.42 times brighter than Beta Crucis, or 42% brighter (Figure 10.7).

The joint brightness of the two members of a double star, seen as one star, is found by adding the brightnesses of the individual members. Alpha Crucis is a double star, the magnitudes of its members being 1.41 and 1.88.

Assume the joint magnitude is m. Thus

$$\frac{1}{2.512^m} = \frac{1}{2.512^{1.41}} + \frac{1}{2.512^{1.88}}$$

i.e. $\quad 2.512^{-m} = \dfrac{1}{3.6646} + \dfrac{1}{5.64985}$

$$= 0.449876$$

$$\therefore -m \log 2.512 = \log 0.449876$$

$$\therefore -m(0.4) = -0.3469$$

$$\therefore m = 0.867$$

This can be rounded off to 0.87, as given in Table 10.3. In the case of the double star Alpha Centauri, the magnitudes of the two members are 0.00 and 1.39. The joint magnitude works out at −0.266. If this is rounded off to −0.27, it makes the star brighter than it actually is. It must therefore be rounded off to −0.26.

Some of the stars which had been classified as magnitude 1 were found to have magnitudes less than 1, according to the Pogson scale. The star Vega (Alpha Lyrae) was taken as standard and assigned the magnitude 0.00. The magnitude of the brightest star, Sirius, worked out at −1.46. At its brightest, the planet Venus has a magnitude of −4.4; the second brightest star, Canopus, −0.72; the full Moon, −12.5, and the Sun, −26.72!

The standardisation of magnitudes of stars was based on a group of stars within 20 degrees of the North Pole and of magnitudes between 2 and 20. These stars are called the North Polar series.

Absolute Magnitude

The brightnesses of the stars are affected by their distances from the Earth. Brightness is inversely proportional to the square of the distance, that is, if the distance is doubled, the brightness becomes $(\frac{1}{2})^2$, namely one-quarter; if the distance is halved, the brightness becomes $(2)^2$, namely four times.

The visual magnitude does not indicate what the intrinsic brightness of a star really is, because it does not take distance into consideration. To compare the intrinsic brightnesses of stars, the system of absolute magnitudes is used. In this system, the brightnesses of stars are compared as if they were all situated at the same distance. The distance used is 10 parsecs, namely 32.6 light years. The absolute magnitude is therefore the visual magnitude that a star will have if it were situated at a distance of 32.6 light years from the Earth.

If a star of brightness l has a visual magnitude m at a distance d, then its brightness will be L and its magnitude M at a distance of 32.6 light years. Then

$$\frac{L}{l} = \left[\frac{d}{32.6}\right]^2 = \frac{\frac{1}{2.512^M}}{\frac{1}{2.512^m}} = 2.512^{(m-M)}$$

Instead of the distance d, in light years, we can write $\frac{3.26}{P}$, where P is the parallax in arc seconds. Therefore

$$\left[\frac{3.36}{P} \times \frac{1}{32.6}\right]^2 = 2.512^{(m-M)}$$

Thus $\left[\dfrac{1}{10P}\right]^2 = 2.512^{(m-M)}$

$$\therefore 2 \log\left(\frac{1}{10P}\right) = (m - M) \log 2.512, \text{ and}$$

$$2(\log 1 - \log 10 - \log P) = (m - M) \log 2.512$$

Thus $2(0 - 1 - \log P) = (m - M)(0.4)$

$$\therefore -2 - 2\log P = (m - M)(0.4)$$

Divide each term by 0.4:

$$\therefore -5 - 5 \log P = m - M$$

$$\therefore M = m + 5 + 5 \log P \qquad (1)$$

This equation gives the absolute magnitude, M, in terms of the apparent, visual magnitude, m, and the parallax, P, in arc seconds.

Because the parallax P is inversely proportional to the distance D in parsec, we can substitute $\frac{1}{D}$ for P. Equation (1) then becomes:

Distance Modulus

$$M = m + 5 + 5 \log \frac{1}{D}$$
$$= m + 5 + 5 \log D^{-1}$$
$$\therefore M = m + 5 - 5 \log D \quad (2)$$

This equation gives the absolute magnitude, M, in terms of the apparent, visual magnitude, m, and the distance, D, in parsecs. If the distance is in light years, it has to be divided by 3.26.

Distance Modulus

The difference between the apparent, visual magnitude, m, and the absolute magnitude, M, is known as the distance modulus, that is $(m - M) = 5 \log D - 5$.

The formula can be transformed into:

$$5 \log D = m - M + 5$$

so that $\log D = \dfrac{m - M + 5}{5}$

Table 10.2 gives the distance in parsec and in light years for various values of the distance modulus.

The apparent, visual magnitude of Canopus is -0.72 and its absolute magnitude is estimated as -8.5. Its distance can be calculated as follows:

$$\log D = \frac{m - M + 5}{5} = \frac{-0.72 - (-8.5) + 5}{5}$$
$$= 2.556$$

D is the antilog of $2.556 = 360$. Therefore the distance of Canopus is 360 parsecs, which equals $360 \times 3.26 = 1174$ light years. (Later, we shall see how the absolute magnitude may be determined without knowledge of the distance of the star.)

The absolute magnitude, M, can be calculated from equation (1) if the apparent magnitude and the parallax are known; and from equation (2) if the apparent magnitude and the distance in parsecs are known. If the distance is in light years, it can be converted to parsecs by dividing by 3.26.

The apparent magnitude of Sirius (Alpha Canis Majoris) is -1.46 and its parallax is $0.376''$. The absolute magnitude of Sirius can then be found as follows:

$$M = m + 5 + 5 \log P \quad (1)$$

i.e. $M = -1.46 + 5 + 5 \log(0.376)$
$$\therefore M = 3.54 + 5(-0.4248)$$
$$= 3.54 - 2.12$$
$$= 1.42$$

Table 10.2 Distance modulus

Modulus $m - M$	Distance Parsec	Distance Light years
0	10	32.6
0.5	12.59	41.0
1.0	15.85	51.7
1.5	19.95	65.0
2.0	25.1	81.8
2.5	31.6	103
3.0	39.8	129.75
4.0	63.1	205.7
5.0	100	326
6.0	158.5	517
8.0	398	1 297
10.0	1 000	3 260
15.0	10 000	32 600
20.0	100 000	326 000

Dividing Sirius' distance of 8.67 light years by 3.26 gives its distance as 2.66 parsecs. Now, using equation (2):

$$M = m + 5 - 5 \log D \quad (2)$$
$$\therefore M = -1.46 + 5 - 5 \log 2.66$$
$$= 3.54 - 5(0.4249)$$
$$= 3.54 - 2.12$$
$$= 1.42$$

To calculate the absolute magnitude of the Sun, its distance must be converted into parsecs:

$$149\,597\,870 = \frac{149\,597\,870}{9.46 \times 10^{12}} \text{ light years}$$
$$= \frac{149.597\,870 \times 10^6}{9.46 \times 10^{12} \times 3.26}$$
$$= 4.848 \times 10^{-6} \text{ parsecs}$$

Substitute this value for D and (-26.72) for the Sun's apparent magnitude in equation (2):

$$\therefore M = m + 5 - 5 \log D$$
$$= -26.72 + 5 - 5 \log(4.848 \times 10^{-6})$$
$$= -21.72 - 5(0.685\,56 - 6)$$
$$= -21.72 - 3.4278 + 30$$
$$= 4.85 \text{ (correct to two decimal places)}$$

Table 10.3 THE BRIGHTEST STARS

Name of star	Abbr.	Co-ordinates 2000.0 Right ascension (RA)	Co-ordinates 2000.0 Declination (Dec)	Magnitude Visual m	Magnitude Absolute M	Distance (light yrs)*	Spectral type
1. Sirius	αCMa	06 45 09	−16 43.0	−1.46	1.42	8.67	A1 V
2. Canopus	αCar	06 23 57	−52 41.7	−0.72	−8.5	1174	F0 Ia
3. Alpha Centauri A	α¹Cen	14 39 37	−60 50.0	0.00 ⎫ −0.26	4.4	4.29	G2 V
Alpha Centauri B	α²Cen			1.39 ⎭	5.7		K1 V
4. Arcturus	αBoo	14 15 40	+19 11.0	−0.04	−0.2	37	K2 IIIp
5. Vega	αLyr	18 36 56	+38 47.0	0.03	0.5	26.4	A0 V
6. Capella	αAur	05 16 41	+45 59.9	0.08	0.3	45	G8 III
7. Rigel	βOri	05 14 32	−08 12.1	0.12	−7.1	912	B8 Ia
8. Procyon	αCMi	07 39 18	+05 31.5	0.38	2.6	11.4	F5 IV
9. Achernar	αEri	01 37 43	−57 14.2	0.46	−1.6	85	B5 IV
10. Betelgeuse	αOri	05 55 10	+07 24.4	0.5	−6.0	650	M2 Iab
11. Agena	βCen	14 03 49	−60 22.4	0.61	−5.1	452	B1 II
12. Altair	αAql	19 50 47	+08 52.1	0.77	2.2	17	A7 IV–V
13. Aldebaran	αTau	04 35 55	+16 30.6	0.85	−0.3	55	K5 III
14. Alpha Crucis A	α¹Cru	12 26 36	−63 05.9	1.41 ⎫ 0.87	−3.9	360	B1 IV
Alpha Crucis B	α²Cru			1.88 ⎭	−3.2		B3 IIn
15. Antares	αSco	16 29 24	−26 25.9	0.96	−4.7	442	M1 Ib
16. Spica	αVir	13 25 12	−11 09.7	0.98	−3.5	258	B1 V
17. Pollux	βGem	07 45 19	+28 01.6	1.14	0.2	50	K0 III
18. Fomalhaut	αPsA	22 57 39	−29 37.3	1.16	2.0	22	A3 V
19. Deneb	αCyg	20 41 26	+45 16.8	1.25	−7.5	1800	A2 Ia
20. Beta Crucis	βCru	12 47 43	−59 42.3	1.25	−5.0	580	B0 III
21. Regulus	αLeo	10 08 22	+11 58.0	1.35	−0.6	85	B7 V
22. Adhara	εCMa	06 58 38	−28 58.3	1.50	−4.4	490	B2 II
23. Castor A	α¹Gem	07 34 36	+31 53.3	1.99 ⎫ 1.58	1.2	46	A1 V
Castor B	α²Gem			2.85 ⎭	2.1		A5 V
24. Gamma Crucis A	γ¹Cru	12 31 10	−57 06.8	1.63 ⎫ 1.62	−0.5	87	M3 III
Gamma Crucis B	γ²Cru	12 31 17	−57 06.9	6.42 ⎭	4.3		A2 V
25. Shaula	λSco	17 33 36	−37 06.2	1.63	−3.0	274	M3 III

Source: A. Hirshfeld and R.W. Sinnott, *Sky Catalogue 2000.0* (1982).
*Some distances adapted to allow for reddening due to interstellar gas and dust.

This means that, at a distance of 10 parsecs (= 32.6 light years), the Sun would be a dim star of magnitude 4.85. The Sun is therefore much fainter than Sirius, whose absolute magnitude is 1.42.

$$\frac{\text{Brightness of Sirius}}{\text{Brightness of Sun}} = 2.512^{(4.85-1.42)}$$

$$= 2.512^{(3.43)} = 23.55$$

Sirius is thus, in reality, more than 20 times brighter than the Sun. Canopus, whose absolute magnitude is −8.5, is 218 900 times brighter than the Sun! The temperature of such a star must be very high indeed and the nuclear reactions in its nucleus must be very rapid.

Table 10.3 contains data on the 25 brightest stars. The brightnesses of double stars in the list have been combined to reveal their visual brightnesses.

Naming the Stars

Today, we still use the system of naming the stars which was introduced by J. Bayer in 1603, in which he used the letters of the Greek alphabet to name the stars in each constellation, according to their degrees of brightness: alpha for the brightest, beta for the second brightest, and so on. These letters are used with the genitive of the Latin name of each constellation. The brightest star in the constellation of Orion, for example is Alpha Orionis (Betelgeuse); the second brightest, Beta Orionis (Rigel). Since the Greek alphabet contains 24 letters, only the 24 brightest stars in each constellation could be named in

The Greek Alphabet					
alpha	α	iota	ι	rho	ρ
beta	β	kappa	κ	sigma	σ
gamma	γ	lambda	λ	tau	τ
delta	δ	mu	μ	upsilon	υ
epsilon	ϵ	nu	ν	phi	ϕ
zeta	ζ	xi	ξ	chi	χ
eta	η	omecron	o	psi	ψ
theta	θ	pi	π	omega	ω

this way. Fainter stars were numbered according to the Flamsteed numbers, such as 61 Cygni, etc.

More comprehensive naming systems are used in the large catalogues, such as the *HD Catalogue* (after Henry Draper), in which all the stars are numbered in sequence from west to east according to their right ascensions. (The stars in Appendix 3 which have numbers are from the *HD Catalogue*.)

Stars have also been named by using a letter as prefix to indicate a certain astronomer, such as L for Luyten, Lac for Lacaille, Σ for F.G.W. Struve, 0Σ for Otto Struve, β for S.W. Burnham, I for R.T.A. Innes, B for W.H. van den Bos, etc.

Stars can very conveniently be named according to their co-ordinates. Sirius (Alpha Canis Majoris) can be indicated by 6 45 09 − 16 43 0, because it is located at the position having right ascension 6 hours 45 minutes, 9 seconds and declination south 16 degrees, 43 minutes 0 seconds. Vega (Alpha Lyrae) is 18 36 56 + 38 47 0, that is, the star situated at the point having right ascension 18 hours, 36 minutes, 56 seconds and declination north 38 degrees 47 minutes. (The nomenclature used for variable stars will be dealt with in Chapter 11.)

Stellar Spectra

The intrinsic brightness of a star will, in large measure, depend on its temperature. The temperature will have a considerable effect on the star's spectrum.

In 1866, A. Secchi produced the first classification of stellar spectra. He divided the spectra into three classes:

Class I: white stars having strong hydrogen lines, such as Vega (Alpha Lyrae);

Class II: yellow stars having strong metallic lines, such as Arcturus (α Boötis);

Class III: red stars having broad, fluted bands of molecules, such as titanium oxide (TiO), for example Antares (Alpha Scorpii).

The progress in stellar spectroscopy was so rapid that Secchi's very useful classification was superseded in the year 1900, by a more extensive system. The Harvard College Observatory system gained general approval. It provided for six spectral types: B, A, F, G, K and M. Subsequently the system was expanded by the addition of two types, W and O, before type B and by the addition of side branches to types G and K:

$$W - O - B - A - F - G \begin{matrix} \nearrow R-N \\ \searrow C \end{matrix} K - M \begin{matrix} \\ \searrow S \end{matrix}$$

99% of the stars in the *HD Catalogue* fall within types B to M.

Each spectral type is further subdivided into ten classes: 0, 1, 2, ... 9. Classes 1, 2, 3 and 4 of type O were removed to type W and called W5, W6, ... W9.

Stars belonging to the types on the left-hand side of the series are called early or young stars; those on the right-hand side are called late or old stars. The reason for this is that the stars on the left-hand side have very high temperatures and they cannot endure as long as those on the right-hand side, which have lower temperatures.

There are two main groups in type W: WC and WN, which have strong carbon and nitrogen lines respectively in their spectra.

Side branches of type G are types R and N, being orange and reddish in colour; as well as type C, which contains much carbon. Type K, class 5 has a side branch S.

The temperatures of the stars decrease consistently from W to M. This is known because the degrees of ionisation of atoms are highest in type W; then type O, and type B. The high degrees of ionisation in the earlier types require temperatures in the range 80 000 to 25 000 K..

The stars in types M, S, R and N have broad bands in their spectra, caused by molecules such as titanium oxide. These molecules can exist only if the temperature is low enough, around 3000°C.

Properties of the Spectral Types

Type W These are also known as Wolf–Rayet stars, named after their discoverers, C.J.E. Wolf and G.A.P. Rayet. Such stars are very rare and have temperatures of 80 000 K; their spectra are dominated by broad, emission lines of hydrogen and helium. Those stars that also have emission bands of carbon and nitrogen are classed as WC and WN. These stars are found at the centres of "planetary nebulae" – they are stars that have cast off their surface layers and are thus surrounded by shells of glowing gas. In small telescopes they have the appearance of planets, hence their name, but they bear no relationship to planets. A good example is the Ring Nebula M57, in Lyra, situated at 18 517 + 3258.

Gamma Velorum (08 09.5–47 20) is a typical W8 star. Its absolute magnitude is –4.1.

Type O The temperatures of this type range between 40 000 and 25 000 K. Because of their high temperatures, these stars have spectral lines of singly and doubly ionised helium, HeII and HeIII respectively (neutral helium is HeI). The spectra also contain lines of doubly and trebly ionised oxygen, OIII and OIV; as well as doubly and trebly ionised nitrogen, NIII and NIV. Prominent spectral lines are CIII 4650 Å, HeI 4471 Å, HeII 4541 Å, Hγ 4340 Å, SiIV 4089 Å and NIII 4097 Å. These stars are known as helium stars, for example, Zeta Puppis (08 03.6–40 00) an O5 of absolute magnitude –7.1; Sigma Orionis (05 38.7–02 36), an O9 of absolute magnitude –4.4.

Type B These have temperatures between 25 000 and 11 000 K. The spectra contain no lines of ionised helium, but do contain lines of singly ionised oxygen, OII, and nitrogen, NII. As the series proceeds from B0 to B9, the oxygen and nitrogen emission lines become fainter, showing that the temperature decreases steadily. Absorption lines of hydrogen begin to appear and become stronger through the series. Some of these stars also have emission lines of hydrogen, showing that they have a very rarefied atmosphere. (The presence of emission lines in a spectrum is indicated by the letter e, such as B 2e.) Prominent spectral lines are HeI 4471 Å, HeII 4541 Å, CIII 4540 Å, SiIV 4089 Å, SiII 4128 Å and Hδ 4102 Å. Although these stars are proportionately not very abundant, their brightness makes them visible from afar. They tend to appear in swarms. Examples are the Orion stars: Beta Orionis, Rigel (05 14.5–08 12) which is a B8 Ia star, that is a super-giant, of absolute magnitude $M = -7.1$; Gamma Orionis (05 25.0–01 12), a B2 III giant, having $M = -3.6$; and Epsilon Orionis (05 36.2–01 12), a B0 Ia star, having $M = -6.2$.

Type A The temperature of this type ranges between 11 000 and 7500 K; they are plentiful. Hydrogen absorption lines are strongest in types A0 and A1, but weaken thereafter. The spectra contain no helium lines. Lines of singly ionised calcium, CaII, begin to appear. Prominent lines are: Hα 6563 Å, Hβ 4861 Å, Hγ 4340 Å, Hδ 4102 Å and Hε 3970 Å – these five lines are part of the Balmer series; the *K* line 3934 Å and the *H* line 3968 Å, both of ionised calcium; the MgII line 5173 Å, which is very bright; TiII at 4303 Å; and neutral iron at 4299 Å and 4303 Å.

Examples of A type stars are: Vega, Alpha Lyrae (18 36.9 + 38 47), an A0 V star having $M = 0.5$; Castor, Alpha Geminorum (07 34.6 + 31 53.3), which is an A1 V star having $M = 1.2$; Fomalhaut, Alpha Piscis Austrini (22 57.7–29 37.3), an A3 V star having $M = 2.0$.

Type F The temperatures of this type are between 7500 and 6000 K. The earlier classes F0 and F1 have equally strong hydrogen and singly ionised calcium, CaII, absorption lines. The CaII lines are in the ultra-violet portion of the spectrum. As the series progresses from F0 to F9, the Balmer lines of hydrogen weaken and the calcium lines become stronger. Many fine absorption lines of metals begin to appear. The *H* and *K* lines of calcium and the Hδ and Hε lines are strong. Lines of neutral calcium, CaI, 4227 Å, and a band at 4370 Å also appear.

Examples are: Canopus, Alpha Carinae (06 24–52 41.7)

which is an F0 Ia super-giant, having $M = -8.5$; Procyon, Alpha Canis Minoris (07 39.2 + 05 31.5), an F5 IV with $M = 2.6$; also the star Wezen, Delta Canis Majoris (07 08.6 – 26 24), an F8 Ia super-giant, of $M - 8.0$.

Type G The temperatures of this group are between 6000 and 5000 K. The hydrogen absorption lines continue to weaken but are easily visible. Metallic lines get stronger from G0 to G9. Lines of ionised atoms are barely visible. Two groups of G stars can be distinguished: (a) those with fine absorption lines in the spectra are giants having low pressures in their atmospheres – that is why the lines are so fine; and (b) those that have broader absorption lines – they are dwarf stars with high pressures in their atmospheres. These two groups are designated as g, for giant, and d, for dwarf, before the spectral type. The Sun is a typical dwarf G2, namely dG 2; and Capella, Alpha Aurigae (05 16.7 + 45 59.9) a typical giant, gG 8 or G8 III.

The neutral calcium line 4227 Å and the $H\delta$ and $H\gamma$ lines are strong.

Type K These have temperatures between 5000 and 3500 K. Metallic absorption lines are very strong, especially those of iron. Ionised calcium, CaII, lines are at their strongest in type K1, having temperatures 4000 K to 5000 K among the dwarfs and 3500 K to 4500 K among the giants. Later types have absorption lines of molecules such as titanium oxide (TiO) and zirconium oxide (ZrO).

Examples are: Arcturus, Alpha Boötis (14 15.7 + 19 11), a K2 IIIp, with $M = -0.2$ (the letter p indicates that there is something peculiar about the spectrum). The Main Sequence star Epsilon Eridani (03 22.4–09 29.8) is a K2 V star, having $M = 6.1$. The reddish Aldebaran, Alpha Tauri (04 35.9 + 16 30.6), is a K5 III, with $M = -0.3$.

Strong spectral lines are: the K and H lines of ionised calcium, CaII, at 3934 Å and 3968 Å; the g line of CaI at 4227 Å; neutral iron at 4383 Å and 4325 Å; and ionised iron at 4172 Å. The hydrogen gamma line is still visible.

Type M Temperatures of this type range from 3500 down to 3000 K. The spectra contain fluted bands of titanium oxide (TiO), which become fainter on the red side of the spectrum. There are still many metallic lines. Hydrogen lines are almost invisible, except in certain variable stars, when they become visible in certain phases of the variability.

Examples are: Betelgeuse, Alpha Orionis (05 55.2 + 07 24.4), an M2 Iab super-giant with $M = -6$; Antares, Alpha Scorpii (16 29.4–26 25.9), an M1 Ib, with $M = -4.7$; Gamma Crucis (12 31.3–57 06.9), a M3 III, having $M = -0.5$.

Besides lines of CaI at 4227 Å and bands of titanium oxide, the Balmer lines appear as emission lines.

Type R These have temperatures from 3000 down to 2000 K, and are very rare. The violet and blue portions of their spectra are fairly bright. These stars are not as red as the M and N types.

Examples are: variables such as R Coronae Borealis (15 48.6 + 28 09), which really has a G type spectrum; V Arietis (02 15.0 + 12 14), an R4 type which varies in magnitude between 7.3 and 8.3.

Type N Temperatures of this group range from 3000 down to 2000 K. The fluted bands of carbon compounds are fainter in the violet and fainter than the R types. Most of these stars are irregular variables and very red. An example is W Canis Majoris (07 08.1–11 55), an N3 and one of the reddest stars known.

Type C These stars are rich in particles of carbon in their atmospheres. They are very red and are sometimes classified as type N stars.

Type S Temperatures of this group range between 2000 and 3000 K. Spectra are similar to M types but have bands of zirconium oxide (ZrO) and titanium oxide (TiO). They are all giants and are usually classified as long-period variables. An example is: R Andromedae (00 24.0 + 38 35), which is an S6e. Its apparent magnitude varies between 5.0 and 15.3, a brightness range of 13 000 times!

IN the G, K and M types, g and d are used to indicate giants and dwarfs.

The letter D is used to indicate white dwarfs.

Figure 10.8, (*overleaf*) by W.C. Rufus and R.H. Curtis, of Michigan University, shows how the spectra are graded. Through the types B, A, F, G, K and M, there is an almost imperceptible, but gradual change in brightness and prevalence of lines of various atoms.

The following additions are tagged on to the spectral types and classes:

e emission lines present;
p something peculiar in the spectrum;
m strong metallic lines;
s sharp spectral lines;
ss very sharp spectral lines;
n nebulous lines on account of the spin of the star;
nn lines very nebulous;
k absorption lines of interstellar calcium present.

Ia brightest super-giants;
Ib less bright super-giants;
II bright giants;
III ordinary giants;
IV sub-giants (small giants);
V main sequence stars, which are the vast majority;
VI small dwarfs (red dwarfs).

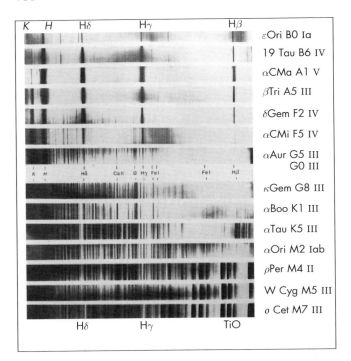

Figure 10.8 Stellar spectra

The Hertzsprung–Russell Diagram

The first astronomers to plot a graph showing the relationship between spectral types of stars and their absolute magnitudes (or luminosities) were E. Hertzsprung and H.N. Russell. The diagram is known as the Hertzsprung–Russell diagram; in short, the H–R diagram (Figure 10.9).

Sky Catalogue 2000.0 by A. Hirshfeld and R.W. Sinnott (1982) contains the data relating to 45 269 stars, down to as faint as apparent visual magnitude 8.05. From this catalogue, 380 stars of all types and classes were selected to cover the entire range of absolute magnitudes. To this group was added 24 faint stars, the data of which were supplied by J. Churms of the South African Astronomical Observatory. The data concerning the 404 stars are contained in Appendix 3. Many of the distances listed have been calculated according to the spectral types, and others have been estimated. Many of the absolute magnitudes are actually bolometric magnitudes which reflect the total radiation of a star over all wavelengths and not merely the radiation at the visual wavelength of 5400 Å. Adjustments have also been made for the colours of the stars, and reddening due to interstellar gas and dust has been taken into account. The resulting absolute magnitudes are the best obtainable.

The data of absolute magnitudes against spectral types and classes have been plotted in Figure 10.9. Black dots indicate stars of which the distances have been determined trigonometrically so that their absolute magnitudes could be calculated accurately, but still taking into account the necessary corrections. The letter t indicates that the distance has been determined trigonometrically; and the letter s indicates that the distance has been adjusted according to the spectrum.

Interwoven with the black dots are open circles for those cases where distances have not been determined by trigonometric methods.

Each dot or circle on the graph represents many stars of equal absolute magnitude and similar spectral type and class. The dot representing K0 and absolute magnitude 0.2 actually represents 40 stars in the catalogue, all of which have the same spectral type and class and the same absolute magnitude.

Under type F, class 1, the star Epsilon Sculptoris (εScl) has the same absolute magnitude as the star HD77370, whose distance has been determined trigonometrically as 24 parsecs. Because these two stars have identical spectral type and class, and because their visual apparent magnitudes are almost identical, their distances must be equal. The plotted point is thus both a dot and an open circle. The absolute magnitudes of the two stars must also be equal. The apparent magnitude of Epsilon Sculptoris is 5.31 and its absolute magnitude is 2.8. Its distance can be calculated as follows:

$$\log D = (m + 5 - M) \div 5$$
$$= (5.31 + 5 - 2.8) \div 5 = 1.5$$

Therefore D equals the antilog of 1.5, that is, 31.6 parsecs. The distance of εScl can thus safely be put down as 30 parsecs.

When one looks at the series of ordinary giants (III), it is apparent that black dots and open circles alternate. The four open circles under spectral classes G4, G5, G6 and G7 must therefore have the same absolute magnitude of 0.3 as the star under spectral class G8. This is the star Alpha Aurigae whose distance, 13 parsecs, is well known. According to calculation, the distance of Alpha Aurigae works out at 9 parsecs. The discrepancy is due to the reddening of the star's light by gas and dust in the Milky Way galaxy.

The visual magnitudes of the four stars under classes G4, G5, G6 and G7 are: ζPyx 4.89, μVel 2.69, v^1Eri 4.51 and

Hertzsprung-Russell Diagram

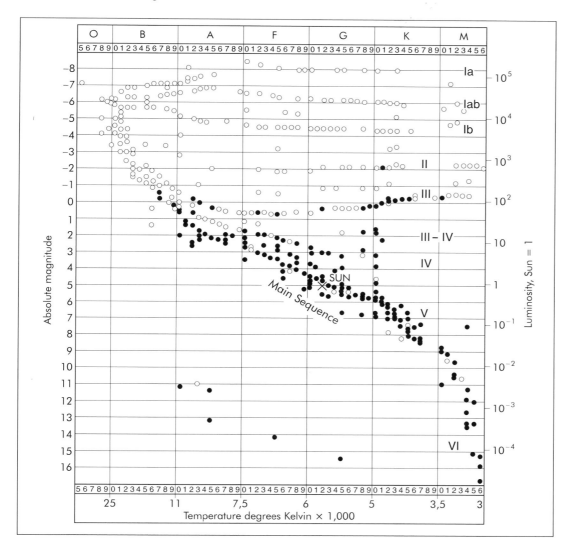

Figure 10.9
The Hertzsprung–Russell diagram: luminosity against spectral type

λPyx 4.69. They all have the same absolute magnitude of 0.3 and their distances work out at 83, 30, 69 and 76 parsecs respectively. In *Sky Catalogue* their distances are given as 75, 30, 63 and 76 parsecs, which agree very well with the calculated distances. In the cases of ζPyx (75 instead of 83 pc) and υ′ Eri (63 instead of 69 pc), adjustments have been made according to the spectra.

It is therefore possible to deduce the absolute magnitude of a star from its spectrum. This has been done in most cases of distances beyond 30 parsecs. (Later, we shall see that other methods of distance determination are also helpful in this regard.)

Because the absolute magnitudes in each spectral class spread over a range, the idea took root that some stars must be giants in order to have luminosities such as −2, −4.5, −6.2 and even −8. If a star has a large surface, it will appear to be brighter than a smaller star of the same temperature. If the temperature is the same, the spectral type and class will be the same and the greater brightness will require a greater area of radiation. The temperatures of the various spectral types are indicated at the bottom of the graph.

Luminosities in terms of the Sun's luminosity (= 1) are given on the right-hand side of the graph. The absolute magnitude, M, corresponding to any given value of the luminosity, say 100 times that of the Sun, whose absolute magnitude is 4.85, may be calculated as follows:

$$\frac{1}{2.512^M} \div \frac{1}{2.512^{(4.85)}} = 100$$
$$\therefore 2.512^{(4.85-M)} = 100$$
$$\therefore (4.85 - M)\log(2.512) = \log 100$$
$$\text{i.e. } (4.85 - M)(0.4) = 2$$
$$\therefore M = 4.85 - (2 \div 0.4) = -0.15$$

A star of absolute magnitude −0.15 is thus 100 times brighter, or more luminous, than the Sun. One of absolute magnitude −7.65 is 100 000 times more luminous than the Sun (= 10^5 times). At the other end of the scale, there are stars such as Proxima Centauri, the nearest star, having an absolute magnitude of 15.1, which means it is more than 500 000 dimmer than the Sun. The faintest star known is Van Biesbroeck's star (19 14.5 + 05 04) in the constellation of Aquila, which has an absolute magnitude of 19.3, making it 602 950 times fainter than the Sun.

Most of the stars congregate in the Main Sequence, which snakes across the graph from top left to bottom right, forming a seemingly impenetrable barrier to the lower limits that the luminosities of stars can have. This is due to the effects of temperature: a given temperature implies a certain luminosity for a certain radiating surface area. Generally, stars of greater mass also have greater luminosities.

If a star becomes larger by expanding, its luminosity will increase, provided its temperature stays the same. During the course of a star's evolution, it can change its position on the H–R diagram. It seems certain that stars move from the Main Sequence to the areas occupied by the giants: IV, III, II and I.

Since there are so many countless billions of stars, one could expect that there would be stars of all sorts of spectral types and luminosities. The fact that this is not the case shows that stars cannot reside long in the vacant areas of the H–R diagram; that is why none or few stars are found in the vacant areas. The absolute magnitudes of the giant stars tend to values such as: Ia −8; Ib −4.5; II −2; and III at 0. The Main Sequence spreads over absolute magnitudes from −6 to +16.

Colour Index

At present, the magnitudes of stars are determined by means of photo-electric photometers and by means of the charge coupled device. Magnitudes may be determined photometrically by measuring the sizes of the images on photographic plates. The photographic magnitude is not equal to the visual magnitude, because the photographic plate is more sensitive to blue light than to yellow light. Stars that radiate strongly in the blue will therefore have lower magnitude numbers than equally bright stars that are more yellowish or reddish.

The difference between the photographic (or blue) magnitude of a star and its photovisual (or yellow) magnitude is called the colour index of the star. Since the blue magnitudes (B) are greater (smaller numbers) and the visual magnitudes (V) are smaller (larger numbers), bluish stars will have negative or small colour indices. Reddish stars will have positive or larger colour indices. (The colours of the stars are not really blue and red.)

Use is made of filters that permit only certain wavelengths of light to pass through. Photographic magnitudes are usually measured at the wavelength 4250 Å in the blue part of the spectrum; and photovisual magnitudes at the wavelength 5280 Å in the yellow part of the spectrum. The colour index is thus equal to $B - V$, that is, blue magnitude minus visual magnitude. Magnitudes measured in the ultra-violet are designated as U. These three colour magnitudes are termed the UBV system. It is further extended by measuring magnitudes through red and infra-red filters, thus giving rise to the $UBVRI$ system.

The zero on the scale of colour indices has been so chosen that Main Sequence stars of spectral class A0, such as Vega (Alpha Lyrae), have a colour index of zero, that is, $U - B = B - V = 0$.

In the UBV system devised by H.L. Johnson and W.W. Morgan, the magnitudes are measured through three filters: ultra-violet (U) through a filter which permits light at a wavelength of 3600 Å to pass through; a blue magnitude (B) through a filter allowing the wavelength 4200 Å to pass through; and the visual magnitude V, through a filter which allows the wavelength 5400 Å to pass through.

H–R Colour Index Diagram

If absolute magnitude is plotted against colour index (Figure 10.10), a distribution similar to the Hertzsprung–Russell diagram is obtained. If this graph is compared with the H–R diagram of absolute magnitude against spectral type, the correspondence will be apparent. There is thus a regular relationship between the colour indices of stars and their temperatures and spectral types and classes.

Figure 10.10 can be used to read off the absolute magnitude of a star, if its colour index is known, that is, if the value of its blue magnitude minus its visual magnitude is known. For example, the star Kappa Ceti is a main sequence star, having a $B - V$ colour index of 0.68. From the graph, it is seen that its absolute magnitude must be about +5. The star Epsilon Hydrae is a G0 III giant; its colour

H-R Colour Index Diagram

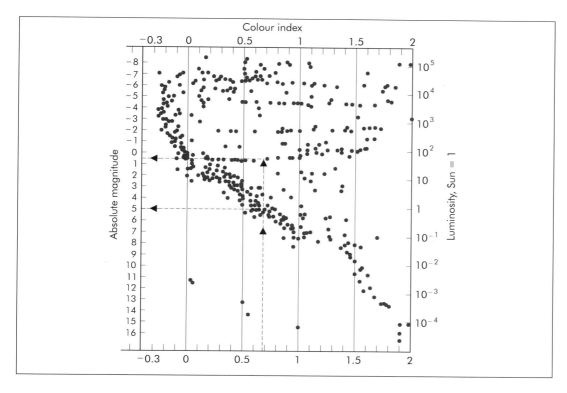

Figure 10.10 Absolute magnitude plotted against colour index

index of 0.68 indicates that its absolute magnitude is about 0.6. The distances of such stars can then be easily calculated from the formula $\log D = (m + 5 - M) \div 5$.

The light emitted by a star therefore carries all the secrets of the star: its chemical composition, its temperature, its absolute magnitude, its distance, its speed of motion along the line of sight, its mass and its diameter. G.E. Satterthwaite (*Encyclopaedia of Astronomy*, Hamlyn, London) gives the relationships between spectral type, colour index and temperature (Table 10.4, *overleaf*). The bolometric corrections, when applied to the visual absolute magnitudes, yield the bolometric magnitudes. The bolometric magnitude of a star is its absolute magnitude over all the wavelengths of the entire spectrum. It is a measure of the total energy flux of the star.

Table 10.3 (page 164) of the brightest stars contains five super-giants (Ia, Ib), three bright giants (II), seven ordinary giants (III), four sub-giants (IV) and ten Main Sequence stars.

The 45 269 stars down to magnitude 8.05 in *Sky Catalogue 2000.0* are not representative of the stars as a whole. The largest telescopes reveal stars as faint as visual magnitude 22. Because the stars of spectral types B and A are so bright, these seem to preponderate. The majority of type M stars are fainter than magnitude 8.05 and do not appear in the catalogue. Of the nearest stars, listed in Table 10.1 (page 161), 22 out of the 32 are type M. The stars in the catalogue comprise 13% type B, 20% type A, 16% type F, 14% type G, 32% type K and 4% type M. It is possible that classes F5, F6, ... F9, G0, ... G9 and K1, ... K5 have habitable planets. These stars comprise 40% of all the stars in the catalogue, that is 18 000 stars.

According to size, 1.3% are bright super-giants, 1.1% bright giants, 35% ordinary giants, 9% sub-giants and 42% Main Sequence stars.

The most distant stars listed are about 8000 parsecs (26 000 light years) from Earth. Within this sphere, of radius 26 000 light years, somewhat less than half of all the stars in the Milky Way galaxy are to be found, namely some 40 000 million.

Stellar Diameters

To measure the diameter of even a giant star may seem impossible, since, in the largest telescopes, stars are mere points of light and do not reveal discs. This was certainly one of the most difficult problems that astronomers had to face. However, the solution was found in the nature of light itself, in the fact that light waves undergo interference, that

Table 10.4 Colour index, temperature and bolometric correction

Spectral type	Visual colour	Colour index		Temperature		Bolometric correction		
		Main Seq.	Giants	Main Seq.	Giants	Main Seq.	Giants	Super-giants
O5	white	−0.6		79 000		−5.6		
B0	white	−0.33		25 200		−2.7		
B5	white	−0.18		15 500		−1.58		
A0	white	0.00		10 700		−0.72		
A5	white	0.20		8 500		−0.31		
F0	yellow	0.33		7 500		0.00		
F5	yellow	0.47		6 500		−0.04	−0.08	−0.12
dG0	yellow	0.57		6 000		−0.06		
gG0	yellow		0.67		5 200		−0.25	−0.42
dG2	yellow	0.63		5 800		−0.08	−0.32	−0.53
dG5	yellow	0.65		5 400		−0.10		
gG5	yellow		0.92		4 600		−0.39	−0.65
dK0	orange	0.78		4 900		−0.11		
gK0	orange		1.12		4 200		−0.54	−0.93
dK5	red	0.98		3 900		−0.85		
gK5	red		1.57		3 600		−1.35	−1.86
dM0	red	1.45		3 500		−1.43		
gM0	red		1.73		3 400		−1.55	−2.2

is, the amplitudes of the waves are either strenghthened or wiped out, causing alternate brightness and darkness.

On a photographic plate, a bright star makes a sizeable spot. This is called a false disc, because it is not a disc of the star itself, but the dispersion of light around the point image, owing to the interaction of lightwaves on each other. If the visual image of a star is greatly enlarged, a false disc, known as an Airy disc (Figure 10.11) is formed. There is a bright spot in the centre, surrounded by alternate rings of brightness and darkness.

This phenomenon is known as interference. When two lightwaves are in phase, so that the crests of the two waves coincide, the two waves will maximally strengthen each other, so that the light will be bright. If the crest of one wave coincides with the trough of another, the two waves will wipe each other out and there will be darkness.

To illustrate this principle, two wave sources, A and B, can be generated in a trough of mercury (Figure 10.12).

In the figure, solid circles represent crests, and broken circles represent troughs of the two wave sources. When two crests, or two troughs, coincide (at the black dots), both the crests and the troughs are strengthened. When a crest and a trough coincide (at the open circles), one wave is cancelled out by the other. When strengthening takes place, the amplitude of the wave is increased. When wiping out takes place, there will be no wave. In this way, stationary patterns of amplified crests and cancelled-out waves are produced. This illustrates the interference between waves.

A.-J. Fresnel (1788–1827) was the first to obtain

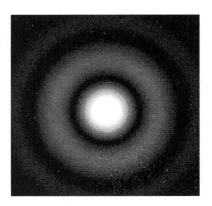

Figure 10.11
Airy disc

Interference of Light Waves

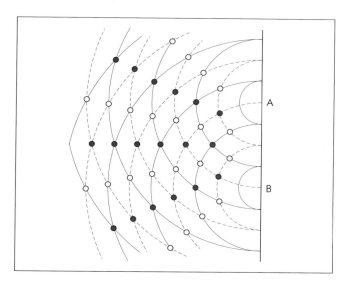

Figure 10.12 Interference between waves to form stationary patterns

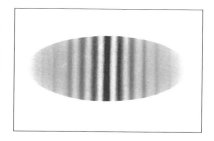

Figure 10.14 Interference fringes

interference fringes experimentally. He used two plane mirrors to obtain two rays of sunlight, from which he was able to show that light is propagated in waves.

The formation of a false disc is illustrated in Figure 10.13, according to an experiment by T. Young in 1802.

A and B represent two very narrow slits (at right angles to the plane of the paper), very closer together, at a small distance d apart. Sunlight is sent through a collimating lens, so that the rays falling on the slits A and B are parallel and in phase.

PCP' is a ground-glass screen on which the light can be seen. The point P is so situated that BP is one-half wavelength ($= BG = \frac{1}{2}\lambda$) longer than AP. At P the two waves are therefore out of phase and will cancel each other, forming a dark band.

At C, which is equidistant from A and B, the crests of the two waves will coincide and strengthen each other to form a bright band or fringe.

Beyond P there will be another bright fringe, followed by a second dark fringe, and so on (Figure 10.14). Similar fringes will be formed between C and P'.

Since the distance AB ($= d$) is very small, arc AG, with P as centre, can be considered to be a straight line. The arc AG thus forms right angles with both GP and AP, because these lines are radii of the circle having P as centre.

The two triangles BGA and PCF are similar, having two angles of the one equal to the corresponding angles of the other. Therefore the lengths of sides opposite equal angles are proportional (Appendix 1, section 1.4). That is:

$$\frac{CP}{BG} = \frac{FP}{AB} \quad \text{(opposite angles PFC and PCF)}$$
$$\quad \text{(opposite angles BAG and BGA)}$$

$$\therefore \frac{x}{BG} = \frac{D}{d}, \text{ or } \frac{x}{\frac{1}{2}\lambda} = \frac{D}{d}, \text{ or } \frac{\frac{1}{2}\lambda}{x} = \frac{d}{D}$$

Distance d is known and distances x and D can be measured, so that the wavelength λ can be calculated:

$$\frac{\frac{1}{2}\lambda}{x} = \frac{d}{D}, \text{ so that } \lambda = \frac{2xd}{D}$$

If D is 5 metres and d 1 mm, that is, 1×10^{-3} metre, the wavelength will be:

$$\lambda = \frac{2x(1 \times 10^{-3})}{5} \text{ metres}$$

If the separation between the slits, x, is set at 1.5 mm ($= 1.5 \times 10^{-3}$ metre), λ will be given by:

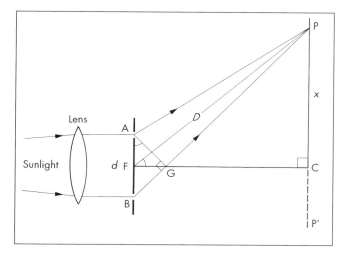

Figure 10.13 Generating a false disc by means of interferometry

$$\lambda = \frac{2(1.5 \times 10^{-3})(1 \times 10^{-3})}{5} = 0.6 \times 10^{-6}$$
$$= 6000 \times 10^{-10} \text{ metre} = 6000 \text{ Å}$$

Alternately, if the wavelength, λ, is known, the distance, d, between the sources of the light rays can be calculated, if x and D are measured. The distance, d, could be the diameter of a star, with points A and B on opposite sides of a diameter. If the distance, D, of the star is known, the star's diameter will be given by:

$$d = \frac{\frac{1}{2}\lambda D}{x}$$

In practice, the angular diameter APB of the star is being measured by this method of interferometry. If the distance of the star is known, the actual angular diameter of the star will be known.

In 1920 the new $2\frac{1}{2}$-metre reflector on Mount Wilson was used to measure the diameter of the star Betelgeuse. A 6-metre long beam was erected across the top of the telescope (Figure 10.15). Two plane mirrors, A and B, were placed 3 metres apart on the beam, making angles of 45 degrees so as to reflect the rays of light from the extremities of the star to the plane mirrors C and D, from where the rays fall on the $2\frac{1}{2}$-metre mirror. A shield was so placed that no light from the star could fall directly on the main mirror.

By altering the distance between the mirrors A and B, moving them symmetrically, interference fringes were

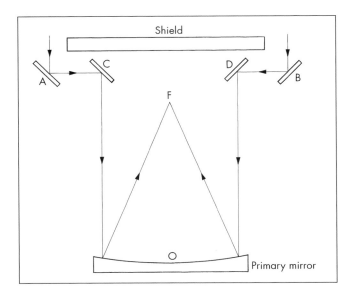

Figure 10.15 Principle of the interferometer

Figure 10.16 Airy diffraction disc

obtained from the two sources A and B, in the form of a false disc (Figure 10.16).

The plane mirrors A and B now took the places of the narrow slits of Figure 10.13. A microscope, having a magnification of 1600 times, was used to measure the distances between the interference fringes. This distance was found to be 0.005 mm. The plane mirrors A and B were then systematically moved further apart, while the distance between mirrors C and D was kept at 114 cm. The separation of A and B was continued until the interference fringes just vanished. If the alignment of the mirrors had been disturbed, the fringes would also have vanished. To ascertain whether this had perhaps happened, the telescope was directed on to another star. Reappearance of the interference fringes proved that the optical train had remained in alignment.

The disc of a star is not identical to the narrow slits. Experiments done in the physics laboratory showed that the angular measure of a disc works out to $1.22\lambda/d$ (known as the Rayleigh limit), where d is the distance between the two mirrors A and B, and λ is the wavelength.

IN the case of Betelgeuse, the fringes just vanished when the mirrors A and B were 307.34 cm apart.

Before the angular diameter, θ, could be calculated, the wavelength of the light from the star had to be determined. The distance x between the interference fringes was found to be 0.0005 cm. The distance CD = a was found to be 114 cm, and the focal length, f, of the main mirror (= FO in Figure 10.15) was 1000 cm. The wavelength, λ, was given by: $\lambda = ax/f$, that is:

$$\lambda = 114(0.0005) \div 1000$$
$$= 0.000\,057 = 5700 \times 10^{-8} \text{ cm}$$

Stellar Diameters

The angular diameter of Betelgeuse is therefore given by:

$$\theta = 1.22\lambda/d$$
$$= \frac{5700 \times 10^{-8} \times 1.22}{307.34} = 0.000\,000\,226\,2 \text{ rad}$$

In arc seconds, this angle is equal to:

$$0.000\,000\,226\,2 \times 206\,265 = 0.046\,67''$$

The angular diameter of Betelgeuse must therefore be at least 0.046 arc second. At an estimated distance of 200 parsecs, namely 652 light years, the diameter of Betelgeuse is therefore $0.046 \times 200 = 9.2$ AU. This is equal to $9.2 \times 149\,597\,870$ km $= 1376$ million km. This is 995 times the Sun's diameter. The radius of Betelgeuse, 4.6 AU, means that if Betelgeuse were placed at the Sun's centre, its surface would almost reach the orbit of Jupiter!

The brightness of Betelgeuse changes from time to time, but not by very much: this is due to its expanding and shrinking, as many other variable stars do.

IN order to obtain a greater separation between the incident rays from the opposite sides of a star, R.H. Brown built a circular rail track. On this he mounted two compound mirrors, each consisting of 252 hexagonal mirrors, 38 cm in diameter, thus obtaining a total diameter of 6.5 metres. By moving the compound mirrors on the rail track, he obtained a separation of 188 metres. The incident rays were focused on to photo-electric cells and a computer analysed the electron flux and thereby obtained an interference pattern. By 1970, Brown and his colleague J. Davis had succeeded in measuring the diameters of 27 stars, some as small as 0.001 arc second.

THE diffraction pattern in a large telescope is inclined to flicker on account of vibrations in the atmosphere. To overcome this, A. Labeyrie, D.Y. Gezare and R.V. Stachnik developed the method of speckle interferometry. The flickering patterns are registered at great speed and combined again by a computer to form analysable interference patterns. This is one of the most modern developments in astronomy.

Diameters of Stars By Means of Occultations

D.S. Evans, T.G. Barnes and C.H. Lacy made use of occultations of stars by the Moon to obtain interference patterns. When the Moon occults a star, the star's light is seen by the unaided eye to vanish instantaneously, but when the light from the star is monitored by means of high-speed photo-electric devices, interference patterns are obtained. The varying intensities of the rays of light grazing the limb of the Moon during the last fraction of a second, before the star vanishes, are converted by a photo-electric cell into varying electric impulses, having time intervals between 0.001 and 0.008 second. Figure 10.17 shows the varying light intensities in the diffraction fringes produced by the light from a single star.

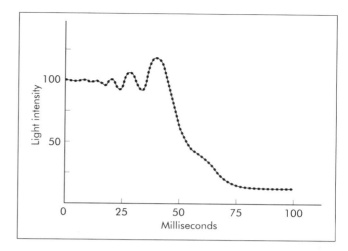

Figure 10.17 Diffraction during occultation. Graph by P. Bartholdi, Haute Provence Observatory

It is easy to measure the distances between the diffraction fringes and then to calculate the angular diameter of the star.

This method also serves well to indicate the existence of double stars and to measure their angular separations. The method also works in cases where the stars are so close together that they cannot be resolved by means of a telescope.

Table 10.5 Angular diameters of stars (McDonald Observatory)

Star	Spectral type	Angular diameter
R Leonis	M6.5e/M9e	0.076
α Scorpii	M1 Ib	0.041
μ Geminorum	M3 III	0.0135
TX Piscium	C	0.0098
X Sagittarii	K2 III	0.0044
ξ^2 Sagittarii	K1 III	0.0030
42 Librae	K4 III	0.0025

Good results have been obtained by the McDonald Observatory, Texas, angular diameters as small as 0.0025 arc second having been measured. The stars listed in Table 10.5 are all giants, but the method works equally well for smaller stars, provided they are bright enough.

Diameter of Star By Means of Colour Index

The light intensity received from a star depends on its angular diameter. It ought to be possible to deduce the surface temperature of a star from its angular diameter.

If the magnitude of a star, in terms of its total radiation over all wavelengths, that is, its bolometric magnitude, is known, it ought to be possible to calculate its angular diameter.

To determine the bolometric magnitude, the visual magnitude, V, has to be adjusted. This adjustment is known as the bolometric correction.

Evans, Barnes and Lacy (mentioned above) realised that, while they were busy measuring the angular diameter of a star, they were actually measuring a combination of the star's temperature and its bolometric correction. They called this the parameter for surface brightness F_v. F_v must, to some extent, correlate with the colour index. They found a fair correlation with the $B - V$ colour index, but the correlation with the $V - R$ colour index was very good.

The Barnes–Evans formula which they derived for the parameter for surface brightness F_v is as follows:

$$F_v = 4.2207 - 0.1V - 0.5 \log \psi'$$

where 4.2207 is a constant derived from calibration of the Sun with regard to F_v, V is the visual magnitude, and ψ' the angular diameter of the star in arc seconds.

If F_v can be derived from the colour index and V is measured, then ψ' is the only unknown in the formula and can be calculated.

Figure 10.18 contains the data for stars of which the angular diameters were determined by means of occultations.

The graph shows a regular relationship between the $V - R$ colour index and the visual surface brightness parameter F_v. To find the value of $V - R$, the apparent visual magnitudes of the stars are determined through yellow and red filters. When this is known, the value of F_v can be read off the graph. By substituting this value of F_v in the Barnes–Evans formula, the angular diameter ψ' can be calculated in milli-arc seconds.

We can test this on the well-known star Epsilon Eridani (03 32 23 –09 29.8), the tenth nearest star to the Earth. Its

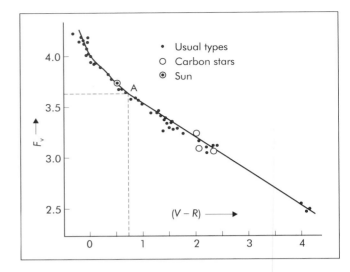

Figure 10.18 Relation between $(V - R)$ colour index and visual surface brightness parameter

apparent visual magnitude is 3.73 and its $V - R$ colour index is 0.72.

Draw the vertical above the point 0.72 on the $V - R$ axis to cut the line on the graph at A. From A, draw the horizontal line to cut the F_v axis; it cuts this axis at 3.65. Substitute this value for F_v and 3.73 for V in the Barnes–Evans formula:

$$F_v = 4.2207 - 0.1V - 0.5 \log \psi'$$
$$\therefore 3.65 = 4.2207 - 0.1 \times 3.73 - 0.5 \log \psi'$$
$$\therefore 0.5 \log \psi' = 4.2207 - 0.1(3.73) - 3.65$$
$$= 0.1997$$
$$\therefore \log \psi' = 0.1997 \div 0.5 = 0.3954$$
$$\therefore \psi' = \text{antilog } 0.3954 = 2.485$$

Thus, the angular diameter of Epsilon Eridani is 2.485 milli-arc seconds, namely 0.002 485 arc second. This is equal to 0.002 485 ÷ 206 265 radians = R radians, which in turn is equal to the sine of the angle of 0.002 485 arc second. This angle is subtended by Epsilon Eridani, whose distance is 3.30 parsecs. This is equal to:

$$3.30 \times 206\,265 \times 149\,597\,870 \text{ km} = A \text{ km}$$

The diameter of Epsilon Eridani is thus equal to R times A, namely 1 226 777 km. It can therefore safely be stated that the diameter of Epsilon Eridani is at least 1 200 000 km. This is seven-eighths of the Sun's diameter of 1 392 530 km. Since Epsilon Eridani is a K2 star, of absolute magnitude 6.1

(compared with the Sun's absolute magnitude of 4.85), it can be expected that it will be somewhat cooler and smaller than the Sun. The Barnes–Evans theory therefore agrees with measurements.

The size of a star will certainly be affected by its mass. To find out how the masses of the stars are determined, we shall have to examine double stars, that is, stars that revolve around each other. While they revolve around each other, they are bound in each other's gravity, which depends on the masses of the stars.

2 Double Stars

In Tables 10.1 and 10.3, stars such as Alpha Centauri, 61 Cygni, Alpha Crucis, etc. are indicated as being double. To the naked eye they appear as single points of light, but a telescope resolves them into separate points. This may be due to their being nearly in the same direction, without any physical connection. Such stars are called optical doubles. They usually maintain their relative positions, even for centuries, but if the stars of a double are gravitationally connected, it will be seen, after some years of observation, that they revolve around each other. More correctly, they revolve around their common centre of mass, their barycentre.

The first star which was telescopically discovered to be double was Mizar (Figure 10.19), namely Zeta Ursae Majoris (13 23.9 + 54 56), the discoverer having been G.B. Riccioli, in 1651.

People with good eyesight were able to distinguish the two stars Mizar and Alcor, in the tail of the Great Bear, but it required a telescope to resolve Mizar itself into a double.

Alpha Crucis was the first star in the Southern Hemisphere which was identified as double, by G. Tachard in 1685, at Cape Town. The two members of Alpha Crucis are very bright, having absolute magnitudes of −3.7 and

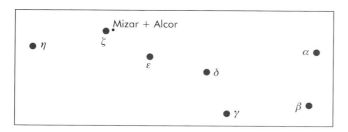

Figure 10.19 The double star Mizar, in the constellation Ursa Major

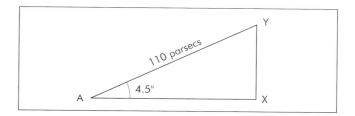

Figure 10.20 Distance between Alpha Crucis A and B

−3.2. They are thus 2600 and 1600 times brighter than the Sun, whose absolute magnitude is 4.85. Their spectral types, B1 IV and B3 II, show that they are bright giants. Since 1862 their separation has decreased from 5.4″ to 4.5″, although their position angle has changed only slightly. Their distance from the Earth is estimated at 110 parsecs, very nearly 360 light years.

In Figure 10.20, X and Y represent the two members and A represents the Earth. The angle XAY, of 4.5 arc seconds, is equal to 4.5 ÷ 206 265 = 0.000 021 82 radians. This means that sin XAY = XY ÷ 110 = 0.000 021 82. Therefore:

$$XY = 0.000\,021\,82 \times 110 = 0.0024 \text{ parsec}$$
$$= 0.0024 \times 206\,265 = 495 \text{ AU}$$

The two members of Alpha Crucis are thus 495 AU apart. We can roughly estimate their masses as each being twice that of the Sun, and their barycentre as being 247.5 AU from each. Now, by applying Kepler's third law:

$$T^2 = A^3 \div (2+2) = 247.5^3 \div 4$$
$$\therefore T = 1947 \text{ years}$$

They therefore take almost 2000 years to complete one revolution. That is why there has been no appreciable change in their relative positions since 1862.

THE monitoring of double stars is very interesting and also very useful. All that an amateur astronomer needs is a telescope with a mirror of 150 mm, and either a filar micrometer in the eyepiece, or a 35 mm camera, attached to the telescope, so as to be able to photograph the double stars from time to time and then to measure their positions on the photographic plates.

Table 10.6 (*overleaf*) gives the position angles and the angular separations of the double star 70 Ophiuchi (18 05.5 + 02 30), for the years 1970 to 1980.

The data of Table 10.6 are represented in Figure 10.21 (*overleaf*). The primary star, A, is kept at the origin and the position angles occupied by the B component are indicated by the vectors AB, etc. The position angles are measured

Table 10.6 The double star 70 Ophiuchi

Date	Position angle	Separation
1970	52.5°	2.6″
1972	37.5°	2.4″
1974	18.0°	2.3″
1976	353.0°	2.3″
1978	334.0°	2.5″
1980	317.0°	2.6″

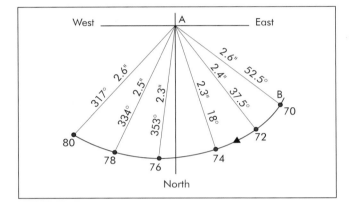

Figure 10.21 Position angles and separations of 70 Ophiuchi A and B, 1970–1980

from north through east. This star, which was discovered as a double by William Herschel in 1779, is 16.5 light years from the Earth. The greatest separation between the members, their apastron, was reached in 1845 and again in 1933. Their periastron, closest approach to each other, took place in 1895 and 1982. The plane of the orbits, of eccentricity 0.5, is inclined by 121 degrees to the plane of the sky. The data concerning the components of 70 Ophiuchi are given in Table 10.7.

Another double which has a beautiful appearance in the telescope is Alpha Centauri. Its A component has a visual magnitude of 0.00 and absolute magnitude 4.4 (compared with the Sun's 4.85). The intrinsic brightness of Alpha Centauri A is 1.5 times greater than that of the Sun:

$$2.512^{4.85-4.4} = 2.512^{0.45} = 1.5 \text{ times}$$

The absolute magnitude of component B is 5.7 and it has only 0.457 times the brightness of the Sun.

N.L. de Lacaille discovered Alpha Centauri to be double in 1752. The period of revolution is 80 years. The semi-major axis of the orbit of B around A is 17.6″ and the apparent separation varies between 2″ and 22″. The pair were at apastron in 1995 and periastron in 1955.

Figure 10.22 shows the apparent orbit of component B around component A, as seen from the Earth. The plane of the orbit is inclined by only 11 degrees to the line of sight, making it appear very eccentric, while the actual eccentricity is 0.52, as shown in Figure 10.23. The distance between the two components varies from 11 to 35 AU.

When measuring the position angle between the members of a double star, the crosswires in the eyepiece must be set so that the vertical wire passes through the brighter of the pair and points north–south, such as AN in Figure 10.22. The horizontal wire must then be turned so that it passes through both stars, for example AB_1 or AB_2,

Table 10.7 Double star 70 Ophiuchi

| Component | Magnitude | | Spectrum | Mass | Diameter |
	Visual	Absolute		Sun = 1	
A	4.2	5.8	K0V	0.82	0.9
B	5.9	7.5	dK6	0.6	0.7

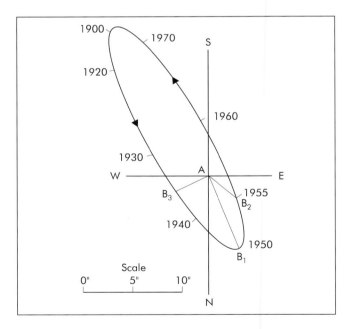

Figure 10.22 Apparent orbit of Alpha Centauri B around Alpha Centauri A

through east to AB_3. When the fainter member is at B_1, the position angle is NAB_1, when it is at B_2, the position angle is NAB_2, and so on.

When the time of revolution, the period, is known, the mean distance, A, between the two members can be calculated by using Kepler's third law. If T and A are in years and AU, $A^3 = T^2$.

The average period of Alpha Centauri is 80 years. Therefore $A^3 = 80^2 = 6400$. Thus

$$A = \sqrt[3]{6400} = 6400^{1/3} = 18.566 \text{ AU}$$

The length of the semi-major axis of the elliptical orbit is therefore 18.566 AU, nearly as large as the average distance of Uranus from the Sun, 19.19 AU. It was found by G.P. Kuiper that the average of the average separations of double stars is 20 AU.

The actual orbits of the two members of Alpha Centauri, when viewed at right angles to the plane of the orbits, are ellipses of eccentricity 0.52 (Figure 10.23). When the more massive component, A, is at points A, B, C and D, the lighter component, B, is at points 1, 2, 3 and 4 respectively. The straight line through the centres of the two stars always passes through the barycentre, S, for example A–S–1, B–S–2, C–S–3 and D–S–4.

The measurements by means of the filar micrometer depict the massive member as being stationary at A, one of the foci of the large ellipse YZ1X, and the less massive member as revolving in the large ellipse, YZ1X.

The distance of Alpha Centauri A from the barycentre is always 43/55 times the distance of Alpha Centauri B from the barycentre. From this it follows that:

$$\frac{\text{Mass of component A}}{\text{Mass of component B}} = \frac{55}{43} = 1.279$$

The mass of component A is therefore 1.279 times that of component B.

If the masses of the two members are M_1 and M_2 then the gravitational force which M_1 exercises on M_2 is

$$\frac{G_1 M_1 M_2}{r^2}$$

where r is the mean distance between the two masses M_1 and M_2, that is, the semi-major axis of the elliptical orbit 1–X–Y–Z of component B around component A.

The gravitational force which M_2 exercises on M_1 is

$$\frac{G M_2 M_1}{r^2}$$

These two gravitational forces are equal, but act in opposite directions. They illustrate Newton's third law: every action is accompanied by an equal but opposite reaction.

The two forces which the two stars exercise on each other cause accelerations towards each other, which are each equal to the force divided by the mass being accelerated. Thus the acceleration of M_1 on M_2 is

$$\frac{G M_1 M_2}{r^2} \div M_2 = \frac{G M_1}{r^2}$$

So, too, the acceleration of M_2 on M_1 is $\dfrac{G M_2}{r^2}$

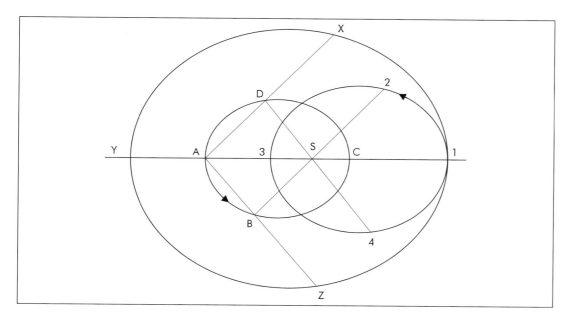

Figure 10.23
Actual orbit of Alpha Centauri A is A–B–C–D; that of Alpha Centauri B 1–2–3–4. Vectors AX and B2 are parallel and equal. Vector AZ is parallel and equal to vector D4. S is the barycentre

The total acceleration is the sum of the two accelerations. It is equal to:

$$\frac{GM_1}{r^2} + \frac{GM_2}{r^2} = \frac{G}{r^2}(M_1 + M_2)$$

Since G/r^2 is a pure number, we can say:

> The total acceleration is proportional to the sum of the masses of the two stars $(M_1 + M_2)$.

In the case of the Sun, the total mass of all the planets combined is infinitesimal in comparison with the mass of the Sun, so the masses of the planets can be ignored. But we see that, in the cases of double stars, the sum of the two masses must be considered.

If A is the mean distance between the members of a double star, a the mean distance of the Earth from the Sun, T the period of the double star and t the period of the Earth ($= 1$ year), then combining Newton's law of gravitation with Kepler's third law:

$$\frac{M_1 + M_2}{M_{Sun} + M_{Earth}} = \frac{A^3}{T^2} \div \frac{a^3}{t^2}$$

We now take the Sun's mass as equal to 1; a and t for the Earth are equal to 1 AU and 1 year, respectively. Therefore

$$\frac{M_1 + M_2}{1} = \frac{A^3}{T^2} \div \frac{1^3}{1^2}$$

i.e. $$M_1 + M_2 = A^3 \div T^2$$

We see that the sum of the masses of the two stars, in terms of the Sun's mass equal to 1, is equal to the cube of the mean distance in AU between the two stars, divided by the square of the mean period of revolution.

In order to apply this formula, T and A have to be measured. T can easily be determined by observing the pair of stars for some years. A can be measured in arc seconds ($= x$). To convert this to AU, the distance of the star must be known, that is its parallax p.

If the distance is d, then $\sin p = 1/d$. If the mean distance between the stars is r, then $\sin x = r/d$ (see page 158).

$$\therefore \frac{\sin x}{\sin p} = \frac{r}{d} \div \frac{1}{d} = r \quad \text{and} \quad r = \frac{x}{p}$$

Because the angles x and p are very small, $\sin x = x$ and $\sin p = p$.

$$\therefore M_1 + M_2 = \frac{r^3}{T^2} = \left[\frac{x}{p}\right]^3 \div T^2$$

In the case of Alpha Centauri, the semi-major axis $x = 17.66''$ and the parallax $p = 0.76''$.

$$\therefore M_1 + M_2 = \left[\frac{17.66}{0.76}\right]^3 \div 80^2$$

$$= (23.237)^3 \div 6400 = 1.96$$

The sum of the masses of the components of Alpha Centauri is therefore equal to 1.96 solar masses.

Since the size of the ellipse of component B (1-2-3-4) (Figure 10.23) is 55/43 times the size of the ellipse of component A (A-B-C-D), the mass of A is 55/43 times the mass of B.

$$\therefore M_1 = \frac{55 M_2}{43} \quad \text{and} \quad M_1 + M_2 = 1.96$$

$$\therefore \frac{55 M_2}{43} + M_2 = 1.96$$

i.e. $$\frac{98 M_2}{43} = 1.96$$

$$\therefore M_2 = 0.86 \quad \text{and} \quad M_1 = \frac{55}{43}(0.86) = 1.1$$

The mass of Alpha Centauri A is therefore equal to 1.1 solar masses and that of Alpha Centauri B is 0.86 solar masses.

Double stars therefore provided the secret key whereby the masses of the stars could be determined. The values of the masses of the stars proved that the supposition that was made from their spectra, that they were bodies like the Sun, was correct.

Alpha Centauri A is of the same spectral type and class, G2 V, as the Sun. That its mass is almost identical to that of the Sun shows that there is a close relationship between spectral class and mass. Its absolute magnitude of 4.4 shows that it is somewhat brighter than the Sun, being 1.5 times more luminous.

Among the nearer stars, we find that the masses of the two members of Procyon (in Canis Minor) are 1.78 and 0.65 solar masses; 61 Cygni has masses of 0.63 and 0.60 solar masses; UV Ceti, 0.044 and 0.035 solar masses. The Sun is thus 22.7 and 28.5 times more massive than the two members of UV Ceti.

In general, most stars have masses between 0.1 and 10 times that of the Sun. Masses greater than 10 times are not unknown. In 1922, J.S. Plaskett discovered a super-giant double star (HD 47129) situated at 03 34 + 06 11 in the constellation of Monoceros, 1.25 degrees south-west of the Rosette Nebula, NGC 2237. The masses of the two members could be 30 and 40 solar masses, although Plaskett claimed one member to have a mass 60 times that of the Sun. Appendix 4 contains the data for about 40 double stars.

William Herschel was the first astronomer to announce the existence of double stars that have a physical relationship and that revolve around each other. In 1802 he published the orbital positions of Castor A and B, which

Double Stars

he had monitored since 1779.

If the stars are counted separately, at least 50% are found to be double. The *I D S Catalogue* contains the data about 65 000 multiple systems that are close enough to be studied through Earth-based telescopes.

For the study of double stars, a telescope of long focal length is necessary. The 68-cm refractor at the Johannesburg Observatory has a focal length of 10.917 metres. It was used for 40 years by R.T.A. Innes, H.E. Wood, W.H. van den Bos and W.S. Finsen.

In 1912, R.T.A. Innes discovered that Alpha Centauri A and B had a third member, about 13 000 AU from their barycentre. It is on the Earth's side of Alpha Centauri A and B and is therefore the nearest star; it carries the name Proxima Centauri. Its parallax is 0.762″ and is 0.02 light years closer than Alpha Centauri, namely 4.28 light years distant. It is only one-twentieth the size of the Sun and its absolute magnitude is 15.1. It is a flare star and periodically suddenly brightens by a magnitude or so – a star most unsuitable to sustain habitable planets.

Among the noteworthy doubles, we find the beautiful pair Xi Boötis (14 51.4 + 19 06), which W. Herschel discovered in 1780. It is one of the nearer doubles, being only 22 light years distant. Its period is 150 years and mean separation 5″. The two stars have a beautiful colour contrast: one is yellow (G8 V) and the other reddish-purple (a K4 V star). Their apparent visual magnitudes are 4.7 and 6.8, and they are easily visible in a small telescope. The two members were at their furthest apart (at apastron) in 1984.

The red giant Antares is situated in Scorpius (16 29.5 −26 25.9). It is coloured orange or saffron-red and has a green companion, which was discovered in 1819 by Professor Burg, during an occultation by the Moon. It is not easy to see the companion, because of the glare of Antares. The companion is of spectral type dB4 and it is actually 50 times more luminous than the Sun. The luminosity of Antares is 9000 times that of the Sun.

Eta Coronae Borealis (15 22.9 + 31 33) has completed four revolutions since its discovery in 1826 by F.G.W. Struve; its period of 41.56 years is therefore very well known. The two members are very similar to the Sun, with regard to mass, size and luminosity. The apparent visual magnitudes are 5.7 and 6.0, easily visible in a telescope, although their greatest separation is only 1 arc second.

Omicron-2 Eridani (04 15.2–07 39) is an exceptional double. It is actually a triple star, since the companion itself is a double, consisting of a white dwarf (DA5) and a red dwarf (dM 4e). These two have a period of 248 years and their average separation is 34 AU. They were at apastron in 1990, when they were 9″ apart. The data, from *Burnham's Celestial Handbook*, regarding the two members of the

Table 10.8 Omicron-2 Eridani

	Member B	Member C
Visual magnitude	9.7	10.7
Absolute magnitude	11.2	12.3
Luminosity (Suns)	0.0027	0.0008
Spectrum	DA5	dM 4e
Mass (Suns)	0.44	0.20
Diameter (Suns)	0.019	0.43
Density (g cm^{-3})	65 000	1.8

companion of Omicron-2 Eridani, also known as 40 Eridani, are given in Table 10.8. The data are very accurate because Omicron-2 Eridani is very close, being only 4.9 parsecs from the Earth.

Alpha Geminorum (Castor) situated at 07 34.6 + 31 53.3, in Gemini, has an A component which consists of two white stars, A1 V and A2 V, that revolve around each other in only 9.2128 days. They are each about twice the size of the Sun and twelve times as bright. The B component also consists of two stars, B1 and B2, that revolve in 2.9383 days! They are each 1.5 times the Sun's size and 6 times brighter. The two pairs take 400 years to complete one revolution. At a distance of 72.5″, that is, 1000 AU, a third member which has two sets of lines in its spectrum is also a double. It is known as YY Geminorum and it takes thousands of years to complete one revolution around the barycentre of the A and B components. Castor is therefore a very exceptional group of stars.

70 Ophiuchi (18 05.5 + 02 30) has two members of types K0 V and dK6, of apparent magnitudes 4.2 and 5.9: one is yellow and the other is orange. Their mean separation of 4.2″ is 23 AU and their period is 87.85 years. Situated at a distance of 17.3 light years, 70 Ophiuchi is near enough to be studied in detail. It seems that there are inexplicable deviations in the orbits of the two members. These could be caused by an invisible body, or bodies, of mass about 1% of the Sun's mass. Possibly it is a large planet, having 10 times the mass of Jupiter. It could also be a very faint star.

During the years 1834 to 1844, F.W. Bessel found that there were irregularities in the proper motion of Sirius, against the background stars. Sirius did not move in a straight line. Bessel ascribed this to the gravitational influence of an invisible companion (Figure 10.24, *overleaf*). He calculated that the invisible companion has a period of revolution of 50 years, approximately (Figure 10.25). Double stars which are discovered by measurement are called astrometric binaries.

In 1851, C.H.F. Peters calculated the orbit of the invisible

Figure 10.24 Sirius and companion

companion, using Bessel's measurements. The letters *a*, *b*, *c*, *d* and *e* in Figure 10.25, on the solid curve, indicate Sirius' positions, and the calculated positions of the companion are indicated by the numerals *1, 2, 3, 4* and *5*.

In 1862, A.C. Clark, using a 47-cm refractor, succeeded in spotting the companion (Figure 10.24). Its magnitude was found to be 8.65; it is therefore at least 11 000 times fainter than Sirius (magnitude −1.46)!

$$2.512^{8.65-(-1.46)} = 2.512^{10.11}$$
$$= 11\,071$$

That is why it is lost in the glare of Sirius, making it difficult to see and to photograph. Figure 10.24 was made by R.B. Minton of Marietta Aerospace by printing the image through a red filter on to black and white film. The distance of the companion from Sirius varies between 3″ and 11.5″. The period of revolution is 49.98 years, a remarkable confirmation of Bessel's prognostication.

The two members were at their furthest apart (apastron) in 1875 and again in 1975. When Bessel made his measurements in 1834–1844, Sirius A and B were at positions *b* and *2* (Figure 10.25). Sirius' deviation from the straight line was then changing most rapidly and that made it easier for Bessel to spot the deviation – strange how good luck plays into the hands of the devoted worker!

When the orbits of Sirius A and B were known, it was easy to calculate the masses of the two stars: Sirius A was found to be 2.35 solar masses and B 0.98 solar masses. How is it possible that a star, having a mass almost identical to that of the Sun, can be so faint? At a distance equal to that of Sirius, the magnitude of the Sun would be 0.34. How is it then that the magnitude of Sirius B is 8.65? The only feasible explanation was that Sirius B had to be very small – a dwarf star. But dwarf stars are usually red, while the spectrum of Sirius B is type A, which has a temperature of about 9000 K, compared with the Sun's 6000 K. To be white, but also faint at that distance, its diameter could be no more than 30 000 km, that is, 1/46 of the Sun's diameter and only $2\frac{1}{3}$ times that of the Earth. Sirius B must therefore be a white dwarf. Many more white dwarfs have since been discovered. Their absolute magnitudes range from 9 to 16, and the fainter ones require large telescopes.

With a solar mass packed into 1/46th of the Sun's diameter, the density of Sirius B has to be very high. It works out at 88 kg cm^{-3}, 8000 times that of lead. This was the first indication that matter could be compressed by a factor of more than 1000 times.

Proper Motion of Stars

Although the starry skies appear to be immutable, the stars do possess motions, called proper motions. The proper motion of a star is the angular measure in arc seconds which a star moves in one year, in the plane of the heavens, at right angles to the line of sight from the Sun. Barnard's star (17 57.8 + 04 41.6) has the largest proper motion known: 10.25″. This amounts to 1 degree in 351 years: it is a dM5 red dwarf, of apparent magnitude 9.53 and absolute magnitude 13.4. At 5.98 light years, it is the second nearest star to the Earth, if the three members of Alpha Centauri are counted as one.

Stars also have motions towards or away from the Earth (or rather, the Sun). This is known as radial velocity and it is measured by means of the Doppler effect: it is listed as km s^{-1}, along the line of sight from the Sun.

If V is the radial velocity and c the velocity of light, then a wavelength λ will be displaced by a small amount $\Delta \lambda$ in the spectrum, given by:

$$V = c\frac{\Delta \lambda}{\lambda}$$

If the hydrogen alpha line, 6563 Å, is displaced by 1 Å, the radial velocity causing the shift is equal to:

$$V = c\frac{1}{6563} = \frac{300\,000}{6563} = 45.7 \text{ km s}^{-1}$$

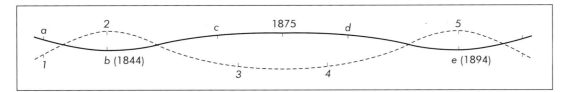

Figure 10.25 Proper motion of Sirius (solid curve). Path of companion (broken curve)

Double Stars

Table 10.9

Star	Proper motion (arc seconds per year)	Radial velocity (km s^{-1})
Barnard's star	10.25	− 67.5
Kapteyn's star	8.72	+151.3
Alpha Centauri	3.67	− 13.8
Sirius	1.32	− 5.0
Aldebaran	0.20	+ 54.1
Canopus	0.03	+ 20.5

Table 10.9 contains the annual proper motions and radial velocities of a few stars. The radial velocities are correct to one-tenth of a km per second. If the motion is away from the Sun, it is indicated as positive; towards the Sun as negative.

The real velocities of the stars in space can be resolved into two components: the transverse velocity, at right angles to the line of sight; and the radial velocity, along the line of sight.

In Figure 10.26, B represents Barnard's star, and A the Sun. The radial velocity of Barnard's star towards the Sun is 67.5 km s^{-1}, and its proper motion is 10.25″ per year. BD is the resultant into which the two velocities can be compounded.

The angular measure, 10.25″, is equal to 10.25 ÷ 206 265 = 0.000 049 69 radians. At a distance of 6 light years, Barnard's star therefore moves through a distance of 6(9.46)(10^{12})(0.000 049 69) km in one year. Its transverse velocity is thus this value divided by 365.25(24)(3600) seconds, namely 89.4 km s^{-1}. This is represented by BC. The velocity towards the Sun, 67.5 km s^{-1}, is represented by BA. The value of the resultant, BD, can be calculated by means of the parallelogram of velocities.

Since ABCD is a rectangle, CD = AB. To solve triangle BCD, the value of angle θ must be calculated:

$$\tan \theta = \frac{DC}{BC} = \frac{67.5}{89.4} = 0.7550$$
$$\therefore \theta = 37°$$

$$\sin \theta = \sin 37° = 0.6018 = \frac{DC}{BD}$$

$$\therefore BD = \frac{67.5}{0.6018} = 112 \text{ km s}^{-1}$$

Therefore, the real velocity in space of Barnard's star is 112 km s^{-1}, in the direction BD, obliquely past the Sun. It will be at its closest to the Sun (and the Earth) at point E.

Let AB now represent the distance of Barnard's star, 6 light years. Therefore:

$$AE = 6 \cos 37° = 6(0.7986) = 4.8$$

The closest that Barnard's star will get to the Sun is 4.8 light years; it will never be as close as Alpha Centauri is now.

At its velocity of 112 km s^{-1}, it will take Barnard's star 9662 years to reach that closest point; then it will speed further away.

Since 1916, Barnard's star has been carefully monitored by means of the 61-cm refractor at Sproul Observatory. P. van de Kamp found that there are regular oscillations around its mean motion in a straight line. The deviations have periods of 11.7 and 20 to 26 years. These can be ascribed to the gravitational attractions of two companions, 0.8 and 0.4 times the mass of Jupiter, revolving at distances of 2.7 and 3.8 AU from the star. According to its luminosity, the mass of Barnard's star is only 0.14 solar mass. A planet

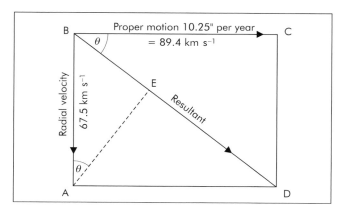

Figure 10.26 Motion in space of Barnard's star

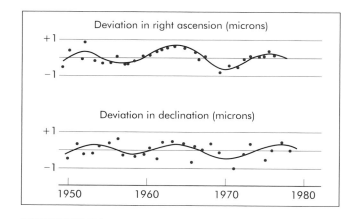

Figure 10.27 Deviations in the proper motion of Barnard's star

of 0.8 Jupiter mass could therefore have an appreciable effect on the straight line course of the star. Figure 10.27 (*previous page*) shows the deviations in right ascension and in declination of the star, as published in *Sky and Telescope*, March 1979.

Canopus, the second brightest of the stars, has a proper motion of 0.03″ per year and a recession velocity of 20.5 km s^{-1}. One could perhaps think that Canopus is moving directly away from the Sun, but at its distance of 1174 light years, a proper motion of 0.03″ means a transverse velocity of 51 km s^{-1}. The nearest that Canopus ever got to the Sun was 1090 light years, and that was 8.7 million years ago.

Spectroscopic Binaries

Some double stars are so close together that they cannot be resolved optically into separate points of light. It may be that the system is very far away, or that the stars are really very close together. The spectroscope, however, reveals two sets of lines, meaning that there must be two sources of light. Each set of lines reveals the spectral type and class of each star. If the plane of the orbits of the two stars is somewhat inclined to the plane of the sky, the spectral lines will shift alternately from the red to the blue and back again, corresponding to the period of revolution of the stars. These stars are called spectroscopic binaries.

The first spectroscopic binary to be discovered was Mizar (Zeta Ursae Majoris), by E.C. Pickering in 1889. Table 10.10 contains some examples of spectroscopic binaries.

The time that the spectral lines take to shift from the red towards the blue, and back, is a measure of the period of revolution of the binary. From the varying radial velocities, the inclination of the plane of the orbits can be calculated, as well as the separation between the members. Then the sum of the masses can be calculated, as in the case of ordinary double stars.

The ratio between the means of the radial velocities of the two stars gives the ratio of the sizes of the elliptical orbits. The inverse of this ratio is the ratio of the masses of the individual stars. From the sum and the ratio of the masses, the individual masses can be calculated, as was done in the case of ordinary doubles.

Multiple Stars

Theta Orionis, the four bright stars in the form of a trapezium, in the heart of the Great Nebula in Orion, M42, provides most of the light which causes the nebula to glow. The trapezium is a very interesting object, easily visible in a small telescope (Figure 10.28).

The brightest of the four stars is C, magnitude 5.4, a super-giant O6 star. Second brightest is D, a B0 star; then follows A, an A7; and B is the faintest, an eclipsing binary, of period 6.471 days. It is also known as BM Orionis.

In 1975, member A was also identified as an eclipsing binary of period 65.432 days, according to R. Burnham (*Celestial Handbook*, Dover Publications Inc., New York). The primary eclipse lasts 20 hours and the secondary eclipse $2\frac{1}{2}$ hours.

Within a radius of 5′ (one-sixth of the Moon's diameter) of the Trapezium, there are no less than 300 stars brighter than magnitude 17. K.A. Strand found that the group is dispersing and that the outward motion is no more than 300 000 years old. This group of stars is therefore very young. The nebula in which these stars are located is the raw material out of which new stars are continually being born.

There are many groups of stars which have a physical relationship, insofar as having a common motion in space. There are many such clusters occurring in or near to the plane of the Milky Way; they are called galactic clusters.

Table 10.10 Spectroscopic binaries

Star	Co-ordinates	Magnitude V	Spectrum
τ Tauri	04 42.2+22 57	4.3	B3 V
ζ Aurigae	05 02.5+41 05	3.75	K4 Ib
β Aurigae	05 59.5+44 57	1.9	A2 IV
β Leonis Min.	10 27.9+36 42	4.20	G8 III
ι Librae	15 22.2−19 47	4.53	B9 IV
ξ Scorpii	16 04.4−11 22	4.16	F6 V
α Herculis B	17 14.7+14 23	5.4	G0 II
τ Sagittarii	19 16.9−27 40	3.32	K1 III
ξ Aquarii	21 37.7−07 51	4.68	A7 V
ω Piscium	23 59.3+06 52	4.01	F4 IV

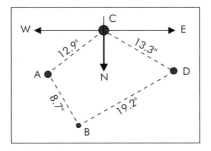

Figure 10.28
Theta Orionis – the Trapezium

Star Clusters

Figure 10.29 Galactic cluster NGC 2437 (or M46)

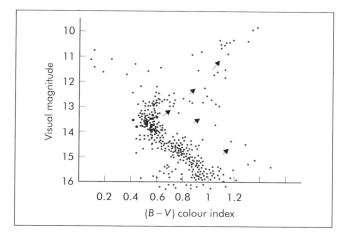

Figure 10.30 H–R diagram of M67

About fifty have been selected from *Burnham's Celestial Handbook*, and are listed in Appendix 5.

The constellation Puppis is richly endowed with galactic clusters: one of the most beautiful of these is NGC 2437 (or M46) at 07 41.8−14 49 (Figure 10.29).

The photograph of M46 was taken by Lowell Observatory, through a 33-cm telescope. There are about 150 stars within a space of 25′, a bit less than the size of the full Moon. At the top of the densest part of the cluster is a star with a faint halo – a planetary nebula, which is not part of the cluster.

NGC 4755, in Crux (12 53.6−60 20), namely Kappa Crucis, is a feast for the eyes. It is rightly called the Jewel Box.

A characteristic of galactic clusters is that the stars in some of them are of the same types. NGC 6405 (M6) in Scorpius (17 40.1−32 13) consists almost exclusively of B stars, the majority of them being in the Main Sequence. This gave rise to the surmise that the stars in a galactic cluster must have had a common origin.

Because the stars in a galactic cluster are so densely packed, they can all be considered as being equally far from the Earth. They must therefore have a linear relationship between their apparent magnitudes and the logs of their luminosities. An H–R diagram of their colour indices and apparent magnitudes can therefore be drawn, without any knowledge of their absolute magnitudes. Figure 10.30 shows that most of the members of the galactic cluster NGC 2682 (M67) in Cancer (08 50.4 + 11 49) belong to the Main Sequence. A considerable number have, however, left the Main Sequence and moved into the giant branch, as is indicated by the arrows. These giants can be identified by their spectra and their brightness. M67 is fairly far from the plane of the Milky Way, 1500 light years, while most of the galactic clusters occur in the plane of the Milky Way. M67 also differs from the other galactic clusters since the latter do not have nearly as many stars that have moved away from the Main Sequence. All the stars in the cluster NGC 2362, in Canis Major, located at (07 18.8−24 57), for example, belong to the Main Sequence.

NGC 6475 or M7 (17 53.9−34 49) is just visible to the naked eye and shows up very well in binoculars. The cluster covers an area of 60′, twice the Moon's diameter. It contains about 50 bright stars, of magnitudes 7 to 11, consisting of types A and B, all in the Main Sequence.

(Rather than spend a large sum of money on a big telescope, an amateur astronomer would do better with a 15 or 20 cm reflector, fitted with a photo-electric photometer. He could then do professional work, determining the apparent visual magnitudes of the stars in a cluster of his choice, through a blue filter (4200 Å) and a yellow filter (5400 Å). The difference between these two magnitudes is the $(B - V)$ colour index. Plotting these against visual magnitudes will yield an H–R diagram of the cluster. He would then be able to determine the distance and the age of the cluster.)

One of the best known among open clusters is the Pleiades, or Seven Sisters, situated in Taurus at 03 47.0 +

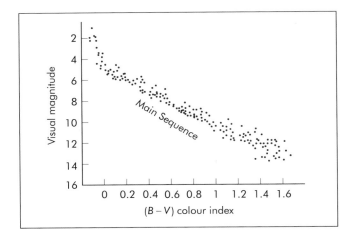

Figure 10.31 H–R diagram of the Pleiades

24 07. The cluster covers an area of 100′ (more than three lunar diameters) and contains about 200 stars. Most of them are bright B types in the Main Sequence of the H–R diagram (Figure 10.31). The figure shows that the stars have not yet left the Main Sequence and therefore they must still be very young, not more than a hundred million years old.

The brightest star of the cluster is Alcyone (Eta Tauri, or 25 Tauri), which is 1000 times brighter than the Sun, and probably 10 times the Sun's diameter. Its apparent magnitude is 2.86 and, at a distance of 410 light years, its absolute magnitude must be −2.64. By contrast, the faintest star in the cluster is about one-hundredth of the Sun's brightness.

The Pleiades are known by their Flamsteed numbers in Taurus (Figure 10.32).

The Pleiades (Figure 10.33) are shrouded in nebulosity, which shows up well in a long time-exposure photograph. The stars which have the densest nebulae are Merope, Alcyone, Maia and Celaeno. It is probable that the nebulae are the remains of the primaeval nebula, consisting of gas and fine dust, out of which the cluster condensed by the process of accretion. The fact that the nebula is still visible also indicates that the stars must be very young. New, young stars have very strong stellar winds. Pleione, in particular, is very active in this regard. In 1888 its spectrum showed that it was surrounded by an expanding shell of gas. By 1903 it had disappeared. In 1937 a second shell of gas and dust appeared, and by 1951 it had also vanished.

V.M. Slipher found that the spectrum of the nebulae in the Pleiades cluster was identical to that of the stars themselves. This shows that the nebulae reflect the radiation which they receive from the stars. The nebula around Merope is the largest, being $2\frac{1}{2}$ light years across. It is listed as a nebula (NGC 1435).

The cluster is moving in a south-south-easterly direction at 5.5″ per century. At its distance of 410 light years, or 126 parsecs, this implies a transverse velocity of 32.8 km s^{-1}. The radial velocities are of the order of

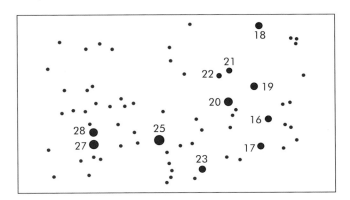

Figure 10.32 The Pleiades.
- 16 Celaeno B7 IV
- 17 Electra B6e III
- 18 B8 V
- 19 Taygeta B6 V
- 20 Maia B7 III
- 21 Asterope B8e V
- 22 Asterope B9 V
- 23 Merope B6 IV
- 25 Alcyone B7e III
- 27 Atlas B8 III
- 28 Pleione B8ep

Figure 10.33 The Pleiades (Hans Vehrenberg photo)

+7.2 km s^{-1}. The real velocity in space works out at 33.6 km s^{-1}. In 30 000 years, the cluster will have moved through a full lunar diameter.

The Hyades

Also in Taurus, 13 degrees south-east of the Pleiades, we find the open cluster of the Hyades (04 27 + 16 25) which has the appearance of a letter V on its side. The V depicts the forehead of the Bull. The brightest star in the V, Aldebaran, is not a member of the cluster, lying halfway between the cluster and the Earth.

Among the brightest stars in the cluster are fourteen type A and seven type F.

The whole custer is moving towards the point 6 h 08 m +9°06′, not far from the bright star Betelgeuse in Orion. Figure 10.34 (from an article by Paul Hodge, editor of the *Astronomical Journal*, which appeared in *Sky and Telescope*, February 1988) shows the directions in which the stars of the cluster are moving. The directions are parallel to each other, but they appear to converge because the cluster is receding and has an effect of perspective.

In order to determine the transverse velocity of the cluster, its distance from the Earth must be determined. Trigonometrically, the parallax of the cluster has been set at 0.02″. The distance is thus 1 ÷ 0.02 = 50 parsecs, namely 163 light years. A difference of only 0.005 arc seconds in the parallax means a difference of between 10 and 16 parsecs in the distance. It therefore became necessary to determine the distance of the cluster by other methods, because the Hyades is one of the nearest galactic clusters and, if its distance can be determined with accuracy, it would affect the whole distance scale of the Milky Way.

L. Boss pointed out in 1908 that the cluster is close enough for the parallel paths of the stars to show perspective. It therefore seems as if the stars are converging to a point. With the passage of time, the cluster will appear to have become smaller, because of its recession. This effect can be used to determine the distance of the cluster. The principle is illustrated in Figure 10.35.

In reality, all the stars are moving in the direction of the arrow, but because the distance from the Earth is increasing, angle θ_2 is smaller than θ_1.

If the radial velocity is V, the cluster will move through a distance Vt in time t.

Because angles θ_1 and θ_2 are very small, $\theta_1 = R/r$ and $\theta_2 = R/(Vt + r)$, where r is the distance of the cluster and R is the width. R remains unchanged.

Therefore $R = \theta_1 r$ and $R = \theta_2(Vt + r)$

so that $\theta_1 r = \theta_2(Vt + r) = \theta_2 Vt + \theta_2 r$

giving $\theta_1 r - \theta_2 r = \theta_2 Vt$

i.e. $r(\theta_1 - \theta_2) = \theta_2 Vt$

Thus, the distance of the cluster is given by:

$$r = \frac{\theta_2 Vt}{\theta_1 - \theta_2}$$

In order to determine the distance r, the two angles θ_1 and θ_2 must be measured at the beginning and end of the lapse of time t. The radial velocity V must also be measured.

L. Boss obtained a value of 129 light years for the distance r, and this was confirmed by H.G. van Bueren who obtained a value of 132 light years by an adaptation of this method.

However, in 1966 G. Wallerstein found, when he took the Wilson–Bappu effect into consideration, that the distance

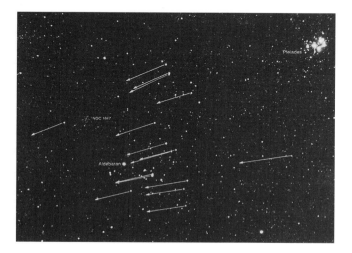

Figure 10.34 The Hyades moving cluster

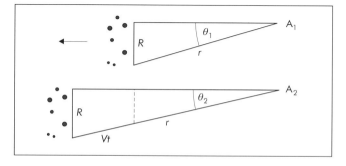

Figure 10.35 Moving cluster appears to become "smaller"

differed considerably from 130 light years. The Wilson–Bappu effect relates the widths of the *H* and *K* absorption lines of calcium in the spectrum to the absolute magnitudes of the stars: the wider the lines, the greater is the luminosity. This provided another method of determining absolute magnitude and therefore of determining distances. In the cases of dwarf stars like the Sun, the *H* and *K* lines become weaker as the star ages. A.H. Vaughan designed a photometer which counted the photon flux for the frequencies of the *H* and *K* calcium lines. The instrument also revealed the fact that some stars undergo cyclical changes, similar to the Sun's sunspot cycle. Fifteen stars, studied by A.H. Vaughan and G.W. Preston, showed an 8.5% variation over periods of 7 to 14 years (compare this with the Sun's 11-year sunspot cycle).

A study of the binary stars in the Hyades cluster showed that their luminosities did not tally with their masses. In order to tally, their luminosities had to be somewhat greater. In other words, their distances had to be more than 130 light years. Applying the Wilson–Bappu effect, the distances had to be 150 light years.

In 1975, R. Hanson studied the fainter members of the Hyades cluster, by measuring their proper motions on photographic plates. From this, he calculated a distance of 157 light years.

Another study, based on proper motions, by T.E. Corbin, yielded a distance of 144 ± 6 light years.

A refinement of the parallaxes of the Hyades gave a distance of 147 ± 6 light years.

Interferometric measurements of the orbits of binaries among the Hyades, and checking these by means of occultations by the Moon, yielded a distance of 145 light years.

One of the Hyades (the star HD 27130, a spectroscopic binary) was found to be an eclipsing binary, which means that the plane of the orbits of the pair lies in the line of sight, so that the stars alternately eclipse each other. Their masses and luminosities could then be determined. From these data, R. McClure derived a distance of 162 light years.

The weight of probability therefore ascribed a distance of 155 ± 5 light years to the Hyades.

The middle part of the Hyades covers an area of 8 light years in diameter, but this is only a small fraction of a much larger cluster, 80 light years across, which is called the Taurus moving cluster.

The Relationship Between Mass and Luminosity

In 1924, A.S. Eddington derived a relationship between the mass and luminosity of stars. Eddington compared the masses of double stars, derived from Kepler's third law, with their luminosities. He found that there exists a very high correlation between the mass and luminosity of Main Sequence stars: the luminosity is proportional to 3.5th power of the mass. Sub-giants and ordinary giants also fit into this relationship, but super-giants and white dwarfs do not fit the relationship.

In 1958, M. Schwarzschild published a list of Main Sequence spectroscopic binaries in his book *Structure and Evolution of the Stars* (Dover Publications Inc., New York) (see Appendix 1, section 1.12).

The data are the averages for each pair of stars, based on measurements by Z. Kopal (*Transactions of the International Astronomical Union*, Vol. IX, 1955). Appendix 1, section 1.12 also contains the data of visual doubles, monitored by P. van de Kamp, S.L. Lippincott and E.F. Flather, as published in the *Astronomical Journal*. There are also measurements by A.A. Strand, R.G. Hall, K.O. Wright and D.M. Popper, as published in the *Astrophysical Journal*. In cases where the spectral types do not correspond to those published elsewhere, this is because the magnitudes used here are bolometric. The bolometric correction for the Sun is −0.08. Therefore the Sun's bolometric magnitude is 4.85 − 0.08 = 4.77, which is rounded off to 4.8. The luminosity of a star is then 2.512 raised to the power of 4.8 minus the bolometric magnitude of the star. In Appendix 1, section 1.12, the masses of the stars have been derived from the logs of the masses, using a pocket calculator.

The coefficient of correlation between the logs of the luminosities and the logs of the masses, worked out in Appendix 1, section 1.12 is almost perfect, giving a value of 0.98 (1.00 being perfect). Appendix 1, section 1.11 shows how the formula for calculating the coefficient of correlation is derived.

This very high coefficient of correlation of 0.98 forms the basis of the mass–luminosity law, which states: the logarithms of the luminosities of stars are proportional to the logarithms of the masses; or the luminosities are proportional to the masses raised to the power n. In the cases of Main Sequence stars, $n = 3.5$; for massive stars, $n = 4$; and for solar type stars, $n = 2.5$.

Figure 10.36 is the graph of the logarithms of the luminosities against the logarithms of the masses; or, more directly, bolometric magnitudes against masses.

The slope of the straight line of best fit is calculated in Appendix 1, section 1.12. It works out at 3.14 for the value of b. For the stars plotted on the graph, b is found to be equal to 3, which is very close to the calculated value of 3.14.

In Appendix 1, section 1.12, the standard deviation and the 95% confidence limits are calculated for the case of the straight line of best fit. The 95% confidence limit for

Mass–Luminosity Law

log(luminosity) against log(mass) is 0.8. This has been indicated on the graph. We see that this limit does encompass all the plotted points, except those for the white dwarfs. Drawing a vertical line above the point 4, we see that it cuts the straight line of best fit at 1.9. Thus a star of 4 solar masses has a log (luminosity) of 1.9 plus or minus 0.8, that is between 1.1 and 2.7, whose antilogs are 13 and 500, so that the star's luminosity is between 13 and 500 times that of the Sun.

In Appendix 1, section 1.12, the 95% confidence limits of log(mass) against log(luminosity) is 0.25. From the graph, we see that a star whose log(luminosity) is −1 has a log(mass) of −0.26 ± 0.25, that is between −0.51 and −0.01, whose antilogs are 0.33 and 0.98. In 95% of cases, therefore, a star whose log(luminosity) is −1 (having bolometric magnitude 7.3) will have a mass of 0.33 to 0.98 solar masses.

A luminosity of 1 (that of the Sun) has a log equal to 0, and this corresponds to a bolometric magnitude of 4.8.

Careful examination, however, shows that the plotted points tend to cluster around a cubic curve, rather than a straight line. From the cubic curve which has been drawn, the values shown in Table 10.11 have been read off.

In Appendix 1, section 1.13, these values of L and M are substituted in the general form of the cubic equation:

$$aL^3 + bL^2 + cL = M$$

in order to find the values of the coefficients a, b and c. The equation of the cubic works out very neatly to:

$$3L^3 - 5L^2 + 57L = 250M$$

Table 10.11

Point on graph	A	B	C	D	E
Luminosity L	3	2	1	−1	−2
Mass M	0.83	0.47	0.22	−0.26	−0.63

Any value of L, the log(luminosity), can be substituted into this equation to find the corresponding value of the mass M.

Suppose a star has a bolometric magnitude of 1.4. Its luminosity is equal to:

$$2.512^{(4.8-1.4)} = 2.512^{3.4} = 22.9$$

The log(luminosity) = log 22.9 = 1.36.

Substitute $L = 1.36$ in the equation:

$$3L^3 - 5L^2 + 57L = 250M$$
$$\therefore 3(1.36)^3 - 5(1.36)^2 + 57(1.36) = 250M$$
$$\therefore 250M = 75.82 \quad \therefore M = 0.303 = \log(\text{mass})$$
$$\therefore \text{Mass } M = \text{antilog } 0.303 = 2 \text{ solar masses}$$

The mass of the star is thus twice the mass of the Sun. The absolute magnitude of Sirius is 1.42 and therefore its mass, according to the mass–luminosity law, must be about twice the Sun's mass. On page 182 we saw that the mass of Sirius is 2.35 solar masses by applying Kepler's third law to the orbits of Sirius A and Sirius B. The mass–luminosity law needs no further corroboration.

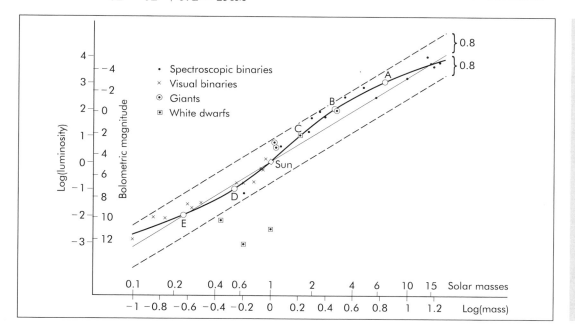

Figure 10.36 The relationship between mass and luminosity

When A.S. Eddington derived the mass–luminosity law in 1924, based on the radiant energies of stars, he concluded that the luminosity of a Main Sequence star is proportional to the 3.5th power of the mass. If we consider only the portion of the graph of Figure 10.36 (*previous page*) between the points E and A, we see that L increases from –2 to +3 (= +5), while M increases from –0.63 to 0.83 (= +1.46). The slope of this portion of the straight line is therefore $5 \div 1.46 = 3.4$, which is in excellent agreement with Eddington's 3.5th power.

The mass–luminosity law is a mighty weapon in the armoury of astronomers. Simply, by measuring the total energy flux of a Main Sequence star, over all the wavelengths, that is, the bolometric magnitude, the mass of the star can be found!

The bolometric magnitude can also be derived from the visual absolute magnitude by applying the bolometric correction (Table 10.4). In the case of the Sun, its absolute magnitude of 4.85 is corrected by the amount –0.08, yielding the value 4.77, rounded off to 4.8 for the Sun's bolometric magnitude.

Bolometric Magnitude and the Diameter of a Star

The bolometric magnitude is a measure of the total energy flux from the whole surface of the star. This flux will depend, in large measure, on the temperature. If the star's temperature can be determined, its bolometric magnitude will be an indicator of the size of the surface area, from which the diameter can be calculated.

In order to do this, M.K.E.L. Planck (Max Planck) studied the radiation of a black body. A black body is a body which is in thermal equilibrium with its environment and which absorbs all the radiation which falls on it, and re-radiates it without loss. These conditions prevail in the nuclei of stars. The radiant energy from the nucleus of a star is speedily absorbed and re-radiated in all directions (that is isotropically). The conditions on the surface of a star begin to deviate from those required for black body radiation, but the star's surface is still approximately a black body.

In Planck's laws of radiation, there is a relationship between the radiation, or emissivity, and the temperature. Planck's relationship between the colour index, CI, of a star and its colour temperature T_C is:

$$\text{CI} = \frac{7270}{T_C} - 0.53 \quad (1)$$

The colour index is the difference between the visual magnitude through a blue filter (4200 Å) and through a yellow filter (5400 Å). When the colour temperature, T_c, is known, the bolometric correction, B_C, can be calculated from Planck's second equation:

$$B_C = 42.60 - \frac{28\,400}{T_C} - 10 \log T_C \quad (2)$$

When the bolometric correction is known, the bolometric magnitude, M_{bol}, can be calculated. Knowing the colour temperature, T_C, and the bolometric magnitude, M_{bol}, the star's diameter can be calculated in terms of the Sun's diameter = 1:

$$M_{bol} = 42.35 - 10 \log T_C - 5 \log D \quad (3)$$

where D is the ratio between the diameter of the star and that of the Sun.

Let us apply Planck's formulae to a well-known Main Sequence star, Sirius, of which the visual absolute magnitude 1.42 (= M_V) is known. Sirius is an A1 star and therefore its colour index = 0.0.

(a) Sirius' colour temperature:

$$T_C = \frac{7270}{\text{CI} + 0.53} = \frac{7270}{0 + 0.53} = 13\,700 \text{ K}$$

(b) Calculate the bolometric correction:

$$B_C = 42.60 - \frac{28\,400}{T_C} - 10 \log T_C$$
$$= 42.60 - \frac{28\,400}{13\,700} - 10 \log 13\,700$$
$$= -0.84$$

Therefore Sirius' bolometric magnitude is:

$$M_{bol} = M_V - 0.84 = 1.42 - 0.84 = 0.58$$

(c) Calculate Sirius' diameter, D:

$$5 \log D = 42.35 - M_{bol} - 10 \log T_C$$
$$= 42.35 - 0.58 - 41.37 = 0.40$$
$$\log D = 0.08, \text{ and therefore } D = 1.2$$

Although the mass of Sirius is 2.35 times that of the Sun, its diameter is only 1.2 times the Sun's diameter.

Let us try Planck's formulae on a giant star not on the Main Sequence, namely the giant Arcturus, Alpha Boötis, a K2 IIIp star, having a parallax of 0.09″. Its visual magnitude m_V is –0.04 and colour index 1.23.

The visual absolute magnitude M_V of Arcturus is given by:

$$M_V = m_V + 5 + 5 \log P \quad (P \text{ is the parallax})$$
$$= -0.04 + 5 + 5(\log 0.09)$$
$$= -0.04 + 5 + 5(-1.046) = -0.27$$

$$T_C = \frac{7270}{1.23 + 0.53} = 4130 \text{ K}$$

$$B_C = 42.6 - \frac{28\,400}{4130} - 10 \log 4130 = -0.44$$

$$M_{bol} = -0.27 - 0.44 = -0.71$$

$$5 \log D = 0.71 + 42.35 - 10 \log 4130 = 6.90$$

$$\therefore \log D = 1.38 \text{ and therefore } D = 24$$

By the method of interferometry, the diameter of Arcturus subtends an angle of 0.02″. Let us use this value to calculate Arcturus' diameter and compare the result with the value given by Planck's laws (Figure 10.37).

The mean distance of the Earth from the Sun is 149 597 870 km. Dividing this by the Sun's diameter, 1 392 800 km, we find that the Earth's mean distance from the Sun is 107 times the Sun's diameter (= 107S). In Figure 10.37:

$$\frac{D}{d} = \tan a = a \text{ and } \frac{107S}{d} = \tan p = p$$

$$\therefore \frac{D}{d} \div \frac{107S}{d} = \frac{a}{p}$$

$$\therefore D = 107S \times \frac{a}{p} = 107S \times \frac{0.02}{0.09} = 23.78S$$

This value must be taken as identical to the result (24) obtained from Planck's laws. (The author expresses his thanks to Girsch Pulik, former Chairman of the Johannesburg Centre of the Astronomical Society of Southern Africa, for his assistance with the Planck calculations.)

In the case of the super-giant Betelgeuse, Planck's formulae give a diameter of 1000 times that of the Sun. By the method of interferometry (page 175) we obtained a diameter of 995 times the diameter of the Sun for the diameter of Betelgeuse.

Thus, in spite of the fact that Betelgeuse is a super-giant, far from the Main Sequence of the H–R diagram, Planck's formulae give the correct result for its diameter.

We therefore find very good agreement between the values of the diameters of stars obtained by the methods of:

(a) interferometry,
(b) occultations by the Moon,
(c) visual absolute magnitudes and colour indices, and
(d) Planck's radiation laws.

These results show that the physical laws are universally applicable.

The white dwarfs do not fit into the mass–luminosity law because their surface areas are too small, being about the size of the Earth's surface, which is 1/10 000th that of the Sun's surface. Their luminosities are too low and they fall below the line of best fit in Figure 10.36. The three giants that are plotted on the diagram do fall within the 95% confidence limits of probability. Super-giants are outside the limits, and also do not fit into the mass–luminosity laws.

The close connection between mass and luminosity can be taken further, by calculating the temperature for any given mass or luminosity of the stars.

In 1900, Max Planck devised the quantum theory. This postulates that energy does not flow in a continuous stream, but that it consists of discrete parcels. It requires one quantum of energy to displace an electron from one orbital around the nucleus of an atom to the next higher orbital. One quantum of energy is radiated when an electron falls from one orbital to the next lower orbital, nearer to the nucleus.

The energy radiated is proportional to the frequency, ν, of the radiation, that is, $E = h\nu$, where h is a constant, equal to $6.626\,196 \times 10^{-34}$ joule-seconds, or $6.626\,196 \times 10^{-27}$ erg-seconds.

The higher the frequency of radiation, the shorter is the wavelength and the greater is the amount of energy emitted. The shorter the wavelength and the higher the frequency, the higher is the temperature of the maximum flux of the radiation. The curve of the intensity of radiation against wavelength for a higher frequency is entirely above that for a lower frequency. The higher-frequency curves have higher maximum temperatures (Figure 10.38).

Although the stars are not perfect black body radiators, they approximate to that condition and their energy fluxes are in good agreement with the black body curves. By measuring the radiation of a star at all wavelengths, the curve which best fits can be taken as the indicator of the star's temperature. This works very well for the middle wavelengths, 4000 Å to 6000 Å.

Because of the absorption of ultra-violet and infra-red wavelengths by the Earth's atmosphere, measurement of these wavelengths can best be done by probes in orbit outside the atmosphere.

The fact that the maxima of the intensities of radiation are displaced towards the violet for higher frequencies gave rise to the law formulated by W. Wien. This states that the

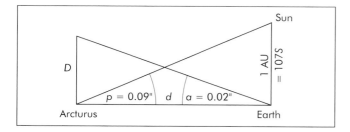

Figure 10.37 Diameter of Arcturus

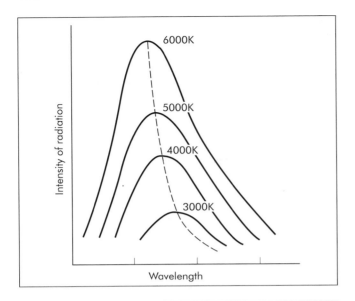

Figure 10.38 Black body radiation curves

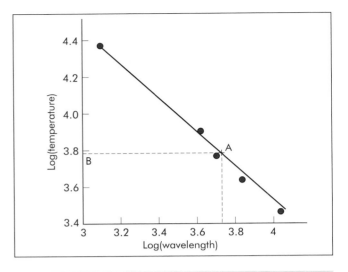

Figure 10.39 Wien's law

wavelength at which maximum energy is radiated is inversely proportional to the absolute temperature of the radiating body. As the temperature is increased, so the intensity of the radiation is displaced more to the violet part of the spectrum. All that needs to be done is to measure the wavelength, λ, of the maximum energy flux, to obtain the temperature T of the star. Table 10.12 gives the temperatures appropriate to the wavelengths. The wavelengths are those of the hydrogen lines in the solar spectrum.

The graph of the logarithms of the temperatures against the logarithms of the wavelengths (Figure 10.39) is a straight line with slope equal to −0.97.

Stars such as the Sun, which have their maximum radiation at approximately 5200 Å, in the yellow part of the spectrum, have a temperature of about 6000 K. The calculation is done as follows.

The log of 5200 is 3.72; mark 3.72 on the x-axis [log(wavelength)] and draw the vertical to cut the straight line at A. Now draw a straight line horizontally to cut the y-axis [log(temperature)] at B. The value of B is 3.78. This is the log of the temperature. The temperature is the antilog of 3.78, namely 6025 K.

The Stefan–Boltzmann Law

The Stefan–Boltzmann law makes use of the total energy radiated by 1 square centimetre of the star's surface; thus, the total surface area of the star must be known. The law states that the total energy radiated is proportional to the area under the black body curve for the star (Figure 10.38). This can be calculated mathematically by means of the integral calculus; it turns out that the energy radiated is proportional to the 4th power of the Kelvin temperature.

The Stefan–Boltzmann equation to determine the radius of a star is derived from Planck's law:

$$\log R = \frac{5900}{T} - 0.2 M - 0.016$$

This equation connects the radius R and the temperature T in degrees Kelvin to the absolute bolometric magnitude (M).

How the three laws compare is seen in the values they give for the temperature of the Sun, as shown in Table 10.13.

The differences may be due to the deviation of the Sun

Table 10.12 Relationship between temperature and wavelength of radiation

λ	log λ	T	log T
1 215	3.085	23 400	4.37
3 970	3.60	7 900	3.90
4 861	3.69	5 900	3.77
6 563	3.82	4 400	3.64
10 050	4.02	2 900	3.46

Table 10.13 The Sun's temperature

Law	Temperature of Sun (K)
Planck (shape of energy curve)	5600–6150
Wien (maximum of energy curve)	6025
Stefan–Boltzmann (area under curve)	5750

from being a black body.

Spectrum Lines and Temperature

As the temperature of a star rises, more and more electrons are removed from the atomic nuclei. This is the process of ionisation. The electron carries a negative electric charge. Having lost an electron, the remaining nucleus and electrons will be positively charged on the whole. The hydrogen atom has only one electron and it can thus undergo only one degree of ionisation. The oxygen atom has eight electrons: 2 in the inner shell, 2 in the second shell and 4 in the outer shell. To wrench one electron from the oxygen atom in its ground state requires 35 eV. The remaining positive oxygen ion is indicated by OII. To free two electrons from the oxygen atom requires 55 eV. The oxygen atom is then doubly ionised and is written as OIII. To free three electrons requires 77 eV, and the oxygen is then trebly ionised: OIV.

If the lines in a star's spectrum indicate a certain degree of ionisation, the energy required to bring about that degree of ionisation can be calculated, and from that the temperature.

The very hot O-type stars have spectra showing lines of doubly and trebly ionised oxygen (OIII and OIV). To bring about this degree of ionisation on a large scale requires a temperature of 25 000 to 40 000 K.

B-type stars reveal spectral lines of singly ionised oxygen (OII) and nitrogen (NII), and their temperatures are between 11 000 and 25 000 K.

A-type stars have spectra showing lines of singly ionised calcium (CaII) and their temperatures are 7500 to 11 000 K.

F-type stars have ionised lines only in the ultra-violet. In G-type stars, there is very little ionisation to be found. In K and M-types, lines of molecules appear. These molecules cannot exist at temperatures higher than 3500 K.

A star's spectrum is therefore a very good indication of its temperature.

Chapter 11
Variable Stars

1 Eclipsing Binaries

Besides the method of naming stars, described on page 165, it was decided that capital letters before the genitive of the Latin name of the constellation be used only from A to Q. Letters from R to Z were to be reserved for variable stars, such as R Coronae Borealis, U Geminorum, etc.

When Z was very quickly reached, it was decided to use two capital letters, beginning with RR, then RS, RT, ... RZ, followed by SS, ST, ... SZ; then TT, ... TZ. When ZZ was reached for the 54th variable in a constellation, it was decided to continue with AA, AB, ... AZ, followed by BB, BC ... BZ; up to QZ, which would link up with RR. The letter J was not used, but I was. In this way provision was made for 334 variables in any constellation.

When this became inadequate, astronomers proceeded with V335, V336, ... followed by the abbreviation of the name of the constellation, such as V337 Aql ... .

Today there are catalogues containing many thousands of variable stars, for example, the *Russian General Catalogue of Variable Stars*, edited by P.N. Kholopov, N.P. Kukarkin *et al.*, which contains 28 435 variable stars. The constellation Sagittarius alone contains more than 4000 variables.

If the plane in which the orbits of a double star lie is in the line of sight, or nearly so, each star of the pair will periodically move across the line of sight in front of the other and then move behind it, thus causing eclipses and being eclipsed. The eclipses result in the light intensity varying. When the two stars are next to each other, the light will be at a maximum. As one star proceeds to eclipse the other, the brightness will decrease until the eclipse has reached its maximum phase. Then the light intensity will increase again.

Although these eclipsing binaries do not undergo intrinsic variations in the stars themselves they are nevertheless classified as variable stars and are named as set out above.

The first eclipsing binary which was studied telescopically was Algol, or Beta Persei (03 08.2 + 40 57). Its variations in light intensity must have been known since antiquity, because its name, which means "demon" in Arabic, was used by the Babylonians. It was different from other stars, and therefore it had to be a demon.

The deaf-and-dumb John Goodricke (1764–1786) studied the light variations of Algol and, in 1782, noticed the regularity in the light curve. The time lapse between two primary minima remained constant at $2\frac{7}{8}$ days. Today's value is 2.867 31 days, or 2 days, 20 hours, 48 minutes and 55 seconds. At the minima the magnitude of the star was 3.40, compared with its brightest of 2.12. The duration of the eclipse is 10 hours, while the eclipsing star moves from B_1 to B_3 (Figure 11.1, *overleaf*).

When the star reaches B_1, it just starts to eclipse member A. The light curve is rounded off at B_1, because the limbs of the stars are darker than their discs, as in the case of the Sun.

When star B moves in front of the star A, there is a sudden drop in brightness. The brightness is a minimum at B_2, the primary minimum. The light curve is sharp at B_2 because the eclipse of component A is either momentarily total or partial. There is only one moment of minimum brightness.

As star B moves from B_2 to B_3, the brightness increases up to B_3, where there is a rounding-off because of limb-darkening. From this moment onwards both stars are visible, but there is still a slight increase in brightness, from B_3 to B_4 because of light from the brighter star A being reflected from the surface of the dimmer star B. The facing surface of B is at its brightest just before it moves behind star A. It is then 1.7 times brighter than in position B_1, B_2

195

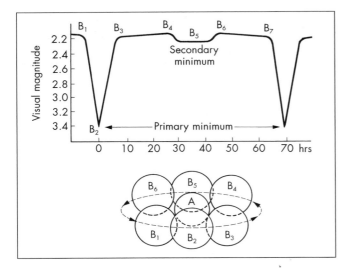

Figure 11.1 Light curve and model of eclipsing binary, Algol

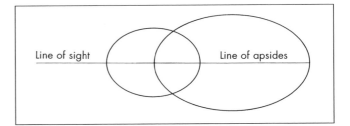

Figure 11.2 Secondary eclipse mid-way between primary eclipses

and B_3.

When star B moves behind star A, the secondary eclipse takes place. The secondary eclipse B_4–B_5–B_6 is very shallow in the light curve, because star B is fainter per square metre of surface than star A. The flat secondary minimum shows that star B is larger than star A and a measurable amount of time elapses while A covers the same area of B. At B_7 the whole process is repeated.

John Goodricke was the first to explain the changes in brightness as being due to mutual eclipses of two stars which revolve around each other – a very "revolutionary" idea in 1782!

Spectroscopic analysis shows that the primary star A is a B8 V. It has 100 times the Sun's luminosity and nearly three times its diameter. Its mass is 3.5 to 4 times that of the Sun. The spectrum of the fainter companion B could not be satisfactorily obtained because of the glare of the primary. The calculated size of the secondary indicates that it is a G8 or K0 star, of mass about equal to that of the Sun; its diameter is more than 3.4 times that of the Sun. It is therefore a giant of luminosity class III.

Although the telescope cannot resolve the two stars, the light curve reveals a mass of evidence. The period of revolution (the time from one primary eclipse to the next) is very well known. By means of Kepler's third law, the mean distance between the two stars can be calculated, firstly in AU and then in km. The mean distance of the bright, primary star from the barycentre is 2.57 million km, and that of the secondary 9 million km. Therefore their masses are in the ratio of 9:2.57, or 3.5 to 1. The eccentricity of the orbits is 0.033 – closer to circles than the orbits of six of the Sun's planets.

If the position of the secondary eclipse is exactly half-way between the two primary eclipses in the light curve, the orbits are either circles or, if they are ellipses, their line of apsides must coincide with the line of sight from the Earth, as is illustrated in Figure 11.2. In all other cases the secondary eclipse will be nearer in time to one primary eclipse than to the other. The further the secondary minimum is removed from the mid-point between the two primary minima, the greater is the eccentricity of the orbits and the nearer the line of apsides is to the plane of the sky. The plane of the orbits of Algol is tilted by only 8 degrees to the line of sight. During a primary eclipse, 79% of the surface of the primary star is eclipsed.

The secondary minimum moves in a period of 32 years from one side to the other vis-à-vis the primary minima. This means that the eccentricity of the orbits varies, or the angle between the line of apsides and the plane of the sky varies in a periodic way. It can be indicative of the line of apsides rotating, which will be the case if the ellipses are not closed (Figure 11.3).

The rotation of the line of apsides can be ascribed to the fact that the ellipses are open. This will be brought about by the gravitational force of a third body, or because the stars are oval in shape. The rotation of the line of apsides is similar to the rotation of the line of apsides of the Moon's orbit around the Earth, caused by the gravitational attraction of the Sun. Mercury's orbit also has a shift of the perihelion. The explanation of the magnitude of this shift was one of the triumphs of Einstein's Theory of Relativity. The shift of the perihelion of Mercury could not be fully accounted for by Newton's law of gravitation.

From the amount of rotation of the ellipses, in Algol's case the mass of the perturbing body could be calculated. The presence of the third member was also revealed spectroscopically. It is an F-type Main Sequence star and is brighter than the dull B-component of the eclipsing pair.

Eclipsing Binaries

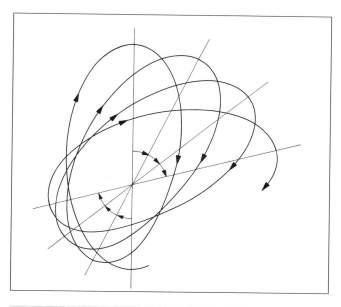

Figure 11.3 Open ellipse has rotating line of apsides

The third member revolves around the eclipsing pair in 1.86 years at a mean distance of 80 million km from the barycentre. The plane of the orbit of the third member is tilted to the line of sight by 37 degrees, and the eccentricity of the orbit is 0.2 – much more eccentric than the orbits of the eclipsing pair.

The elements of the three members of Algol are as shown in Table 11.1.

Algol's distance is not known with certainty. The parallax is between 0.03″ and 0.04″, so that its distance is between $1 \div 0.03 = 33.3$ parsecs (or 108 light years) and $1 \div 0.04 = 25$ parsecs (or 81 light years). The average of these values is 29 parsecs or 94.5 light years.

Algol is very bright and is a good object for an amateur astronomer to use as a start in studying eclipsing binaries. The period is short enough so that all the phases of the eclipse can be measured in one month. The times of Algol's maxima and minima are published monthly in *Sky and Telescope Magazine*.

Appendix 6 contains lists of eclipsing binary stars.

Beta Lyrae Stars

An altogether different type of eclipsing binary is Beta Lyrae (18 49.9 + 33 32); it was discovered in 1784 by John Goodricke. The light curve (Figure 11.4) shows a continual variation with rounded primary minima of magnitude 4.1 and secondary minima at magnitude 3.8, almost half-way between the primary minima.

The atmospheres of the two stars are probably in contact and their shapes have been deformed into ellipsoids by their mutual gravitational attractions. Both stars are giants, being 20 and 10 times the Sun's mass. The eccentricity of the orbits is 0.024 and their average distance apart 35 million km.

From the light curve, it can be seen that one star is 1.27 times the size of the other; their diameters are 15 and 19 times that of the Sun. Their "surfaces" are therefore about 11 million km apart. As is usual with binaries, the smaller star is more massive than the larger star.

Table 11.1 Algol

	A	B	C
Mass (Suns)	3.5	1.0	1.3
Diameter (Suns)	3.0	3.4	1.5
Absolute magnitude	–0.2	3.4	3.2
Luminosity (Suns)	100	3.6	4.4
Spectral type	B8 V	K0 IV	F2 V

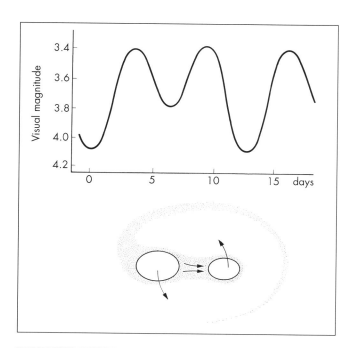

Figure 11.4 Light curve of Beta Lyrae, and model of the system

The distance modulus of the pair is 7.0, that is, 800 light years, and the absolute magnitude works out at −4.0.

The primary star is of type B7 Ve, of luminosity 3000 times the Sun's. The secondary star is an A8p, having a peculiar spectrum.

By applying the Doppler effect, it has been found that the gas streams between the two stars are flowing at 280 km s^{-1}. This flow of gas tends to make the more massive star even more so. This causes a lengthening of the period by 9.4 seconds per year. There is also a stream of gas moving outwards into space, causing a continual mass-loss, which also contributes to the lengthening of the period.

According to theory, the more massive star evolves more quickly away from the Main Sequence of the H–R diagram than the less massive star, and moves to the area of the giant stars.

The life-time of a close binary, such as Beta Lyrae, is necessarily short, because of the great masses of the stars, and very few such binaries will be found.

As the secondary star becomes more massive, it will move further into the area of the giants so that its luminosity will increase, becoming greater than that of the primary.

It is also possible that the invisible secondary star collapsed on its core after becoming a giant, because its outward radiation pressure decreased when the helium in its core underwent atomic transformation to heavier atoms; thus it collapsed so far under its own gravity that it became a gravitational vortex (commonly known as a "black hole"). This is an area in which the density of matter is almost infinite, and thus anything but a "hole". The gravitational force is so strong here, and the escape velocity so high, that neither matter nor any radiation can get out because the escape velocity exceeds the velocity of light. According to Einstein's theory of relativity, the mass of a body increases as its velocity increases and, at the velocity of light, its mass becomes infinite so that it requires an infinite force to move it. An infinite force cannot, of course, exist.

Because no radiation can escape from a gravitational vortex, it cannot be observed visually. But the masses of gas which gravitate into it will be heated so much by friction and acceleration tthat they will radiate X-rays, before vanishing into the vortex. X-rays have been observed as coming from Beta Lyrae, and it is thus possible that a gravitational vortex exists there.

Near to Beta Lyrae in direction is the star Gamma Lyrae, of magnitude 3.25. It can be used as a standard by an amateur observer while monitoring Beta Lyrae, whose magnitude varies between 3.4 and 4.1. Who knows what rare phenomenon may not at some time appear there?

Appendix 6, section 6.2 contains a list of Beta Lyrae eclipsing variables.

Among the Beta Lyrae eclipsing stars, UW Canis Majoris is noteworthy because it consists of two giant stars. One is an O7. Their diameters are 18.6 and 14.8 times that of the Sun, and their masses 19 and 32 times respectively. UW Canis Majoris also displays gas streams and the eclipses are partial.

The majority of Beta Lyrae stars are of type B.

W Ursae Majoris Stars

Another type of eclipsing binary stars consists of dwarfs of the Sun's size, and these have very short periods. The prototype is W Ursae Majoris (09 43.8 + 55 57). Figure 11.5 shows the light curve. The period is only 8 hours, actually 8 hours 27 seconds! The curve of primary eclipse is somewhat flattened and lasts 20 minutes. The secondary eclipse is 0.1 magnitude brighter – magnitude 8.6 against 8.7 of the primary eclipse. The secondary eclipse is absolutely central between the primary eclipses, showing that the orbits are circles, or that the line of apsides lies in the line of sight.

The rounded-off maxima in the light curve show that the atmospheres of the stars are in contact. The diameters of the two stars are 1 580 000 and 1 150 000 km, respectively; their diameters are 1.14 and 0.83 times that of the Sun, and their mean distance apart is 1.6 million km. Table 11.2 shows the properties of the two stars. The two members show marked resemblances to the Sun.

Measurements based on the Doppler effect show that the velocities of the two stars in their orbits are 120 and 240 km s^{-1}, and that the system is shrouded in gas-streams. The stars revolve so close together that their

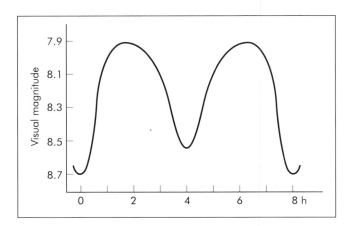

Figure 11.5 Light curve of W Ursae Majoris

Table 11.2 W Ursae Majoris

	A	B
Mass (Suns)	1.0	0.6
Diameter (Suns)	1.14	0.83
Absolute magnitude	4.4	4.75
Luminosity (Suns)	1.45	1.00
Spectral type	dF 8	dF 6

shapes are deformed into ellipsoids.

The distance of W Ursae Majoris is about 200 light years.

According to H. Shapley, this type of eclipsing binary is 20 times more prevalent than all the other types put together. This is probably because of their closeness to each other, since if the eclipsing binaries are far apart, the line of nodes must be very close to the line of sight for eclipses to be visible from Earth.

Appendix 6, section 6.3 contains data about some W Ursae Majoris stars.

TW Ceti and Iota Boötis are W Ursae Majoris stars with very short periods, 0.3169 days (= 7 hours 36 minutes 20 seconds) and 0.267 days (= 6 h 24 m 29 s), respectively. They both complete a whole cycle in less than one night! They dim from magnitudes 8.8 to 9.6 and from 6.5 to 7.1. Iota Boötis has the distinction that both members are G2 V stars, the same class as the Sun.

The spectral types of the members of UV Leonis are G0 and G2.

RW Tauri undergoes an eclipse from magnitude 7.6 to 12.0. It dims by a factor of 57 times!

o^1 Cygni is unique because its primary eclipse lasts 63 days!

Some eclipsing binaries do have very long periods (Appendix 6, section 6.4): VV Cephei has a period of 20.35 years. The K3 member of Epsilon Aurigae has a period of 27 years; the eclipsing star in this case is probably a giant, oval mass of gas. VV Cephei and Epsilon Aurigae are probably not Algol types.

V444 Cygni consists of a WN5 Wolf–Rayet star and an O6 star, of masses 18 and 32 times that of the Sun. Their visual absolute magnitudes are −3.1 and −4.8 and their luminosities 1450 and 6900 times that of the Sun.

When the light of the eclipsed star passes through the atmosphere of the eclipsing star, an absorption spectrum of the atmosphere of the latter can be obtained, enabling the chemical composition, temperature and pressure, as well as the magnetic fields, to be determined. Advanced technology is necessary for this type of measurement.

If the plane of the orbits of an eclipsing binary does not lie in the line of sight, the eclipses will be partial and both primary and secondary minima will be sharp.

If one member is considerably larger than the other, both minima will be flattened, because some time will elapse while the smaller star is behind the larger, or while it transits it. Examples of eclipsing binaries with two flattened minima are U Cephei (01 02.3 + 81 53), very near to the north celestial pole, and U Sagittae (19 18.8 + 19 37) which is 10 degrees north of the bright star Altair in Aquila.

U Sagittae

The lapse of time between eclipses of U Sagittae is 3.380 619 33 days (= 81 h 8 m 5.5 s). Totality of the primary minimum lasts 1.48 hours (= 1 h 28.8 m). Both minima are flattened because eclipse of the bright star is total and the dim star is transited by the bright star. The plane of the orbits lies in the line of sight.

The primary star is a B8 V and the secondary a G2 III, that is, Main Sequence and giant, respectively. Data obtained by A.H. Joy, D.H. McNamara and K.A. Feltz, as printed in Burnham's *Celestial Handbook*, are as shown in Table 11.3.

Because of gas-streams leaving the stars, loss of mass takes place, coupled with lengthening of the period. The light curve shows that the B component reflects light from the A component, just before it passes behind the A component and just after it emerges from eclipse. That is why the light curve is not horizontal before and after a secondary eclipse (Figure 11.6).

The radial velocities are determined when the stars are furthest from the eclipsing phase, because the stars then move along the line of sight. The radial velocity of the secondary of U Sagittae is 209 km s^{-1}. In the time between two primary eclipses (= about 81 hours) the secondary completes one circuit, which we can consider to be a circle, approximately. At 209 km s^{-1}, the secondary covers a distance of 81(3600)209 km in 81 hours. This distance is the circumference of its orbit, considered to be a circle of

Table 11.3 U Sagittae

	A	B
Mass (Suns)	3.5	1.4
Diameter (Suns)	2.52	3.32
Absolute magnitude	− 0.4	1.8
Luminosity (Suns)	120	15
Spectral type	B8 V	G2 III

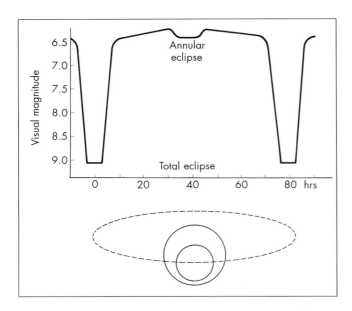

Figure 11.6 Light curve and model of U Sagittae

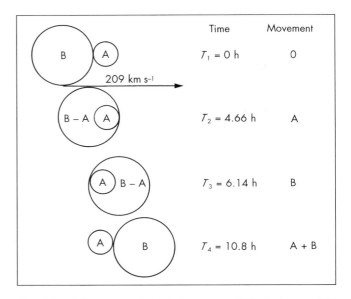

Figure 11.7 Times of primary eclipse of U Sagittae-A by U Sagittae-B. The primary star is considered to be stationary; and the four moments of contact are depicted

radius equal to R km.

$$\therefore 2\pi R = 81(3600)209 = 60\,944\,400 \text{ km}$$
$$\therefore R = 60\,944\,400 \div 2(3.141\,59)$$
$$= 9\,699\,911, \text{ namely } 9\,700\,000 \text{ km}$$

The diameter of the orbit of the secondary is twice the radius = 19 400 000 km.

If the times of first and last contact are known for each star, the diameters of the stars can be calculated, if we assume the eclipse to be centrally orientated. If the diameters of the stars are A and B, we can see from Figure 11.7 that:

movement A takes 4.66 hours, and
movement B takes 6.14 hours

In 4.66 hours, B covers a distance equal to the diameter of A (= 4.66 (3600)209) km which equals 2.52 solar diameters.

In 6.14 hours, B covers a distance equal to its own diameter (= 6.14(3600)209) km which equals 3.32 solar diameters.

The radial velocity of the primary is 84 km s^{-1}. Its mass compared with the mass of the secondary is 209 ÷ 84 = 2.5, that is, the two masses are 3.5 : 1.4, as stated in Table 11.3.

The light curve can also be used to calculate the separate magnitudes of the two stars.

Suppose the magnitudes of B and A are x and y, respectively. The depth of the primary minimum to the depth of the secondary minimum is as 29 is to 2 = 14.5 to 1. (This is measured from the graph.) Therefore the magnitude of x divided by the magnitude of y = 14.5.

$$\therefore 2.512^{(x-y)} = 14.5$$
$$\therefore (x-y) \log 2.512 = \log 14.5$$
$$\therefore (x-y)(0.4) = 1.16, \text{ and } (x-y) = 2.9$$

From the graph, we can see that the joint magnitude of the two stars, taken together, is equal to 6.3 (6.5 minus the depth, 0.2, of the secondary minimum).
Therefore:

$$\frac{1}{2.512^x} + \frac{1}{2.512^y} = \frac{1}{2.512^{6.3}} = \frac{1}{331}$$
$$\text{or } \frac{1}{a} + \frac{1}{b} = \frac{1}{331}$$

We know that $\dfrac{a}{b} = \dfrac{14.5}{1}$ so $\dfrac{1}{b} = \dfrac{14.5}{a}$

$$\therefore \frac{1}{a} + \frac{14.5}{a} = \frac{1}{331}, \quad \therefore a = 5130$$

i.e. $2.512^x = 5130$, from which $x = 9.3$

But $x - y = 2.9$
$$\therefore y = x - 2.9 = 9.3 - 2.9 = 6.4$$

Intrinsic Variables

Thus, the magnitudes of the two members of U Sagittae are 6.4 and 9.3.

Appendix 6 contains data about the various kinds of eclipsing binary stars that can be used by an amateur astronomer for monitoring.

Among the Algol types, ST Carinae is exceptional because of its very short period of less than 1 day.

Delta Orionis was the first star of which the spectrum revealed lines of stationary calcium; this proved the existence of calcium in interstellar space. It was carried out by Potsdam Observatory in 1904.

RW Tauri has the greatest difference between maximum and minimum of all the stars listed, 7.6 against 12.0, and has a convenient period of 2.7688 days.

2 Intrinsic Variables

Although the magnitudes of eclipsing binaries do vary, they are not true variable stars since their changes in brightness are due to their orientations with regard to the line of sight. There are no physical changes in those stars themselves, except in the cases of very close binaries whose atmospheres intermingle. We shall now deal with variable stars that undergo inherent variations.

Delta Cepheid Stars

Stars that undergo intrinsic variations show a great variety of light curves: from the very regular to the irregular. The periods of the variabilities range from a few minutes to hundreds of days.

One of the most important of the intrinsic variables is the class of stars called the Cepheid variables. The prototype of this class of variable is the star Delta Cephei (22 29.2 + 58 25) which was discovered by John Goodricke in 1784. The light curve is very regular and has a period of 5 days, 8 hours, 47 minutes, 31.9 seconds, or 5.366 341 days. The photographic magnitude varies from 4.63 to 3.78. During the change in brightness the spectrum changes from F5 to G2; this shows that intrinsic changes take place in the star itself and that its temperature also changes. The star therefore alternately becomes yellower and whiter, as its temperature changes. No form of eclipse can explain the light brightness changes. Goodricke was absolutely correct when he explained the variations as due to pulsations of the star – that it alternately expands and contracts.

The lapse of time of 1.66 days between minimum and maximum brightness is shorter than that between maximum and minimum, 3.7 days. The slope of the rising part of the graph of magnitude against time is steeper than the falling part of the graph (Figure 11.8).

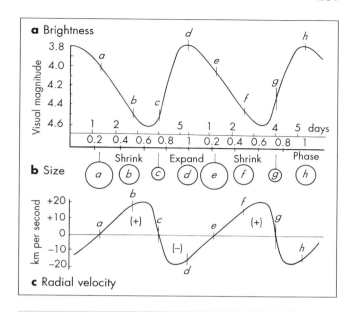

Figure 11.8 Delta Cephei: **a** Light curve; **b** Variation in size; **c** Radial velocity of surface

By applying the Doppler effect, it is evident that the radial velocity of the star's surface is positive while the star contracts – the surface then moving away from the Earth (*a* to *c* on the graph) – and negative while the star expands – the surface then moving towards the Earth (*c* to *e* on the graph).

According to calculations by H. Shapley and A.S. Eddington, Delta Cephei is at its brightest one-quarter of a period before it reaches its maximum size, when its radial velocity is zero, at *a* and *e*.

Maximum brightness is indicated by the point *d* on the graph of brightness against time.

When the star shrinks, from *a* to *c*, the brightness decreases because the radiating area decreases. Just before the star reaches its minimum size at *c*, the increase of pressure caused by the shrinking brings about a rise in temperature, causing the brightness to increase, even before the star starts expanding. Then the increase in brightness is supplemented by the increase in surface area, after point *c* has been passed on its way to greatest brightness at *d*, which is one-quarter of a period before greatest volume is reached at point *e* on the graph. After the point *d* has been passed, the brightness decreases because the increasing surface area causes a drop in temperature, which makes the star get dimmer. After maximum size at *e*, contraction, accompanied by positive radial velocity, decrease the size of the radiating area, so that the temperature drops further and the dimming goes on. The increased pressure raises the temperature (between f and g) and the next cycle commences.

The mean magnitude, x, of Delta Cephei is that magnitude which divides the area below the light curve into equal halves. A good approximation is arrived at by taking the arithmetic mean between the maximum and minimum magnitudes of 3.78 and 4.63, namely 4.20.

Astrophysical calculations show that the difference between extremes of brightness are brought about by a 10% to 12% change in volume during expansion and contraction. RR Lyrae expands by 7.2%; T Vulpeculae by 15.2%; η Aquilae by 9.1%; ζ Geminorum by 8.5% and Delta Cephei, itself, by 11.9%.

Some Delta Cepheids have secondary periods of 0.7 times the main period (Appendix 7, section 7.1).

According to the theory set out by C. Hoffmeister, G. Richter and W. Wenzel, on page 47 of their book *Veränderliche Sterne* (Springer-Verlag, Berlin), the pulsating of the Cepheids is due to absorption of the energy flux from the star's centre, by doubly ionised helium, HeIII, at a depth of a few hundred thousand km below the surface. An undamped oscillation then causes the surface to expand and the temperature to drop. The drop in temperature decreases the degree of ionisation of the helium, causing contraction to take place. The contraction then raises the temperature, thus increasing the ionisation and causing the process to be repeated. The critical factor is the amount of helium present. Periods of oscillation, calculated on the basis of this theory, agree well with those of well-monitored Cepheids.

The sizes of Delta Cephei depicted in Figure 11.8 are somewhat exaggerated.

Today, thousands of Cepheids are known; their periods vary from less than one day to a few hundred days. The light curves are not all regular, which is as expected, considering the huge forces generating the expansion and contraction.

The Cepheids are giants of masses 3 to 16 times that of the Sun. The spectra of Delta Cep. stars are of type F (Appendix 7, section 7.1).

The variable stars which are grouped together as "Classical Cepheids" are F and G giants, having temperatures between 5500 and 6500 K. Their absolute magnitudes lie between -2 and -7, making them 500 to 50 000 times brighter than the Sun. The amplitudes of their changes in magnitude vary between 0.5 and 1.7 magnitudes; and their periods vary from 3 to 40 days. Cepheids of very short periods, as well as those of long periods, are placed in separate classes.

RR Lyrae Stars

The periods of these stars are very short, being less than 1 day (Appendix 7, section 7.2). The prototype, RR Lyrae (19 25.5 + 42 47) has a period of 0.5668 days. Its maximum magnitude is 7.06 and minimum 8.12.

These stars are all type A or late type B and are bright giants. They all have the same absolute magnitude of 0.8, that is, 40 times brighter than the Sun. If a variable can be, identified as being an RR Lyrae star, it can be assumed that its absolute magnitude is 0.8 and, if its apparent magnitude is determined, its distance can be calculated. In this respect, these stars are valuable.

RR Lyrae periods range from 0.2 to 1.2 days and the amplitudes of the variations are 0.2 to 2 magnitudes. The maxima of their negative radial velocities, during expansion, coincide with the maxima in the light curves.

Delta Scuti Stars

The periods of these stars are even shorter, being of the order of 0.3 day, somewhat overlapping those of the RR Lyrae stars. The amplitudes of the variations are 0.2 to 0.7 magnitude. A photo-electric photometer is therefore necessary to measure their magnitudes. Delta Scuti stars can be considered as dwarf Cepheids. Some have secondary periods of 0.76 times that of the main period. Their maximum expansions take place within 0.1 period of their maximum brightnesses. The graph of radial velocity is therefore a mirror image of the light curve. The spectral types vary from A0 to F5 III, and they show the H and K emission lines of singly ionised calcium CaII (Appendix 7, section 7.3).

W Virginis Stars

These are similar to the Cepheids, but the descending part of the light curve contains a hump, or wave (Figure 11.9). These stars are 0.7 to 2 magnitudes fainter than the Cepheids of equal periods.

While the majority of the Classical Cepheids have periods of 5 to 7 days, W Virginis variables have periods ranging from 1 to 22 days (Appendix 7, section 7.4).

Cepheids are Population I stars and are rich in metals. The W Virginis stars are Population II stars and poor in metals.

The amplitudes of their light curves vary by 0.3 to 1.2 magnitudes. These variables occur mostly in globular star clusters as well as in the plane of the Milky Way.

Intrinsic Variables

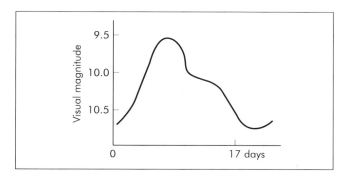

Figure 11.9 Light curve of W Virginis

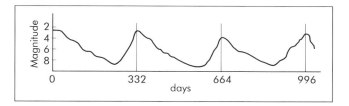

Figure 11.10 Light curve of Omicron Ceti, Mira

Long-Period Variables

Variables of very long periods of some hundreds of days are giants, occupying the top right-hand corner of the H–R diagram. The star Omicron Ceti, called Mira, the Wonderful, was the first star of this class to be discovered, telescopically in 1638 by Holwarda. Its spectrum has molecular lines and bands of titanium oxide (TiO), zirconium oxide (ZrO), carbon monoxide (CO), titanium hydride (TiH_2), methane (CH_4) and zirconium hydride (ZrH_2). These lines and bands become broader as the stars get fainter, until the spectrum vanishes. The spectra have emission lines which appear one quarter-period before maximum and which vanish before minimum.

Although the brightnesses of long-period variables can change by a factor of thousands of times, for example, χ Cygni changes in magnitude from 3.3 to 14.2, a factor of 23 000 times in brightness, the bolometric magnitudes of these stars at maximum are only twice those at minimum. This shows that the total energy flux does not vary much.

These long-period variables are late M-type stars and even C and S types; they are red giants and have very low densities. Because they expand greatly, their surfaces cool down to temperatures as low as 2500 K.

The reddest of the Omicron Ceti variables is R Leonis (9 47.6 + 11 26), an M8 IIIe star, unsurpassed in beauty.

T Centauri (14 41.8–33 36) is an easy object to observe, its magnitude varying between 5.5 and 9.0 over a period of 90.4 days. The period of Omicron Ceti itself is almost 332 days. Its co-ordinates are 02 19.3–02 59. Other examples of Mira types are listed in Appendix 7, section 7.5.

The light curve of Mira is shown in Figure 11.10. The curve displays continual wave effects as if the star is vibrating.

Semi-Regular Variables

Long-period variables with irregular periods (Appendix 7, section 7.6) are classed as semi-regular variables. Their amplitudes are smaller, being about 2 magnitudes. Some have secondary periods. These stars are divided into four groups:

SRa These are giants and super-giants, of types K, M, C and S. Their light curves are similar to those of the long-period variables, but the amplitudes are less. Examples are: Z Aqr and AM CrA.

SRb At times the periods of these become shorter; at other times they appear to be constant. These stars display emission lines and belong to types K, M, C and S. R Dor and S Lep are good examples.

SRc These have small amplitudes of about 1 magnitude and long secondary periods. They belong to types G8 to M6. Examples are α Ori and α Her (Figure 11.11).

SRd These are of types F, G and K, and the spectra have strong emission lines. They are bright giants and super-giants. The minima of the light curves are alternately deep and shallow. Examples are: S Vul and SX Her.

In general, the emission lines in the spectra of semi-regular variables vanish for shorter periods than in the case of the long-period variables. The emission lines originate in

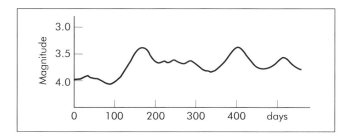

Figure 11.11 Light curve of semi-regular variable Alpha Herculis

hot atoms just below the surface; when these hot atoms erupt into the chromosphere and corona, they cool down and the emission lines vanish.

The majority of semi-regular variables have periods between 100 and 200 days.

Irregular Variables

There is large group of variables that have no connection with periodicity; they are the irregular variables. The spectral lines are broad, showing that the stars spin rapidly; some cast off shells of matter. Intense disturbances caused by localised supersonic shockwaves in the shells of matter are the cause of the irregularities in the light emission.

The irregular variables usually have small amplitudes, for example U Del and V Aql. The latter star is deep red and classified by some astronomers as semi-regular.

External factors can play roles in the light intensity which a star emits. Stars shrouded in nebulosity are known as nebular variables. The nebulae can be dark or bright. J.S. Glasby in *The Nebular Variables* (Pergamon Press, Oxford) distinguishes three groups of nebular variables:

1. RW Aurigae variables that have sudden maxima and minima of large amplitudes (Figure 11.12; see also Appendix 7, section 7.8).
   RW Aurigae experiences four or five maxima in a period of about 40 days. These rapid changes can be succeeded by long periods during which the brightness changes by hardly $\frac{1}{2}$ magnitude, probably because the star then happens to be in a nebula-free region of space. A considerable number occur in the nebulae of Orion. Their spectral type is dG, that is, dwarfs at temperatures of about 6000 K.
2. T Orionis variables (Appendix 7, section 7.9) which brightly illuminate the nebulae in which they are located. They are B8 to A3 in type, but mostly A0 at temperatures between 11 000 and 20 000 K. Their amplitudes comprise 2 to 3 magnitudes.
3. T Tauri variables (Appendix 7, section 7.10). These stars have smaller amplitudes, although T Tauri itself has an amplitude of 5 magnitudes.

Their spectra show that they are red dwarfs and are usually shrouded in dark nebulae. The variability in the gas and dust envelopes is greater than that in the stars themselves. Probably, small variations in the brightness of T Tauri stars bring about larger variations in their infra-red and ultra-violet radiations, and these can have large effects on the molecules and gas of the nebulae.

The light curves have portions that look like sine curves. The stars are of type dG, somewhat like the Sun, but their spectra have about 100 times as many lithium lines as the Sun's spectrum. The absorption lines are broad, showing that T Tauri stars spin rapidly, much more rapidly than other stars (like the Sun) that occupy their region of the H–R diagram.

The infra-red portions of their spectra show that the stars are shrinking and that they have discs of dust around them. These discs may be planetary systems in the making. Large telescopes are required to see these stars

In 1988 it was found that the star Beta Pictoris has a disc of dust which radiates in the infra-red, showing that the dust must be finely divided. The dust particles will eventually accrete to form planets.

R Coronae Borealis Variables

Another type of very irregular variable is R Coronae Borealis (15 48.6 + 28 09).The absolute magnitudes of this type (Appendix 7, section 7.11) are between −4 and −5; they are bright giants. The light curve of R CrB (Figure 11.13) shows sudden diminutions in brightness after long periods

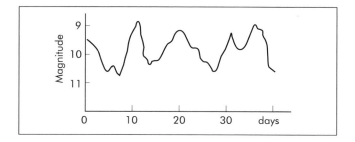

Figure 11.12 Light curve of RW Aurigae

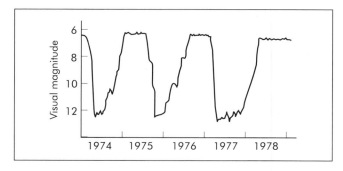

Figure 11.13 Light curve of irregular variable R Coronae Borealis

The Carbon–Nitrogen–Oxygen Cycle

of steadiness. The sudden decreases in brightness are caused by shells of carbon dust which are cast off by the stars. High above the chromosphere, and above the corona, where the temperature drops below 7000 K, the carbon condenses to fine particles which cut off the light of the star. After a lapse of 200 to 300 days, the brightness suddenly recovers and then remains constant for a considerable time. This happens when the stellar wind blows the carbon particles away.

These stars belong to the rare R type although R Coronae Borealis itself is a G-type star. RY Sagittarii (19 16.3–33 32) reaches a minimum magnitude of 15 and brightens to magnitude 6. It also casts off shells of carbon and displays emission lines in the spectrum.

The C–N–O Cycle

Besides the nuclear reactions which take place in the Sun, and which also occur in stars, the carbon–nitrogen–oxygen cycle is of great importance in the production of stellar energy. Carbon atoms, which are plentiful in the R CrB stars, form the basis of a series of nuclear reactions, as shown in Figure 11.14.

The nucleus of the carbon atom serves as catalyst to bring about the fusion of four protons, numbered ⊕1, ⊕2, ⊕3 and ⊕4, to form one helium nucleus.

In the core of a star (and the Sun), where the temperature is of the order of 15 million degrees K, the sub-atomic particles are under great pressure and move at tremendous speeds.

In the first step of the reactions, at the top of the figure, a fast-moving proton, ⊕1, enters the nucleus of a carbon atom. The normal carbon nucleus consists of 12 nucleons: 6 protons ⊕, and 6 neutrons O. When the first proton enters the nucleus, the nucleus has 7 protons and 6 neutrons and thus has become a nucleus of nitrogen, $^{13}_{7}N$, consisting of 13 nucleons, of which 7 are protons. This nitrogen nucleus is radioactive and it emits a positive electric charge, a positively charged electron (or positron) e^+, as well as neutrinos with an energy of 1.2 MeV, and very high frequency gamma rays.

Because a positive charge has been lost, the nucleus has only 6 positive charges remaining, but the mass of the nucleus is still 13. This means that the 7th proton has become a chargeless neutron. The nucleus thus has 6 protons and 7 neutrons. It has therefore again become a carbon nucleus, $^{13}_{6}C$.

When a second proton, ⊕2, enters this nucleus, it will have 7 protons and 7 neutrons. It therefore has become a normal nitrogen nucleus $^{14}_{7}N$.

When a third proton, ⊕3, now enters this nucleus, a positron is shot out, as well as gamma rays and neutrinos, having an energy of 1.74 MeV. Because a proton has been added, the mass of the nucleus is now 15, namely 8 protons and 7 neutrons. This is an oxygen nucleus and is radioactive. Having lost a positron, the charge has decreased to 7 positive charges, that is 7 protons, but the mass is still 15 nucleons: 7 protons and 8 neutrons, a proton having changed into a neutron. It is therefore again a nitrogen nucleus $^{15}_{7}N$, although its mass is 15 a.m.u.

When a fourth proton, ⊕4, enters this nucleus, it contains 16 nucleons, of which 8 are protons and 8 neutrons. This is a normal oxygen nucleus, $^{16}_{8}O$, but a rearrangement of the nucleons takes place and the very stable nucleus of helium, $^{4}_{2}He$, is formed, containing 2 protons and 2 neutrons, and is shot out. The 12 nucleons which remain consist of 6 protons and 6 neutrons, that is the original carbon nucleus, $^{12}_{6}C$, has been formed again. And the the whole process is repeated.

In actual fact, the four penetrating protons have been formed into a helium nucleus, $^{4}_{2}He$, which is 0.028 63 a.m.u. lighter than the total mass of four protons. This loss of mass is converted into energy, according to Einstein's famous equation:

$$E = mc^2$$

where m is the mass lost, in grams, and c is the speed of light, in cm s^{-1}. This works out to 615.5 × 10^9 calories of heat per 4 grams of stellar matter, which has been converted into helium.

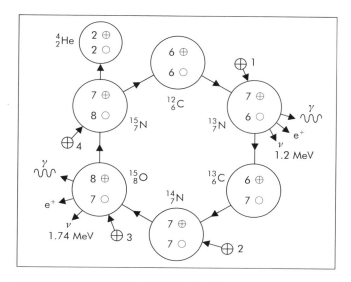

Figure 11.14 The carbon–nitrogen–oxygen cycle

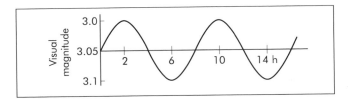

Figure 11.15 Light curve of Beta Cephei

RV Tauri Variables

The spectra of these irregular variables have sharp hydrogen absorption lines, showing that the corona contains much hydrogen. During the deep minima of these stars, bands of titanium oxide appear in the spectrum; this means that the temperature must drop to 3000 K. The maxima are sharp and the amplitudes reach 5 magnitudes. Amplitudes of 3 to 4 magnitudes are common. Periods are 30 to 150 days.

An appreciable percentage of metals (nuclei heavier than hydrogen and helium) shows that RV Tauri stars belong to Population I stars. The spectral types are F and G at maximum and K and M at minimum.

Beta Cephei Variables

This is an exceptional group of pulsating variables, with amplitudes of only 0.1 magnitude; the pulsations are non-radial. Spectral types are B0.5 to B2 and the luminosity classes IV and III (Appendix 7, section 7.13).

The periods are very short: 3 to 7 hours. The light curves (Figure 11.15) are almost identical to sine curves. These variables are also classed with Canis Majoris stars.

Besides the irregular variables treated above, there are other variables that are peculiar in one way or another.

Peculiar Variables

Eta Carinae (10 45.1−59 41) is a very peculiar sort of star. In the 17th and 18th centuries it was of magnitude 2, but early in the 19th century its magnitude started to fluctuate. After a very sharp increase in brightness in 1838, it rose to magnitude −0.8 by 1843 – about as bright as Canopus, the second brightest of all the stars. The fluctuation in brightness continued until 1867, when it suddenly dropped to magnitude 7.6 (Figure 11.16).

Eta Carinae is shrouded in dense nebulosity. Doppler

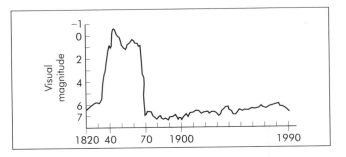

Figure 11.16 Light curve of Eta Carinae

effect measurements show that the nebulae are expanding at 480 km s^{-1}. The computed distance is 6400 light years and the mass can be as much as 80 solar masses. However, K.H. Hofman and G. Weigelt, using the 3.6-metre telescope of the European Southern Observatory and the method of speckle extinction, found, according to their article in *Sky and Telescope*, that Eta Carinae may really be a quadruple star: Eta itself, and three companions within 0.1 and 0.2 arc second, and about 2.7 magnitudes dimmer than Eta. It therefore has to be doubted that the mass of Eta Carinae is 80 solar masses.

Since 1870, Eta Carinae has displayed magnitudes ranging between 6 and 7.

Among the most well-known peculiar variables, are the following:

1. UV Ceti (Appendix 7, section 7.14), which is a flare star. At irregular intervals, it suddenly brightens from magnitude 13 to 6.5, but usually brightens by 1 to 1.5 magnitudes. UV Ceti is one of the nearer stars, at 2.74 parsecs. The nearest star, Proxima Centauri, is also a flare star, brightening by 1 to 1.5 magnitudes at times; within minutes, these stars become $2\frac{1}{2}$ to 4 times brighter. Flare stars are small, being about 0.4 solar mass and have low luminosities.
2. Certain white dwarfs, of spectral type DA, with surface temperatures of 12 000 degrees and densities of 1000 kg cm^{-3}, undergo sudden sharp brightening of 0.3 magnitude (Figure 11.17). Their cycles range from 1 to 15 minutes. The prototype is ZZ Ceti and the stars are also called ZZ Ceti stars. The figure shows the light curve of ZZ Piscis, according to measurements by B.R. Pettersen (*Astronomical Journal*, 1969, 3.173). These stars also have secondary periods.
3. Rotating variables (Appendix 7, section 7.15) are stars whose surfaces are not equally bright all over, because chemical and temperature differences exist on account of strong magnetic fields. The differences in magnitude

U Geminorum Variables

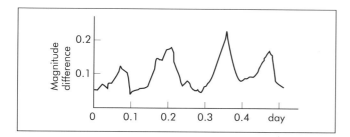

Figure 11.17 Light curve of white dwarf ZZ Piscis

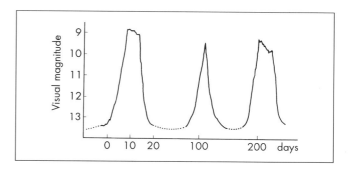

Figure 11.18 Light curve of U Geminorum

are small, 0.01 to 0.1. The periods range from 0.5 to 160 days. The tracking down of these minute changes in magnitude shows how efficient modern photo-electric photometers are. Examples of this kind of variable, are: α^2Canum Venaticorum, of spectral type B9.5Vp to A0p, and γArietis, an A1p star.

4. BY Draconis stars are K and M dwarfs whose spectra have emission lines. Their oscillations in magnitude are small.
5. FK Comae Berenices stars are rapidly spinning G and K giants. Their spectra have broad emission lines of ionised calcium, CaII. Sometimes they have emission lines of hydrogen. Most are spectroscopic binaries whose periods of variation of a few days are equal to their periods of revolution. It is possible that they also have irregular blotches.
6. Symbiotic variables are stars that, as it were, "live on each other" by the to-and-fro exchange of material. They are of late spectral types, K and M. Besides absorption lines, they also have emission lines of ionised helium and oxygen. Their amplitudes can be as much as 3 magnitudes (Appendix 7, section 7.16).
7. Cataclysmic variables, or eruptive variables (Appendix 7, section 7.17) are dwarf stars that undergo sudden changes. The best known example is U Geminorum (07 54.8 + 22.01). In contrast with R CrB stars that suddenly become several magnitudes dimmer, the eruptive variables suddenly become 2 to 6 magnitudes brighter.

In 1855, U Geminorum suddenly became 9 magnitudes brighter and disappeared after two weeks. A few months later N.R. Pogson found that it had returned to its former brightness and then it slowly faded. Because the sudden brightening gives these stars the appearance of being new stars, they are called novae. Those such as U Geminorum and Z Camelopardalis, which usually have small eruptions, are called dwarf novae. Figure 11.18 shows the light curve of U Geminorum.

There is a linear relationship between the amplitudes and the periods of the dwarf novae: the longer the period, the greater is the amplitude.

Spectroscopically, it has been found that the majority of these stars are close binaries, in which streams of gas flow from a cool, solar type star to a white dwarf companion. The stars also eclipse each other. Figure 11.19 shows the light curve during one revolution.

The gas-streams from the larger star form an accretion disc around the white dwarf. At the point where the gas-stream meets the accretion disc, a bright spot is generated and this is responsible for the sudden increase in brightness. As the two stars revolve around each other, this spot moves and it is seen at its brightest just before eclipse takes place.

Eruptions of dwarf novae take place at random and the brightening of the accretion disc can vary according to the temperature of the impinging gas from the larger component (Figure 11.20). This figure is a tracing of the diagram by E.L. Robinson in the *Astrophysical Journal.*.

In the case of SS Cygni, the roles of the two stars are reversed and it is the white dwarf which is losing material.

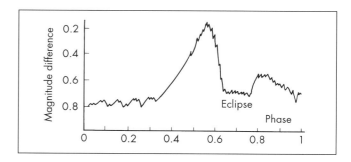

Figure 11.19 Differences in magnitude of U Geminorum during one revolution

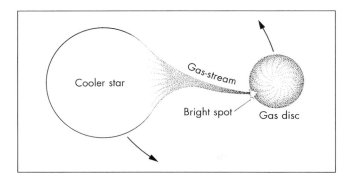

Figure 11.20 Cataclysmic variable star

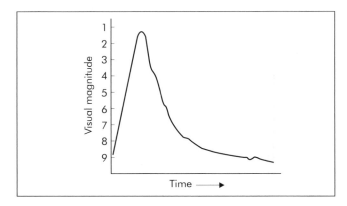

Figure 11.21 Light curve of typical nova

The orbits of the close binaries are usually circles, but some do have eccentric orbits, for example RX Andromedae and Z Camelopardalis.

Most U Geminorum stars have emission lines in their spectra, as they approach maximum brightness. This is due to the high speed of more than 3000 km s^{-1} at which the accretion disc rotates.

There is a generic link between dwarf novae and repetitive novae. W Ursae Majoris stars may be the progenitors of dwarf novae.

Usually, the more massive member of a close pair is the first to evolve away from the Main Sequence of the H–R diagram. While so doing, it expands and sheds its upper layers. These outer layers of gas are absorbed by the lighter member, so that it begins to gain mass. After the member which first left the Main Sequence has lost a certain amount of its upper layers, it shrinks to a white dwarf, stripped of its upper layers. By this time the originally less massive member may start moving away from the Main Sequence, and will also begin to expand. The roles of the primary and secondary members then become reversed and the gas-streams will flow in the opposite direction, from the cooler, yellow star, to the white dwarf, exactly as in the case of the dwarf novae.

Novae

Novae are not new stars, but stars which have suddenly become considerably brighter. Many of them are found in the plane of the Milky Way.

Novae have amplitudes of 8 or 9 magnitudes. After such a sudden increase in brightness there is a rapid decrease of 3 or 4 magnitudes, followed by a gradual drop in brightness over a period of some hundreds of days. The light curves of novae are very similar (Figure 11.21).

DQ Herculis (1934) and V732 Sagittarii (1936) both showed a gradual fall in magnitude for about 100 days, but then there followed a sudden decrease in brightness for 30 days and a recovery of 4 or 5 magnitudes over a period of 60 days, before the gradual dimming resumed (Figure 11.22). The drop in brightness, after 100 days, must have been caused by an eclipse.

Many novae were missed in the past, but today there exists a world-wide monitoring patrol, the members of which regularly photograph the same sections of the sky. Any flare-up will be seen immediately.

Novae are classified according to the speed of the change of the brightness of the star.

Na This typifies rapid novae which vary between 11 and 13 magnitudes. After the rapid change, there is a drop in brightness of 3 magnitudes over a period of 110 days. Absolute magnitudes work out at −9. Examples are: GK Persei (1901); V603 Aql (1918); V476 Cygni

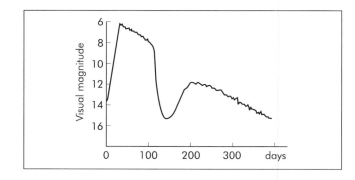

Figure 11.22 Light curve of V732 Sagittarii

Planetary Nebulae

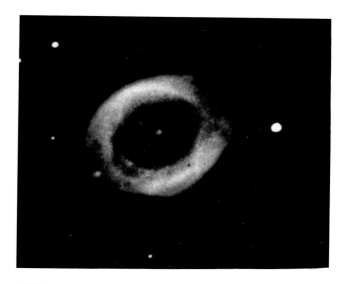

Figure 11.23 The Ring Nebula, M57 in Lyra

(1920); and XX Tauri (1927).

Nb These are slower: amplitudes between 8 and 11 magnitudes and the fall of 3 magnitudes takes longer than 110 days. Sometimes they linger at maximum for months on end. They may also have secondary brightenings during the dimming phase. Examples are: HR Delphini (1967); and V732 Sagittarii (1936) (Figure 11.22).

Nc These novae are very slow: RT Serpentis (1915) brightened to magnitude 10.5, and maintained it for 10 years before it slowly dimmed, reaching magnitude 14 in 1942. Examples are: V407 Cygni (1936); FU Orionis (1937); V939 Sagittarii (1914); and V941 Sagittarii (1910).

Nr These are repetitive novae (Appendix 7, section 7.18). It is possible that many, if not all, novae will become repetitive. VY Aquarii brightened by 8 magnitudes in 1907, remained dormant for 55 years and then erupted in 1962. U Scorpii has undergone no less than four eruptions: in 1863, 1906, 1936 and again in 1979. Be on the lookout for another between 2018 and 2022, or perhaps earlier! The co-ordinates of U Scorpii are 16 h 22 m 30 s − 17°53′. At each eruption, U Scorpii brightened through 9 magnitudes − a factor of 4000 times!

The brightest repetitive novae can attain absolute magnitudes of −7 or −8 during the maxima.

The spectra of novae are typified as type Q. Those novae whose spectra were known before eruption were stars of types A and B. However, Nova Delphini (1967) was an O9 star (almost a B).

At maxima, novae have continuous spectra with the greatest intensity in the ultra-violet. Often there are broad emission and absorption lines. After maximum, the continuous spectrum fades and the hydrogen lines become stronger. Then the hydrogen emission lines fade and nebulous bands appear. The nebulae may consist of the gases cast off during eruption. Eventually, the spectrum shows a dull continuum and strong Wolf–Rayet bands of HeI, CII, CIII, CIV and OII to OV, as well as NIII, NIV and NV. These lines of ionised atoms show that the temperature must be of order of 35 000 K.

The stars at the centres of planetary nebulae, such as M57, the Ring Nebula in Lyra (18 53.6 + 33 02) are probably white dwarfs (Figure 11.23).

The shells of gas of the planetary nebula are cast off when the star reaches the giant stage. The casting off of the gas contributes to the brightening.

Another beautiful example of a planetary nebula is the Helix Nebula, NGC 7293 in Aquarius (22 29.6−20 48) (Figure 11.24). It is easily visible in a small telescope, using a low power. A time-exposure photograph is the best way to reveal details of the expanding shells.

When a star casts off material, the remainder shrinks very forcibly to a size calculated to be 2% of the Sun's size. Not only are these residues very dense, but their temperatures reach 100 000 K!

The gases are illuminated by the ultra-violet radiation from the star which occupies the mid-point of the nebula. The typical greenish-blue colour of the gas shells is due to doubly ionised oxygen, OIII. A photograph through a red filter shows striations of outward-moving gases. Planetary nebulae may be very slow novae.

Because the brightest novae reach absolute magnitudes of −9, they can be used to determine distances, by measuring their apparent magnitudes and comparing these

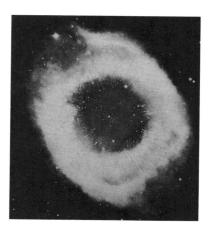

Figure 11.24 The Helix Nebula in Aquarius

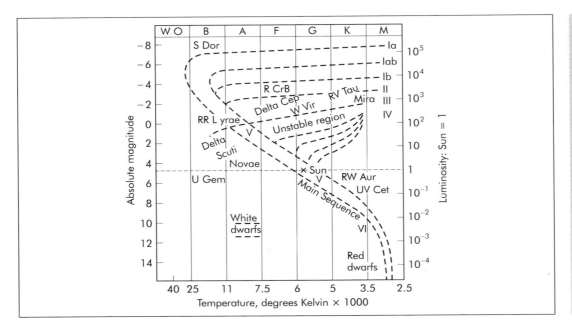

Figure 11.25
Distribution of stellar types in the H–R diagram

with the absolute magnitude of −9.

In Figure 11.25, the various types of variable stars are depicted in the positions they occupy in the H–R diagram. It is noteworthy that the pulsating variables occupy the region of instability. The RR Lyrae stars occupy the horizontal strip because they all have the same absolute magnitude of 0.8. Novae, U Geminorum stars and white dwarfs fall below the Main Sequence. S Doradus, purported to be the brightest star, is situated in the top left-hand corner and is a B or O giant. It may, however, be a multiple star.

Supernovae

In August 1885, there appeared a bright star in the heart of the Andromeda nebula, which rose to magnitude 6 and then faded in typical nova manner. The star was dubbed S Andromedae. There was no certainty as to whether it belonged to the Andromeda nebula, M31, or whether it merely lay on the line of sight.

In 1917, G.W. Ritchey, using the $2\frac{1}{2}$-metre telescope on Mount Wilson, found novae on photographs of the nebula, which appeared to be similar to novae of the Milky Way. At maximum, their magnitudes were about +15. If the absolute magnitudes of novae in the Andromeda nebula were also −9, as in the Milky Way, then the distance of the Andromeda nebula had to be:

$$\log D = (m + 5 - M) \div 5$$
$$= (15 + 5 + 9) \div 5 = 5.8$$
$$\therefore D = \text{antilog } 5.8 = 630\,957$$

namely 630 957 parsecs, or 2 057 000 light years. Such a distance was unknown in 1917. If S Andromedae was a rapid nova, which dims through 13 magnitudes, it had to go down to magnitude 19. Long time-exposures showed no sign of S Andromedae, not even down to magnitude 23. It could, therefore, safely be assumed that the star must have dimmed through 23 − 6 = 17 magnitudes. This is much more than the 13 magnitudes through which a bright nova dims.

If S Andromedae was nearer than the Andromeda nebula, its remnant would have been visible at magnitude 19. Since there was no sign of any such object, the star could not be nearer than M31.

It was remarkable that S Andromedae had reached a brightness which rivalled that of M31 itself. Later, we shall see that M31 consists of at least 100 thousand million solar masses, and if S Andromedae was one of those stars, it must have been a most exceptional nova.

If the calculated distance of M31 was correct, then the absolute magnitude of S Andromedae had to be −18:

$$M = m + 5 - 5\log\frac{2\,057\,000}{3.26}$$
$$= 6 + 5 - 29 = -18$$

Supernovae

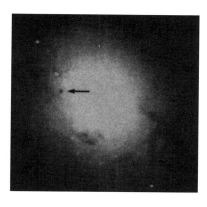

Figure 11.26 Remnant of S Andromedae in light of iron – 3860 Å

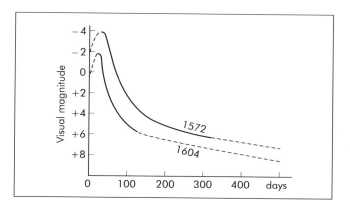

Figure 11.27 Light curves of supernovae of Brahe, 1572, and Kepler, 1604

Astronomers recoiled from accepting this, because they had no experience of an absolute magnitude of −18 – more than two thousand million times the brightness of the Sun!

In 1985, Robert A. Fesen and his colleagues succeeded in tracking down the remnant of S Andromedae by using a charge coupled device with the 4-metre reflector at Kitt Peak. It shows as a small black dot against the core of the Andromeda nebula. The photograph (Figure 11.26) was made in the light of iron at the wavelength of 3860 Å, and was published in *Sky and Telescope* of March 1989, page 247. The size of the dot is 0.3 arc second. At that distance it represents about 1 light year and cuts off the light of a few hundred thousand stars. From point size in 1885, this means a speed of expansion of 4000 to 5000 km s^{-1}. The wavelength of 3860 Å was very suitable because of the great mass of iron which is formed in a supernova explosion.

Searches through photographs of other similar nebulae revealed examples similar to this type of nova, of unknown brilliance. It was therefore decided to call this very brilliant nova a supernova.

Attention was again paid to the "new stars" which Tycho Brahe and Johannes Kepler had monitored in 1572 and 1604. The light curves which they obtained are depicted in Figure 11.27. The two curves which show the fading of the stars are very similar, with a difference of about two magnitudes. Brahe and Kepler could monitor the stars down to magnitude 6. Extrapolation to magnitudes 13 to 15 shows that the two stars must have faded through 17 magnitudes. Probably, the new star which appeared in the time of Hipparchus was a supernova.

Records of Chinese astronomers mention "visiting stars" in 1006, 1181 and 1054, the latter in the constellation of Taurus. At this point in the sky there is a nebula M1, the Crab nebula (05 34 + 22 01). The gases in this nebula are approaching the Earth at a radial velocity of −1000 km s^{-1}. Extrapolating backwards, this means that these gases must have left the point of explosion 900 to 950 years ago, that is, between 1040 and 1090, which fits in well with the date 1054. According to the records, the star, when at its brightest, was as bright as the planet Jupiter. At a distance calculated to be 6000 light years, the absolute magnitude of the star must have been −15 (Figure 11.28).

In the centre of the nebula, there is a star of magnitude 16, which radiates radio and optical pulses at a frequency of 30 per second. It has been called a pulsar.

Using the 2½-metre telescope at Mount Wilson, F. Zwicky was able to distinguish different types of supernovae in distant galaxies:

Type I   These supernovae have absolute magnitudes

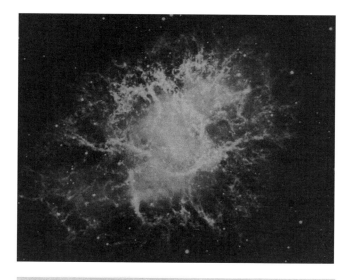

Figure 11.28 The Crab nebula

reaching −19.5 and their light curves have sharp maxima. Their spectra reveal strong lines of iron.

Doppler effect measurements ascribe velocities of −10 000 km s^{-1}, that being the speed of expansion, and it is much greater than those of ordinary novae. The changes in the spectra are very regular, and backward extrapolation gives accurate times for the maxima in the light curves.

Type II   These have much broader maxima in the light curves. The fading after maximum is rapid but it soon slows down and, after 60 days, there is a steady decline in brightness. Their absolute magnitudes reach −16 to −18. The Crab nebula was probably formed by a Type II supernova. Spectra of Type II show similarities to those of novae.

While novae concentrate in the plane of the Milky Way, supernovae occur anywhere, and especially in the spiral arms of galaxies, where there are many massive stars.

The light and electro-magnetic radiation of a supernova constitute only a small portion of the total energy emitted. The kinetic energy (energy of motion) is ten times greater. Still, the brightness of a supernova can rival that of the whole galaxy of thousands of millions of stars. A great amount of energy is whisked away by the mass-less neutrinos.

As far as is known, there have been only 6 or 7 supernovae in the Milky Way galaxy since 1006 (Table 11.4).

The light curves of the supernovae of 1572 and 1604 indicate that they were Type I. The estimated magnitude of −15 for the supernova of 1054 shows that it must have been a Type II.

Research in depth by D.H. Clark and F.R. Stephenson into Chinese records of the visiting star of 1181 shows that the radio source 3C 58 occupies the identical position in the sky.

About 10 supernovae are discovered per year in the arms of spiral galaxies, outside the Milky Way.

The first astrophysical theory about supernovae was compiled by F. Hoyle, Margaret Burbidge, G.R. Burbidge and W.A. Fowler. Their theory suggests that:

1. When a massive star has used up a certain amount of its hydrogen fuel, in the process of converting it into helium, the sudden diminution in radiation, which had been assisting the gas pressure to resist the crushing power of gravity, results in a collapse of the overlying layers of the star, on to its centre.
2. The energy released by the collapse expels the greater proportion of the star's mass into space.
3. All that remains is a neutron star, of unimaginable density, brought about by the pressure forcing the electrons into the protons, thus converting them into neutrons.

During the lifetime of a star, a balance is maintained between the outward pressure and the inward force of gravity (Figure 11.29). The Sun has been in such a state of balance for at least 4 to 5 thousand million years, despite losing mass at the rate of 4 million tons per second, and it can go on at the same rate for another five thousand million years. And then? Then the Sun will suddenly start collapsing on its nucleus, while hydrogen fusion to helium will go on in its overlying layers. The collapse will raise the temperature of the Sun's core to 100 million degrees. This high temperature will cause the helium which had been formed in the core by the fusion of hydrogen, itself to undergo fusion – the helium flash – to form heavier nuclei, such as carbon, oxygen, etc. The tremendous amount of energy liberated will blast the overlying layers away. The Sun will thus expand and leave the Main Sequence to

Table 11.4 Historic supernovae

Date	Constellation	Co-ordinates RA	Dec	Type	Distance (light years)	Magnitude
1006	Lup/Cen	15 02.8	− 41 57	I	3 300	?
1054	Taurus	05 34.5	+ 22 01	II	6 500	−4 to +15 (Crab)
1181	Cassiopeia	02 02.0	+ 64 37	II?	26 000	0
1572	Cassiopeia	00 25.3	+ 64 09	I	10 000	−4.5 (Tycho)
1604	Ophiuchus	17 30.6	− 21 09	I	16 000–30 000	−2.5 (Kepler)
1680?	Cassiopeia	23 23.0	+ 58 45	II?	9 100	?
1885	Andromeda	00 42.7	+ 41 16	I	2 200 000	+6.0
1987	Doradus	05 35.8	− 69 18	II	166 000	+2.4

Stellar Evolution

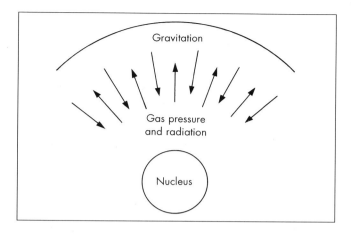

Figure 11.29 Balance between gravity and gas pressure plus radiation

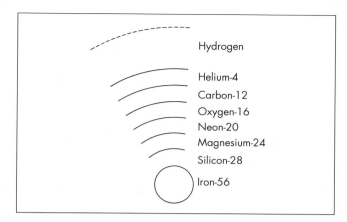

Figure 11.30 Onion-like stellar nucleus

become a red giant. During this expansion, the Sun will gobble up the inner planets. After reaching its maximum size, the radiation will decrease and the Sun will commence to make its final collapse, ending up as a white dwarf.

The career of a more massive star is somewhat different. Because of its greater mass, the pressure on the core is greater, making its temperature higher. This higher temperature will make the fusion reactions go faster, so that a massive star will not sojourn as long in the Main Sequence before becoming a red giant. A star 10 times as massive as the Sun will consume its hydrogen 1000 times as fast. The helium fusion will last about half a million years. Three helium nuclei fuse to form one carbon nucleus:

$$_2^4He + {_2^4}He + {_2^4}He \rightarrow {_6^{12}}C$$

The radiation from this last reaction will help to resist the collapse to a white dwarf. But when the collapse does recommence, the raised temperature will convert the carbon nuclei into oxygen nuclei, with another release of energy and the inexorable pressure of gravity will successively form layers of neon-20, magnesium-24, silicon-28, sulphur-32, argon-40, calcium-40 and titanium-48. The atomic masses of these nuclei are all multiples of four and they are very stable – these nuclei probably consist of groupings of helium nuclei.

The star now has the form of an onion – shell upon shell – as in Figure 11.30. By the time silicon-28 is formed, it begins to fuse directly into iron-56. This iron nucleus is very tightly bound – it has the largest packing fraction of all nuclei. Fusion into nuclei heavier than iron requires energy to be supplied – energy is not liberated. Besides iron-56, cobalt-56, which is very radioactive, is also formed and this adds to the total amount of energy liberated. Now the gravitational force becomes irresistible and catastrophic collapse takes place. The irresistible force of gravity comes up against the incompressibilty of the core of the star and a shockwave rips the star apart. The speeds of expansion of the order of $10\,000$ km s^{-1} lead to an indescribable explosion. All that is left is a small core of neutrons – a mass equal to that of the Sun, compressed into a diameter of 10 km. This is how a neutron star comes into existence.

How massive must a star be to follow this course of events, rather that that leading to the formation of a white dwarf? The answer was found in the 1950s by S. Chandrasekhar of the University of Chicago. When matter is compressed, the density is increased because the molecules are pressed closer together. The electrons surrounding the nuclei of the atoms offer resistance because they are all negatively charged, and like charges repel each other. Chandrasekhar found that there is a limit to the degree of compression which the electrons can withstand. In a star the balance can be maintained if the mass is less than 1.44 solar masses. This is called the Chandrasekhar limit.

A white dwarf of less than 1.44 solar masses can go on existing indefinitely. Strange to say, Type I supernovae, the more violent type of supernova, come about when the white dwarfs blow themselves to smithereens. This happens because white dwarfs which go supernova are members of binaries. If a white dwarf is one of a close binary, it will strip matter from the companion, because of its highly concentrated gravity. The white dwarf will then become more massive and exceed the Chandrasekhar limit. The increased pressure on the core will raise the temperature so that fusion of heavier nuclei will take place. When the energy flux exceeds 20 times the binding energy of the star, it blows itself apart, the overlying layers shooting off at speeds of $10\,000$ km s^{-1}. During the explosion, at least one solar mass of the unstable nickel-56 is formed. It decays to cobalt-56 which is radioactive and in its turn decays to iron-56. The speed of decay of the cobalt-56 is exactly enough to explain the fading of a Type I supernova.

Apparently, no remnant is left when a white dwarf goes supernova.

To become a Type II supernova, a star must have at least 8 times the mass of the Sun. Such massive stars do not linger long in the Main Sequence before expanding to red giants. When silicon fusion produces iron-56 in its core, the speed of this reaction accelerates, so that within one day, the iron core itself exceeds the Chandrasekhar limit, so that complete collapse of the iron core takes place. So great is the pressure on the core that electrons are forced into the protons, forming neutrons. This happens together with the escape of neutrinos, which carry off tremendous amounts of energy. According to calculation, the neutron core reaches a density of 4×10^{11} g cm^{-3}. Now the neutrinos cannot escape so easily and they collide with atomic nuclei.

The decrease in the number of electrons has the effect of lowering the Chandrasekhar limit to a value between 1.44 and 1.2 solar masses. This mass is then the greatest mass which can collapse as a unit. Now a pressure wave on the surface of the collapsing core raises the density to 2.7×10^{14} g cm^{-3}. This is equal to the density of an atomic nucleus. The separate nucleons then fuse together to form a single massive nucleon. This is the ultimate limit of compressibility. Now a rebound sends out shockwaves at speeds of $30\,000$ to $50\,000$ km s^{-1}. The shockwaves burst through the onion-like layers of the core and blast the matter into space.

Independently of each other, L.D. Landau and J.R. Oppenheimer formulated the concept of super-dense stars, consisting largely of neutrons. In 1938, F. Zwicky suggested that such super-dense stars would be formed in supernova explosions. The density of such stars is equal to that of an atomic nucleus. Ths gravitational collapse which precedes a supernova explosion will also generate very strong magnetic fields. A neutron star will therefore possess a strong magnetic field. Electrons coming from the nucleus will spiral around the lines of force of the magnetic field, at relativistic speeds, thus generating synchrotron radiation.

The mass of the star plays a large role in determining the precise course of events. If the mass of the star exceeds 20 solar masses, the compression during the collapse phase will be so great that the remnant neutron star will get crushed down to unimaginable densities. It will become a gravitational vortex (black hole), from which nothing can escape, not even electro-magnetic radiation, because the velocity of escape from the surface of the gravitational vortex is greater than the velocity of light, and so great a velocity cannot exist.

While the shockwaves of a supernova explosion are ripping the star apart, all manner of atomic transformations will take place so that helium and all the various atomic nuclei will be formed. Many of these nuclei are radioactive and decay, with emission of gamma rays and neutrinos. Cobalt-56 is a very important product of the explosion; its radioactivity is enormous. If a supernova explosion takes place close enough to the Earth, it ought to be possible to detect cobalt-56 in the spectrum of the exploding star.

Astronomers had hoped for such an event to take place, and they were rewarded on the night of 23–24 February 1987. During the night of 24 February 1987, the amateur astronomer Gordon Garradd noticed a bright star $\frac{1}{2}$ degree south of the Tarantula nebula, 30 Doradus, in the Large Magellanic Cloud, at 05 35.8–69 18, which had not been there before. Ian Shelton of Toronto University had been photographing this area nightly, from Chile, by means of a 250-mm telescope. The photograph of 23 February 1987

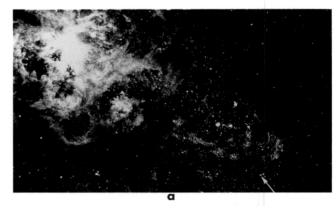

a

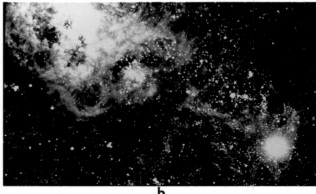

b

Figure 11.31 Supernova 1987A: **a** Before the explosion; **b** After the explosion

showed nothing extraordinary, but the photograph of 24 February 1987 showed a bright star of magnitude +3. This was the long-hoped-for supernova! Figure 11.31 shows the before and after photographs of the supernova, Sanduleak −69°202.

The star, Sanduleak −69°202, was known to be of magnitude 12 and spectral type B3. At the distance of the Large Magellanic Cloud, its absolute magnitude was:

$$12 + 5 - 5 \log \frac{166\,000}{3.26} = -6.5$$

It was, therefore, a blue super-giant of mass about 10 solar masses – an excellent candidate for a Type II supernova!

During the early morning of 24 February 1987, its apparent magnitude was +3.2. Overnight, it had brightened by almost 9 magnitudes. Over the next five days, the magnitude dropped to +4, showing that the surface layers had begun cooling. Then the brightness gradually increased for the next 71 days, when it reached a maximum brightness of +2.4. At that stage, its absolute magnitude was

$$2.4 + 5 - 5 \log \frac{166\,000}{3.26} = -16$$

It stayed at maximum for 16 days – this broad maximum being typical of Type II supernovae.

The brightening through 9 magnitudes was on the low side for a Type II supernova and the absolute magnitude of −16 was also slightly low.

The next 18 days saw a rapid drop in magnitude from 2.4 to 3.3. After that the steady decline, lasting several hundred days, commenced.

The light curve shown in Figure 11.32, by the South African Astronomical Observatory, Sutherland, Cape Province, is the first light curve of a supernova showing a drop in brightness immediately after the original brightening. The rest of the curve is typical of a Type II supernova.

A remarkable fact about this supernova is that neutrinos were detected by the Kamiokande and Irvine–Michigan–Brookhaven detectors, at the precise moment that the light was received. The 19 neutrinos were electron antineutrinos, which are the easier type to detect. Since neutrinos can penetrate a thickness of light years of lead, without interacting with a particle, the result was noteworthy.

A spectrum taken by the Kuiper Airborne Observatory showed strong emission lines of singly ionised iron, nickel and cobalt.

The brightness of SN 1987A, and the rate of decline after maximum, showed that at least 20 000 Earth-masses of nickel-56 were ejected. The nickel-56 very rapidly decayed to cobalt-56. The energy of the cobalt-56 radiation was the power behind the visual and infra-red radiation from the supernova.

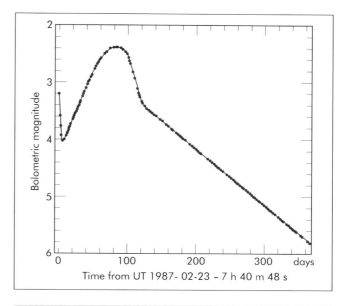

Figure 11.32 Light curve of SN 1987A by the South African Observatory, Sutherland

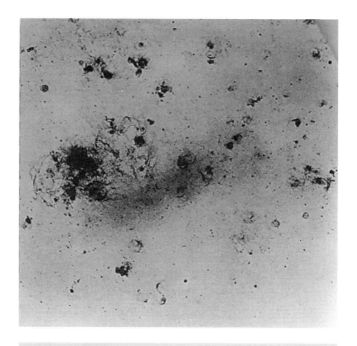

Figure 11.33 Red filter photograph of the Large Magellanic Cloud, showing supernova remnants

This supernova provided astronomers with a golden opportunity. Because the distance of 166 000 light years of the Large Magellanic Cloud is well known, all measurements of magnitudes and energy fluxes could be made with great accuracy. The astrophysical theories about supernovae could now be tested on the basis of accurate measurements.

This supernova is only one of many that had formerly taken place in the Large Magellanic Cloud. Figure 11.33, taken by the Royal Observatory, Edinburgh, using a red filter, shows many supernova remnants, in the form of small "bubbles". The Large Magellanic Cloud contains many bright stars of spectral type B. More supernovae can be expected to occur among these massive stars.

Pulsars

When Tycho Brahe saw his "new star" in 1572, he thought that it was the birth of a star. Today, we know that exactly the opposite is true: a supernova is the death of a star, or at least the end of its normal period of radiation.

No optical remnant of Tycho's new star (SN 1572) has been found. Radio telescopes have, however, found a remnant. Figure 11.34 is a tracing of a radio chart by Baldwin, as it appeared in *IAU Symposium No. 31*, page 352. The lines on the chart are lines of equal radio strength. Where the lines crowd together, there is a sharp gradient in the radio strength caused by rapidly moving gases in the form of an expanding shell. When gases move at great speed, radio waves and X-rays are generated. Space probes which have been launched into orbits outside the Earth's atmosphere are able to detect radiation, which cannot penetrate the atmosphere.

As early as 1964, an orbiting satellite discovered X-rays as coming from M1, the Crab nebula in Taurus (05 31.5 + 21 59). Because of the X-ray emission, M1 is also known as Taurus X-1. In the Cambridge list of pulsars, it is known as 3C-41.

The Einstein satellite's X-ray image of the Crab nebula shows synchrotron radiation which is responsible for 96% of the X-rays. The energy of the electrons which spiral around the magnetic lines of force is of the order of a million million (10^{12}) electron volts. The other 4% of the X-rays radiate from a star in the middle of the Crab nebula, left over as a supernova remnant of the explosion of 1054.

Measurements of the radial velocities of the gases in the nebula, made by E.P. Hubble, indicated to him that the nebula had to be the remnant of the Chinese "visiting star" of 1054.

Because there is a remnant, this explosion could not have been a Type I supernova – it had to be Type II.

All stars spin and, when a star collapses, the spinning rate has to increase in order to maintain angular momentum; in the same way as a skater spins faster when he draws his arms into his body. Angular momentum can be transferred but cannot be destroyed.

The star in the centre of the Crab nebula is a neutron star. It emits pulses of light and X-rays at a frequency of 30 per second, because the neutron star spins at a rate of 30 times per second. Once during each spin, the Earth receives a pulse of radio waves, X-rays and light. That is why the name pulsar was coined. Figure 11.35b shows a series of 10 photographs, taken by a swiftly acting camera of the National Optical Astronomy Observatories, USA. Eighteen photographs per second were taken. Photos 2 and 3 show the star at its brightest. Photos 7 to 10 show the reappearance of the light flash, with a secondary maximum in photos 8 and 9. Figure 11.35a shows the light curve.

All pulsars that have been discovered since 1967 show similarities. The periods of the pulsars listed in Appendix 7, section 7.19 vary from 3.745 497 ... to 0.001 557 8 .. seconds. At 30 revolutions per second, the Crab pulsar has a period of 0.333 3 ... seconds.

Pulsars radiate their energy in pulses because it is

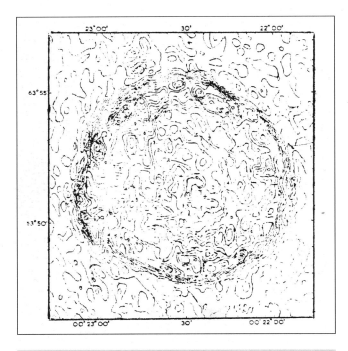

Figure 11.34 Radio chart of SN 1572 at the wavelength of 21.3 centimetres

Supernova Remnants

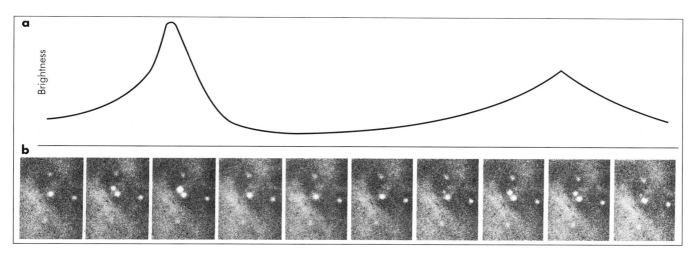

Figure 11.35 a The light curve of the Crab Pulsar, Taurus X-1. **b** Ten photos of the Crab Pulsar taken at intervals of $\frac{1}{18}$ second, show the variation in the light intensity coming from the central star

synchrotron radiation, in which swiftly moving electrons gyrate around the magnetic lines of force. The synchrotron axis is inclined to the axis of spin. Every time that the synchrotron axis points towards the Earth, we receive a pulse of radiation, consisting of radio waves, X-rays and light.

The synchrotron radiation must have the effect of polarising light, and this is exactly what happens to the light from the pulsar.

The rate at which the Crab pulsar is losing rotational energy, measured by the lengthening of its period, is equal to the rate of the synchrotron radiation, proving that the synchrotron radiation has its origin in the pulsar.

Figure 11.28 (page 211) – the Crab nebula, by Dr Hans Vehrenberg – gives a very good impression of the texture in the gases expanding from the pulsar.

The Einstein satellite, HEAO-2, which was launched in 1978, worked very well for two years; it spotted many sources of X-rays. From the remnant of Tycho's supernova it received X-rays, coming from a spherical shell of gases that fit the remnant very well. The absence of synchrotron radiation from this remnant shows that there is no pulsar there. The supposition that SN 1572 was a Type I supernova is borne out by this finding, because a white dwarf will leave no pulsar as a remnant.

In the constellation Vela, there is an extended nebula that is possibly the remnant of a supernova. A neutron star lurks in the coils of the expanding gases, although it is not in the centre. Figure 11.36 was taken by Richard West of the European Southern Observatory.

Another extended nebula which is probably a supernova

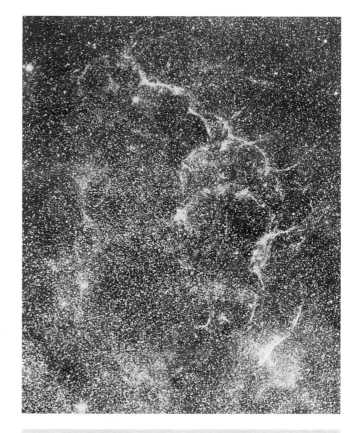

Figure 11.36 The nebula in Vela

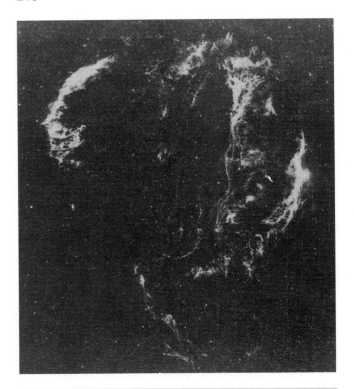

Figure 11.37 The Veil nebula in Cygnus

mately at 20 h 48 +31° (Figure 11.37). No central star or pulsar has been found there. This nebula must therefore be the remnant of a white dwarf supernova.

The remnants of SN 1006 and SN 1604 (Kepler's supernova in Ophiuchus) have no neutron stars or pulsar's. The possibility that there may be gravitational vortices there cannot be excluded.

The spectra of the Crab nebula and Tycho's remnant differ widely. The X-ray spectrum of the Crab is a continuum, but that of Tycho shows thermal emission lines of silicon, sulphur and argon. These elements were formed in the explosion.

From the X-ray image of Tycho's remnant, F.D. Seward, P. Gorenstein and W.H. Tucker calculated the remnant to have a mass of 1.5 solar masses. This fits the theory that Tycho's supernova was a white dwarf explosion, since the mass just exceeds the Chandrasekhar limit.

With the lapse of time, the nebular remnants will dissipate into space. Of the 92 pulsars listed, many do not have any nebulae. Even if these nebulae glow for 100 000 years, this is astronomically speaking a very short time.

SS 433

A very strange object, brought to the attention of the astronomical world by N. Sanduleak and C.B. Stephenson, lurks in Aquila at the point 19 h 08 m +05°.

Sanduleak and Stephenson placed a large prism in front of the objective lens of their telescope and obtained spectra of all the stars in the field of view. All the stars formed spectra which appeared as smears, except one object, whose spectrum appeared as a single dot (Figure 11.38).

All the spectra had several lines, but this one consisted only of the hydrogen alpha line at 6563 Å.

In 1978, D.H. Clark and P. Murdin found emission lines of ionised hydrogen and helium in the spectrum. The object was also found to radiate X-rays and radio waves. The spectrum also has unidentified lines that could possibly have been hydrogen lines, which had undergone a large Doppler shift from 6563 Å to 7400 Å. This supposes a speed of recession of 40 000 km s^{-1}, which is 13% of the speed of light, a speed far in excess of the velocity of escape from the Milky Way! It was also found that the wavelengths of the redshifted lines changed by 1% after a few days, a change in the radial velocity of 5000 km s^{-1}. While some lines underwent redshifts, other lines underwent blueshifts. In a period of 30 days, the radial velocity changed from 20 000 to 50 000 km s^{-1}! This tremendous change had to be ascribed to streams of gas, some having negative and others positive radial velocities.

Then it was found that the hydrogen and helium lines were trebly split, showing both red- and blueshifts. The redshift indicated a velocity of 27 000 km s^{-1} and the blueshift 6000 km s^{-1}. The maximum redshift indicated a velocity of 50 000 and the maximum blueshift 30 000 km s^{-1}. The radial velocity indicated by the lines between the shifted lines was constant at 12 000 km s^{-1}. This is still 30 times the velocity of escape from the Milky Way!

Figure 11.38 Single-line spectrum of SS 433 – the dot in the middle

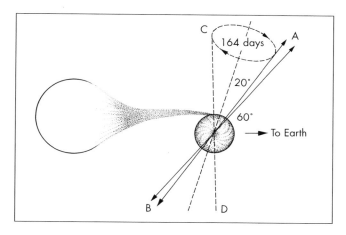

Figure 11.39 Model of pulsar SS 433

The changes in velocity had a period of 164 days: every 82 days the red- and blueshifts changed over. This could not be ascribed to a revolving body. At such speeds a binary pair would fly apart; to hang together would require a mass of 10^9 solar masses.

The solution, suggested by B. Margon, was that the object must be a pulsar, which is casting off two streams of gas, one approaching the Earth obliquely and causing the blueshift, and the other receding obliquely and causing the redshift. The axis of the pulsar is tilted by 80 degrees to the line of sight, and the streams of gas make angles of 20 degrees with the spin axis. The approaching gas-stream thus makes an angle of 60 degrees with the line of sight (Figure 11.39). The real speed of stream A must therefore be $40\,000 \div \cos 60° = 80\,000$ km s^{-1}, or 26% of the speed of light! Stream B, which causes the redshift, moves in the opposite direction at the same speed as stream A.

When the spectral lines cross over, there is still a remaining constant speed of 12 000 km s^{-1}. This is due to time dilation. According to Einstein's Theory of Relativity, time slows down as speed increases. At a speed of 26% of the speed of light, time dilation works out at 4% of the speed of light. This is exactly 12 000 km s^{-1}, namely the residual speed when the spectral lines cross over. This is another proof of Einstein's Theory of Relativity.

The streams of gas precess in a period of 164 days. In this period of time the spectral lines change from maximum redshift to maximum blueshift, and back again: BA gyrates to DC and then to BA, thus describing a cone in space.

The swiftly moving streams of hydrogen cannot have their origin in the pulsar. There must be a close companion from which the pulsar is drawing off matter. This matter, largely hydrogen, spirals in on the pulsar, forming an accretion disc around it. The rapidly moving electrons spiralling around the magnetic lines of force of the pulsar, generating synchrotron radiation, blast the hydrogen into space at the speed of 26% of the speed of light. This speed is close to the escape velocity from the surface of a neutron star.

The question arises whether there are other examples like SS 433. If other examples cannot be found, it may be due to the fact that the life of such a gas-spouting pulsar is very short, perhaps no longer than 10 000 years.

Stellar Variety

The varieties of the stars are almost infinite. Some have temperatures higher than 50 000 K; some are very steady, while others flare up and even explode; their masses vary from one-twentieth that of the Sun to 10 times, and more, and their sizes vary by a factor of 688 million!

Red giant	1 376 000 000 km
The Sun	1 392 000
White dwarf	13 830
Neutron star	2
Gravitational vortex	2/3

Chapter 12
The Milky Way Galaxy

On a clear, moonless night, away from city lights, one can see a broad streak of light spanning the heavens. During the northern winter months, it stretches from south to north, eastwards of the well-known constellation Orion. In the summer months, it stretches from the Southern Cross, through the tail of Scorpius, northwards to Cygnus. This faint streak of light is the Milky Way. The Greek word for it, *Galaxias*, is derived from "*galaktos*", like milk. From this the word Galaxy has been derived. The Galaxy includes all the stars, planets, pulsars, nebulae and much more.

The Milky Way, as seen from Earth, is the view that we get from inside the Galaxy, from our vantage point, situated about two-thirds of the radius of the Galaxy from its centre. Galileo was the first person to see, through his 2.5-cm telescope, that the Milky Way consists of countless numbers of stars.

At the northern end of the Milky Way there is a split, or bifurcation, just about devoid of stars, in the constellations of Aquila and Cygnus. We know today that this apparent lack of stars is due to dark matter obscuring the stars.

By the year 1800, William Herschel had made a diagrammatical representation of the Galaxy, based upon his many years of careful charting of the heavens. He did not know the distances of the stars, and considered the brighter stars as being nearer and the dimmer ones as being further away. In the direction of Cygnus, he depicted the bifurcation as a region without stars (Figure 12.1).

Because the Milky Way stretches right across the sky, Herschel concluded that the Sun and Solar System must be situated near the middle of the Galaxy.

By means of the trigonometrical method, it is possible to measure the parallax of a star, down to 0.01 arc second. This indicates a distance of 100 parsecs, or 326 light years. Within this distance, there are about 5000 stars. Photographs show that there are countless millions of stars in the Galaxy. They must therefore be further than 300 light years. But how much further?

There was no other method of measurement that could be applied. Astronomers resorted to the brightness of stars as indicators of distance and this could only be approximately correct.

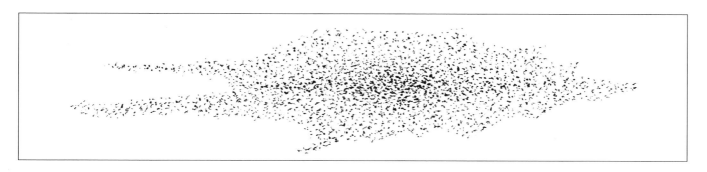

Figure 12.1 Herschel's representation of the Milky Way Galaxy

The Period Luminosity Law

Early in the 20th century, Miss Henrietta S. Leavitt, of Harvard College Observatory, concentrated on a study of the Cepheid variables in the two Magellanic Clouds. By 1912 she had made the remarkable discovery that there exists a relationship between the apparent magnitudes of the Cepheids in the Magellanic Clouds and their periods of variation: the longer the period, the brighter is the star and the smaller is its magnitude number.

Figure 12.2 is a tracing of the actual graphs that Leavitt drew, showing the apparent magnitudes against (a) the periods of their variations, and (b) the logarithms of the periods. Her publication, *Periods of 25 Variable Stars in the Small Magellanic Cloud*, appeared in the Harvard Observatory circular No. 173 of 1912.

Leavitt (see Figure 12.3) pointed out that the stars in the Small Magellanic Cloud are all approximately equally far away, provided that the Cloud was very far away. This could be asserted with confidence because the stars in the Cloud have magnitudes of 13 and fainter. The Cloud is so far away that there is no appreciable difference between the distances of the stars on the far side compared with the near side of the Small Magellanic Cloud.

Leavitt formulated the following law: at a common distance, the Cepheid variable stars have a linear relationship between their apparent magnitudes and the logarithms of their periods. The linear relationship can be seen in graph b where straight lines fit the distributions of the maxima and minima of the variables.

The period of a variable star can be measured very accurately. The apparent magnitude can then be read off from graph b. If, for example, the period of a Cepheid is 10 days, then log 10 = 1. The vertical line above the point 1 on the log-axis cuts the straight lines of best fit at 14.8 and 13.6.

Leavitt pointed out that a constant difference must exist between apparent and absolute magnitudes for the Cepheids of the Small Magellanic Cloud, because they are all at more or less the same distance from the Earth. The period–luminosity law can therefore be used to derive the absolute magnitude of a Cepheid variable, provided a starting point on the scale of absolute magnitudes could be found. This uncertainty is indicated by means of question marks (?) on the right-hand vertical axis of graph 12.2b.

If there was one Cepheid variable close enough to determine its distance trigonometrically, its absolute magnitude could be calculated from the formula:

$$M = m + 5 - 5 \log D$$

where m is the apparent magnitude and D the distance in parsecs.

But there was no Cepheid variable close enough to determine its distance trigonometrically, and this presented a very difficult problem.

To overcome the problem, the spectral types and classes of Cepheids were compared with those of stars of known distances and absolute magnitudes.

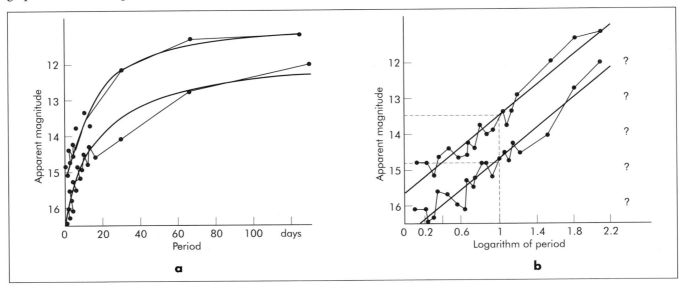

Figure 12.2 Relationship between apparent magnitude and (a) period, and (b) logarithm of period of Cepheid variable stars

The Period Luminosity Law

Figure 12.3 Henrietta S. Leavitt

Another method was to compare the Cepheids with the members of clusters of known distances.

Harlow Shapley made the assumption that the parallaxes of nearby Cepheids could be determined on the basis that their average transverse velocities must be equal to their average radial velocities with regard to the Sun's neighbourhood. From the statistic values of the parallaxes, the mean value of $(m - M)$ could be found. From the measured values of the apparent magnitudes, m, the absolute magnitudes, M, could then be calculated. These values of the absolute magnitude, M, were then ascribed to the Cepheids of the Small Magellanic Cloud having the same periods as the nearby Cepheids.

Shapley was assisted by M.L. Humason and E.P. Hubble of Mount Wilson Observatory. Eventually, in the 1920s, a scale of absolute magnitudes corresponding to the scale of apparent magnitudes was decided upon. From this, it was found that the Magellanic Clouds are more than 100 000 light years distant. This corroborated Leavitt's supposition about the distance of the Magellanic Clouds.

Astronomers now had a mighty weapon to attack the problem of distances greater than 100 parsecs (326 light years). With one leap, measurable distances reached beyond 100 000 light years.

By measuring the periods and apparent magnitudes of Cepheid variables all over the Milky Way, it was found that the diameter of the Milky Way must be at least 50 000 light years. The Magellanic Clouds, at a distance of 100 000 light years, must be outside the Milky Way Galaxy; hence they were separate galaxies or clusters. This was the first indication that the Universe does not consist only of the Milky Way, but that the Milky Way is only one of the building blocks of the Universe. Besides the Magellanic Clouds, there could be other systems of stars.

All over the Milky Way there are nebulae. Some of these were being resolved into systems of stars, by the $2\frac{1}{2}$-metre telescope of Mount Wilson. It resolved the Andromeda Nebula, M31, or NGC 224 (00 h 42.1 m + 41° 12.4′) into a system of separate stars, of unmistakable spiral form (Figure 12.4, *overleaf*). E.P. Hubble discovered Cepheid variables in the Andromeda Galaxy. By measuring their periods and apparent magnitudes, he found that the Andromeda Galaxy is at least 900 000 light years distant, far, far beyond the confines of the Milky Way Galaxy.

When W. Baade studied the Andromeda Galaxy through the 5-metre telescope of Mount Palomar, he found no trace of RR Lyrae variables. If the distance of M31 is 900 000 light years, the apparent magnitude of the RR Lyrae variables whose absolute magnitudes are 0.8 had to be 23:

$$0.8 = m + 5 - 5\log\frac{900\,000}{3.26}$$
$$\therefore m = 0.8 - 5 + 27.2 = 23$$

The 5-metre telescope could easily see down to magnitude 24. It was unthinkable that a giant galaxy such as Andromeda would have all sorts of stars, but no RR Lyrae stars. M31 thus had to be further than 900 000 light years away. Baade felt that the Cepheid scale needed revising.

Baade distinguished two classes of stars in the Andromeda Galaxy:

1. Bright, blue stars of small and even negative $B-V$ colour indices, occurring mainly in the spiral arms, similar to the bright stars in the Sun's neighbourhood. These he called Population I stars. They also have high percentages of atoms heavier than hydrogen and helium.
2. Stars in the nucleus of M31, mainly dimmer, yellowish stars, of large colour indices and low content of atoms heavier than hydrogen and helium. These he called Population II stars.

The spectra of Population I stars are rich in metallic lines and those of Population II are poor in metallic lines.

Baade also found that the Cepheid variables in the two populations differed in respect of luminosities and periods. The Population I Cepheids are, on average, 1.5 magnitudes brighter than those of Population II. The scale of absolute magnitudes, as applied to the Cepheids prior to 1952, was therefore in error and the distances determined according to that scale had to be multiplied by a factor of 2. The more accurate value for the distance of the Andromeda Galaxy now came to 2 200 000 light years. This value agrees with the findings based on supernova S Andromedae.

Baade suggested separate period–luminosity relationships for: (a) the very short-period RR Lyrae stars; (b) the classical Cepheids; (c) the long-period Cepheids. The RR

Figure 12.4
The Andromeda Galaxy, M31 or NGC 224. Lower left is NGC 205, right M32, or NGC 221

Lyrae stars are of type A, with absolute magnitudes of 0.8; while the classical Cepheids are F and G giants, of absolute magnitudes −2 to −6. The long-period Cepheids are M types. Temperatures are also taken into account in calculating absolute magnitudes.

A.R. Sandage and G.A. Tammann published the absolute magnitudes and periods of classical Cepheids in the *Astrophysical Journal*, 1969, 157.683. They are listed in the first three columns of Table 12.1. Appendix 1, section 1.14 explains from first principles how the equation of the straight line of best fit is derived.

The data of absolute magnitudes against logs of the periods are plotted in Figure 12.5. The co-ordinates of the points A and B on the line are found by substituting 0.3 and

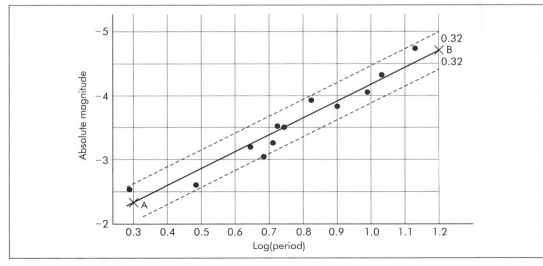

Figure 12.5
The period–luminosity law for classical Cepheid variables

The Period-Luminosity Law

1.2 for log P in the equation of the straight line. The ordinate of A = −2.29 and that of B = −4.72.

The last three columns of the table are used to calculate the standard deviation and 95% confidence limits (= ± 0.32), which apply in a normal distribution.

Take as an example, a Cepheid whose period is 10 days. The log of 10 is 1. When 1 is substituted for log P, $M = -1.48 - 2.7(1) = -4.18$. With 95% confidence, it can be said that the absolute magnitude of a Cepheid of period 10 days is −4.18 ± 0.32, that is, between −3.86 and −4.50.

Of Delta Cephei, the prototype of the Cepheids, we know the period (5.366 days) and the photographic visual magnitude, which varies between 4.63 and 3.78, with a mean of 4.2. Using the equation for the straight line of best fit, we can calculate the absolute magnitude of Delta Cephei:

$$M = -1.48 - 2.70 \log P$$
$$= -1.48 - 2.7 \log 5.366$$
$$= -1.48 - 2.7(0.729\,65)$$
$$\therefore M = -3.45$$

The distance of Delta Cephei, is therefore:

$$M = m + 5 - 5 \log D$$
$$\therefore -3.45 = 4.2 + 5 - 5 \log D$$
$$\therefore \log D = \frac{4.2 + 5 + 3.45}{5} = 2.53$$
$$\therefore D = \text{antilog } 2.53 = 338.8$$

The distance of Delta Cephei is therefore 338.8 parsecs, namely 338.8(3.26) = 1100 light years. The author worked out this result in 1989, for the Afrikaans version of this book. Delta Cephei was listed as being 630 light years

Table 12.1 Period–luminosity law

	Sandage, Tamman data			Equation to line of best fit by method of least squares						
Cepheid	Abs. mag. $M = y$	Period P (days)	$\log P = \ell$	$\ell - \bar{\ell} = x$ ($\bar{\ell} = 0.766$)	Product xy	x^2	Substitute $x = \log P$ in $a + bx = c$	Subtract $y - c = d$	d^2
SUCas	−2.54	1.95	0.290	−0.476	+1.209	0.227	−2.26	−0.28	0.0784
EVSct	−2.62	3.09	0.490	−0.276	+0.723	0.076	−2.80	+0.18	0.0324
CECasb	−3.20	4.48	0.651	−0.115	+0.368	0.013	−3.24	+0.04	0.0016
CFCas	−3.07	4.87	0.688	−0.078	+0.239	0.006	−3.34	+0.27	0.0729
CECasa	−3.27	5.14	0.711	−0.055	+0.180	0.003	−3.40	+0.13	0.0169
UYPer	−3.54	5.36	0.729	−0.037	+0.131	0.001	−3.45	−0.09	0.0081
VYPer	−3.51	5.53	0.743	−0.023	+0.081	0.001	−3.49	−0.02	0.0004
USSgr	−3.93	6.74	0.829	+0.063	−0.248	0.004	−3.72	−0.21	0.0441
DLCas	−3.84	8.00	0.903	+0.137	−0.526	0.019	−3.92	+0.08	0.0064
SNor	−4.03	9.75	0.989	+0.223	−0.899	0.050	−4.15	+0.12	0.0144
XVPer	−4.34	10.89	1.037	+0.271	−1.176	0.073	−4.28	−0.06	0.0036
SZCas	−4.71	13.62	1.134	+0.368	−1.733	0.135	−4.54	−0.17	0.0289
$n = 12$	−42.60 = Σy		12)9.194 $\bar{\ell} = 0.766\,16$	+0.002	−1.651	0.608		−0.01 $\Sigma d = 0$	0.3081 = Σd^2
	Corrections: 12(0.000 16):			−0.002 $\Sigma x = 0$	+0.007 $\Sigma xy = $ −1.644	0.000 $\Sigma x^2 = $ 0.608		Variance = $(Sd)^2 = \frac{\Sigma d^2}{n} = \frac{0.3081}{12}$	

Equation to straight line of best fit: $y = a + bx$
$a = \frac{\Sigma y}{n} = \frac{-42.60}{12} = -3.55$; $b = \frac{\Sigma xy}{\Sigma x^2} = \frac{-1.644}{0.608} = -2.70$
$\therefore y = M = -3.55 - 2.70x = -3.55 - 2.70 \log(P - 0.766)$
$\therefore M = -1.48 - 2.70 \log P \pm 0.32$

$= 0.025\,675$
$\therefore Sd = \sqrt{0.025\,675} = 0.160$
$=$ Standard deviation
$2Sd = 0.32 = 95\%$ Confidance limits

distant. However, in November 1993, *Sky and Telescope* reported that G. Gatewood and his assistants at Allegheny Observatory, using the Multichannel Astrometric Photometer, measured Delta Cephei's parallax with unprecedented accuracy and got a value of $0.003'' \pm 0.001$. This implies a distance of 333 parsecs, or 1100 light years.

Amateur astronomers could do valuable work by measuring the periods of Cepheids and their apparent magnitudes and then calculating their absolute magnitudes and distances.

The most accurate and most recent determinations of the distances of the Magellanic Clouds give distances of 166 000 light years for the Large Cloud and 200 000 light years for the Small Cloud.

Starstreaming

To determine the shape and size of the Milky Way Galaxy, it was not sufficient to measure the distances of the stars, but also necessary to determine their motions.

Early in the 20th century, J.C. Kapteyn, in collaboration with David Gill of Capetown Observatory, discovered that the stars move in streams. Statistically, the nearer stars, within 200 parsecs, move in two opposed directions.

Later, J.H. Oort examined the mean values of proper motions and radial velocities of stars more than 2 kiloparsecs from the Sun, and discovered a differential rotation in the Galaxy (Figure 12.6).

This can best be explained by means of parallel, revolving streams. The stars in the Sagittarius arm (arc AB), nearest to the centre of the Milky Way, do not move faster than those in the Orion arm (CD) or those in the Perseus arm (EF) but, because the angular measure of arc AB is greater than those of arcs CD and EF, it appears as if the stars in AB are moving faster. The Milky Way Galaxy

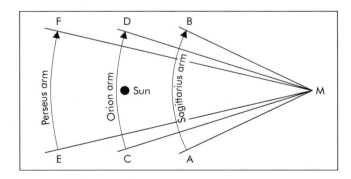

Figure 12.6 Starstreams in the Milky Way

Figure 12.7 M74 (NGC 628) in Pisces

must therefore have a spiral structure, like the Andromeda Galaxy and many others. M74, or NGC 628 (01 36.7 + 15 47), is a very good example of a spinning pinwheel galaxy, its plane of revolution being at right angles to the line of sight (Figure 12.7).

In the 1950s, W.W. Morgan, D.E. Osterbroek and S.L. Sharpless delineated the positions of blue-white O and B super-giant stars, as belonging to various spiral arms: Perseus arm, Orion arm, Sagittarius arm and the Carina arm, a side-branch. This work corroborated the findings of radio astronomy. Figure 12.8 is an idealised representation of the spiral arms in the Milky Way and shows the position of the Sun.

The small arrow shows the direction in which the Sun is moving; the Sun's Way. Jupiter's orbit is magnified 10^{12} times. The line through the small ellipse shows the line of cut between the Ecliptic and the plane of the Galaxy's spiral arms. The Ecliptic is inclined at 55 degrees to the plane of the spiral arms. The Sun moves at about 20 km s^{-1} towards the point 18 h +34° in the constellation of Hercules. This is the Apex of the Sun's Way.

Radio astronomy applied the Doppler effect to measure the speeds and directions of motion of masses of hydrogen.

Figure 12.9 depicts the graphs of velocities in the Galaxy at various distances from the centre. The graphs are by M. Schmidt (solid curve) and L. Blitz (broken curve). At the

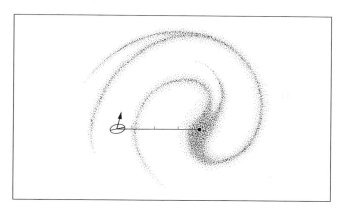

Figure 12.8 The Sun's situation in the spiral arms of the Galaxy

centre of the Milky Way, the speed of rotation is zero, but at 1500 light years from the centre, the speed of rotation is 240 km s^{-1}. For the next 10 000 light years, the speed decreases to 200 km s^{-1}; then it increases again up to a distance of 30 000 light years from the centre, where the speed is 250 km s^{-1}. The Sun happens to be at about this distance from the centre, so that the Sun's speed around the hub of the Galaxy is 250 km s^{-1}. Further from the centre, the speed decreases again. L. Blitz, however, found that there is an increase in speed beyond 40 000 light years from the centre. Schmidt's curve has a better chance of being correct if the mass of the dark matter beyond the spiral arms is small; if the mass of the dark matter is large, Blitz's curve will be more correct.

This curve, of speeds of revolution around the centre of the Galaxy, differs widely from that of the speeds of the planets in their orbits around the Sun (Figure 6.10, page 67). That is so because, in the case of the Solar System, almost all the mass is concentrated in the Sun at the centre, whereas in the case of the Galaxy, most of the mass is spread in and between the spiral arms.

Both Schmidt and Blitz agree on a speed of 250 km s^{-1} for the Sun and the stars in its neighbourhood. If we consider the Sun's orbit around the centre of the Galaxy to be a circle, the distance covered by the Sun in one revolution is the circumference of the circle with radius 30 000 light years. If we divide this distance by the Sun's speed of 250 km s^{-1}, we find that it takes the Sun 226 million years to complete one revolution, provided its distance from the centre and its speed do not change. In the 5000 million years of the Sun's existence, it has thus completed about 22 laps.

The radial velocities of hydrogen masses of gas will be a maximum at points where they move along the line of sight. By measuring these speeds and locations, the structure of the Milky Way is found to be spiral in shape.

Shape of the Galaxy

From the neighbourhood of the Sun, the starstreams in the spiral arms give the impression that the stars are spread right round the expanse of the sky. The Galaxy must therefore have a flattened shape. In the direction of the constellation of Sagittarius, there is a concentration, or hump, of stars. This must be the centre of the Galaxy. J.C. Kapteyn isolated 206 regions of the sky and made statistical counts of the numbers of stars, according to their increasing faintness, in order to gain a picture of the structure of the Milky Way, depicted in Figure 12.10.

Later work has shown that this representation is fairly correct, except that the hump is somewhat peanut-shaped.

The diameter of the Galaxy is at least 100 000 light years (somewhat more than 30 000 parsecs). The thickness of the lenticular portion, containing the spiral arms, is 2600 light years at the outer rim and 10 000 light years in the central hump. The stars concentrate in the plane of the "lens" and in the central hump.

The stars in the hump are old stars, poor in heavy atoms. They are surrounded by an attenuated matrix of gas and dust. Most of the gas and dust in the hump has been blown away by supernova explosions, which must have taken place in the Galaxy's centre. The stars near the centre radiate strongly in the infra-red. On the outskirts of the hump, there is a ring-shaped girdle containing giant molecular complexes. The expansion of this ring indicates that it started moving away from the centre some

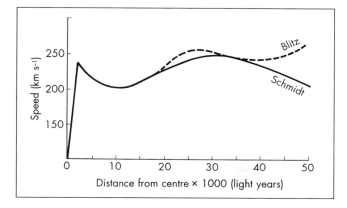

Figure 12.9 Graph of speed against distance from centre of the Galaxy

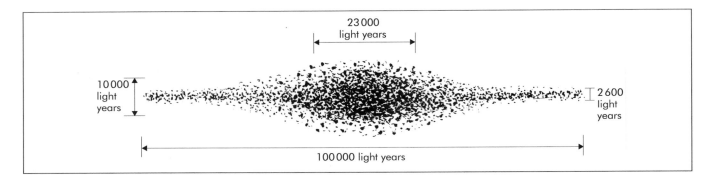

Figure 12.10 Side view of a model of the Milky Way Galaxy

30 million years ago, as the result of concerted supernova explosions.

Within 10 light years of the centre, there are several million stars, embedded in ionised gas. Each of these clouds contains about a solar mass. They revolve around the centre in 10 000 years, compared with the Sun's 226 million years. In the centre there has to be a super-massive object. Measurements by B. Ballick and R.L. Brown of the radiation from the centre indicate that the central object is a mere 10 astronomical units in size (about as large as the space within Jupiter's orbit around the Sun). Its mass is about 50 million solar masses. It must therefore be a giant gravitational vortex or black hole. It could have formed as a result of supernova explosions of some 20 million stars of 2.5 solar masses each. E.E. Becklin and G. Neugebauer found that the centre of the Galaxy radiates strongly in the infra-red. This must be secondary radiation from the gas and dust around the centre. The expanding gas and dust, followed by shock waves and magnetic waves, would have condensed to a second generation of stars by the process of accretion, thus forming stars richer in heavy atoms, which were formed in the supernova explosions.

At a distance of 30 light years from the centre, there is a revolving ring of hydrogen at a temperature of 5000 K.

One thousand light years from the centre, there is a ring of atomic and molecular hydrogen and fine dust. Here and there the hydrogen is ionised, that is, the protons are stripped of their electrons so that the temperature must be 10 000 K. Clusters of newly formed blue-white super-giant stars are found here.

Ten thousand light years from the centre, there is a girdle of neutral hydrogen. This was discovered in 1974 by J.H. Oort and G.W. Rougoor. Doppler measurements show that this ring is also revolving and expanding at speeds between 50 and 135 $km\,s^{-1}$. This ring could also have originated as a giant smoke ring puffed out from the centre.

Such a casting-out process seems to be continuing.

Nebulae in the Milky Way

There are many diffuse nebulae in the Milky Way Galaxy: some of them have spectacular appearances. One of the largest and most beautiful is the Great Nebula in Orion (M42 or NGC 1976) situated in the sword of Orion at 5 h 35.4 m right ascension and declination $-5°27'$ (Figure 12.11). This was the first nebula to be photographed – by H. Draper in 1880. Long time exposures reveal the

Figure 12.11 M42, the Great Nebula in Orion. Photograph by Jason Ware, Frisco, Texas

Table 12.2 Elements in the Orion Nebula

Element	Number of atoms m^{-3}
hydrogen	882 500
helium	88 250
carbon	529
oxygen	220
nitrogen	176
sulphur	32
neon	9
chlorine	1
fluorine	0.1

Figure 12.12 The Horsehead Nebula close to the star Zeta Orionis

extent of the nebula as being 65 minutes of arc.

Colour index determinations attach a distance of 1600 light years to the stars embedded in the nebula. The nebula therefore stretches over 1600 sin(65 ÷ 60) = 30 light years. There is enough matter in the nebula to form 10 000 stars of the Sun's mass.

The nebula fluoresces on account of the ultra-violet radiation from the quadruple star, Theta Orionis, the Trapezium, in its heart. The faint greenish sheen of the nebula is due to emission by doubly-ionised oxygen OIII, at the wavelengths of 4957 and 5007 Å. The spectral lines of these wavelengths are two of the forbidden lines, being radiation from gases at very high temperatures, but very low pressure.

As in the case of other nebulae, hydrogen is the most abundant of the gases. Spectroscopic analysis has revealed the abundances of gases in M42 as detailed in Table 12.2, according to R. Burnham, in his *Celestial Handbook*.

It is clear from the table that hydrogen is very much more abundant than the other elements. The oxygen content is only 0.02% compared to the hydrogen content.

The whole constellation of Orion swarms with nebulae. Here and there, dark blotches are to be found; they are gas and dust clouds, so dense that no light from the stars beyond can penetrate. The Horsehead Nebula in Orion is a beautiful example (Figure 12.12). Photos taken in the infrared show that there are stars present in the dark regions. The red glow of the nebula is due to hydrogen emission at the wavelength of the hydrogen alpha line.

Many dark areas are to be seen in the Rosette Nebula, NGC 2237 in Monoceros (also containing the open cluster of stars, NGC 2244) situated at 6 h 32.3 m + 4°54′ (Figure 12.13).

The Cone Nebula in Monoceros (06 40.9 + 9 54) is another striking example of bright and dark nebulosity mixed together. A bright star illuminates the gas and dust of the nebula, but the denser part of the nebula casts a conical shadow. Figure 12.14 is by Dr Hans Vehrenberg.

Sagittarius contains many nebulae, among others: the Trifid Nebula (18 02.6–23 02). Also known as M20 or NGC 6514, it has dark lanes against a brightly illuminated background (Figure 12.15).

Another example is the Lagoon Nebula, or M8 (18 03.8–24 23), and also the Swan, or Omega Nebula (18 20.8–16 11). These nebulae have masses of glowing hydrogen. Thirty of the most striking nebulae have been listed in Appendix 8, section 8.1.

Most of the diffuse nebulae occur in the plane of the

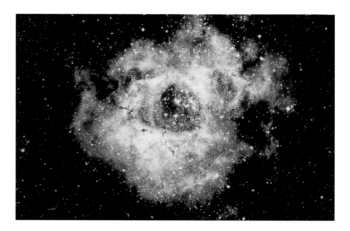

Figure 12.13 The Rosette Nebula in Monoceros and the open cluster NGC 2244

Figure 12.14 The Cone Nebula in Monoceros

Milky Way and they contribute considerably to the mass of the Galaxy.

Four astronomers, A. Vidal-Madjar, C. Laurent, P. Bruston and J. Audouze, have discovered a nebula of hydrogen gas at a distance of 933.12×10^9 km; that distance is only 0.03 parsec, or 36 light days. It is

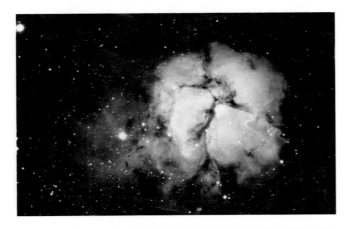

Figure 12.15 The Trifid Nebula M20 in Sagittarius

approaching the Sun at a speed of 15 to 20 km s^{-1}, and will reach the Solar System in about 2000 years. Depending on the density of the cloud, the Earth's climate could undergo changes if the Sun's temperature were to increase on account of gas streaming onto the Sun.

Interstellar Molecules

In 1904, J.F. Hartmann suggested that the absorption line of ionised calcium at the wavelength of 3967 Å, in the violet part of the spectra of certain very bright stars, is actually due to calcium ions in interstellar space and not to the stars themselves. The Potsdam Observatory found that the calcium line did not fit in with the rest of the spectrum, and it did not display the Doppler effect of the other spectral lines of stars with high radial velocities.

Later, it was found that neutral sodium was another constituent of interstellar space. Glowing nebulae, in the neighbourhood of hot stars, also showed the presence of ionised hydrogen.

The first interstellar molecule, apart from atoms, was discovered in 1937, namely the methylidine radical (CH$^+$). It is a molecule consisting of one carbon and one hydrogen atom. In its ionised form, it appears, together with cyanogen (CN), in the spectra of blue-white O and B type stars. The temperatures of these stars are actually too high for molecules to exist. These molecules had, therefore, to be constituents of interstellar space. In the cases of stars with measurable radial velocities, it was found that the spectral lines of these molecules did not participate in the Doppler effect of the other spectral lines.

The dawn of radio astronomy radically affected this branch of astronomy. Radio astronomy had its birth when Karl Jansky found that he was receiving radio noise from a certain part of the sky. It recurred with a periodicity of 24 hours. The direction from which the noise came was found to be the direction of the constellation of Sagittarius in the plane of the Milky Way. The radio noise was received whenever Sagittarius was in the line of sight of the radio telescope. This point in the sky is now called Sagittarius-A.

By 1940, Grote Reber had built the first radio telescope, with which he did pioneering work. In 1942, J.S. Hey discovered that the Sun radiates radio waves.

After the Second World War, thousands of radio sources were discovered. Jupiter was found to emit radio waves which reached a maximum whenever Jupiter's satellite, Io, was in a certain position in its orbit around Jupiter.

Other radio sources were found in various galaxies, in supernova remnants and in nebulae consisting of atomic hydrogen, radiating on a wavelength of 21 cm.

Proofs were forthcoming that certain molecules radiate

in the radio waveband. The hydroxyl radical (OH) radiates strongly on the wavelength of 18 cm. Other molecules tracked down were water (H_2O), radiating on 1.3 cm, formaldehyde (HCHO) on 6.2 cm, and ammonia (NH_3) on 1.3 cm.

During the 1970s a veritable flood of molecules was discovered: carbon monoxide (CO) at 2.6 mm, hydrocyanic acid (HCN) at 3.4 mm, methyl alcohol (CH_3OH) at 35.9 cm, formic acid (HCOOH) at 18.3 cm, and silicon oxide at 2.3 mm, (Appendix 8, section 8.2 contains a list of 58 molecules that have already been discovered in nebulae and interstellar space. It is taken from *Larousse Astronomy*, Hamlyn, London. The list covers discoveries up to 1985.)

The list reveals that the molecules consist of the following atoms: hydrogen, nitrogen, carbon, oxygen, sulphur, silicon and chlorine.

Apart from hydrogen, all these atoms had been manufactured in the cores of stars, where the nuclear reactions fuse hydrogen into heavier atoms. The molecules had been distributed in space by supernova explosions. The diffuse nebulae must have been born in supernova explosions.

The stars in the spiral arms of the Galaxy also contain these atoms. These stars are of Population I; their spectra show appreciable amounts of atoms heavier than hydrogen and helium. These stars must have condensed out of gas and dust nebulae, and they are therefore second or third generation stars. While the oldest stars consist mainly of hydrogen, with a 10% admixture of helium, the younger Population I stars contain fair amounts of atoms heavier than hydrogen and helium.

The Milky Way has been intensively surveyed at the wavelengths that are peculiar to the molecules: hydroxyl (OH), formaldehyde (HCHO) and carbon monoxide (CO). These molecules are strongly concentrated in the plane of the Milky Way and more so near its centre. Most of the other molecules occur in patches.

Within 1000 light years of the Sun, there are 12 dark clouds of gas and dust. If this is average, the Galaxy must contain at least 3000 such clouds. The dust in these clouds comprises only 1% of the total mass of the nebulae. The heavy atoms in the Sun also constitute 1% of the total mass of the Sun. This supports the theory that stars, like the Sun, must have condensed out of such nebulae.

The dark nebulae are very cold, having temperatures below 25 K (−240°C). The matter in the dark nebulae tends to coagulate into globules, called Bok globules, after their discoverer Bart Bok. The globules are visible as tiny black dots against the background of the glowing nebulae. It is presumed that the globules are the last stages in the coagulation, or accretion process, leading to the formation of proto-stars. Further accretion will increase the pressure on the centre of a globule and raise the temperature so that it will start radiating in the infra-red. The Great Nebula in Orion M42 is a very good example of a gas and dust cloud, in which new stars are being formed by the process of accretion.

These stars are called T Tauri stars. They are subject to irregular outbursts of stellar winds. In the dense, dark area behind the bright nebula, infra-red radiation is to be found, and this is typical of young stars.

Hydrogen atoms, which radiate on 21 cm, are to be found all over the spiral arms of the Galaxy. The hydroxyl radical and the carbon monoxide molecule are very resistant to the disintegrating effect of light waves, and they also occur all over in the Galaxy.

Radio telescopes are able to determine the temperatures of dark gas and dust clouds. One of the hottest of these clouds occurs near the centre of the Galaxy; it is known as Sagittarius-B2. Its temperature is about 100 K and its density is very high, 10^8 particles per cm^3. This cloud, which is nearly 20 light years in diameter, is almost twice the diameter of other dark clouds.

Surveys made by radio telescopes on ionised clouds of hydrogen have shown that the Galaxy consists of spiral arms. These spiral arms are delineated by the many blue-white super-giant stars. The distribution of bright gas and dust clouds fits into this structure. The ages of these stars are still very low. Some have been estimated to be only 10 million years old. These gas and dust clouds usually occur on the inside of the spiral arms.

According to C. Lin and F.H. Shu, the dark clouds are an indication of a density wave which compresses the gas and dust, as the density wave moves outwards from the centre of the Galaxy. Lin and Shu found that this density wave moves at two-thirds of the rotational speed of the Galaxy, so that it gets overtaken by the stars in the spiral arms.

That part of the Galaxy far from the spiral arms is called the halo. In the halo, there is a streak of stars and dust, stretching from the centre of the Galaxy towards the north galactic pole. This is known as the North Polar branch or spur. It is a strong radio emitter.

Besides single stars in the halo, there are also concentrations of stars, called globular clusters.

Globular Clusters

Among the 103 objects which C.J. Messier (1730–1817) listed as objects to avoid, in his search for comets, there are several which look like the nuclei of comets in a small telescope. Larger telescopes reveal them as aggregates of hundreds of thousands of stars, packed tightly into spherical or globular clusters.

Figure 12.16 Globular cluster Omega Centauri

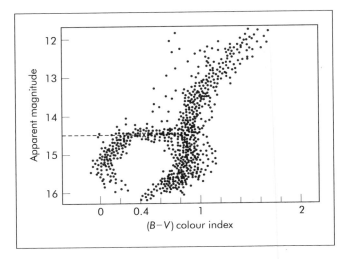

Figure 12.17 Apparent magnitude–colour index H–R diagram of Omega Centauri

H. Shapley (1885–1972) applied the period–luminosity law to determine the distances of the Cepheid variables which he found in these globular clusters. The clusters were found to be 10 000, and more, light years distant. The largest of these clusters is Omega Centauri, NGC 5139 (13 26.8–47 29). Figure 12.16 is by C. Papadopoulos, author of the *Celestial Atlas* (see page 79).

Omega Centauri is just visible to the naked eye, although its distance is 17 000 light years. It forms an equilateral triangle with Gamma and Epsilon Centauri. It is undoubtedly the most spectacular of the globular clusters, containing more than one million stars. It spreads over a disc 30' in diameter, as large as the full Moon, which at that distance means 150 light years. The largest telescopes give it a diameter of 70' and a width of 346 light years. At its centre, the density of the stars is 25 000 times that of the Sun's neighbourhood, the stars there being, on average, one-tenth of a light year apart. The nearest star to the Sun is 43 times further away. In a sphere of volume 333 light years centred on the Sun, there is only the Sun and Alpha Centauri. In the same volume in Omega Centauri, there are 7000 stars!

From a colour index H–R diagram, the distance modulus works out to be 14.3 magnitudes, but a correction of 0.7 magnitude has to be made to allow for absorption of light by the gas and dust in the plane of the Milky Way, making the correct distance modulus 13.6. Therefore:

$$13.6 = 5 \log D - 5$$

$$\therefore D = 5248 \text{ parsecs} = 17\,000 \text{ light years}$$

The integrated spectrum of the cluster is that of F7. After correcting for absorption of light, the total luminosity comes to 1 million times the luminosity of the Sun. If the average stellar mass in Omega Centauri is 1 solar mass, it must contain 1 million stars.

Many of the stars in Omega Centauri are red giants of absolute magnitude −3. By means of the 188-cm telescope of the Radcliffe Observatory, Pretoria, accurate values were obtained for the apparent magnitudes of 7500 stars in the cluster. The H–R diagram of apparent magnitude against colour index (Figure 12.17) shows that a large number of stars have left the Main Sequence and moved to the giant branch. (Absolute magnitudes are not necessary, because the stars are equally far away – there is a constant difference between absolute and apparent magnitudes.)

We see that the apparent magnitude of the RR Lyrae variables in the horizontal branch of the graph is 14.5, which, corrected for light absorption, comes to 13.8. From this, the distance works out to 13 000 light years.

More than 200 Cepheid variables have been discovered in Omega Centauri, as well as many R R Lyrae stars, whose periods are shorter than one day. H. van Gent discovered one with a period of $1\frac{1}{2}$ hours!

The H–R diagrams of other globular clusters are similar to that of Omega Centauri. They must all therefore be in the same stage of evolution, and equally old.

Ages of the Globular Clusters

Many of the stars in the globular clusters are red giants.

Ages of the Globular Clusters

They are therefore older than Main Sequence stars, such as the Sun. During its sojourn in the Main Sequence, a star converts hydrogen into heavier atoms, such as helium, carbon, etc. If f is the fraction of the star's mass which is converted into heavier atoms, then the total energy liberated during that time is given by $E = fmc^2$, where m is the mass which has been converted and c is the velocity of light.

The energy E is also equal to the product of the luminosity L and the time T, that is, $E = LT$.

Therefore $LT = fmc^2$, so that
$$T = fc^2(m \div L)$$

The value of f is approximately the same for all stars. Therefore T is proportional to $(m \div L)$.

According to the mass–luminosity law, L is proportional to the (3.5)th power of the mass. Thus we can write m raised to the power 3.5, for L. The lifetime is then proportional to $\dfrac{m}{m^{3.5}}$, namely $\dfrac{1}{m^{2.5}}$.

Thus, the more massive the star, the shorter is its stay in the Main Sequence.

Table 12.3 gives the times of sojourn in the main sequence, for stars of various masses, taking the Sun's sojourn time as 5×10^9 (five thousand million years).

Although many of the stars in Omega Centauri are giants, their masses do not differ much from that of the Sun. Compared with the Sun's stay of 5×10^9 years in the Main Sequence, that of a star of 1.2 solar masses is:

$$(5 \times 10^9)(1 \div 1.2^{2.5}) = (5 \times 10^9)(1 \div 1.578)$$
$$= (5 \times 10^9)(0.6337) = 3.17 \times 10^9 \text{ years}$$

that is, three thousand million years. Because many of the stars of Omega Centauri left the Main Sequence some time ago, the age of the cluster must be considerably more than that of the Sun, about 8 to 10 thousand million years. The Sun can continue shedding matter at its present rate of 4 million tons per second for another few thousand million years, before it finally leaves the Main Sequence.

Icko Iben writes in the *Scientific American* collection

Figure 12.18 47 Tucanae, NGC 104, at a distance of 16 000 light years

"New Frontiers in Astronomy" that he found very few metallic lines in the spectra of globular clusters; less than a tenth of that in the stars in the spiral arms of the Galaxy – the further they are from the centre of the Galaxy, the poorer they are in those lines. The heavy atoms that they do contain must have come from the centre of the Galaxy, because very few supernova explosions could have taken place in the globular clusters – they are Population II stars, on average, and hence only slightly more massive than the Sun.

While Omega Centauri is the largest and most spectacular of the globular clusters, 47 Tucanae (00 24.1–72 05), in Tucana, does not lag far behind. Nor does M13 (16 41.7 + 36 28) in Hercules. The diameter of 47 Tucanae is 210 light years, and its integrated spectrum is G3. Thus, it contains more metals than the other clusters. Figure 12.18 is by the American National Optical Astronomy Observatories and Figure 12.19 is by Hans Vehrenberg, *Atlas of Deep Sky Splendors*.

Table 12.3 Time spent in the Main Sequence

Solar masses = m	$\dfrac{1}{m^{2.5}}$	No. of years spent in Main Sequence
2	$\dfrac{1}{5.657}$	883×10^6
5	$\dfrac{1}{55.9}$	89×10^6
8	$\dfrac{1}{181}$	28×10^6
10	$\dfrac{1}{316}$	16×10^6

Figure 12.19 Globular cluster M13, NGC 6205, in Hercules. Integrated spectral type F5

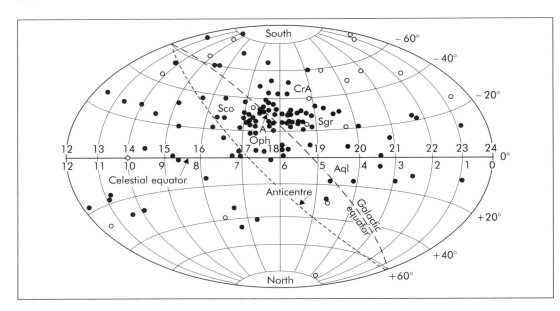

Figure 12.20
Positions of the 137 globular clusters. Open circles are situated on the "back" of the paper. Direction of centre of the Milky Way is shown by A

Distribution of the Globular Clusters

Of the 137 globular clusters, listed in the *Astronomical Ephemeris*:

  13 are of right ascension 16 to 17 hours;
  26 are of right ascension 17 to 18 hours;
  25 are of right ascension 18 to 19 hours.

The 3 hours of right ascension, 16 h to 19 h, comprising $12\tfrac{1}{2}\%$ of the equatorial arc of the sky, contain 46.7% of the total number of globular clusters. They are thus concentrated largely in one part of the sky (Figure 12.20). They also cluster around declination 30 degrees south. The arrow A marks the centre of the densest concentration of stars, in Sagittarius. It is also the location of Sagittarius-A, and is considered to be the direction of the centre of the Milky Way Galaxy. The co-ordinates of this point are

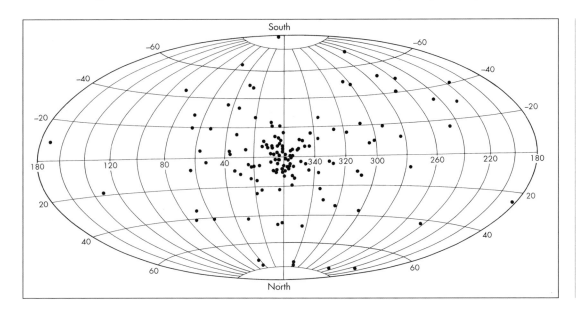

Figure 12.21
Distribution of globular clusters according to galactic latitude and longitude (Appendix 9)

Position of the Sun in the Galaxy

17 h 42 m right ascension, and −28°55′ declination. Between 17 h and 19 h right ascension and −20 and −40 degrees declination, there are 45 globular clusters, comprising one-third of the total number of 137.

On the other side of the sky, between 0 h and 12 h (that is 180 degrees), there are only 19 globular clusters. They are indicated by means of open circles and are situated on the "back" of the paper. A few of these clusters are also very far from the Galaxy and belong to outer space.

In Appendix 9, the column headed ($m - M$) denotes distance modulus. The nearest cluster is NGC 6397 (17 40.7–53 40) in Ara. Its distance modulus, 12.3, means it is 9400 light years away from the Sun. The furthest away cluster, PAL-2 (04 46.1 + 31 23), between Auriga and Perseus, has a distance modulus of 20.9, giving it a distance of 493 000 light years. It is situated far, far beyond the outskirts of the Milky Way Galaxy. The vast majority of the globular clusters have distances between 25 000 and 60 000 light years from the Sun.

The globular clusters must therefore be an integral part of the Milky Way. How they are located around the centre of the Milky Way is indicated in Figure 12.21. In order to draw this figure, the co-ordinates of the clusters had to be adapted from right ascension α and declination δ, to galactic co-ordinates, namely galactic longitude l, and galactic latitude b. The author expresses his thanks to G.C. Jacobs and J.J. Franken for having done these calculations on their computers.

Figure 12.21 reveals that the globular clusters are distributed more or less symmetrically around the centre of the Galaxy, with a concentration near the centre.

The globular clusters are thus caught in the gravitational field of the Milky Way, and must describe orbits around it.

Position of the Sun in the Galaxy

According to their right ascensions and declinations, the

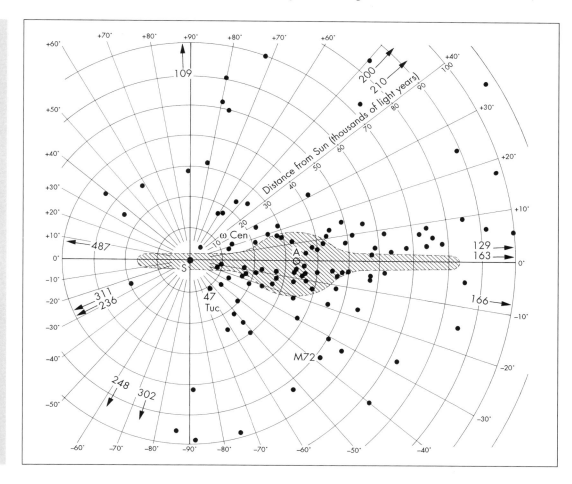

Figure 12.22 Projection of positions of globular clusters on the plane through the Sun (= S), the centre A, and the pole of the Galaxy

globular clusters are mostly in one part of the sky. It is unthinkable that they could be distributed in any other way than around the Galaxy. If the Sun's position is moved to one side of the Galaxy, it will have the effect of moving the clusters to the other side of the sky. The positions of the 117 clusters whose distances are known have been projected in Figure 12.22 on the plane through the Sun and the centre and pole of the Galaxy. The last two columns of Appendix 9 contain the calculated co-ordinates: A' is the projected distance from the Sun, and $k°$ is the angle between the directions of the clusters and the centre of the Galaxy. As an example: M72, NGC 6981, has galactic longitude 35.6 degrees and latitude −33.1 degrees. Its projected distance works out to 50.6 thousand light years, and the angle $k°$ comes to −38.7 degrees. It is indicated in the figure as M72. 47 Tucanae and ω Centauri are also indicated.

The Milky Way Galaxy, as seen along the plane of the spiral arms, is shaded in. Its centre is indicated by circle A, situated about 32 thousand light years from the Sun.

The clusters that fall outside the diagram are indicated by arrows and their projected distances. Those with projected distances 302, 248, 236, 311 and 487 do not seem to belong to the Milky Way, but are interlopers from intergalactic space.

The mean of the declinations of the globular clusters is −30 degrees, which is very close to the declination, −28°55′, of the centre of the Milky Way Galaxy.

The apparent positions of most of the globular clusters, between right ascensions 16 h and 19 h, provided the key whereby the position of the Sun in the Galaxy could be determined. The Sun is nowhere near the centre of the Galaxy, although the continuous arch of the Milky Way across the heavens gives the impression that the Earth and Sun are in the centre of it.

The dynamics of the revolving spiral arms, and their velocities, enabled calculations to be made of the total mass of the Milky Way. It turns out to be 10^{11} solar masses. Since the Sun is an average star, we can say that the Milky Way Galaxy contains 10^{11} stars, if we assume that the average star has a mass equal to that of the Sun. The Milky Way Galaxy therefore contains at least one hundred thousand million stars! This figure includes the stars in the halo, as well as those in the globular clusters.

What lies outside the Milky Way Galaxy? We shall see in Chapter 13.

Chapter 13
Beyond the Milky Way

In our imagination we can now leave the Milky Way. Imagine that we go off at an angle of some 30 degrees to the plane of the Galaxy and travel in the direction of the constellation of Orion, going through the Orion arm and the Perseus spur. If we now look backwards, we get an overview of the spiral arms wound around the nucleus of the Galaxy. We see the nucleus in the far distance, with much dark matter among the hosts of stars. If this dark, obscuring matter were absent, the nucleus of the Galaxy would be blindingly bright. Interspersed there are glowing clouds of ionised hydrogen. They have enabled astronomers to demarcate the spiral arms. There is also much dark matter between the spiral arms, especially on their inside edges.

All over there are bright nebulae (Figure 13.1), such as M1 (the Crab), M42 (the Great nebula in Orion), NGC 2264 (the Cone nebula), Eta Carinae, NGC 6990 (the Veil nebula in Cygnus), M57 (the Ring planetary nebula in Lyra) and

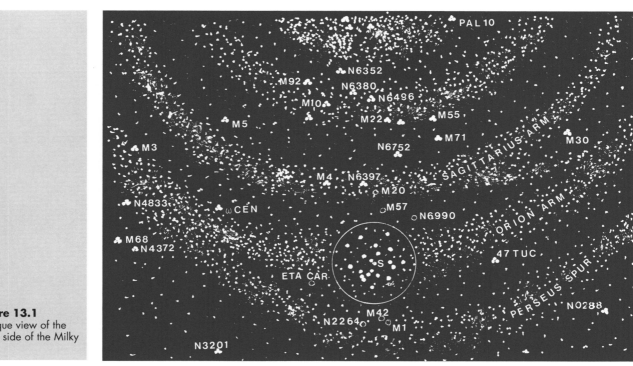

Figure 13.1
Oblique view of the Sun's side of the Milky Way

237

M20 (the Trifid nebula in Sagittarius).

The twenty brightest stars in the neighbourhood of the Sun (marked S) have been somewhat magnified. Among these there are many dwarf stars, smaller and dimmer than the Sun. The circle having the Sun as centre has a radius of 5000 light years. The distance from the Sun to the centre of the Galaxy is 30 000 light years, the Sun being about three-fifths of the way from the centre to the outer rim.

Many of the objects on the far side of the nucleus are obscured from our view by the swarms of stars and dark matter in the nucleus.

Standing out like beacons are the globular clusters. Those on the Sun's side of the centre of the Milky Way are designated by their M and NGC numbers. Closest to the Sun is NGC 6397, in the constellation of Ara, 9.4 thousand light years distant. M4 in Scorpius is second nearest, being 11.6 thousand light years from the Sun. 47 Tucanae is 14 000 light years distant and Omega Centauri, the largest cluster, is 17 000 light years away.

Most of the globular clusters are firmly held in the gravitational field of the Milky Way Galaxy. They describe orbits around the centre of the Milky Way. Some of the globular clusters are more than 200 000 light years from the Sun and even greater distances from the Milky Way, such as PAL-2 (523 000 light years away) and PAL-1 (361 000 light years). These two clusters were discovered by means of the 5-metre telescope on Mount Palomar. It is to be doubted whether the four furthest globular clusters are permanently caught in the gravitational field of the Milky Way. Most probably, they describe orbits which alternately circle the Milky Way and then some other, nearby galaxy. In the neighbourhood of some giant galaxies, points of light occur, which are most probably globular clusters.

The Magellanic Clouds

Among the nearest systems of stars, we find the two Magellanic Clouds. They are named after the leader of the first expedition around the globe of the Earth. In 1519, while Magellan's fleet was sailing southwards in the Atlantic Ocean, the crew nightly saw two hazy patches in the southerly direction. They could not be atmospheric clouds. Some astronomers have considered them as fragments of the Milky Way.

The large Magellanic Cloud (LMC), situated in Dorado between 5 h and 6 h right ascension and 69 degrees south declination, is at a distance of 50 920 parsecs, or 166 000 light years (Figure 13.2); and the Small Magellanic Cloud (SMC), situated at right ascension 1 h and declination 72 degrees south, is at a distance of 62 880 parsecs, or 205 000 light years (Figure 13.3). These distances

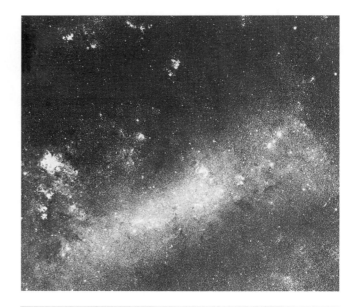

Figure 13.2 The Large Magellanic Cloud

are known fairly accurately, thanks to the large number of Cepheids which occur there.

Countless millions of stars occur in the two clouds. The brightest stars have magnitudes of +10. Suppose the absolute magnitude of these stars is x, then this value

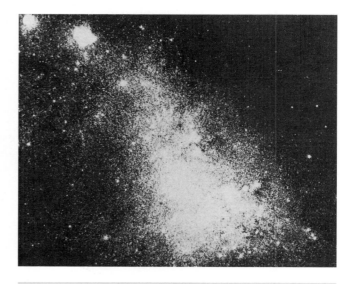

Figure 13.3 The Small Magellanic Cloud

can be calculated as follows (assuming a distance of 166 000 light years for the LMC):

$$\therefore \frac{2.512^x}{2.512^{10}} = \left(\frac{32.6}{166\,000}\right)^2 = \left(\frac{1}{5\,092}\right)^2$$

$$\therefore 2.512^x = 2.512^{10} \div (5092)^2$$

$$\therefore x \log 2.512 = 10 \log 2.512 - 2 \log 5092$$

$$\therefore x(0.4) = 10(0.4) - 2(3.7)$$

$$\therefore x = -8.5$$

This value agrees very well with the absolute magnitude of the brightest stars in the Milky Way Galaxy. These bright stars, wherever they occur, can be used to measure distances.

Most of the stars in the Magellanic Clouds have apparent magnitudes of 13, 14 ... . At that distance the Sun's magnitude would be 23.3 – invisible in all but the very largest telescopes. The two Clouds must have many such stars.

The Clouds are too distant to show proper motions, but radial velocities can be determined with ease, by measuring the redshift and blueshift in the spectra. The LMC is receding at 260 km s^{-1} and the SMC at 150 km s^{-1}. The Sun's speed of 250 km s^{-1} in the Galaxy must be taken into account. The SMC thus appears to be coming closer, while the LMC is receding from the Milky Way.

The two Clouds thus revolve around each other at a distance of 80 000 light years apart, while also revolving in the gravitational field of the Milky Way.

Calculations show that the total mass of the LMC is 10^{10} solar masses – between one-tenth and one-twentieth of the mass of the Milky Way; the mass of the SMC is between 1/6 and 1/7 that of the LMC.

Both Clouds are irregular in shape, although the LMC does show traces of a bar-spiral form.

The Clouds contain many red and blue giant stars, and Cepheids.

The colour indices of the stars in the LMC indicate that it consists largely of Population I stars; while those of the SMC indicate that the SMC consists of equal numbers of Population I and II stars. It must be borne in mind that stars as faint as the Sun are invisible and it is only the bright stars that we are concerned with.

The LMC contains a star, S Doradus (05 18.2–69 15), which is probably the brightest star known, having an absolute magnitude of -9. There are many bright nebulae in the LMC and it is possible that S Doradus may be a close binary, enveloped in the nebula NGC 1910, thus adding to its brightness.

The Tarantula Nebula in the LMC, NGC 2070 or 30 Doradus (05 39.1–69 34), is probably the largest ionised nebula.

The LMC contains many dark gas and dust clouds, but the SMC contains fewer.

The LMC also contains many clusters of stars including globular clusters, such as NGC 1978 and NGC 1866. H. Shapley identified 700 open clusters, some so large that he named the nebula NGC 1936, Constellation I. It spans 500 light years and contains 24 000 solar masses in stars, and the nebula 60 000 solar masses.

There are many supernova remnants in and around the LMC (Figure 11.33, page 215). The photograph has at least 30 bubble-shaped objects, which must be supernova remnants. The LMC has many massive stars, so that it was no surprise that a supernova erupted there on 24 February 1987 (Figure 11.31). Coincident with one of these remnants, a gamma ray burster was discovered by the probe Venera 11, on its way to Venus. Venera 12 and the solar probe Helios-2 also registered the gamma ray burst and so did other satellites. The location of the gamma ray burster could thus be pinpointed as the nebula N9, coincident with a supernova remnant in the LMC. Because the distance of the LMC is known, the strength of the burster could be assessed. It was found to be 100 000 times more intense than other gamma ray bursters in the Milky Way. One explanation for this is that the bursters may originate in galaxies far beyond the Milky Way.

The Milky Way's Nearest Neighbours

Besides the two Magellanic Clouds, there are a number of smaller stellar systems in the neighbourhood of the Milky Way. In 1979 the International Astronomical Union published a list of them. There are two spherical dwarf systems, 200 000 and 270 000 light years from the Milky Way. Another dwarf system lies at 326 000 light years, and four others between 326 000 and 750 000 light years. Among the latter are the Sculptor and Leo systems, or dwarf galaxies. They are very faint and it requires long time-exposures with the largest telescopes to obtain their images, their magnitudes being 17.5.

These spherical and elliptical galaxies are about 50 times bigger than a large globular cluster, but their stars are 100 times more scattered. There may be many more of these systems too faint to be identified. Cepheids and RR Lyrae variables in these galaxies made the determination of their distances possible. Table 13.1 (*overleaf*) contains a list of the dwarf galaxies in the neighbourhood of the Milky Way Galaxy.

Between the Milky Way and the Magellanic Clouds, there

Table 13.1 Galactic systems in the neighbourhood of the Milky Way

Name	Co-ordinates RA Dec	Distance (thousand light yrs)	Morphological type	Diameter (thousand light yrs)	Mass (solar masses)
The Milky Way			SA(s)ab	100	1×10^{11}
LMC	05 23.6 − 69 45	166	SB(s)m	26	1×10^{10}
SMC	00 52.7 − 72 50	205	SB(s)mp	16	15×10^{8}
Sculptor	00 59.9 − 33 42	254	E3	8	1×10^{6}
Fornax	02 39.9 − 34 32	613	E0p	20	2×10^{7}
Leo I	10 08.4 + 12 18	717	E3	6	4×10^{6}
Leo II	11 13.5 + 22 10	750	E0	4	2×10^{6}
NGC 6822	19 44.9 − 14 48	1500	IB(s)m	6	1×10^{9}

is a streak of hydrogen gas. The Sculptor and Fornax systems move in an extension of it.

In its region of space, the Milky Way is accompanied by about ten or so smaller systems and by 137 globular clusters.

The Andromeda Galaxy

The Andromeda Galaxy is situated at a distance of 2 200 000 light years; its co-ordinates are (00 42 + 41 12) and it is a giant spiral system, at least twice the size of the Milky Way. It is also accompanied by a number of smaller systems (Table 13.2). Dwarf systems would be invisible at that distance, so it is not strange to find that Andromeda is escorted by systems larger than those around the Milky Way.

The largest satellite galaxy to Andromeda is M33, NGC 598 in Triangulum (Figure 13.4). It contains twice as

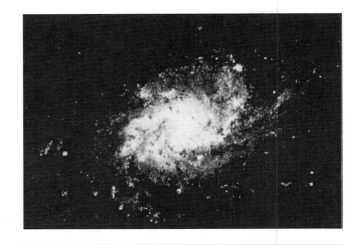

Figure 13.4 Spiral galaxy M33, NGC 598 in Triangulum

Table 13.2 Andromeda galaxy and satellite galaxies

Name	Co-ordinates RA Dec	Distance (thousand light yrs)	Morphological type	Diameter (light yrs)	Mass (solar masses)
Andromeda, M31, NGC 224	00 42.7 + 41 16	2200	SA(s)b	170	3×10^{11}
NGC 147	00 32.2 + 48 30	2200	E5p	8	9×10^{9}
NGC 185	00 39.0 + 48 20	2200	E3p	9	1×10^{10}
NGC 205	00 40.4 + 41 41	2100	E5p	14	8×10^{9}
NGC 221, M32	00 42.7 + 40 52	2200	cE2	7	4×10^{9}
NGC 598, M33	01 33.9 + 30 39	2350	SA(s)cd	60	2×10^{10}
IC 1613	01 04.8 + 02 07	2400	IAB(s)m	15	4×10^{8}

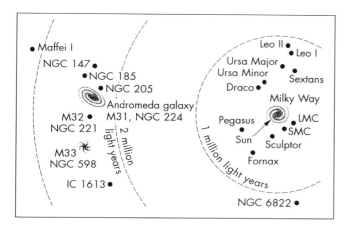

Figure 13.5 The Local Group of galaxies

much matter as the Large Magellanic Cloud and one-fifth of that of the Milky Way. There are another four galaxies nearly as big as M33: NGC 147, NGC 185, NGC 205 and M32 (or NGC 221). The total amount of matter in the neighbourhood of the Andromeda Galaxy is much greater than that in the region of the Milky Way Galaxy.

Taken together, the two groups of galaxies are called the Local Group. Figure 13.5, which depicts the distribution of the galaxies in the Local Group, is derived from *Galaxies* by T. Ferris, published by Stewart, Tabori and Chang, New York. Appendix 10 contains 85 of the brightest galaxies.

In 1854, William Parsons (Lord Rosse) constructed a speculum mirror of 183 cm – the largest up to that time. It enabled him to prove that some of the round and oval nebulae had spiral structures, especially M51 in Canes Venatici (13 29.9 + 47 12), which showed a magnificent full-face spiral. An extension of one of the spiral arms is in contact with a smaller nebulous galaxy (NGC 5195), which probably is elliptical in shape. The radial velocities of M51 and its satellite are 340 and 390 km s^{-1}. Dustlanes are visible in M51.

M51 extends over 10 minutes of arc – one-third of the diameter of the full Moon. Its distance is 35 million light years, 16 times that of the Andromeda Galaxy. It contains at least $1\frac{1}{2}$ times as many stars as the Milky Way, and is somewhat more than 100 000 light years across. The firm Zeiss, in Germany, built a camera which photographed the galaxy in the infra-red. From this the positive print of Figure 13.6 was made.

Classification of the Galaxies

There are apparent differences and similarities between the thousands of galaxies. E.P. Hubble made a classification according to their morphologies, that is shapes and structures.

Some galaxies are round or oval, some have spiral arms, some have a bar in the middle, and others are irregular in shape.

Hubble typified the oval galaxies as elliptic, and graded them from E0 to identify the spherical shapes, to E7 for the most flattened lens shapes. Figure 13.7 shows M87, an E0 elliptical galaxy in Virgo (12 30.8 + 12 24) or NGC 4486. It is surrounded by hundreds of globular clusters, the dots around the periphery of the galaxy – a very good example of the distribution of such clusters around a galaxy; five hundred globulars have been counted there. M87 is in the centre of a great swarm, or cluster of galaxies in Virgo – in his *Celestial Handbook*, R. Burnham lists no less than 217 in the constellation of Virgo. M87 is a strong emitter of radio

Figure 13.6 The Whirlpool galaxy M51, NGC 5194, in Canes Venatici, interacting with NGC 5195

Figure 13.7 E0 elliptical galaxy M87 or NGC 4486 in Virgo

waves; its radial velocity is +755 km s^{-1}. Hubble classified NGC 205, next to the Andromeda galaxy and measuring 8′ by 3′, as an E5 elliptical.

Among spirals, Hubble separated the ordinary spirals from those with a bar through the central bulge, as S and SB galaxies respectively.

Galaxies having large central bulges and tightly wound spiral arms are classed as Sa; as the bulge gets smaller and the spiral arms more loosely spread, his classification proceeded to Sb and Sc. Later, class Sd was added.

Similarly the barred spirals were classified by Hubble as SBa, SBb and SBc. To this class SBd was added.

The lens-shaped galaxies were classed as S0 and SB0.

Galaxies of irregular shape were classed as Irr.

Hubble's classification (Figure 13.8) did not imply an evolutionary process: it was simply a morphological classification, according to the shapes and structures of the galaxies.

The Whirlpool Galaxy (M51) was classed as Sa; and M81, NGC 3031 in Ursa Major (09 55.6 + 69 04) as Sab, being between classes Sa and Sb (Figure 13.9).

A beautiful example of an SBb barred spiral is NGC 1300 in Eridanus (03 19.7–19 25) (Figure 13.10). NGC 2997 in Antlia (09 45.6–31 11) is a fine example of an Sc, having a small central bulge and spiral arms well spread out (Figure 13.11). M83, or NGC 5236 in Hydra (13 37.0–29 52) is an SBc barred spiral (Figure 13.12).

Next there is NGC 4565 in Coma Berenices (12 36.3 + 25 59), seen almost edge-on, its plane in which the spiral arms revolve being tilted by only 4 degrees from the line of sight. It comes very close to being a true picture of the Milky Way, seen edge-on. It is probably an Sb or Sc galaxy (Figure 13.13).

Later, Hubble's classification was extended to include in-between types as well as more details, for example, the

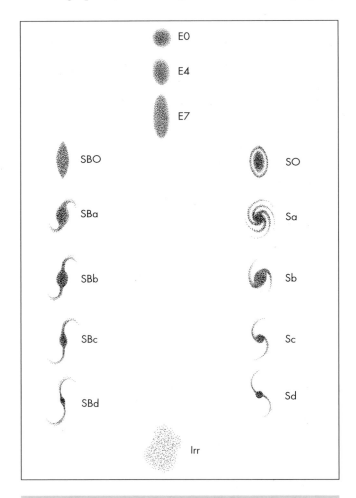

Figure 13.8 Hubble's classification of galaxies

Figure 13.9 M81, NGC 3031 in Ursa Major has a huge central bulge

Figure 13.10 NGC 1300 in Eridanus with well-developed bar and spiral arms

Figure 13.12 M83, NGC 5236 in Hydra, an SBc galaxy

Figure 13.11 NGC 2997 in Antlia, an Sc galaxy

Figure 13.13 NGC 4565 in Coma Berenices – almost edge-on

point where the spiral arms originate: in the bulge (s) or in a ring of matter around the bulge (r).

A.R. Sandage and G. de Vaucouleurs extended Hubble's classification, by including in-between types in a three-dimensional diagram. The ordinary spirals are classified as SA and the barred spirals as SB. If the spiral arms originate in the bulge, the type is SA(s) or SB(s); if the spiral arms originate in a ring outside the bulge, the types are SA(r) and SB(r). They also added the d types and subdivided the irregulars into Sm and Im. Sm indicates slight but recognisable signs of spiral arms, as for example the Large Magellanic Cloud, which is classified as SBm. In the Im group there are no signs of spiral arms, such as in the Small Magellanic Cloud, which may, however, have a semblance of a bar.

De Vaucouleurs extended the S0 and SB0 types into $S0^-$, $S0^o$ and $S0^+$. Today, these types are indicated by the letter L, which suits their lens shapes better.

In 1960, S. van den Bergh found that a relationship exists between the morphology of the Sb, Sc and Irr galaxies and their luminosities: most of the bright galaxies have long, well-developed spiral arms. He was able to subdivide the Sc galaxies into five luminosity classes: I for the brightest and V for the faintest. Among Sb galaxies, he distinguished classes I, II and III.

Among the bright galaxies, spirals account for 75% of the total. Elliptic and S0 galaxies make up 20%. There are

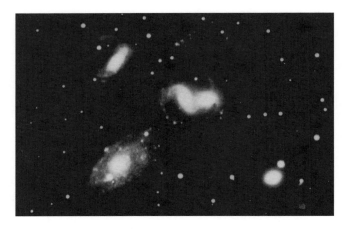

Figure 13.14 Stephan's Quintet – four interacting galaxies

many dwarf galaxies that appear to be elliptical. If they are included, the percentage of ellipticals reaches 60% of the total. Among spirals, the numbers of SA, SB and SAB galaxies are about equal.

There are many interesting cases among the Irregulars. Maffei I, discovered by P. Maffei, is almost in the same direction as NGC 147, but at a distance of 3.3 million light years can be reckoned as a distant member of the Andromeda group. It radiates strongly in the infra-red, which is very strange for a galaxy.

NGC 5195 and one spiral arm of M51 are distorted by their interaction. In the ultra-violet NGC 5195 is invisible. This means that it contains only old stars which radiate on long wavelengths.

NGC 3077 (10 03.3 + 68 44) is in the gravitational influence of M81 in Ursa Major, which is at a distance of 10 million light years.

NGC 4038 and 4039 (12 01.8–18 52), in Corvus, known as the ringtail galaxies, or the Antennae, are in grazing contact: each has a spiral arm which has been flung out into space.

Stephan's Quintet, NGC 7317, 7318 A and B, 7319 and 7320 in Pegasus (22 36.0 + 33 58), actually has four galaxies interacting with each other (7320 does not seem to belong to the group as its radial velocity is entirely different) (Figure 13.14).

The giant galaxy Centaurus A, which is 16 million light years away, has some very peculiar properties. It is calculated to be three times as massive as the Milky Way. It radiates intensely in X-rays, infra-red and radio wavelengths. The greatest part of its radiation stems from its core, but it also has two "clouds" on the projection of its axis of spin, 16 thousand and more than a million light years from the core. These clouds were discovered by radio telescopes; they consist of gases which must have been hurled out of the core by means of a gigantic explosion. The galaxy is elliptical, but in the foreground there is a gas and dust streak diagonally across the centre of the galaxy (Figure 13.15).

Usually, elliptical galaxies do not contain gas and dust. In the gas and dust lane of Centaurus A, bright, young stars appear, while the elliptical core appears to consist entirely of old, red stars. The young stars in the dust lane must have been formed *in situ*. If the gas and dust lane had been cast out of the core, the shock waves, following on, would have triggered the formation of new stars by compression exerted on the gas and dust.

Alternatively, the gas and dust lane could be the spiral arm of another galaxy, just in front of Centaurus A. It might also be a galaxy being cannibalised by Centaurus A. The set-up appears to be enigmatic.

A short-exposure photograph of the giant galaxy M87 shows a dense jet of material coming from the centre.

Usually, the galaxies near the "top" of Figure 13.8 are referred to as "early" types and those at the "bottom" as "late" types, although there seems to be no evolutionary process involved. There are systematic differences between the galaxies, which depend upon the relative numbers of bright, blue stars and the systematic increase of gas and dust. This may be indicative of the ages of the galaxies. All galaxies that contain masses of old, developed stars must have more or less the same ages. These ages probably vary greatly, estimates between 10 and 18 milliard (18×10^9)

Figure 13.15 Centaurus A, NGC 5128

years being the order of the day.

Secondly, elliptic galaxies are more massive than spirals. This difference in mass cannot be explained by any evolutionary process. The mass of a galaxy plays a big role in its morphological development, by which its shape is determined. The mass of a galaxy depends on the amount of matter that was available in the region of its formation.

Thirdly, there are differences in the angular momentum of various types of galaxies, which cannot be explained on any evolutionary basis.

Redshift

In the nineteenth century, when the galaxies were first discovered, there was no certainty as to whether they were within the Milky Way or beyond its confines. The period–luminosity law enabled the distances of the closer galaxies to be determined by means of the Cepheids which they contain. The galaxies were found to reside beyond the confines of the Milky Way – they form the separate building blocks of the Universe.

Most of the work in determining the distances of the galaxies was done by H. Shapley, E.P. Hubble (Figure 13.16), M.L. Humason and V.M. Slipher, using the 2.5-metre telescope on Mount Wilson, California.

It was found that the galaxies were more or less evenly distributed in space, outside the Milky Way. This was the first indication that the Universe is isotropic – that its structure is similar in all directions. But, most importantly, Slipher and Hubble found that the spectral lines of most of the galaxies are shifted towards the red end of the spectrum, and that the fainter the galaxies are, the greater is the redshift. This means that the fainter galaxies have greater speeds of recession. It was safe to assume that the fainter galaxies are further away. Therefore, the further

Figure 13.16 E.P. Hubble

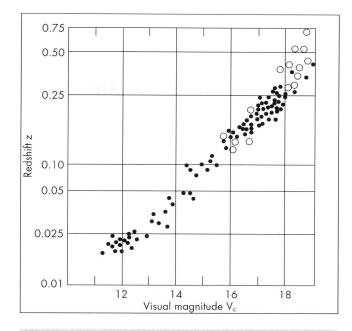

Figure 13.17 Hubble's diagram. Recent additions: open circles

away a galaxy is, the greater is its speed of recession. Figure 13.17 is a tracing from an article by W. Somerville in the *Encyclopaedia of Astronomy*, Hamlyn, London. The figure shows the relationship between redshift and faintness, or greater distance, corresponding to higher magnitude numbers. This is known as Hubble's first diagram. From it we can see that the redshift, z, is proportional to visual magnitude, V_c. From this, in 1929 Hubble was able to formulate a law as follows: the speed of recession, V, of a galaxy is proportional to its distance, D, in megaparsecs. Mathematically, it can be stated as follows:

$$V = H_0 D$$

where H_0 is the Hubble constant.

This provided a new method for determining distances. The redshift, z, of the spectral lines could be measured very accurately. The velocity of recession, V, is then equal to cz, where c is the velocity of light, and therefore $H_0 D = cz$. This method does not depend on Cepheid variables, which become invisible in the more distant galaxies.

The constant H_0 is the gradient of the straight line which best fits the data on the graph of speed of recession and distance in megaparsecs.

To determine the value of Hubble's constant, it was necessary to determine the distances of a sufficiently large number of galaxies by means of the Cepheid variables and

RR Lyrae variables which are visible in the galaxies. These distances could then be plotted against the redshifts.

The redshift, z, is the fraction by which the wavelength of a spectral line is increased, compared with the wavelength of emission. If λ is the wavelength received, and λ_0 the wavelength of emission, then the redshift, z, is given by:

$$z = \frac{\lambda - \lambda_0}{\lambda_0}$$

The value of z can be determined for any particular wavelength, but the H and K lines of ionised calcium in the violet end of the spectrum are very suitable, since there is room for the redshift to move right across the width of the spectrum, from the violet end to the red end.

Hubble's first value for the constant H_0, in 1935, was 530 km s^{-1} per megaparsec. (One megaparsec = 1 million parsecs = 3.26 million light years.)

On account of the revision of the period–luminosity law, based on the recalibration of the absolute magnitudes of Cepheids, and the consequent increase in the distances derived from the law, today's values of the Hubble constant vary between 50 and 100 km s^{-1} Mpc^{-1}.

In the *Astrophysical Journal*, Vol. 196 (1975) pp. 313–328, A.R. Sandage and C.A. Tammann present their findings for:

1. 37 Sc galaxies whose distances were determined from HII ionised nebulae;
2. groups of galaxies for which the mean distances were used;
3. 75 galaxies of type ScI to ScIII, whose distances were derived from their absolute luminosities.

In all three cases, they found that the Hubble constant H_0 is about 55 km s^{-1} Mpc^{-1}.

Their graph for the three groups of galaxies, of speed of recession in km s^{-1} against distance in megaparsecs, is shown in Figure 13.18.

All the galaxies fall within an acute triangle AOC. The largest value that the Hubble constant can have, is the slope of the boundary line AO = AD ÷ OD = 600 ÷ 60 = 100 km s^{-1} Mpc^{-1}. The minimum value is CD ÷ OD = 2000 ÷ 60 = 33 km s^{-1} Mpc^{-1}. The statistical average is given by the slope of the line OB = BD ÷ OD = 3300 ÷ 60 = 55 km s^{-1} Mpc^{-1}.

Recent determinations of the value of the Hubble constant, based on the study of Cepheids in one galaxy, have given values of 80 to 87 km s^{-1} Mpc^{-1}. The high values for H_0 obtained by these researchers give an age for the Universe that is less than the ages of globular clusters; the age of the Universe being the reciprocal of the Hubble constant.

Hubble's law gave rise to the concept of the expanding Universe: that all parts of the universe are moving away

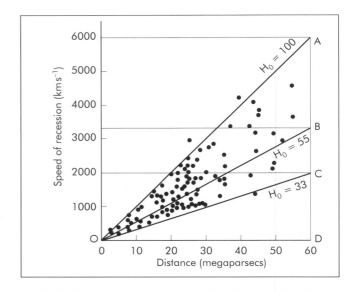

Figure 13.18 Graph by Sandage and Tammann to determine the Hubble constant

from each other, and the greater the distance, the greater is the speed of recession. This involves important cosmological principles.

The galaxies were found to occur in clusters. The Local Group consists of some 20 galaxies, most of them small. The cluster in Virgo contains hundreds at a mean distance of about 55 million light years. Almost all the objects in Figure 13.19 are galaxies in the Coma Berenices cluster.

The cluster in Hercules (Figure 13.20) contains thousands of galaxies. A distant cluster of galaxies in Hydra has

Figure 13.19 Cluster of galaxies in Coma Berenices

Methods of Determining Distances

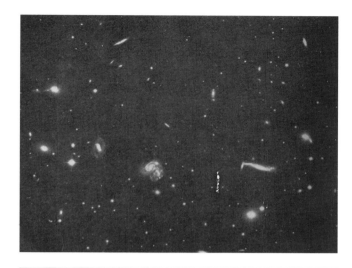

Figure 13.20 Cluster of galaxies in Hercules

at least 60 members which are listed in R. Burnham's *Celestial Handbook*.

Distance determinations had by this time reached to the realm of the galaxies, scattered beyond the confines of the Milky Way. We shall now group them together.

Methods of Determining Distances

The distances to which the following methods reach are shown in Figure 13.21. The new telescopes of the 1990s extend these limits somewhat.

1. Trigonometrically, distances can be measured as far as 100 parsecs, or 326 light years (= 0.0001 megaparsecs).
2. Radial velocities and proper motions of open clusters which are receding give distances up to 1630 light years.
3. The absolute magnitudes of stars in the Main Sequence of the H-R diagram are well known, and also their distances. When the absolute magnitude, read off from the H-R diagram, is compared with the apparent magnitude, the distance follows from the distance modulus $(m-M)$.
4. RR Lyrae stars all have the same absolute magnitude of 0.8. When the absolute magnitude is subtracted from the apparent magnitude, the distance modulus will yield distances of these stars up to half a million light years away.
5. Cepheid variables are brighter than RR Lyrae stars, and can yield distances up to 13 million light years.
6. The brightest novae reach absolute magnitudes of -9 and can be used as beacons to measure distances up to 30 million light years.
7. The brightest stars have absolute magnitudes of -8.5 and they can be used in the same way as novae.
8. The total integrated magnitudes of globular clusters also yield distances up to 30 million light years.
9. Clouds of ionised hydrogen, HII, are assumed to be of equal size, on average. Their apparent sizes can be measured by means of radio telescopes, yielding distances up to 40 million light years.
10. The brightness of ScI galaxies are very similar. Their apparent magnitudes can be used to measure distances up to 320 million light years.
11. Supernovae reach absolute magnitudes of -19.5. They can be used as beacons up to distances of 1000 million light years.

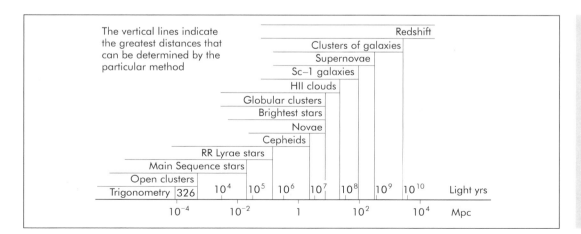

Figure 13.21 Limits of methods of distance determination. From W. Somerville, *Encyclopaedia of Astronomy*

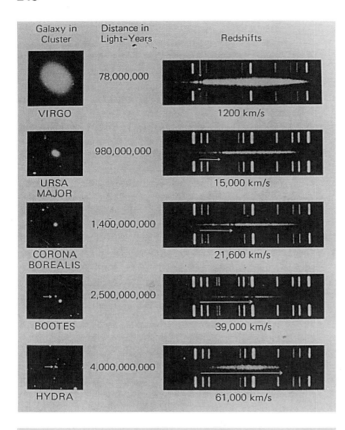

Figure 13.22 Redshifts of H and K calcium lines of distant galaxies

Relativistic Velocities

The further away the bodies are, the greater are their velocities of recession. If a body is so far off that its velocity of recession is equal to the velocity of light, it will be invisible because its light could never reach the Earth. This distance is known as the observable limit of the Universe. The simple formulae for redshift

$$z = \frac{\lambda - \lambda_0}{\lambda_0} \quad \text{and} \quad cz = H_0 D$$

are not applicable for velocities that are an appreciable percentage of the velocity of light. When velocities come close to the velocity of light, the equations of Einstein's Theory of Relativity have to be used. The formula for redshift is

$$1 + z = \sqrt{(c+v) \div (c-v)}$$

where c is the velocity of light and v is the velocity of the receding body. Table 13.3 gives the velocities of recession

Table 13.3 Redshift and recession velocity

z	log z	v = %c
0.5	−0.30	38
0.7	−0.17	49
1.0	0.00	60
2.0	0.30	80
3.0	0.48	88
4.0	0.60	92

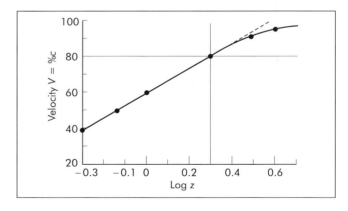

Figure 13.23 Velocity of recession against log of redshift

12. By assuming average sizes for galaxies in clusters, the apparent sizes give distances up to 12 000 million light years.
13. The redshift of the spectral lines can be used in conjunction with Hubble's law to determine distances up to the limits of the observable Universe. Figure 13.22 shows the spectra of the brightest galaxies in Virgo, Ursa Major, Corona Borealis, Boötes and Hydra clusters. Over the range of distances encompassed by these clusters of galaxies, the H and K lines of ionised calcium shift from the violet towards the red end of the spectrum, indicated by the arrows in Figure 13.22. This illustration, by the Hale Observatories, shows in a spectacular way how large the redshift can actually be, and what great velocities of recession are involved in the cases of the very distant clusters of galaxies.

for various values of the redshift, z, using the formula given above.

Figure 13.23 shows the velocity of recession as a percentage of the velocity of light, against the log of the redshift, calculated from the formula above.

Up to a z-value of 2, of which the log is 0.3, there is a linear relationship between velocity of recession and log of the redshift. For greater z-values, the increase in v falls off as v approaches 100%c. Then the curve approaches the limit asymptotically, never actually reaching the limit. The value of the redshift, z, will have to be infinite for the velocity of recession to reach 300 000 km s^{-1}. The limit of the observable Universe can thus never be reached.

Suppose a galaxy has a speed of recession of 96% of the speed of light; what will its redshift be? Using Einstein's formula:

$$1 + z = \sqrt{(c + 0.96c) \div (c - 0.96c)}$$
$$= \sqrt{1.96 \div 0.04} = \sqrt{49} = 7$$
$$\therefore z = 6$$

The distance of a galaxy having a redshift of 6 and recession velocity of 0.96c can be calculated, using Hubble's formula: $H_0 D = V$. If H_0 is taken as 55 km s^{-1} Mpc^{-1}:

$$55D = 0.96(300\,000)$$
$$\therefore D = 5236 \text{ Mpc}$$
$$= 17\,070 \times 10^9 \text{ light years}$$

namely 17 thousand million light years.

If H_0 is taken as 100 km s^{-1} Mpc^{-1}, the distance works out to 9388 million light years, say 10 thousand million light years.

Depending on the value of the Hubble constant, we can say that the observable limit of the Universe is between 10 and 17 thousand million light years approximately.

The Expanding Universe

One could well ask: "Is the Universe infinitely large or is it finite, and if so, what is its size?" Einstein's special theory of relativity shows that speed has an upper limit, namely the speed of light (c). No material body can move at a speed greater than c relative to any observer. We have seen that the furthest observable objects are moving at speeds approaching c, relative to our own galaxy and these objects are at distances of the order of 17×10^9 light years. Are these bodies therefore getting close to the limit of the Universe?

According to Einstein's general theory of relativity, mass causes a warp in the space-time continuum, giving it a strange "curvature". One of the effects of this curvature is that each observer seems to be at the centre of the Universe, according to his observations, based on the Doppler effect, which indicates that all clusters of distant galaxies are receding from our own galaxy and from each other. Although it is impossible to try to visualise this, it seems that the Universe is finite in size (although expanding) but with no boundaries.

The questions "What is outside the Universe" and "What existed before the Universe was formed?" therefore have no meaning. There is no "outside" and there was no "before". Time could not have existed without space and matter.

Will the Universe keep on expanding for ever? This depends on whether there is sufficient matter in the Universe to slow down the expansion, by its force of gravity, and then bring about contraction.

Exploding Galactic Nuclei

In the neighbourhood of the giant galaxy, M81 in Ursa Major, there is an exploding galaxy M82 (09 55.8 +69 41) at a distance of 10 million light years (Figure 13.24).

Radio telescopes show that matter is streaming out of the nucleus of M82 at speeds of 1000 to 1100 km s^{-1}. Extrapolating backwards, this explosion must have commenced 2 million years ago.

What may be happening is that millions of massive stars in the nucleus of the galaxy have reached the supernova stage of their development and started exploding 2 million years ago.

When a galaxy condenses out of gas and dust, much matter will concentrate at its centre, so that the stars which form there by accretion will be very massive. They will

Figure 13.24 Exploding galaxy M82 in Ursa Major

constitute the first generation of stars. Because of their great masses, their temperatures will be much higher than those of average stars. Their nuclear reactions will proceed much faster and they will, very quickly, after only a few hundred million years, reach their supernova stages. The explosions of these stars will be more or less simultaneous. "Simultaneous" on the time-scale of the Universe could comprise a period of millions and even hundreds of millions of years. Astronomically and cosmologically speaking, this is a short period which can be considered as simultaneous. Because the explosion is of short duration, such exploding galaxies will not be plentiful at any instant of time. Other examples of exploding galaxies are NGC 253 (0 47.6–25 17) in Sculptor, appearing to be simmering; and NGC 1275 (3 19.8 + 41 31).

P. Morrison, in an article in *Sky and Telescope*, January 1979, gave an analysis of M82, in which he comes to the conclusion that the galaxy is engaged in a collision with a giant gas and dust cloud, and that its strange appearance is due to bursts of star formation in the perturbed gas and dust.

There are indications that other galaxies, such as the Seyfert, which have particularly bright nuclei, may also be in the stage of collective supernova explosions in their nuclei. NGC 1566 in Dorado (4 20.0–54 56) is a Seyfert galaxy with a very bright nucleus (Figure 13.25).

The Seyfert galaxies have very compact nuclei. Their emission spectra show that they contain swiftly moving masses of gas, which could be in the process of being cast out of the core. The continuum spectra of their cores are intense in the blue and also vary in intensity in short periods. This shows that the portion which is undergoing variation must be comparatively small, not more than one-third of a light year in size. These galaxies also show outbursts in their radio emission, and they also radiate strongly in the infra-red. This comes about on account of the absorption of radiation by dust clouds in the cores followed by re-radiation at longer wavelengths. In this respect, the Seyfert galaxies are similar to quasars. Quasars are star-like objects having very large redshifts and are thus very distantly located. They must thus be inherently very bright, even hundreds of times brighter than the core of a galaxy. The Seyfert galaxy which most nearly rivals the quasars is 3C120, which is only 20 times fainter than the brightest quasar, 3C273. Generally, Seyfert galaxies are a few hundred times fainter in the wavelengths of visible light. Their spectra have broad emission lines from highly ionised active atoms. The forbidden lines are usually very thin.

All galaxies that have a sufficiency of matter in and around their cores will undergo combined supernova explosions. The material which is cast out, consisting of gas and dust, containing elements heavier than hydrogen and helium, such as carbon, oxygen, magnesium, aluminium, silicon and iron, as well as elements heavier than iron, comprises the material from which the spiral arms are formed. Outward-moving shock waves and magnetic waves serve as triggers to set the accretion process going, to form the second generation of stars. Among these stars, there will be bright giants, which are found in the spiral arms of galaxies.

Radio Astronomy

A great expansion of knowledge regarding the Universe beyond the limits of the Milky Way came about as a result of the discovery in 1931–1932 by Karl Jansky, of Bell Telephone Laboratories, that the radio noise which he was receiving reached a maximum every 23 hours 56 minutes (sidereal day), and that it appeared to come from the direction of Sagittarius. This was the birth of radio astronomy. The radio noise from Sagittarius came from the very centre of the Milky Way. This was the first indication that such a central point does exist.

Today, this point is known as Sagittarius A. The radio noise is indicative of the existence of rapidly moving masses of gas at high temperatures in the centre of the Milky Way. The remnants of the supernovae explosions, which must have taken place in the heart of the Milky Way, have compounded together to form a very massive gravitational vortex, or black hole. This concentration of gravity is constantly sucking in gas from neighbouring stars. The gases, spiralling in on the black hole, are heated

Figure 13.25 Seyfert galaxy NGC 1566

by friction of motion and emit thermal radio waves.

In 1942, Grote Reber built the first radio telescope, with which he charted the heavens in radio frequencies. He found concentrations of radio flux in Sagittarius, Cygnus and Cassiopeia. His parabolic reflector consisted of 45 sheets of galvanised sheeting, nailed to a wooden frame. From the focus, the concentrated radio signals were led to a crystal detector. An audio triode amplified the radio waves and converted them into audible signals. Today, a variety of registering systems are used to make traces of the radio signals.

Bodies that absorb electro-magnetic radiation well are able to re-radiate at all frequencies. At low frequencies, the radiation emitted is roughly proportional to the temperature of the body, the total radiation being proportional to the fourth power of the absolute temperature T^4.

According to the Rayleigh–Jeans law, the intensity of radiation is proportional to the square of the frequency, in the range of low frequencies. At low temperatures, maximum radiation takes place on long wavelengths. If the temperature is raised sufficiently, maximum radiation takes place at a dull-red heat and successively at the various colours of the spectrum.

Radio waves are radiations at wavelengths longer than the infra-red, for example microwaves and waves of wavelengths of several metres, such as 31 m, 49 m, etc.

Radio waves have two kinds of origins:

1. thermic radiation in which the intensity, or radio brightness, depends on the temperature of the emitting body;
2. non-thermic radiation, in which the radio waves originate from the rapid movements of electrons, moving in a plasma of positive or negative ions.

Thermic radiation is most intense in the ultra-high frequencies (shortest wavelengths) and non-thermic radiation in the very high frequencies (longer wavelengths).

The intensity of the radio signal obtained from the antenna is proportional to: (a) the flux density of the radio source; (b) the radio frequency bandwidth of the system; and (c) the surface area of the antenna.

A stronger signal is thus dependent on a large surface area of the antenna.

The largest of all radio antennae is that of the radio telescope at Arecibo in Puerto Rico (Figure 13.26). It has a diameter of 300 metres and is constructed in a natural basin between a ring of hills. It cannot be pointed at different parts of the sky and is dependent on the spinning of the Earth to monitor a strip of the sky from west to east. By moving the focal point, which is suspended from three pillars, the width of the strip can be increased.

Resolving Power of the Radio Telescope

The drawback of the radio telescope is its inability to obtain fine resolution. The lobe width, that is, the angle over which the signal is spread, is equal to $\lambda \div D$ radians, or $60\lambda \div D$ degrees, where λ is the wavelength of the radiation and D is the diameter of the dish or antenna.

The wavelengths of radio waves are 10^6 (one million) times longer than visual light waves. When a radio telescope receives waves of wavelength $1\frac{2}{3}$ metre, by means of an antenna of 100 metres in diameter, its bandwidth is equal to:

$$60 \times 1\tfrac{2}{3} \times 100 = 1°$$

compared with the resolving power of the 5-metre telescope on Mount Palomar, which is 1 arc second, or 3600 times better than the above-mentioned radio telescope.

To overcome this problem, two dish-antennae, separated by some distance, are connected to act as an interferometer. Instead of D as divisor in the expression $60\lambda \div D$, the

Figure 13.26 The 300-metre radio telescope at Arecibo, Puerto Rico

Figure 13.27 Twelve of the 27 dishes of the Very Large Array, New Mexico

distance between the two dishes becomes the divisor. To obtain a resolving power of 1 arc second, the two dishes must be 3600 metres apart. Theoretically, the separation can be equal to the diameter of the Earth, or the distance between the Earth and an orbiting antenna.

The Very Large Array in New Mexico has 27 antennae, each of 25 metres, mounted on a rail track in the shape of the letter Y (Figure 13.27), each leg of the Y being 21 km in length. In their most effective positioning, the array has an equivalent width of 27 km. In the late 1950s, Martin Ryle developed the method of aperture synthesis, whereby the separated antennae are synthesised. For this, he and his colleague, Anthony Hewish, received the Nobel prize. On wavelengths of 21 cm, the resolving power of the VLA system is 2.1 arc seconds.

Findings of Radio Astronomy

In 1942, J.S. Hey discovered that the Sun emits radio waves from flares connected with sunspots. The radio telescope derived a radio temperature of 6000 K for the photosphere; 20 000 K for the chromosphere; and 1 500 000 K for the corona.

On 10 October 1946, J.S. Hey and his colleagues registered radio reflections at a wavelength of 5 metres, coming from the Giacobinid meteor shower. These waves originate when the meteors become incandescent, liberating hot gases. They obtained 200 echoes per minute and found the velocities of the meteors to be about 23 km s^{-1}.

In 1948, strong radio sources were discovered by M. Ryle and F.G. Smith, namely Taurus A, Virgo A and Centaurus A. These sources were identified optically as the Crab nebula M1, the extra galactic nebula NGC 4486 or M87, and NGC 5128 respectively.

J.H. Piddington and H.C. Minnett examined the Moon on the 1.25-cm radio waveband. They found that the Moon's temperature lags behind the illumination phase. This indicated that the radio waves which they directed to the Moon must have penetrated the surface before being reflected, showing that the Moon's surface is covered in a layer of dust. The radio temperature of the Moon is thus the temperature of the rocks below the dust layer.

In 1944, H.C. van de Hulst, a 26-year-old assistant at the Leiden Observatory, proved theoretically that radiation from normal hydrogen on a wavelength of 21.3 cm must exist. In 1951, this was found experimentally to be correct. By using this wavelength of 21.3 cm, the occurrence and distribution of clouds of hydrogen in the galaxies and in space could be confirmed. The remnant of Tycho Brahe's supernova (0 25.5 + 64 11) could be charted (Figure 13.28).

The radio source Cassiopeia A was identified as a nebula with rapidly moving filaments, at a distance of 10 000 light years. It is most likely the remnant of a supernova of the year 1700, which was not noticed at the time. The radio

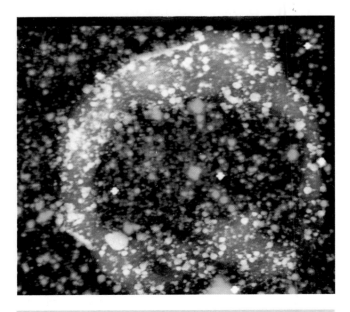

Figure 13.28 Remnant of Tycho's supernova, 1572

source Cygnus A is very exceptional; its redshift gives it a distance of 550 million light years and it is possibly a double galaxy. Its radio flux is a million times that of the Milky Way, and it is therefore called a radio galaxy (Figure 13.29, page 254).

By 1950, it was clear that radio astronomy was a whole new science and that it was destined to affect astronomy radically.

The highest degree of sensitivity of a radio telescope can be obtained only by decreasing the radio noise received to a minimum. For this purpose a maser is used. The amplification of the radio signal is obtained by stimulated emission when the maser is cooled down to a few degrees above zero Kelvin, by means of liquid helium.

B.F. Burke discovered that the radio emission from Jupiter reaches a maximum when the satellite Io is in a certain position in its orbit. Io must therefore be responsible for the perturbations of the magnetosphere and ionosphere of Jupiter.

The clouds which envelop Venus hide its surface from view. Radio waves can penetrate the clouds and the echoes can be picked up. By applying the Doppler effect, it was found that Venus has retrograde rotation in a period of 240 to 250 days. It was found that Mercury is not in synchronous rotation around the Sun, but that it rotates once in 59 days, exactly two-thirds the length of its period of revolution. Later, the Mariner-10 space probe fixed the period of rotation of Venus at 244 days, and that of Mercury at 58.64 days.

Using radar, D.W.R. McKinley was able to prove that meteors move at speeds of less than 72 km s^{-1}. They are therefore caught in the Sun's gravitational field and do not come from outer space.

The Leiden and Sydney Observatories used the 21-cm waveband to chart the positions of the neutral hydrogen clouds in the Milky Way. This revealed a definite spiral pattern, with the Sun in the third arm from the centre and another spiral arm further from the centre than the Sun.

J.H. Oort and G.W. Rougoor found, in 1960, that there is a spiral arm 3 kiloparsecs (10 000 light years) from the centre of the Milky Way, and that it is moving outwards at 50 km s^{-1}. On the outskirts of the Milky Way there is an influx of hydrogen from intergalactic space, being drawn in by the gravitational force of the Galaxy. Most of the hydrogen lies in a thin disc in the plane of the Milky Way. The position of this disc facilitated the determination of the plane of the Milky Way.

The radio flux from the Crab nebula showed that supernova remnants are active in radio frequencies.

A study of flare stars, such as UV Ceti, showed that this star has radio eruptions when it flares up optically. Wherever hot gases occur, radio waves are emitted.

Pulsars

Pulsating radio sources can best be traced on wavelengths of one metre, using a receiver sensitive to rapid oscillations.

At Cambridge, UK, the array of 2048 dipoles is sensitive to the wavelength of 3.7 metres; it occupies 4 acres.

In July 1967, Miss Jocelyn Bell received peculiar radio noises. She and A. Hewish received regularly repeating radio signals from a certain direction in space, with a periodicity of slightly more than 1 second. It seemed as if this could be radio signals from an extraterrestrial intelligence. But the signals were not in code, nor did they show the Doppler effect which a spinning planet would have caused. The pulsating source is CP 1919 + 21 (Cambridge pulsar located at 19 h 19 m + 21°).

Early in 1968, another three of these "pulsars" were discovered, and by 1970 there were more than 50 in the list. They have periods between 0.001 55 and 3.74 seconds (Appendix 7, section 7.19).

T. Gold suggested that the radio pulses are caused by rapidly spinning, very small neutron stars, having strong magnetic fields. According to the theory, the periods must gradually lengthen, and this was proved by observation.

A pulsar having a period of 89 milliseconds was found in the extended radio source Vela X. Soon after, D.H. Staelin and E.C. Reifenstein discovered pulsar NP 0532, with a period of 33 milliseconds in the core of the Crab nebula. This was proof that the pulsations originate in the collapsed remnant of a supernova, that is, a neutron star, as Gold had suggested.

W.J. Cocke, M.J. Disney and D.J. Taylor discovered in 1969 that the Crab pulsar also emits pulses of light, coinciding with the radio pulses (see Chapter 11, page 217). The X-ray satellite found that there were X-ray pulses 100 times more intense than the optical pulses, which, in their turn, are 100 times stronger than the radio pulses.

Recent novae have been found to emit radio waves, for example Nova Delphini (1967) and Nova Serpentis (1970). X-ray stars have also been found, for instance Sco-X1 and Cygnus X-1.

The perfection of the process of making radio spectra led to the discovery of molecules in interstellar nebulae.

Radio Galaxies

Radio studies of extragalactic sources have yielded remarkable results. The Andromeda galaxy has a strong radio flux in its core. The radio flux of Cygnus A is much more intense. The 6-cm wavelength chart of Cygnus A (19 59.5 + 40 44) has a double lobe (Figure 13.29). This

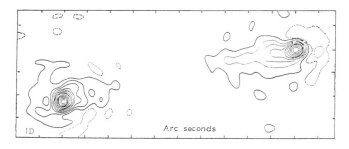

Figure 13.29 Radio chart of Cygnus A at a wavelength of 6 cm

figure is a copy of the chart by S. Mitton and M. Ryle in *The Evolution of Radio Astronomy* by J.S. Hey, Elek Science, London.

The radio flux of Cygnus A is actually concentrated at three points: at the centre and about 5 arc seconds to the sides. Emission lines in the spectrum of Cygnus A, and its peculiar appearance, gave rise to the idea that it consists of two colliding galaxies, or galaxies moving through each other. The probability of such close shaves in space is too small to be able to account for the multiplicity of radio galaxies. It seems clear that in the radio galaxies, matter is being annihilated, with the ensuing release of catastrophic forces, as great as 10^{54} joules.

I.S. Shklovsky suggested in 1960 that the energy of the radio galaxies comes from supernovae explosions in the cores of the radio galaxies. In 1961, E.M. and G.R. Burbidge required a chain reaction of supernovae, and in 1963 W.A. Fowler required the total collapse of a whole galaxy, to explain the enormous amounts of energy which are released.

Quasars

One of the most sensational discoveries of radio astronomy is that of the quasars. The radio source 3C48, which was studied by T.A. Matthews in 1960, was found by A.R. Sandage, after a 90 minute time-exposure with the 5-metre telescope, to be a peculiar type of "blue" star (that is with small $B-V$ colour index). This could be a genuine radio star, but the spectrum, the holder of the secrets of all radiating bodies, showed inexplicable emission lines.

Another quasar, 3C273, underwent occultation by the Moon; this revealed it to be a double source. The 5-metre telescope saw a faint jet, coming from one part. When M. Schmidt studied the inexplicable spectrum, he got the inspiration that the spectral lines had undergone a tremendous redshift, with a z-value of 0.16; that meant a distance of 1400 million light years. At that distance, the optical luminosity must be 100 times that of the core of a whole galaxy!

Similarly, the spectrum of 3C48 could be explained if a redshift of 0.37 was assumed. Its distance would therefore have to be 3000 million light years, using a Hubble constant of 100 km s^{-1} Mpc^{-1}.

The name quasar is derived from "quasi-stellar-radio-source".

The optical counterpart of 3C273 was located on an 80-year-old photograph which showed that the optical brightness was not constant. 3C48 also showed brightness fluctuations with a period of less than one year. Such rapid changes indicate that the body must be very small – not more than one light year in size. How can it then radiate 100 times the energy of a galaxy nucleus, tens of thousands of light years across? It seemed a conundrum.

To study such small objects at such enormous distances, the method of radio interferometry was employed, by using two radio telescopes, a few thousand km apart; their phases had to be synchronised. Two dishes were used, 3074 km apart, on a frequency of 448 MHz (wavelength 67 cm). The separation was thus 4.6 million times the wavelength. This enabled strong interference fringes to be obtained. Some quasars could, however, not be resolved.

The ultra-violet radiation from quasars is usually very strong. The emission lines are broad and often fine absorption lines are present, indicating cooler gas shrouds. All quasars are connected to radio sources of which the bandwidths are very small.

The average visual magnitude of the quasars lies between 14 and 19.5, and their absolute magnitudes between -24.0 and -30.8. An absolute magnitude of -28.4, which is a good mean value, implies a brightness of:

$$2.512^{4.85-(-28.4)} = 2.512^{33.25}$$

that is 2×10^{13} times the luminosity of the Sun. This is 100 times the luminosity of a galaxy the size of the Milky Way. No galaxy is known which has anywhere near the brightness of a quasar. Thus the brightnesses of quasars must be ascribed to something else, besides sheer size. The fact that the brightnesses vary over short periods of less than a year indicates that there must be changes in the radiation itself. Can it be that chain reactions of supernovae explosions are taking place there?

Table 13.4 lists a selection of quasars. The redshift, z, denotes the fraction by which the spectral lines are moved towards the red end of the spectrum. The speeds of recession, v, are given as percentages of the speed of light, calculated from Einstein's formula:

$$1 + z = \sqrt{(c + v) \div (c - v)}$$

Quasars

Table 13.4 Quasars: redshifts, recession velocities, distances

Quasar	Redshift z	Reces. vel. v as %c	Distance ($D \times 10^9$ light years)			
			$H_0 = 50$	$H_0 = 55$	$H_0 = 75$	$H_0 = 100$
PG 0804 + 761	0.10	9.5	1.858	1.689	1.239	0.929
3C 273	0.16	14.7	2.875	2.614	1.917	1.438
3C 48	0.37	30.5	5.965	5.423	3.977	2.983
3C 295	0.46	36.1	7.061	6.419	4.707	3.531
3C 345	0.59	43.3	8.469	7.699	5.646	4.235
PKS 1127 − 14	1.19	65.5	12.812	11.647	8.541	6.406
3C 446	1.40	70.4	13.770	12.518	9.180	6.885
PKS 1610 − 77	1.71	76.0	14.866	13.514	9.910	7.433
3C 9	2.00	80.0	15.648	14.225	10.432	7.824
PHL 957	2.69	86.0	16.822	15.292	11.214	8.411
DHM 0054 − 284	3.61	91.0	17.780	16.181	11.866	8.900
Distance $D \div$ velocity $v =$ Time ($\times 10^9$) years			19.560	17.782	13.040	9.780

$$\text{or } (1 + z)^2 = \frac{(c + v)}{(c - v)}$$

Taking 3C273 as an example, its redshift is 0.16.

$$\therefore (1 + 0.16)^2 = \frac{(c + v)}{(c - v)}$$

$$\therefore 1.3456 \, (c - v) = c + v$$

$$\therefore 0.3456 c = 2.3456 v$$

$$\text{Thus } \frac{v}{c} = \frac{0.3456}{2.3456} = 0.147$$

$$= 14.7\%$$

The speed of recession of 3C273 is thus 14.7% of the speed of light c.

The distances D are calculated from Hubble's formula: $V = H_0 D$. The values of D are multiplied by 3.26, in order to convert them to light years. Various values of H_0 are used, namely 50, 55, 75 and 100 km s^{-1} Mpc^{-1}.

When the distances are divided by the corresponding velocities, constant values are obtained. For example, 3C295: 6.419 ÷ 36.1% = 17.78 thousand million years. The times in the bottom line of the table are the times that have been taken by the radiation to reach the Earth. The ages of the quasars must therefore be 20, 18, 13 or 10 thousand million years, if Hubble's constant is 50, 55, 75 or 100 km s^{-1} Mpc^{-1}.

Taking Hubble's constant to be 55 km s^{-1} Mpc^{-1}, we see that the constant value of the distance column divided by the speed of recession comes to 17.782 thousand million years. In 1995, the most distant galaxy yet discovered, 8C1435 + 635, in Draco, was found to have a redshift of 4.25. The most distant quasar has a redshift of 4.9, making its speed of recession 94.4% of that of light.

Because of the great distances, and therefore great ages of the quasars, we see them by the radiation which left them 10 to 18 thousand million years ago, when they probably were galaxies in the making. Some quasars, such as 3C273 (12 09 06.8 + 02 03 07), have faint auras around them – either gas or outlying parts of galaxies.

On average, the B−V colour index of the quasars is 0.35. When allowance is made for reddening, they must be considered as "blue".

Quasar 3C446 varied by 0.5 to 0.8 magnitude during July 1966, in a period of only 24 hours. Since variations in radiation cannot exceed the speed of light, the core of this quasar cannot be larger than 24(3600)(300 000) km. This is slightly more than twice the diameter of the Solar System – a very small volume indeed.

We observe the quasars by the radiations which left them when the Universe was very young; they may have been the first objects to start radiating.

Properties of Quasars

1. Quasars have star-like appearances.
2. Quasars are usually strong emitters of radio waves.

3. Quasars emit strongly in the ultra-violet.
4. Quasars have large redshifts and are very distant.
5. Quasars vary in short periods in optical and radio frequencies.
6. Their colour indices indicate that they are "blue".
7. Their activity is similar to that of Seyfert galaxies.
8. Several quasars are double, as are some radio galaxies.
9. On account of their synchrotron radiation, their spectra are non-thermal.
10. Visual light from at least one quasar, 3C466, is polarised, as is the gas jet of the elliptical galaxy M87.
11. The radio spectra of some quasars differ from those of radio galaxies, insofar as the energy increases with higher frequencies.
12. Some quasars have shells of absorbent gas.
13. Optical spectra indicate that quasars undergo great turbulence.
14. The spectra of quasars indicate that the atoms of the gases are highly ionised.
15. The abundances of hydrogen, helium, carbon, neon, magnesium, silicon, argon and sulphur correspond to those of many galaxies.
16. The spectral lines of quasars are broad, on account of hot gases, moving at speeds of 2000 to 3000 $km\,s^{-1}$.

Gravitational Lenses

The study of quasars led to the discovery of gravitational lenses, as predicted by Einstein in his Theory of Relativity, and which was observed during the total solar eclipse of 1919. On the cosmic scale, a bright-enough object had to be found, in line with a massive foreground object. The distant object had to be sufficiently bright to enable its double image to be visible.

A photograph of the quasar, located at (10 01 20.7 +55 53 50), had a double image, as seen by the Schmidt camera of Mount Palomar. D. Walsh, R.F. Carswell and R.J. Weymann of Kitt Peak succeeded in obtaining two spectra of the object. Both spectra showed the same redshift of 1.41 and, line-for-line, the spectra were identical and also had the same emission lines and absorption lines, with the same redshift. The gas cloud thus had to be at the same distance as the quasar. The speed of recession was at least 210 000 $km\,s^{-1}$.

The VLA radio telescope, having a resolving power of 0.6″, succeeded in splitting the object. There was no object visible between the two images to serve as gravitational lens. Could the massive object perhaps be a gravitational vortex?

The 5-metre telescope of Mount Palomar yielded two

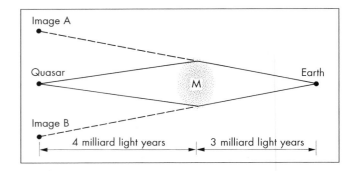

Figure 13.30 Gravitational lens effect of massive object M

images: one circular and the other, slightly oval. Could this oval be the image of a massive galaxy on the line of sight, serving as a gravitational lens? Since the radiation of a quasar is 100 times that of a galaxy, the lensing object would be invisible in the glare of the quasar. There could also be a cluster of galaxies, in line between the Earth and the quasar. The Mount Palomar telescope found a redshift of 0.39 for the elliptic galaxy, which thus had to be at a distance of 3×10^9 light years, taking the Hubble constant as 100 $km\,s^{-1}\,Mpc^{-1}$. On this scale the distance of the quasar must be 7×10^9 light years.

The principle of the gravitational lens is illustrated in Figure 13.30, where M represents the massive object in the form of a cluster of galaxies. The separation between the two images A and B, as seen from Earth, is less than 1 arc second.

If the massive object is slightly off the line of sight, a third image may be obtained. Einstein pointed out that a

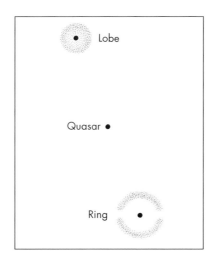

Figure 13.31 Einstein ring and radio lobe of quasar MG 1131 + 0456

ring could be formed of the distant object if the lensing object was exactly on the line of sight (Figure 13.31). Such a ring was discovered by Jacqueline Hewitt and colleagues of Massachusetts Institute of Technology, as reported in *Nature* of 9 July 1988. They used the VLA on a wavelength of 6 cm. The quasar MG 1131 + 0456 in Leo formed one circular lobe and one ring-shaped lobe. Neither the quasar nor the lensing object shows on the radio chart, but the 1.3-metre McGraw-Hill telescope on Kitt Peak obtained images of both the quasar and the lensing object. The redshift of the quasar is 1.74 and that of the massive object 0.254, as measured by observers of Princeton and Caltech, using the 5-metre Mount Palomar telescope.

The distance of the lensing object is 2.2 thousand million light years, and that of the quasar 7.65 thousand million light years, if the Hubble constant is taken at its maximum value of 100 km s^{-1} Mpc^{-1}.

This discovery is a beautiful confirmation of Einstein's Theory of Relativity.

The 3 Degree Background Radiation

Several astronomers, including R.H. Dicke, P.J.D. Peebles, P.G. Roll, D.T. Wilkinson, A.B. Crawford, B.A. Alpher and H. Herman, were on the track of background noise at a radio temperature of 4 to 5 K, which theory showed ought to exist.

G. Gamow had previously postulated a Universe having its origin in an intensely hot fireball at a temperature of 10^{10} K. This idea was based upon G. Lemaître's idea of a primeval atom or "cosmic egg". This fireball then exploded, forming the atoms, largely hydrogen and helium, and proceeded to expand and cool down. At present there ought to be a remnant radiation at a radio temperature of 5 K.

The idea was now in the air.

In 1965, A.A. Penzias and R.W. Wilson, of Bell Telephone Laboratories, were busy trying to eradicate radio noise from their radio receiver. On a wavelength of 7.3 cm they found an ineradicable radio noise at a radio temperature of 3 K. It had no sidereal, solar or direction-determined variations, but was the same in all directions. This finding was

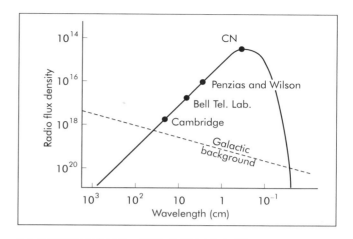

Figure 13.32 The 3 degree black-body curve

published in the *Astrophysical Journal*, No. 142 (1965) page 419, under the title "Measurement of Excess Antenna Temperature at 4080 megacycles per second".

The background radiation has been detected in four different regions of the spectrum.

T.F. Howell and J.R. Shakeshaft, of Cambridge, confirmed the existence of this background radiation at the wavelength of 21 cm.

Microwave measurements have confirmed the existence of the background radiation at a temperature of 2.7 K.

The background radiation is omnipresent and at all wavelengths.

The points on the graph of radio flux density against wavelength, over the range 10^3 to 10^{-1} cm, fit a black-body curve (Figure 13.32).

The existence of the 3 degree background radiation runs counter to the steady-state theory of F. Hoyle, T. Gold and H. Bondi, which states that the Universe is isotropic – generally the same in all directions: always has been and always will be. It therefore does not change with time.

The 3 degree background radiation supports the big bang theory, which postulates a hot beginning to the Universe, followed by expansion and cooling. The 3 degree background radiation is the remnant of the original fireball.

Chapter 14

From the Big Bang to the Present

Space–Time Panorama

The Universe, as revealed by the various instruments at the disposal of astronomers, is in the first instance a panorama of space. The speed of light and other electro-magnetic radiations is not infinitely great – it requires definite amounts of time to reach the Earth: 4.3 years from the nearest star; 2.2 million years from the nearest large galaxy, M31 in Andromeda; 5 to 18 thousand million years from the quasars. We therefore see these objects as they were, 4.3 or 2.2 million or 5 to 18 thousand million years ago. The greater the distance of an object, the further back in time we have to look in order to see it. The Universe is thus also a time panorama.

Chronological Sequence

In the course of thousands of millions of years, all kinds of changes have taken place in the Universe. The Universe is not static, nor changeless: a variety of objects have come into existence from time to time and gone through their evolution. These objects are spread out to our gaze. All that we need to do is to arrange them in the correct chronological order, to have the whole history of the Universe unfold before our eyes. The furthest objects must therefore be the oldest; radiation (light, radio waves, etc.) that we detect now left them when the Universe was much younger. Perhaps we should call the furthest objects the youngest, because we see them by the radiation which left them when they were much younger, that is, when the Universe was much younger. Age and youth are therefore relative terms, which depend on our point of view.

The Test of Observation

Any theories which we may compile to portray the history of the Universe must be subjected to the test of observation. If a theory postulates the existence of a type of object which cannot be confirmed observationally, the theory must be adjusted or dropped altogether. Based on the incontrovertible evidence of observation, we shall now sketch the probable evolution of the Universe, from the big bang, or cosmogenesis, to the present.

The Expanding Universe

The Doppler effect adduces irrefutable evidence that the Universe is expanding: all distant galaxies and quasars are receding from our galaxy and from each other. A two-dimensional analogy would be dots on the surface of a balloon which is being inflated. Although receding from each other and from the "centre", the dots remain constrained to the surface. A better analogy would be foam rubber or soap suds being inflated. Figure 14.1 (*overleaf*) is a chart of the positions of more than a million galaxies, visible from the northern hemisphere. It was made by Professors M. Seldner, B. Siebers, E. J. Groth and P. J. E. Peebles. This chart is a copy from their article in the *Astronomical Journal*, April 1977. It includes galaxies of magnitudes as low as 19. It is a macro representation of the appearance of the universe on the largest possible scale. It is similar to foam, and the bubbles represent voids in space. From the smallest atom to the largest material aggregate in the Universe, one finds much more empty space than matter, but the matter which is present determines the nature of the Universe.

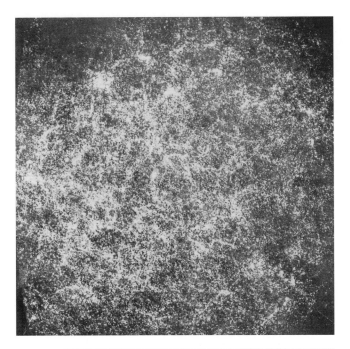

Figure 14.1 Appearance of the Universe on the greatest macro scale – isotropic and homogeneous

The Observable Universe

In Chapter 13 it was pointed out that space-time is curved in such a way that each observer sees himself to be at the centre of the Universe. In principle it would be possible for any observer in the Universe to observe its entirety from each and every position. Although expanding at an enormous rate, it would appear that the universe is at present finite in size, but unbounded.

Olbers' Paradox

Edmond Halley, and others, posed the question: why is the night sky dark? This problem became known as Olbers' Paradox.

H.W.M. Olbers reasoned, in 1823, that if the Universe is infinite, it must contain an infinite number of stars. The cumulative effect of all the stars must be to make the night sky evenly bright, even as bright as the Sun. The glow of all the stars together would have the effect of raising the Earth's temperature to 500 K.

Olbers' solution was that light is absorbed by the intervening background medium, the ether, as it was known. The experiment by A.A. Michelson and E.W. Morley showed that the speed of light is the same in all directions. There can thus be no ether.

The expansion of the Universe gave rise to the idea that the night sky is dark because of the redshift of light from distant receding galaxies. Redshift, however, implies a lowering of frequency and an increase in wavelength, not a decrease of intensity.

Other astronomers cited the limited age of the Universe as the reason for the dark night sky – not enough time has elapsed to make the expanse evenly bright. Both explanations could be correct.

P.S. Wesson (*Astrophysical Journal*, 15 June 1987) compared the relative effects of the two factors. His first finding was that the intensity of the general background illumination, in a static Universe, would be no more than $6 \times 10^{-4}\,\mathrm{erg\,cm^{-2}\,s^{-1}}$. This is faint enough to make the sky appear dark. The expansion of the Universe will lower this intensity of light by half. The age factor is thus more important.

Wesson's finding is therefore that the night sky is dark because the Universe is still young – there has not been enough time to fill all space up to the observable limit with radiating bodies.

This means that the Universe is not infinitely large, nor does it contain an infinite amount of radiating matter.

The Steady-State Theory

A number of theories exist for the origin and structure of the Universe. One of these is the steady state theory of F. Hoyle, T. Gold and H. Bondi.

According to this theory, the observable isotropic and homogeneous appearance of the Universe is an indication that it has always been isotropic and homogeneous and will always remain so. (Isotropic refers to the property that the Universe appears the same in all directions; and homogeneous to the fact that it will have an identical appearance to all observers at any given instant of time).

Small, superficial irregularities in the distribution of mass are observed on a local scale. If, however, the Universe is viewed on a large enough scale, its appearance is isotropic and homogeneous, as shown in Figure 14.1. The isotropic and homogeneous appearance of the Universe is called the "Cosmological Principle".

According to the steady-state theory, there never was an origin for the Universe, a time when it was much smaller. The isotropic and homogeneous state must hold for all time – past, present and future.

To explain the expansion of the Universe, and why the

different parts of the Universe have not dispersed in all directions, the theory postulates a pressure from within. This pressure is brought about by the continuous creation, or coming-into-being, of new matter, in the form of protons and electrons, or atoms of hydrogen, which come into existence throughout space. In the same way as gamma rays give rise to positron–electron pairs, so more energetic rays can give rise to protons and antiprotons.

With the lapse of time, the newly created hydrogen atoms condense to form galaxies. As the receding galaxies get fainter and fainter and vanish at the limit of the observable Universe, new galaxies come into being in the foreground and continue to recede. This is what makes the universe isotropic and homogeneous.

No test of observation could, however, confirm the continuous creation of hydrogen. The theory also cannot explain the 3 degree background radiation.

The Big Bang Theory

This theory requires that one reasons in reverse from the observable fact that the Universe is expanding. This implies that the galaxies must have been closer together in the past, and in the distant past so close that they were all concentrated at one point – designated by Georges Lemaitre as the "cosmic egg". This point in space–time is known as a singularity at which the laws of physics break down. Before this origin, no mass, space or time existed. Even the words "before" and "exist" are invalid in this context because there was no space-time as we know it at present.

Since time did not always exist, there is no way in which the origin of the Universe can be described. Scientists have agreed to describe events from the so-called Planck time, which is defined as 10^{-43} s. The name of this instant was chosen because it is limited by the magnitude of the Planck constant (h), the constant which plays a significant role in quantum physics. From this instant, the big bang and the subsequent events – the cosmo-genesis – can be described in terms of well-known scientific principles.

At Planck time the density and temperature should have been almost infinitely large, according to G. Gamow. Matter, as we know it, did not exist at that time. When the temperature had dropped to the order of 10^{10} K, gamma rays were able to produce electron-positron pairs, as has been demonstrated in the physics laboratory.

The first nucleons, protons and neutrons, were formed shortly after the big bang and these formed the first atomic nuclei, between one second and three minutes after Planck time, when the temperature had dropped to 10^9 K. This was still too hot for the formation of neutral atoms.

A further drop in temperature would have maintained the positron–electron pairs in equilibrium. The original, electrically neutral particles (the neutrons) could then fuse with the positrons to form protons, each proton thus carrying a positive electric charge. The protons are the nuclei of hydrogen atoms. When protons are formed, energetic, mass-less antineutrinos are set free. They move irresistibly at the speed of light and hardly react with matter. The protons could then attract electrons to form neutral hydrogen.

C. Hayashi showed that there would be complete thermal equilibrium between all forms of matter and radiation, at temperatures higher than 10^{10} K. At temperatures lower than 10^{10} K, the neutrons could not maintain equilibrium with the protons, because the formation of positron-electron pairs rapidly diminished. The ratio between the number of neutrons to protons was thus "frozen in", within merely a few hundred seconds. The ratio of neutrons to protons was then 15%.

Simultaneously with the formation of hydrogen atoms, helium was formed by the fusion of two protons and two neutrons to form one helium neucleus. The two isotopes of hydrogen, deuterium ^2_1H and tritium ^3_1H, were also synthesised, but they were subject to rapid decay on account of their radioactivity.

The temperature next dropped rapidly, so that the formation of helium ceased. The ratio of helium to hydrogen was then between 20% and 28%. This is something which can be tested by observation. The abundances of helium to hydrogen in the galaxies, globular clusters, quasars and nebulae have all yielded percentages of approximately 20% helium and 80% hydrogen, which confirms the theory.

We have seen that when a massive star collapses catastrophically on its centre, during a supernova explosion, the remnant is a neutron star. The tremendous contraction overcomes the electro-magnetic forces between protons and electrons, so that they become compressed, resulting in the neutralisation of the electric charges, and a neutron star comes about. There is no electro-magnetic force between the neutrons and they become compressed into what is known as degenerate matter, having a density of no less than 2.7×10^{14} g cm^{-3}.

Cyclic or Pulsating Universe

Could it be that a supernova explosion and the formation of a neutron star is a "laboratory demonstration", on a "small scale", of what can happen to the entire Universe? Could a previously expanding Universe have come to the end of its expansion and then started contracting until, with rapidly increasing pressure and rise in temperature, it became a cosmic egg, consisting solely of neutrons and radiation? The rebound of this contraction would then be the big bang, or rather, the cosmogenesis of the present Universe.

If this is the case, then the Universe is a cyclic or pulsating Universe, which undergoes an infinite series of expansions and contractions.

If the present expansion is to come to an end and be followed by a contraction, there must be a sufficiency of matter in the Universe so that its force of gravitation can halt the expansion and bring about the succeeding contraction.

In Figure 14.2, the moment of the big bang is indicated by O on the time axis when the size of the Universe was zero. The present moment of time is represented by T, some 10×10^9 years after the moment O, taking Hubble's constant at its maximum value of 100 km s^{-1} Mpc^{-1}. The slope of the tangent to the curve, vertically above the point T, gives the present rate of expansion of the Universe. If the rate of expansion was constant over the past 18×10^9 years, the tangent will cut the time axis at O′, 8×10^9 years before O, according to calculations by A.R. Sandage. The time that has elapsed since the big bang is therefore not less than 10×10^9 years and it cannot be more than 18×10^9 years. This agrees with the findings in Table 13.4, page 255.

The present rate of expansion indicates a total expansion time, from O to T′ of 60×10^9 years. Then the expansion will stop, to be followed by contraction from T′ to T″ on the time axis. At first, the contraction will be barely perceptible, but with the lapse of time, the rate of contraction will speed up more and more until, after a further 60×10^9 years, the "great crunch" will come about, with the formation of the cosmic egg all over again – a time of singularity when only neutrons and radiation will exist.

Up to the present, astronomers have not been able to account for a sufficiency of matter in the Universe to

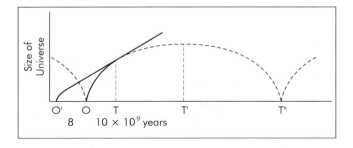

Figure 14.2 Expansion and contraction of a cyclic Universe

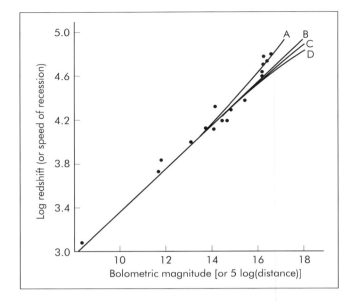

Figure 14.3 The expanding Universe

overcome the expansion which is now going on. The known amount of matter barely accounts for 10% of the required amount. Could this missing matter be locked up in the so-called "dark matter" – brown stars, planets, burnt-out white dwarfs, cold gas and dust clouds, and even gravitational vortices or black holes in the centres of galaxies? In any case, if the Universe is cyclic, there must be a sufficiency of matter present, whether observed or not.

To test whether the expansion will go on for ever, M.L. Humason made a survey of distant clusters of galaxies. He compared their distances by comparing their total luminosities. The fainter clusters must be further than the brighter ones.

Figure 14.3 compares the logarithms of the redshifts with the total magnitudes, or bolometric magnitudes, over the whole spectrum, for 18 clusters of galaxies.

The straight line, C, indicates that the Universe will go on expanding at the present rate for ever.

Curve D fits the steady-state Universe. Curves to the left of C indicate that the expansion will decrease.

Any curve between C and B indicates that the Universe is "open" and infinite.

Curves to the left of B indicate a "closed" and bounded Universe.

The faintest and therefore furthest galaxies fit curve A best. This indicates that the expansion will come to an end and be followed by contraction.

Humason's work seems to show that the steady-state

Measurements of Ages

D.N. Schramm, in his article "The Age of the Elements" in *Scientific American* collection, *New Frontiers of Astronomy*, gives an exposition of the methods employed by physicists in determining ages of radioactive elements. The methods depend on the determination of the amounts of the end-products of radioactive decay, as proportions of the amounts of undecayed elements. The instrument used is the mass spectrometer, in which magnetic fields cause bending of the trajectories of emitted particles from radioactive elements.

The results yield a minimum age for the oldest minerals tested of 7×10^9, and a maximum age of 15×10^9 years. This is in good agreement with Table 13.4, page 255.

The radioactive elements could not have been formed before the big bang. The ages of the elements also agree with the ages of globular clusters (page 233), and with the general age computed for the Universe as a whole.

According to the theory of the formation of hydrogen and helium in the big bang, few if any elements could have been formed in the cosmogenesis. The more massive elements were only formed later, in the cores of stars and in supernova explosions.

Schramm comes to the conclusion that there was a peak, or maximum, in the rate of supernovae about 9×10^9 years ago, at a time of about 11×10^9 years after the big bang.

According to the big bang theory, the continued expansion of the Universe was accompanied by a continual decrease in the temperature, until today there is the omnipresent 3 degree background radiation, which is present in all directions. The presence of this radiation agrees with the theory and with the cosmological principle that the Universe is isotropic and homogeneous. The relative amounts of helium and hydrogen also fit the theory.

The Formation of Galaxies

During the expansion and cooling, the young Universe reached a temperature of 3 000 K after about 300 000 years. At this stage, electrons could bond to atomic nuclei to form neutral atoms. The Universe now became transparent to radiation, namely photons. The photons could now move freely. Some of these photons still pervade the entire Universe as microwaves and they constitute the 3 K background radiation as a relic of the big bang.

Mass is the seat of gravitation. It causes a curvature in the space-time continuum and this exercises a force of attraction on material bodies. In this way matter became coagulated into clumps.

While expanding, clouds of the primeval gases, hydrogen and helium, formed concentrations of mass between 10^8 and 10^{12} solar masses, and with volumes of the order of 100 000 light years. From these concentrations, or condensations, galaxies were born, singly as well as in clusters. Between the clusters, there were great voids, as in the soapsuds' analogy.

The gravity exercised by such a mass of gas would have caused spin to take place and the process of accretion would have concentrated the matter nearer to the centre of such a cloud of gas. The speed of spinning would have increased and this would have caused the cloud of gas to flatten into a lens-shaped form. The first galaxies would therefore have been elliptical in form. Some could have developed the lens-shaped form. These galaxies would therefore contain the oldest stars, stars with low content of atoms heavier than hydrogen and helium. Observation agrees with this.

The evolution of a galaxy depends in great measure on its mass, because the mass determines the strength of the gravitational field exercised by the galaxy.

The distribution of the stars that were subsequently formed determines the masses of those stars, which must have formed by accretion of the gases of the primeval cloud forming the galaxy. The process of accretion will exercise pressure on the centre of gravitation, thus raising the temperature until the gas begins to glow. This would then be a proto-star. Further pressure from the overlying layers would raise the temperature still further, to 15 million degrees K, when the individual protons would fuse to form helium, with the liberation of great amounts of energy – the birth of a star! These first stars were deficient in atoms heavier than hydrogen and helium. They are Population II stars, and are plentiful in globular clusters and in the nuclei of galaxies, as is confirmed by observation.

Observation also confirms that most of the stars in elliptical galaxies consist of 20% helium and 80% hydrogen. Most of the stars in the elliptical galaxies are small stars, of the Sun's mass, with high positive $B-V$ colour indices; they are typical red stars. These stars can remain in the Main Sequence of the H–R diagram for very long periods of time. When these stars eventually leave the Main Sequence, they expand to become red giants, before becoming novae.

The stars in the elliptical galaxies are truly representative of the matter which was formed in the big bang.

Globular Clusters

Most of the original matter formed in the big bang went to form galaxies. In the space between the primeval galaxies, smaller masses of gas, of the order of 10^5 to 10^7 solar masses, condensed to form globular aggregates of stars – the globular clusters, which largely remained caught in the gravitational fields of the galaxies. Out of these smaller masses of gas, stars of the type of Population II were formed. They are smallish stars, of high positive B–V colour indices, of spectral types G, K and M, which are poor in metals. These stars also consist for the most part of 20% helium and 80% hydrogen.

These smaller masses of gas produced 10^5 to 10^7 stars each. There is very little dust in the globular clusters, showing that few supernovae have taken place there.

The globular cluster stars are very old, being of the order of $(13 \pm 3) \times 10^9$ years old, that is, between 10 and 16 thousand million years. These condensations into stars must have taken place 10 and 16 thousand million years ago. Using a Hubble constant of $55 \text{ km s}^{-1} \text{ Mpc}^{-1}$ fixes the age of these stars at 18×10^9 years. The condensation of the first stars, those in the elliptical galaxies and in the globular clusters, must have been completed within the first 2×10^9 years after the big bang.

Some of the condensations to form globular clusters were so far away from the galaxies that they roam about in the intergalactic space.

Massive Stars

Largely at the centres of accreting masses of gas, smaller but denser condensations into stars would have occurred. There are not many stars more massive than 10 solar masses. Stars of 8 to 10 solar masses have a very short stay in the Main Sequence of the H–R diagram – not more than 16 to 28 million years. Then those massive stars will go supernova.

At the centres of the galaxies, where matter is denser, the more massive stars would quickly consume the larger part of their hydrogen, collapse catastrophically on their centres because of the decreased radiation and increased pressure of the overlying layers, reach temperatures of 100 million degrees, and then blow themselves to smithereens in supernova explosions. All sorts of heavier atoms were formed in these explosions and thrown into space, thus enriching interstellar space with heavier atoms. The remnants of stars massive enough would form gravitational vortices, or black holes.

Exploding Galaxies

From 16 to 28 million years after the formation of the first very massive stars near the centres of the galaxies, these massive stars went supernova, singly and in groups. In his article "The Age of the Elements", D.N. Schramm concluded, from his study of the radioactive decay of the elements thorium-232, uranium-238, rhenium-187 and osmium-187, that a peak of supernova explosions was reached about nine thousand million years ago. This is the age of these radioactive elements. The most massive stars were then only 16 to 28 million years old. According to Steven Weinberg in *The First Three Minutes*, the original cooling from the big bang went on for 700 000 years without any exceptional events taking place; this is almost one million years. But another one thousand million years were to elapse before Schramm's peak of supernova explosions took place. There was thus ample time for the condensation of nebulae and the formation of super-massive stars. The evolution of a galaxy in which many supernovae have occurred is very different from that of an elliptical galaxy, which consists of low-mass stars. When the massive stars exploded, the matter was cast out of the galaxy's core at great velocities. In the exploding galaxy M82, the gases are moving outwards at speeds of 1000 to 1100 km s^{-1}. Photographs of M82 reveal it as a galaxy in chaos. There are few other examples of exploding galaxies. Their scarcity implies that the explosions do not last very long, astronomically speaking.

The remnants of the exploding stars formed a massive gravitational vortex, or black hole, in the core of the exploding galaxy. The remnant masses of the millions of stars that exploded are locked up in these black holes in the centres of galaxies. The term "black hole" is a misnomer, because it is not black, nor is it a hole. It is anything but a hole, being a region of almost infinite density. Since this great compression was brought about by gravitation, the term gravitational vortex is more apt. This concept flows directly from Einstein's Theory of Relativity. As long ago as 1805, P.S. Laplace had predicted something similar. The numbers of stars that go to form the gravitational vortices in the cores of galaxies must vary between 10 million and 100 million. These masses could form part of the "missing mass", but only part.

Although a gravitational vortex is invisible, because its escape velocity exceeds the speed of light and all forms of electro-magnetic radiation, none of which can therefore escape from the vortex, the great gravitational attraction of such a vortex will strip matter from nearby stars. The friction undergone by these rapidly moving gases will generate radio waves and even X-rays, before the gases vanish into the vortex. X-rays are observed as coming from

Sagittarius A, in the centre of the Milky Way, proving that it is very likely that a gravitational vortex does exist at the centre of the Milky Way.

Quasars as Exploding Galaxies

The quasars which radiate hundreds of times the energy of a whole galaxy could possibly derive their energy from series of supernovae. This could explain the short periods of their variability, which could be caused by the explosion of smaller or larger groups of massive stars. One supernova radiates an amount of energy equal to that of a whole galaxy. Hundreds of simultaneous supernovae would produce so much more energy. The accretion of gases spiralling in on a massive gravitational vortex would also contribute to the total amount of energy set free. The quasars could thus be galaxies in their formative years, 16 to 28 million years after the condensation of super-massive stars.

P. Morrison determined the distance of M82 to be 10 million light years. How is it that M82 is only exploding in recent times? It could be that it is a slow developer and that galaxies did not all form at the same time, within one thousand million years of the original cosmogenesis.

The matter cast out by supernova explosions contain atoms heavier than hydrogen and helium, which were formed by proton fusion and by the carbon–nitrogen–oxygen cycle in the cores of stars. In the supernova explosion itself, atoms heavier than iron would be formed, such as uranium-238, thorium-232, rhenium-187, etc. – atoms that are radioactive. The rates of their radioactive decay serve as very good chronometers to date the times when they came into existence.

Formation of Spiral Arms

Two main types of spiral galaxies can be distinguished: pure spirals, SAa, SAb, SAc and SAd; and barred spirals, SBa, SBb, SBc and SBd. The barred spirals have spiral arms at the extremities of the bars; these spiral arms vary from short to long, and the bars in inverse ratio.

The bars must be considered as matter that had been cast out of the core of the galaxy. With the passage of time, the outward velocity of this matter would decrease and it would start lagging behind on account of the spin of the core of the galaxy. As time goes by, more and more of the matter in the bar will start trailing behind and the spiral arm will get longer and the bar shorter. Barred spirals will thus tend to become ordinary spirals (Figure 14.4). The speeds of

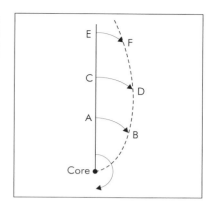

Figure 14.4 Formation of spiral arms

revolution at A and C are approximately equal. The arcs AB and CD are also equal, but because the arc AB has a shorter radius of curvature than CD, point D will lag behind B. At the outer boundary E, the speed of revolution around the core is less than at C or A, so that the matter at F will more readily lag behind. The curve B–D–F thus indicates the beginnings of a spiral arm.

Magnetic fields may also play a part in the formation of spiral arms.

Star Formation in Spiral Arms

The matter in the spiral arms is subject to the formation of clumps by gravitational forces. These condensations form the cores of proto-stars, or of double stars, and in some cases, clusters of stars, depending on the amount of matter present. Observation shows that more than half the stars in the Milky Way are double stars.

The matter from which the stars in the spiral arms condense is enriched by atoms heavier than hydrogen and helium and also heavier than iron. These stars, being the second generation of stars, are members of Population I, like most of the stars in the neighbourhood of the Sun. Population II stars, poor in metals, are largely to be found in elliptical galaxies and in the cores of galaxies. They are old stars and belong to the first generation of stars. They are also plentiful in globular clusters. The Sun, situated in the third spiral arm from the centre of the Milky Way, is a Population I star, containing an appreciable amount of heavier atoms. The Sun was, however, formed by the condensation of gas and dust, which stemmed from a second peak of supernovae explosions, which, according to D.N. Schramm, occurred about five thousand million years ago. The Sun is thus a third generation star. The age of the lunar rocks, 4.6 thousand million years, bears striking evidence to the correctness of Schramm's calculations.

Table 14.1 Variety of Population I stars

Star		Luminosity class	Abs. mag.
Rigel β Orionis	B8 Ia	blue-white super-giant	−7.1
Sirius α Canis Majoris	A1 V	white, Main Sequence	1.4
Procyon α Canis Minoris	F5 V	yellow, sub-giant	2.6
Alpha Centauri A	G2 V	yellow, replica of Sun	4.4
Aldebaran α Tauri	K5 III	orange giant	−0.3
Betelgeuse α Orionis	M2 Iab	red super-giant	−6.0
Barnard's Star	dM 5	red dwarf	13.4

Table 14.1 gives some idea of the variety of stars of Population I to be found in the Sun's neighbourhood – from blue-white super-giants to red dwarfs.

The stars which form in the spiral arms undergo the same processes: initial condensation and increased pressure on the core until the gas begins to glow; followed by further compression until the temperature reaches 15 million degrees K, when fusion of protons begins, accompanied by the liberation of energy on colossal scales. Astrophysicists calculate the gamma rays which are produced in the nuclei of stars have frequencies of the order of 10^{25} Hz. This radiation is rapidly absorbed by particles in the core of the star and just as rapidly re-emitted, but each time at a lower frequency. After one million years have elapsed, the radiation reaches the surface of the star and by that time the frequency has decreased to 6×10^{14} Hz, when the star's photosphere radiates it as visible light at a wavelength of about 5000 Å. Some radiation at the wavelengths of ultra-violet light and X-rays does come through the surface.

The radiation emitted from the core of a star exercises an outward pressure and, together with gas pressure, resists the inward pressure of the overlying layers, thus maintaining a state of balance. The Sun has been in this state of balance for 5000 million years, and can continue for another 5000 million years; then it will leave the Main Sequence of the H–R diagram and proceed to the realm of the red giants. After the red giant phase, the Sun will shrink and eventually become a white dwarf, after having cast off its upper layers.

A star such as the Sun also has convection currents below the surface, which convey heat to the surface. At the surface, some of the convective matter is hurled into space as the solar wind. Some of the matter falls back and forms prominences. The solar and stellar winds pervade interstellar space.

The first perceptible signs of the birth of a star are the infra-red radiations, such as that radiated by the gas and dust clouds around the T Tauri stars.

Because the matter in the spiral arms is more scattered, a variety of stars are formed there. In the Sun's neighbourhood, most of the stars are small, being spectral type M red dwarfs; the more massive stars are scarcer.

Many of the supernovae that occur in galaxies beyond the Milky Way are massive stars in the spiral arms of those galaxies.

Average Stars

When average stars, of mass less than 1.4 solar masses, have reached the stage of having consumed most of their hydrogen, their intensity of radiation will suddenly decrease. They then collapse catastrophically on their centres. This collapse increases the pressure so much that the temperature of the core is raised to 100 million degrees K. This initiates helium fusion – the helium flash – into heavier atoms, such as carbon-12, neon-20, magnesium-24 and silicon-28. The helium flash causes the overlying layers of the star to be blown off and the star becomes a nova. Many novae are constantly being discovered in the plane of the Milky Way, because most of the stars there have masses of less than 1.4 solar masses. Many of these novae are known to become repetitive novae.

The "planetary nebulae", such as the Ring nebula in Lyra, M57, and the Helix nebula NGC 7293, are good examples to show the shells of expanding gases that have been blown off from the star. The remnant of the star is a white dwarf with a temperature in the neighbourhood of 40 000 degrees K.

Astrophysical theories indicate that the white dwarfs are more stable than the average star. They go on cooling down until they radiate in the infra-red and eventually become black cinders – black dwarfs. By that time their gaseous shells will have vanished into space.

The Milky Way possibly contains hordes of white and black dwarfs.

The white dwarfs which are members of binaries tend to strip matter from their companion stars, becoming more massive, until they exceed the Chandrasekhar limit. Then they become Type I supernovae and destroy themselves completely.

The Probability of Planets

In the primeval nebulae from which the stars form by accretion, smaller clumps of gas and dust become the nuclei on which more gas and dust condenses by accretion. These condensations became planets, which are much less massive than the stars.

The Sun has nine acknowledged planets. Besides that, the Solar System contains seven satellites of planets having diameters ranging from 2720 to 5262 km. Two of these satellites are larger than Mercury and six are larger than Pluto. Earth–Moon and Pluto–Charon can be considered as being double planets.

Besides the 16 planetary bodies, there are thousands of lesser moons and planetoids. Lesser bodies are very plentiful.

Since the Sun has such a vast number of lesser bodies in its gravitational field, it is unthinkable that the other stars would not be similarly endowed. By the end of the 20th century, the new generation of telescopes ought to reach finality on the question of the existence of planets revolving around other stars.

In the meanwhile, it has been found that Alpha Lyrae (Vega) and Beta Pictoris each have a disc of matter radiating in the infra-red. These discs are the material out of which planets could condense.

The oscillations in the radial velocities of stars, such as Barnard, Gliese-229, 51 Pegasi and 70 Virginis, among others, can be ascribed to the gravitational attractions of dark bodies, some of planetary size.

Where a massive body forms by accretion, smaller bodies will also accrete. If the Sun had only one planet, one could say that probably half the celestial bodies must be planets. But 16 planetary bodies and vast hordes of lesser bodies indicate that small non-luminous bodies must be at least 10 to 20 times as plentiful as stars. The Milky Way has 10^{11} stars, which probably have 10^{12} planets!

Formation of Planets

While the Sun was forming by accretion of the finely divided dust and gas cast out by supernovae explosions, many separate clumps were formed by the same process of accretion in the neighbourhood of the Sun – not necessarily solid, hard rocks, but also loosely aggregated clumps.

The primeval nebula would have possessed spin, since all bodies spin. The planetesimals that formed in the region of the Sun would therefore have revolved around the Sun from the start, being held in its gravitational field.

G.W. Wetherill (*Scientific American*, October 1969) found by means of a computer simulation that if he started with 100 planetesimals, they would clump together during the lapse of 30.2 million years to form just 22 larger bodies, on account of the gravitational field in which they revolved. After 79 million years, they would have collected into only eleven bodies, and after 100 million years, into four bodies. A similar process could have led to the formation of the Sun's inner four planets and the Moon.

What does observation show?

As the result of the exploration by space probes, we have close-up information on surfaces and atmospheres of all the planets except Pluto. In 1974/1975 Mariner-10 flew past Mercury three times, coming as close as 200 000 km. Its video photographs revealed a surface, uniformly peppered with craters. These craters were formed when loosely aggregated clumps of matter plummeted down on to the gradually growing nucleus, which served as a gravitational centre for the immediate neighbourhood. The surface of Mercury, as we see it, is the end result of the accretion process of material from the primeval nebula. The same process took place throughout the Solar System. The craters on the surfaces of bodies like the Moon and the satellites of Jupiter, Saturn and Uranus could not be eroded away because of lack of water on the surfaces of these bodies. On Mars, wind erosion has covered many of the original craters. On Venus, great craters are extant. On Earth, with its great amount of water, the original craters have all been eroded away. Those craters that still exist were formed very recently.

The ages of the Moon rocks indicate that the accretion process came to an end approximately 3.5 to 4.6 thousand million years ago. The more recent craters on the Moon and Mercury slope conically to the centre and were caused by hard, solid bodies, such as meteorites, and not by loosely aggregated clumps of matter.

The oldest Moon rocks, 4.6 to 6 thousand million years old, are found on the highlands of the Moon. The material of the Moon's maria is only 3.5 thousand million years old. The maria were thus formed 1.1 thousand million years after the greatest density of bombardment came to an end. The bodies that formed the maria were large, solid planetoids, which caused the whole area around the point of impact to melt and cover the surface with lava. The tips of the rims of some craters peep out above the surface of the lava.

Bodies such as the moons of Mars, Phobos and Deimos, have many small craters that must have been caused by meteorites. There is one large crater on Phobos which could almost have resulted in breaking it apart.

Besides its many craters, Jupiter's satellite Callisto has two large systems of concentric ridges, stretching 1500 km from the points of impact where planetoids must have crashed, late in the accretion process. These ring systems are similar to Mare Orientale on the Moon and Caloris Basin on Mercury.

The moons of Saturn also have many craters, most of them having been caused by clumps of ice. These craters are flatter because ice tends to flow.

The moons of Uranus, especially Miranda, are covered with craters. Miranda was almost broken apart, judging by the 20 km high cliffs on its surface.

Triton, Neptune's large moon, also has craters on its icy surface.

In the region of the outer planets, most of the material which plummeted down consisted of clumps of ice – the moons themselves being largely ice.

The accreting material in the inner Solar System was largely rocky material.

There can be no doubt that the planets and their moons were formed by accretion of material from the primeval nebula from which the Sun also condensed.

The material of the primeval nebula was probably fairly homogeneous. But the Moon and the four inner planets are terrestrial, while the outer planets are gaseous and their moons icy. The reason for this difference is because the solar wind, which, as in the case of T Tauri stars, was very strong during the formative years, blew most of the lighter material, such as gases and water, away to the further reaches of the Solar System.

Throughout the Solar System, substances of high melting points, such as rocks and metals, formed the cores on which the accreting material collected. The pressure of the overlying layers caused the rocky material to melt. Dense substances, such as iron and nickel, therefore gravitated down to the nuclei, since they were liquids. In this way the Earth obtained its iron–nickel core.

The water vapour which was blown to the outer reaches froze to form the moons. Saturn's beautiful system of rings is the remnant of an icy moon, which, coming within the Roche limit, was broken up by Saturn's gravitation and spread into rings.

The gases which were blown to the outer reaches gravitated on to the rocky cores to form the planets Jupiter, Saturn, Uranus and Neptune. These gases are largely hydrogen and helium, the most abundant substances in the Universe.

The great pressure exerted by the overlying layers of Jupiter's atmosphere caused the temperature to rise. Despite this, the lower layers of hydrogen liquefied, covering the core with a layer of liquid hydrogen, 25 000 km thick, over a 31 000 km layer of metallic hydrogen. The surface of the liquid hydrogen layer must be very smooth in order that the Great Red Spot, an anticyclone in Jupiter's atmosphere, can endure for centuries.

Gases such as methane and ammonia liquefied and even froze to form fine crystals. These ices form the white spots and streaks in Jupiter's atmosphere and also in the atmospheres of the other outer planets. Jupiter's atmosphere also contains other compounds of hydrogen, as well as fine particles of sulphur and sulphur dioxide.

Jupiter's moon, Io, is exceptional in having a surface consisting largely of sulphur and sulphur dioxide. Jupiter's gravitational force causes churning below Io's surface, thus raising the temperature so much that sulphur and sulphur dioxide are spouted out by several volcanoes on Io's surface. The fine particles of sulphur and sulphur dioxide form a torus, more or less coincident with Io's orbit around Jupiter.

The hydrogen and oxygen of the primeval nebula combined to form water, which is abundant in the Solar System. The Earth's water comes largely from vapour spouted out by volcanoes, the water having been occluded in the rocky material.

Beyond the limit of the planets, the cometary material, consisting of ices of water, methane and ammonia, forms the Oort cloud, from which comets plummet down to the Sun from time to time.

The Earth's Atmosphere

Mixed in with the rocky material that formed the cores of the four terrestrial planets, there were many volatile gases and other substances which had become occluded in the fine, rocky material. These substances had not been blown away by the solar wind. The pressure of the overlying layers forced the volatile substances out of the Earth's interior, through volcanoes. 95% of the gases vented by volcanoes consist of steam, with admixtures of gases such as methane, ammonia, nitrogen and carbon dioxide, and the omnipresent hydrogen and helium. The steam readily condensed to water and fell as torrential rains, eroding away all the original craters on the surface.

Any oxygen that was present would have combined with hydrogen to form water, and with other elements to form oxides and carbonates; it would not have remained free in the atmosphere. The rocks on Earth consist largely of silicon oxide and oxides of metals such as aluminium, magnesium, calcium, iron, etc.

The Earth's Atmosphere

The first atmosphere of the Earth thus consisted largely of methane, ammonia, nitrogen, carbon dioxide and hydrogen. This is known as a reducing atmosphere, since it tends to make substances that combine with it more electropositive.

On each of the planets, the atmosphere which formed depended on the prevailing conditions. Venus and Mercury, so much closer to the Sun, were much hotter. The molecules of the lighter gases moved so rapidly on account of the heat that they readily exceeded the escape velocities of those planets and escaped into space. Mercury's gravity was also too weak to retain any atmosphere whatsoever. Venus, with 81.8% of the Earth's mass, was able to retain the heavy gas, carbon dioxide. Its water vapour was blown away by the solar wind. The carbon dioxide could therefore not combine with water vapour to form carbonic acid, which, in its turn, would have formed carbonates with the rocks. Today, the atmosphere of Venus consists of 96% of carbon dioxide; its atmosphere is 90 times as dense as the Earth's.

Because of the lower gravity on Mars, that planet readily lost its lighter gases to space. Later, when the massive volcanoes on Mars erupted, the water vapour, as steam, readily condensed and fell as rain in torrential downpours. The water easily seeped into the sandy soil, where it froze to permafrost. The dry "riverbeds" on Mars seem to begin nowhere and lead nowhere. The icecaps on the Martian poles do contain frozen water as well as frozen carbon dioxide.

The atmosphere which Mars succeeded in retaining has a pressure of only one-ninetieth that of the Earth, and it consists of 95% carbon dioxide.

In the region of the outer planets, the water vapour froze to ice.

In the region where the Earth formed from the primeval nebula, the temperature was such that water could exist in all three states: solid, liquid and gas. The Earth's gravity was sufficient and the temperature moderate enough to enable it to retain an atmosphere consisting of the gases, methane, ammonia, nitrogen, carbon dioxide and hydrogen.

The steam belched into this atmosphere, and readily condensed to form torrential downpours. The Witwatersrand gold-bearing ores bear witness to a thickness of at least 4000 metres of sedimentary rock that was laid down by water. Radioactive dating of the uranium and thorium that occur in the gold ores gives a date of 3 thousand million years, corresponding to the age of the maria on the Moon. This is also the age of the first semblance of life on Earth.

In this atmosphere at this time, there must have been great lightning discharges in the rain storms that raged. What effect could this lightning have had on the atmosphere as it was at that time? Could the electric discharges have brought about the change from the primeval atmosphere to today's atmosphere, consisting of 80% nitrogen, 20% oxygen and 0.03% carbon dioxide?

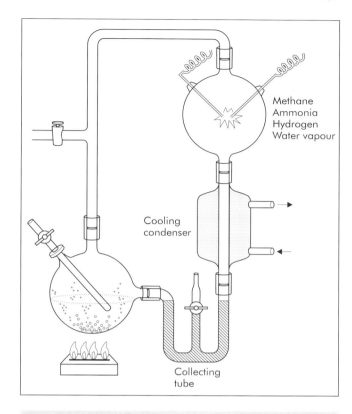

Figure 14.5 Electric discharges form amino acids from the primeval atmosphere

In order to find out, Stanley Miller conducted an experiment in 1952. He sent electric sparks through an atmosphere simulating the primeval atmosphere of methane, ammonia, hydrogen and water vapour (Figure 14.5).

The gases were contained in a flask having two electrodes and connected to a source of water vapour. After the electric sparks had been sent through the gases for a week, Miller found that a brown liquid, containing amino acids, had formed in the collecting tube. The pressure of the gases in the flask had decreased.

Amino acids are essential ingredients that go to build de-oxyribonucleic acid, which is the active substance in the nuclei of living cells.

The first living organisms on Earth absorbed carbon dioxide from the atmosphere and, in the presence of sunlight, the process of photosynthesis took place, whereby starch is formed. Starch is an essential food for living organisms.

The process works as follows. In the presence of sunlight, carbon dioxide combines with water to form the compound CH_2O, and oxygen (O_2) is set free. A more involved reaction between carbon dioxide and water leads to the liberation of oxygen and the formation of glucose. This reaction takes place only in the presence of sunlight and chlorophyll, the green colouring matter in plants and other living organisms:

$$6CO_2 + 12H_2O + \text{light energy} \rightarrow C_6H_{12}O_6 + 6H_2O + 6O_2$$

The glucose ($C_6H_{12}O_6$) is not set free, but combines to form starch.

In this process, amino acids, proteins, lepides, pigments and other organic substances are formed. Nitrogen, phosphorus and sulphur, which take part in these reactions, are derived from minerals in the soil. The complexity of the reactions very soon mounts up.

The process of photosynthesis is fundamental to the existence of life on Earth. The important point is that oxygen is liberated in the process. Through the millennia, the percentage of oxygen in the atmosphere increased, until today it comprises 20%. The original methane and ammonia were totally consumed in forming amino acids. The remaining hydrogen gradually escaped into space. The nitrogen, being inert, did not change much. Lightning discharges caused some nitrogen to combine with oxygen to form nitrogen oxides, which, with water, formed dilute solutions of nitric acid, which, in its turn, combined with minerals to form nitrates. The nitrates are essential as fertilisers.

The Earth thus became able to sustain a variety of life forms, since it is located in an ecosphere where the temperature and other conditions, especially the abundance of water, are just right.

S.H. Dole, in his book *Habitable Planets for Man*, analyses the possible extents of ecospheres around various types of stars. Only stars that have masses between 0.72 and 1.43 solar masses have ecospheres, correctly situated and wide enough to make the existence of habitable planets possible. The stars that comply with these conditions are F2, F3, ... F9, G0, G1, G2, ... G9 and K0, K1, and in some cases M-type stars of mass more than 0.35 solar mass.

Stars of mass more than 1.43 that of the Sun do not reside in the Main Sequence long enough to enable life-supporting planets to develop.

Stars in the above-mentioned classes comprise 25% of all stars. The Milky Way must therefore have 0.25(100 000 000 000) or 25 thousand million stars able to harbour life-supporting planets. If only one out of a hundred of these planets has succeeded in bringing forth life and if one in a hundred of the latter has developed intelligent life, there still must be two and one-half million planets with intelligent life in the Milky Way galaxy.

Then Where is Everybody?

Why have these intelligent beings not visited the Earth, or why have they not made radio contact with us? We must bear in mind that our radio technology is still very primitive. Perhaps the best method of making interstellar contact is by means of laser beams, or by gamma rays, and this we yet have to develop.

We must also reckon with the scale of interstellar space. A radio signal from the nearest star, Alpha Centauri, will take 4.3 years to reach Earth. An immediate reply from Earth will take a further 4.3 years to reach Alpha Centauri. This could hardly be called dialogue; not to mention the time lapses in the case of all the other stars which are further away. What's more, Alpha Centauri is a double star. The luminosities of its two components are 1.5 and 0.64 times that of the Sun. They revolve around each other in 80 years in orbits of eccentricity 0.52, their distance apart varying from 10 to 30 astronomical units. If planetary orbits are possible under these conditions, what would life be like with two suns in the sky, or one during the "day" and one during the "night"? Could life have developed without regular variations between night and day? Would water not boil off into space with two suns in the sky?

We must look further afield, for a star which is single and within classes F2 to K1.

The next nearest candidate is Epsilon Eridani, 10.7 light years distant. It has 32% of the Sun's luminosity. Considering all factors, the probability that Epsilon Eridani has a habitable planet is 3.3% – one out of 30 similar stars will have a planet with life on it. Another single-star candidate is Tau Ceti, 11.94 light years distant. Its luminosity is 53% that of the Sun and the probability of a life-supporting planet is 3.6%. In both cases, the planet would have to be considerably closer to the star so that water can exist in its three states. These two stars do not have the complicating factor of gravitational perturbations, and they will actually reside longer in the Main Sequence of the H–R diagram and thus have many more thousands of millions of years to develop life-supporting planets. To-and-fro communication will take some 20 years.

All other possible candidates are much further away. Delta Pavonis (20 08.7–66 11) is a G5 IV star, 18.6 light years distant. Its absolute magnitude of 4.8 makes it a replica of the Sun. S.H. Dole finds a 5.7% probability that it has a habitable planet – the highest probability of all stars within 22 light years. A radio signal to-and-fro would take 37.2 years.

The age of this star may vary greatly from the Sun's 5 thousand million years. It may be 5 million years younger. Its habitable planet will need another 5 million years to catch up on Earth. If that planet is 5 million years older than the Earth, its technological civilization would have flourished 5 million years ago. How could they have made contact with Earth-dwellers of 5 million years ago?

Thus time and distance are functions of each other.

We could represent time as a straight line subdivided into regular intervals. If we draw a straight line 50 km long, for example from Hyde Park Corner in London to Chatham, or Washington to Baltimore, to represent 5 thousand million years (the lifetime of the Sun), then 1 mm represents 100 years. Life on Earth has been in existence for 3000 million years, that is the last 30 km of our straight line. The 600 million years of vertebrate life is represented by the last 6 km. Pharaoh Cheops lived 4800 years ago, 48 mm from the end of the line. The last two hundred years of our technological civilisation is represented by the last 2 mm.

Figure 14.6 represents approximately the last 12 500 years of the life lines of five hypothetical life-supporting planets. The distances between the lines, namely 1 cm, represent 10 light years.

The endings of the lines do not correspond, because the various planets would not have been born simultaneously, nor would the speeds of their development be equal.

The endings of lines A and B differ by 1 cm, namely 1000 years.

A and C differ by 2 mm, that is 200 years.

A and F do synchronise in time, but they are 50 light years removed from each other. A radio signal from A to F and back will take 100 years. Care to listen?

If a space ship is to travel from A to C (Earth to Delta Pavonis, about 20 light years distant, at one-quarter of the speed of light, 75 000 km s^{-1}) the trip will take $20 \div \frac{1}{4} = 80$ years. Care to come? Actually, the trip will take longer, because of the time required for acceleration and braking at the end of the trip.

It will be dangerous to travel at more than one-quarter of the speed of light. If a spacecraft travels at one-half of the speed of light, the radar waves which it transmits will reach the ship, after reflection from a body on a direct collision course, at the same time as the crash takes place! The spacecraft will not have time to take evasive action.

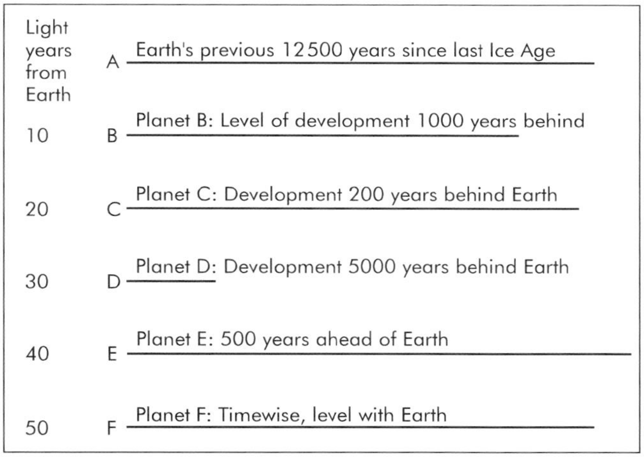

Think of the engines required to boost the ship to 75 000 km s^{-1} with men, fuel and food on board!

Manned travel between stars therefore seems improbable, but perhaps not impossible.

If a space colony, say in a few thousand years, develops to the extent that it becomes independent of the Earth for supplies, and most importantly, independent of the Sun for energy, then such a space colony could undertake an interstellar journey, even if it takes eighty years or more. Since they will not need the Sun for energy, the people could travel at a more leisurely speed. They will undertake the trip with the knowledge that they never will, nor will they want to, return to Earth. At the end of the trip, no one will have any knowledge of the Earth.

But who knows, if Man flourishes for as long as the dinosaurs did – 150 million years – what technology he would not have developed by then to conquer space and time!

Figure 14.6 Ends of 50-km long lines to represent the levels of development of our nearest "imaginary neighbours"

Light years from Earth		
	A	Earth's previous 12 500 years since last Ice Age
10	B	Planet B: Level of development 1000 years behind
20	C	Planet C: Development 200 years behind Earth
30	D	Planet D: Development 5000 years behind Earth
40	E	Planet E: 500 years ahead of Earth
50	F	Planet F: Timewise, level with Earth

Appendices

1	Mathematical Principles	273
	1.1 Trigonometrical Relationships	273
	1.2 Sine Rule	274
	1.3 Area of a Triangle	274
	1.4 Similar Triangles	274
	1.5 Pythagoras' Theorem	275
	1.6 Cosine Formula	275
	1.7 Trigonometric Relationships of Angles greater than 90°	276
	1.8 Circular Measure	276
	1.9 Area of Triangle in Circular Measure	276
	1.10 Logarithms	276
	1.11 Coefficient of Correlation	277
	1.12 Mass–Luminosity Law: Coefficient of Correlation between log(mass) and log(luminosity)	278
	1.13 Mass–Luminosity Law: Equation of Curve	280
	1.14 Equation to Straight Line of Best Fit	281
2	The 88 Constellations	284
3	Stars for the Hertzsprung–Russell Diagram (page 169)	286
4	Masses and Luminosities of Double Stars	292
5	Galactic or Open Clusters	294
6	Selected Eclipsing Binaries	296
	6.1 Algol Type	296
	6.2 Beta Lyrae Type	296
	6.3 W Ursae Majoris Type	297
	6.4 Other Types	297
7	Selected Variable Stars	298
	7.1 Delta Cephei Type	298
	7.2 RR Lyrae Type	298
	7.3 Delta Scuti Type	299
	7.4 W Virginis Type	299
	7.5 Long-Period Type	299
	7.6 Semi-Regular Type	300
	7.7 Irregular Type	300
	7.8 RW Aurigae Type	300
	7.9 T Orionis Type	300
	7.10 T Tauri Type	301
	7.11 R Coronae Borealis Type	301
	7.12 RV Tauri Type	301
	7.13 Beta Cephei Type	301
	7.14 UV Ceti Type	301
	7.15 K ng Type	301
	7.16 Symbiotic Type	302
	7.17 Cataclysmic (U Geminorum) Type	302
	7.18 Novae	303
	7.19 Pulsars	303
	7.20 Planetary Nebulae	304
8	Nebulae	
	8.1 Nebulae in the Milky Way	305
	8.2 Molecules in Interstellar Nebulae	306
9	Globular Clusters (for Figures 12.20, 12.21 and 12.22, pages 234, 235)	307
10	Bright Galaxies	311

Appendix 1
Mathematical Principles

1.1 Trigonometrical Relationships

Let A be any angle. Mark points B, C, D at random. Draw the perpendiculars BE, CF, DG. Measure the lengths BE, CF, DG; AB, AC, AD; and AE, AF, AG.

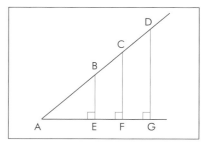

It will be seen that the ratios of any two corresponding sides are constant:

(a) $\dfrac{BE}{AB} = \dfrac{CF}{AC} = \dfrac{DG}{AD} =$ a constant. This constant is called the sine of angle A, namely sin A.

(b) $\dfrac{AE}{AB} = \dfrac{AF}{AC} = \dfrac{AG}{AD} =$ a constant, called the cosine of angle A, namely cos A.

(c) $\dfrac{BE}{AE} = \dfrac{CF}{AF} = \dfrac{DG}{AG} =$ a constant, called the tangent of angle A, namely tan A.

(d) The inverse of the tangent is called the cotangent, namely cot A.

(e) The inverse of the cosine is called the secant, namely sec A.

(f) The inverse of the sine is called the cosecant of angle A, namely cosec A.

THE values of these relationships have been calculated for angles of all sizes, and listed in tables. The values can be obtained directly from a scientific pocket calculator.

What is the value of sin 30°?
Press 3, 0, sin; the answer 0.5 will appear immediately.
Cos 30°? Press 3, 0, cos; the answer 0.86602541 appears.
Tan 30°? Press 3, 0 tan; answer: 0.57735027.
Cosec 30° = 1 ÷ sin 30 = 1 ÷ 0.5 = 2.
Sec 30° = 1 ÷ cos 30° = 1 ÷ 0.86602541 = 1.1547005.
Cot 30° = 1 ÷ tan 30° = 1 ÷ 0.57735027 = 1.7320508.
These inverses can be found simply by pressing the key 1/x.

TO find the cosine of Aristarchus' angle of 89°51′, the 51 minutes must firstly be expressed in degrees, by pressing 51 ÷ 60: Thus: 5, 1, ÷, 6, 0 = 0.85. Press +, 8, 9 = 89.85. Press cos. Answer: 0.00261799. To find the inverse, press 1/x. Answer: 381.97229 = 382, very nearly. The Sun is therefore 382 times as far away as the Moon.

1.2 Sine Rule

ABC is any triangle. Draw AD perpendicular to BC. Then:

$\sin B = \dfrac{AD}{AB}$ and

$\sin C = \dfrac{AD}{AC}$

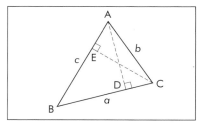

Therefore $\dfrac{\sin B}{\sin C} = \dfrac{AD}{AB} \div \dfrac{AD}{AC} = \dfrac{AC}{AB} = \dfrac{b}{c}$

or $\dfrac{\sin B}{b} = \dfrac{\sin C}{c}$

Draw CE perpendicular to BA, then:

$\sin A = \dfrac{EC}{AC}$ and $\sin B = \dfrac{EC}{BC}$

Therefore $\dfrac{\sin B}{\sin A} = \dfrac{EC}{BC} \div \dfrac{EC}{AC} = \dfrac{AC}{BC} = \dfrac{b}{a}$

Therefore $\dfrac{\sin B}{b} = \dfrac{\sin A}{a}$ and also $= \dfrac{\sin C}{c}$

1.3 Area of a Triangle

The area of a triangle is half the area of the rectangle on the same base and having the same vertical height.

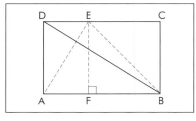

The area of the rectangle ABCD = AB × AD = Area of triangle ADB + Area of triangle DCB. Since the diagonal DB divides rectangle ABCD into two equal halves, the area of triangle ADB = area of triangle DCB.

∴ Area of rectangle ABCD = twice the area of triangle ADB and also twice the area of triangle DCB.

Also: the area of rectangle ABCD = twice the area of triangle ABE, because triangle ABE = triangle AFE + triangle FBE = ½ rectangle AFED + ½ rectangle FBCE = ½ rectangle ABCD.

Therefore: Area of triangle ABE $= \dfrac{AB \times AD}{2}$

The area of a triangle is thus equal to one half of its base times its perpendicular height.

1.4 Similar Triangles

Triangles are similar when the angles opposite corresponding sides are equal, for example the triangles in Appendix 1.1.

Triangles ADE and ABC are similar, because (i) angle A is common to both triangles; (ii) angle ADE = angle ABC because DE is parallel to BC and are cut by the straight line AB; (iii) angle AED = angle ACB, because the parallel lines DE, BC are cut by straight line AC.

$\dfrac{\text{Area of triangle ADE}}{\text{Area of triangle BDE}}$

$= \dfrac{\frac{1}{2} AD \times FE}{\frac{1}{2} BD \times FE} = \dfrac{AD}{BD}$

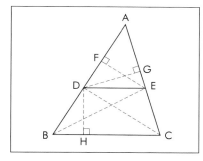

Similarly:

$\dfrac{\text{Area of triangle ADE}}{\text{Area of triangle ECD}} = \dfrac{\frac{1}{2} AE \times DG}{\frac{1}{2} EC \times DG} = \dfrac{AE}{EC}$

But the areas of triangles BDE, ECD are equal, because they have the same base DE, and the same perpendicular height DH.

$\therefore \dfrac{\text{Area } \triangle ADE}{\text{Area } \triangle BDE} = \dfrac{AD}{BD} = \dfrac{\text{Area } \triangle ADE}{\text{Area } \triangle ECD} = \dfrac{AE}{EC}$

$\therefore \dfrac{AD}{BD} = \dfrac{AE}{EC}$ and inversely $\dfrac{BD}{AD} = \dfrac{EC}{AE}$

Add 1 to each side, $\therefore \dfrac{BD}{AD} + \dfrac{1}{1} = \dfrac{EC}{AE} + \dfrac{1}{1}$

Therefore $\dfrac{BD + AD}{AD} = \dfrac{EC + AE}{AE}$

so that $\dfrac{AB}{AD} = \dfrac{AC}{AE}$ or $\dfrac{AB}{AC} = \dfrac{AD}{AE}$

Triangles that are equiangular are called similar and their corresponding sides are proportional to each other.

1.5 Pythagoras' Theorem

In the plane triangle ABC, right-angled at A, draw AD perpendicular to BC. Triangles ABD, CBA have angle B common and angle ADB = angle CAB (each = 90°). Therefore the third angle BAD = the third angle ACD, because the sum of the three angles of a triangle = 180°. Therefore triangles ABD, CBA are similar.

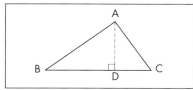

$\therefore \dfrac{AB}{BC} = \dfrac{BD}{AB}$ and $\therefore AB^2 = BC \times BD$

Triangles CAD, CBA are also similar.

$\therefore \dfrac{AC}{BC} = \dfrac{CD}{AC}$ and $\therefore AC^2 = BC \times CD$

Add AB^2 to AC^2:

$\therefore AB^2 + AC^2 = BC \times BD + BC \times CD$
$= BC(BD + CD) = BC(BC)$

i.e. $AB^2 + AC^2 = BC^2$.

Pythagoras' Theorem therefore states:

> In any right-angled, plane triangle, the square on the hypotenuse is equal to the sum of the squares on the other two sides.

The Egyptian Priests (±3500 BC) knew that a triangle with sides 3, 4 and 5 units of length had a right angle opposite the longest side:

$5^2 = 25$ and $3^2 + 4^2 = 9 + 16 = 25$

The angle opposite the side of 5 units long is a right angle. Builders still use this relationship when setting out buildings.

1.6 Cosine Formula

ABC is any plane triangle, with AD perpendicular to BC. According to Pythagoras' theorem:

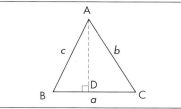

$AB^2 = BD^2 + AD^2$
$= (BC - CD)^2 + AD^2$
$= BC^2 - 2BC \times CD + CD^2 + AD^2$

i.e. $c^2 = a^2 - 2a(CD) + b^2$

But $\dfrac{CD}{b} = \cos C, \therefore CD = b \cos C$

$\therefore c^2 = a^2 - 2a(b \cos C) + b^2$

i.e. $c^2 = a^2 + b^2 - 2ab \cos C$

Similarly:

$a^2 = b^2 + c^2 - 2bc \cos A$, and
$b^2 = a^2 + c^2 - 2ac \cos B$

1.7 Trigonometric Relationships of Angles greater than 90°

If x is an acute angle, then angle MOQ = 180° − x. Triangles OMP, ORQ are congruent. Thus:

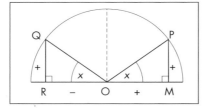

$$\sin \text{MOQ} = \sin(180° - x) = \frac{RQ}{OQ} = \frac{MP}{OP} = \sin x$$

$$\cos \text{MOQ} = \cos(180° - x) = \frac{OR}{OQ} = \frac{-OM}{OP} = -\cos x$$

$$\tan \text{MOQ} = \tan(180° - x) = \frac{RQ}{OR} = \frac{MP}{-OM} = -\tan x$$

1.8 Circular Measure

The diameter of a circle divides a constant number of times into the circumference of a circle. This constant is called pi (π). That is
  Circumference = π times the diameter D
360° correspond to πD.
The radius, r, is half the diameter.
Therefore 180° correspond to πr.
  The value of π is 3.141 592 65 ... . This is very nearly equal to 355 ÷ 113, which equals 3.141 592 9. Roughly, π can be taken as = 22 ÷ 7. We shall use 3.141 59.
  The angle subtended by the radius r, marked off along the circumference of a circle, at the centre of the circle, that is angle AOB, BOC or COD, is called a radian (abbreviated to rad).

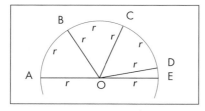

Thus : π radians = 180°

$$\therefore 1 \text{ radian} = \frac{180°}{\pi} = \frac{180°}{3.141\ 59}$$
$$= 57.295\ 828° = 57°17'45''$$

$$1° = \frac{\pi}{180} \text{ radians, and } 15° = \frac{15\pi}{180} \text{ radians}$$

The very small angles that are measured in astronomy are usually expressed in circular measure, for example:

An angle of 57 minutes $(57') = \left(\frac{57}{60}\right)°$

$$= \frac{57}{60} \times \frac{\pi}{180} \text{ rad} = \frac{57(3.141\ 59)}{60(180)}$$
$$= 0.016\ 58 \text{ rad}$$

The angle of 57′ is very small, being less than 1 degree. The sine of such a small angle is equal to the size of the angle in circular measure (radians). For example sin 57′ by means of a pocket calculator:
Press 5, 7, ÷, 6, 0, = (= 0.95); press sin (= 0.016 579 87), which can be rounded off to 0.016 58, which is equal to the size of the angle in radians, as we found above.
  When the angle is only seconds in size, the rule holds even more strongly:

An angle of 1 second (= 1″)
$$= \frac{1}{3600} \times \frac{3.141\ 59}{180} = \frac{1}{206\ 265} \text{ rad}$$

Thus there are 206 265 seconds in 1 radian. This value, 206 265, will often appear.
  Even if the angle is as large as 2°, the sine of the angle equals the size of the angle in radians.

$$2° = \frac{2(3600)(3.141\ 59)}{180(3600)} = 0.034\ 91$$

$$\sin 2° = 0.034\ 90 \text{ rad}$$

1.9 Area of Triangle in Circular Measure

If angle S is very small, angles A and B can be considered as right angles.

$$\therefore \text{Area } \triangle \text{SAB} = \frac{1}{2}(\text{SA})(\text{AB}) = \frac{1}{2}(\text{SA})(\text{AB})\frac{(\text{AB})}{(\text{AB})}$$

$$= \frac{1}{2}(\text{AB})^2 \times \frac{(\text{SA})}{(\text{AB})} = \frac{1}{2}(\text{AB})^2 \div \frac{(\text{AB})}{(\text{SA})} = \frac{1}{2}(\text{AB})^2 \div \sin \text{S}.$$

Both AB and angle S are measured in circular measure.

1.10 Logarithms

(a) The logarithm of a number is the power to which the

base 10 must be raised to equal that number, $1000 = 10^3$. Thus the log of 1000 is 3.

The log of 100 is 2 because $100 = 10^2$.
The log of 10 is 1, because $10 = 10^1$.

Number	As power of 10	Log
1000	10^3	3
100	10^2	2
10	10^1	1
1	10^0	0
0.1	10^{-1}	−1
0.01	10^{-2}	−2
0.001	10^{-3}	−3

Therefore the log of 1 must be 0, in other words $1 = 10^0$. 0.1 is $1 \div 10$, i.e. $\frac{1}{10} = 10^{-1}$. Therefore the log of 0.1 is −1 and therefore the log of 0.01 must be −2, and so on.

The log of a number between 100 and 1000 must therefore be 2 point "something". The "something" can be found from Log tables, or by means of a calculator. Find the log of 365. Press 3, 6, 5, log. The answer: 2.5622929. Four-figure log tables give the log of 365 as 2.5623, rounded off to four decimals.

If the log is known, the number can be found from tables of antilogs, or from a pocket calculator: feed in the log and press INV, or 2nd, according to the make.
(b) When two numbers are multiplied, the logs must be added, for example

$$100 \times 1000 = 10^2 \times 10^3 = 10^{(2+3)} = 10^5$$

that is, the logs of 2 and 3 are added.

Work out: 17.65×0.7052.
17.65 has 2 digits before the decimal. Therefore its log is 1 point "something". Tables give .2467 – this is the mantissa. Therefore the log of 17.65 is 1.2467.
0.7052 has no zeros after the decimal, therefore its characteristic is −1. The mantissa is .8483. Therefore the log of 0.7052 is −1.8483, or $\bar{1}.8483$.

Adding these logs, we get 1.0950. The antilog of .0950 (from tables) is 1245. Since the characteristic is 1, there must be two digits before the decimal, giving 12.45.

By calculator: press 1, 7, point, 6, 5, log: 1.2467447 appears. The calculator gives the log of 0.7052 as −0.1516877, that is, it has subtracted 1 from the mantissa. Therefore subtract .1516877 from 1.2467447 giving 1.090570. Press INV or 2nd: the answer 12.44678 appears. This is more accurate than the 12.45 of the tables, but when rounded off the values agree.
(c) When m has to be divided by n, the log of n is subtracted from the log of m:

$$1000 \div 100 = 10^3 \div 10^2 = 10^{(3-2)} = 10^1 = 10.$$

(d) If a number is raised to a power, the log of the number must be multiplied by the power, for example:

$$(10^3)^2 = (1000)^2 = 1\,000\,000 = 10^6 = 10^{3\times 2}.$$

On a calculator the power key y^x can be used. For example, what is the value of $(13.6)^2$? Press 1, 3, point, 6, y^x, 2, = which gives 184.96.
(e) When a root has to be found, the log of the number is divided by the root. For example, find the cube root of 3375.

Let $x = \sqrt[3]{3375} = (3375)^{1/3}$

$$\therefore \log x = \frac{1}{3} \log 3375 = \frac{1}{3}(3.528\,27) = 1.176\,09$$

$$\therefore x = \text{antilog } 1.176\,09 = 15$$

1.11 Coefficient of Correlation

Three possible distributions of points having X and Y values are shown in the figure. If $\bar{X}$, the mean of X, and $\bar{Y}$, the mean of Y, are subtracted from each of the X and Y values, we obtain $X - \bar{X} = x$ and $Y - \bar{Y} = y$. The co-ordinates of points with respect to new axes through $\bar{X}$ and $\bar{Y}$, the means of X and Y, will be x and y.

In **a** the points fall preponderantly in quadrants I and III. In quadrant I, the x and y values are positive; and in

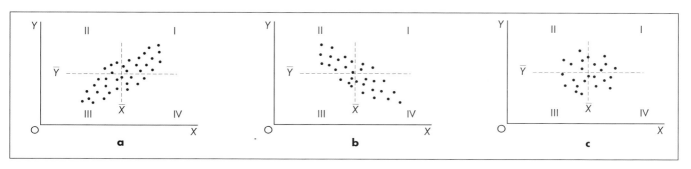

III the x and y values are negative. In both quadrants the product xy will be positive, and the sum of the products $\Sigma xy > 0$. Such a distribution is said to have positive correlation between X and Y, because high values of X correspond to high values of Y, and low values of X have low values of Y.

In **b** most of the points are in quadrants II and IV. In both quadrants the signs of x and y are opposite, so that the products xy are negative and the sum of the products $\Sigma xy < 0$. Such a distribution has negative correlation between X and Y.

In **c** the points are evenly distributed between the four quadrants, so that the positive products of xy in I and III are balanced by the negative products in II and IV. Thus the sum of the products $\Sigma xy = 0$, or close to zero. There is thus no correlation between X and Y.

The mean of the squares of the x values is called the variance in x: $\Sigma x^2/n$, where n is the number of measures. The square root of the variance is the standard deviation: $\sqrt{\Sigma x^2/n} = S_x$. The standard deviation in y is $\sqrt{\Sigma y^2/n} = S_y$.

If each x value is divided by S_x and each y value by S_y, the quotients $x \div S_x$ and $y \div S_y$ will be pure numbers, because the units of measurements cancel out.

When these quotients are multiplied, we obtain the products:

$$\frac{x_1}{S_x}, \frac{y_1}{S_y}; \frac{x_2}{S_x}, \frac{y_2}{S_y}; \text{ and so on}$$

In distribution **a**, the sum of these products is positive; in **b** negative; and in **c** = 0.

The sum of these products, divided by the number of measures n is defined as the Coefficient of Correlation r.

$$\text{Thus } r = \frac{1}{n}\left\{\frac{x_1 y_1}{S_x S_y} + \frac{x_2 y_2}{S_x S_y} + \ldots, \frac{x_n y_n}{S_x S_y}\right\}$$

i.e. $\quad r = \dfrac{1}{n}\sum \dfrac{xy}{S_x S_y} = \dfrac{\Sigma xy}{n S_x S_y}$

Substitute $S_x = \sqrt{\Sigma x^2/n}$ and $S_y = \sqrt{\Sigma y^2/n}$.

$$\therefore r = \frac{1}{n}\frac{\Sigma xy}{(\sqrt{\Sigma x^2/n})(\sqrt{(\Sigma y^2/n)})}$$

n cancels out, so that:

$$r = \frac{\Sigma xy}{\sqrt{\Sigma x^2 \times \Sigma y^2}}$$

The coefficient of correlation, r can never be more than +1 nor less than −1. When $r = +1$, it indicates high positive correlation; $r = -1$, high negative correlation. $r = 0$ indicates no correlation.

1.12 Mass–Luminosity Law: Coefficient of Correlation between log(mass) and log(luminosity) (Stars of Figure 10.36 and Appendix 1.13)

From Table on Page 279

Coefficient of correlation $r = \dfrac{\Sigma xy}{\sqrt{\Sigma x^2 \times \Sigma y^2}} = \dfrac{37.36}{\sqrt{(11.91)(120.54)}} = 0.98$.

Slope of straight line of best fit $= \Sigma xy / \Sigma x^2 = 37.36 \div 11.91 = 3.14$
$\qquad = b$ (in eqn, $y = a + bx$).

Standard deviation in Y: $S_y = \sqrt{\Sigma y^2/n} = \sqrt{120.54 \div 31} = 1.97$.

Standard deviation Y on X: $S_{y.x} = S_y\sqrt{1 - r^2} = 1.97\sqrt{1 - 0.98^2} = 0.392$.

95% confidence limits $= 2 S_{y.x} = 2(0.392) = 0.8$ (rounded off).

Standard deviation in X: $S_x = \sqrt{\Sigma x^2/n} = \sqrt{(11.91) \div 31} = 0.62$.

Standard deviation X on Y: $S_{x.y} = S_x\sqrt{1 - r^2} = 0.62\sqrt{1 - 0.98^2} = 0.123$.

95% confidence limits $= 2 S_{x.y} = 2(0.123) = 0.25$ (rounded off).

Appendix 1.12: Mass–Luminosity Law

Star	log(mass) = X	log(luminosity) = Y	$X - \bar{X}$ = X − 0.16 = x	$Y - \bar{Y}$ = Y − 0.68 = y	xy	x^2	y^2
V 478 Cyg	1.15	3.92	0.99	3.24	3.21	0.98	10.50
Y Cyg	1.24	3.76	1.08	3.08	3.33	1.17	9.49
AH Cep	1.18	3.72	1.02	3.04	3.10	1.04	9.24
V 453 Cyg	1.20	3.60	1.04	2.92	3.04	1.08	8.53
CW Cep	1.00	3.11	0.84	2.43	2.04	.71	5.90
AG Per	.68	2.80	.52	2.12	1.10	.27	4.49
U Oph	.70	2.68	.54	2.00	1.08	.29	4.00
σ Aql	.78	2.43	.62	1.75	1.09	.38	3.06
DI Her	.54	2.43	.38	1.75	.67	.14	3.06
β Aur	.36	1.88	.20	1.20	.24	.04	1.44
AR Aur	.40	1.68	.24	1.00	.24	.06	1.00
RX Her	.30	1.68	.14	1.00	.14	.02	1.00
W W Aur	.26	1.20	.10	.52	.05	.01	.27
TX Her	.28	1.15	.12	.47	.06	.01	.22
ZZ Boo	.23	1.04	.07	.36	.03	.00	.13
VZ Hya	.06	.52	−.10	−.16	.02	.01	.03
YY Gem	−.19	−1.20	−.35	−1.88	.66	.12	3.53
α Cen A	.04	.15	−.12	−.53	.06	.01	.28
α Cen B	−.06	−.32	−.22	−1.00	.22	.05	1.00
η Cas A	−.03	.11	−.19	−.57	.11	.04	.32
η Cas B	−.24	−1.08	−.40	−1.76	.70	.16	3.10
ξ Boo A	−.07	−.24	−.23	−.92	.21	.05	.85
ξ Boo B	−.12	−.77	−.28	−1.45	.41	.08	2.10
70 Oph A	−.05	−.32	−.21	−1.00	.21	.04	1.00
70 Oph B	−.19	−.80	−.35	−1.48	.52	.12	2.19
Kru 60 A	−.57	−1.70	−.73	−2.38	1.74	.53	5.66
Kru 60 B	−.77	−2.08	−.93	−2.76	2.57	.86	7.62
Fu 46 A	−.51	−1.55	−.67	−2.23	1.49	.45	4.97
Fu 46 B	−.60	−1.60	−.76	−2.28	1.73	.58	5.20
Ross 614 A	−.85	−2.08	−1.01	−2.76	2.79	1.02	7.62
Ross 614 B	−1.10	−2.89	−1.26	−3.57	4.50	1.59	12.74
n = 31	+10.40	+ 37.86	+ 7.90	+ 26.88	+ 37.36	+ 11.91	+ 120.54
	− 5.35	− 16.63	− 7.81	− 26.73	= Σxy	= Σx^2	= Σy^2
	31) 5.05	31) 21.23	0.09	0.15	(Corrections: $\bar{X}$ (0.0029) and $\bar{Y}$ (0.0048)		
	0.1629	0.6848	= Σx	= Σy	do not affect the second decimal place)		
	Use $\bar{X}$ = 0.16	$\bar{Y}$ = 0.68					

1.13 Mass–Luminosity Law: Equation to the cubic curve of best fit for the data plotted on the graph of log(luminosity) against log(mass). (Figure 10.36, page 189)

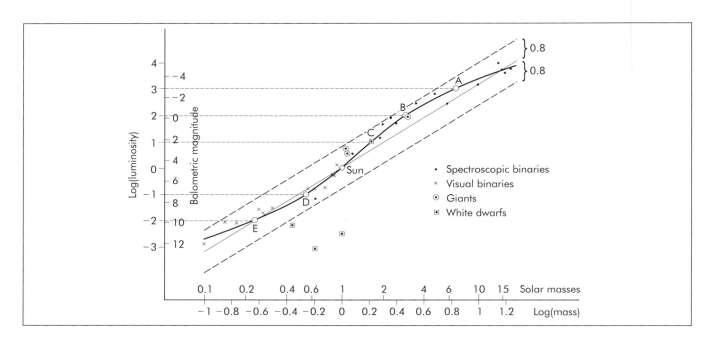

MASS–LUMINOSITY LAW: GRAPH OF LOG(LUMINOSITY) AGAINST LOG(MASS)

The co-ordinates of the points A, B, C, D and E on the cubic graph of log(luminosity) against log(mass) are tabulated below, and the equation to the cubic is derived therefrom.

Points	log(luminosity) = L	log(mass) = M
A	3	0.83
B	2	0.47
C	1	0.22
D	−1	−0.26
E	−2	−0.63

Equation of the cubic curve:

$$aL^3 + bL^2 + cL = M$$

$$27a + 9b + 3c = 0.83 \quad (\times 2): 54a + 18b + 6c = 1.66$$
$$8a + 4b + 2c = 0.47 \quad (\times 3): \underline{24a + 12b + 6c = 1.41}$$
$$a + b + c = 0.22 \quad \therefore\ 30a + 6b = 0.25$$
$$a + b - c = -0.26$$
$$8a + 4b - 2c = -0.63$$
$$\underline{8a + 4b + 2c = 0.47}$$
$$8b = -0.16$$
$$\therefore b = -0.02 \quad \therefore\ 30a + 6(-0.02) = 0.25$$
$$ \quad \therefore\ a = 0.37/30 = 0.012$$

$$a + b + c = 0.22$$
$$\therefore 0.012 - 0.02 + c = 0.22$$
$$\therefore c = 0.228$$

Therefore the equation to the cubic curve is

$$0.012L^3 - 0.02L^2 + 0.228L = M$$

Multiply by 250: $\therefore 3L^3 - 5L^2 + 57L = 250M$ (= equation used on page 189)

1.14 Equation to Straight Line of Best Fit to the plotted data on a graph

Imaginary data, to facilitate the arithmetic, are tabulated in the following table:

Abs. mag. $M = y$	Period P (days)	$\log P = \ell$	$\ell - 0.7 = x$
−2.8	3.16	0.5	−0.2
−3.6	3.98	0.6	−0.1
−3.4	5.01	0.7	0.0
−4.2	6.31	0.8	0.1
−4.0	7.94	0.9	0.2
$-18.0 = \Sigma y$		5)3.5 Mean: $\bar{\ell} = 0.7$	$0.0 = \Sigma x$

The data, absolute magnitude against logs of the period, are plotted in the graph. The points show a steady rise. A straight line that best fits the data has been drawn by eye. We wish to determine the equation to this straight line.

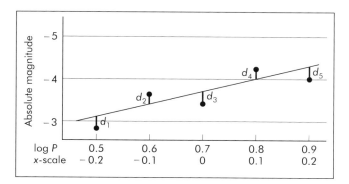

The sum of the l-column in the table is 3.5 and the mean $= \bar{l} = 0.7$. In the fourth column, 0.7 is subtracted from each of the l-values, to obtain the x-scale. This has the same effect as moving the Y-axis to pass through the mean of the l-values (= 0.7).

To determine the equation of a straight line, the values of a and b in the equation $y = a + bx$ must be found.

The line of best fit is the straight line for which the sum of the deviations of the points from the line, d_1, d_2 ... will be zero. But any straight line through the mean of x and y will have deviations that total 0. Therefore, a second stipulation is required: there is only one line for which the *sum of the squares of the deviations will be a minimum*. This will be the line of best fit.

A function is a minimum when the slope of the tangent to the curve is zero, that is when the tangent is horizontal, as at M in the diagram, which represents any function, such as $f(x) = x^2 - 4x + 5$.

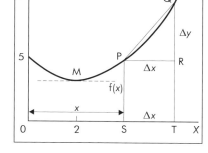

By definition, the slope of the secant PQ is equal to

$$\frac{\text{Delta } y}{\text{Delta } x} = \frac{\Delta y}{\Delta x} = \frac{RQ}{PR}$$

$$\frac{RQ}{PR} = \frac{TQ - TR}{\Delta x} = \frac{TQ - SP}{\Delta x} = \frac{f(x + \Delta x) - f(x)}{\Delta x}$$

$$\therefore \frac{\Delta y}{\Delta x} = \frac{\{(x + \Delta x)^2 - 4(x + \Delta x) + 5\} - (x^2 - 4x + 5)}{\Delta x}$$

$$= \frac{x^2 + 2x\Delta x + (\Delta x)^2 - 4x - 4\Delta x + 5 - x^2 + 4x - 5}{\Delta x}$$

$$= \frac{2x\Delta x + (\Delta x)^2 - 4\Delta x}{\Delta x} = 2x + \Delta x - 4$$

The slope of PQ is therefore equal to $2x + \Delta x - 4$. As Q comes closer to P, the secant PQ tends to become the tangent at P. This will be so when Δx becomes infinitesimally small, or equal to zero, and can thus be neglected without loss of accuracy.

The slope of the tangent at P is thus $2x - 4$. This is known as the derived function or the differential of the function $x^2 - 4x + 5$; it gives the slope at any point. When $x = 2$, the value of $2x - 4$ is zero. The slope is then zero and the tangent is horizontal, as at point M. The function

$x^2 - 4x + 5$ is a minimum at that point.

In general, the derived function of the function ax^n is equal to $n \times ax^{n-1}$.

Line of Best Fit The equation of the straight line of best fit is $y = a + bx$. The deviations $d_1, d_2, \ldots$ are each equal to $y - (a + bx)$. The square of each deviation is $\{y - (a + bx)\}^2$. The sum of the squares of all the deviations is given by $\Sigma\{y - (a + bx)\}^2$. This is equal to:

$$\Sigma(y^2 - 2ya - 2xyb + a^2 + 2xab + b^2x^2)$$

Considering a as the variable and b, x and y as constants, the derived function of the function $\Sigma(-2ya + a^2 + 2xab)$ is $\Sigma(-2y + 2a + 2bx)$, and this must be equal to zero in order to be a minimum.

$$\therefore -2\Sigma(y - a - bx) = 0$$

so that $\Sigma y - na - b\Sigma x = 0 \quad (1)$

When b is taken as the variable, the derived function of $\Sigma(-2xyb + 2xab + b^2x^2)$ is $\Sigma(-2xy + 2ax + 2bx^2)$ and this must be equated to zero for a minimum.

$$\therefore -2\Sigma(xy - ax - bx^2) = 0$$

Therefore $\Sigma xy - a\Sigma x - b\Sigma x^2 = 0(2) \quad (2)$

If $\Sigma x = 0$, as in the table, equation (1) becomes: $\Sigma y - na - b(0) = 0$, so that $na = \Sigma y$. Therefore

$$a = \Sigma y / n \quad (3)$$

Equation (2) becomes $\Sigma xy - a(0) - b\Sigma x^2 = 0$, so that

$$b = \Sigma xy / \Sigma x^2 \quad (4)$$

Equations (3) and (4) give the values of a and b in terms of the number of terms n; the sum of the y terms Σy; the sum of the products Σxy; and the sum of the squares of the x terms Σx^2. For these values of a and b, the sum of the squares of the deviations will be a minimum.

Therefore, extend the table by adding columns of the products xy and the squares of x.

By substituting x values $-0.2, -0.1, \ldots + 0.2$ in the equation $y = -3.6 - 3x$, the y values on the line are obtained. These are listed as c. If the c values are subtracted from the y values in the first column, the deviations of the plotted points from the line, $d_1, d_2 \ldots$ are obtained. The sum of these deviations, $\Sigma d = 0$, as it should be, if the line of fit is correct.

By squaring the deviations and summing them, we obtain $\Sigma d^2 = 0.30$, which is the minimum value of the sum of the squares of the deviations.

The mean of the squares of the deviations, $\Sigma d^2/n$, is the variance.

The standard deviation, S_y, is the square root of the variance, that is

$$S_y = \sqrt{\Sigma d^2/n} = \sqrt{0.3/5} = \sqrt{0.06} = 0.2449$$

In a normal distribution, 95% of the data will fall within the limits $y \pm 2S_y$, that is $y \pm 2(0.2449)$, or $y \pm 0.5$, rounded off. These limits ± 0.5 are indicated by the broken lines, parallel to the straight line of best fit AB.

To draw the straight line of best fit accurately, substitute the x values -0.2 and $+0.2$ into the equation $y = -3.6 - 3x$, to obtain the ordinates of $A \, (= -3.0)$ and $B \, (= -4.2)$.

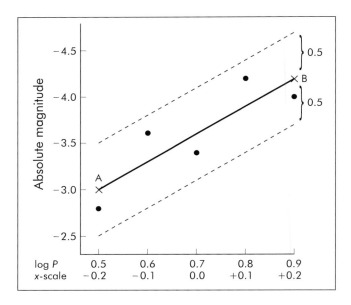

The equation to the straight line that best fits the data, for the Cepheid variables, dealt with in Chapter 12, page 225, was derived in the same way as that outlined here. The equation thus obtained

$$M = -1.48 - 2.70 \, \log \, P \pm 0.32$$

can be used for all Cepheids having periods between 2 and 14 days.

Appendix 1.14

Equation to straight line of best fit by the method of least squares							95% confidence limits		
Abs. mag. $M = y$	Period P (days)	$\log P = \ell$	$\ell - \bar{\ell} = x$	xy	x^2	$y = (a + bx)$ $= c$	$y - c$ $= d$	d^2
−2.8	3.16	0.5	−0.2	+0.56	0.04	−3.0	+0.2	0.04
−3.6	3.98	0.6	−0.1	+0.36	0.01	−3.3	−0.3	0.09
−3.4	5.01	0.7	0.0	0.00	0.00	−3.6	+0.2	0.04
−4.2	6.31	0.8	0.1	−0.42	0.01	−3.9	−0.3	0.09
−4.0	7.94	0.9	0.2	−0.80	0.04	−4.2	+0.2	0.04
−18.0 = Σy $n = 5$		5) 3.5 Mean = $\bar{\ell}$ = 0.7	0.0 = Σx	−1.22 +0.92 −0.30 = Σxy	0.10 = Σx²		0.0 = Σd	0.30 = Σd²

$a = \dfrac{\Sigma y}{n} = \dfrac{-18}{5} = -3.6; \quad b = \dfrac{\Sigma xy}{\Sigma x^2} = \dfrac{-0.3}{0.1} = -3.0$

Equation to straight line: $y = a + bx$

$\therefore y = -3.6 - 3x$

Standard deviation = $\sqrt{\Sigma d^2/n} = \sqrt{0.3/5}$
$= \sqrt{0.06} = 0.2449$

95% confidence limits = 2 × Std. dev. = 0.5

Appendix 2

The 88 CONSTELLATIONS (from *Encyclopedia of Astronomy* by Colin Ronan [Hamlyn])

Latin name, Genitive	Abbr.	English	RA	Dec	Latin name, Genitive	Abbr.	English	RA	Dec
Andromeda, Andromedae	And	Andromeda	1	+40	Dorado, Doradus	Dor	Swordfish	5	−65
Antlia, Antliae	Ant	Pump	10	−35	Draco, Draconis	Dra	Dragon	17	+65
Apus, Apodis	Aps	Bird of Paradise	16	−75	Equuleus, Equulei	Equ	Little Horse	21	+10
Aquarius, Aquarii	Aqr	Water Bearer	23	−15	Eridanus, Eridani	Eri	River	3	−20
Aquila, Aquilae	Aql	Eagle	20	+5	Fornax, Fornacis	For	Furnace	3	−30
Ara, Arae	Ara	Altar	17	−55	Gemini, Geminorum	Gem	Twins	7	+20
Aries, Arietis	Ari	Ram	3	+20	Grus, Gruis	Gru	Crane	22	−45
Auriga, Aurigae	Aur	Charioteer	6	+40	Hercules, Herculis	Her	Hercules	17	+30
Boötes, Boötis	Boo	Herdsman	15	+30	Horologium, Horologii	Hor	Clock	3	−60
Caelum, Caeli	Cae	Chisel	5	−40	Hydra, Hydrae	Hya	Water Monster	10	−20
Camelopardalis, Camelopardalis	Cam	Giraffe	6	+70	Hydrus, Hydri	Hyi	Sea Serpent	2	−75
Cancer, Cancri	Cnc	Crab	9	+20	Indus, Indi	Ind	Indian	21	−55
Canes Venatici, Canum Venaticorum	CVn	Hunting Dogs	13	+40	Lacerta, Lacertae	Lac	Lizard	22	+45
Canis Major, Canis Majoris	CMa	Big Dog	7	+20	Leo, Leonis	Leo	Lion	11	+15
Canis Minor, Canis Minoris	CMi	Little Dog	8	+5	Leo Minor, Leonis Minoris	LMi	Little Lion	10	+35
Capricornus, Capricorni	Cap	Goat	21	−20	Lepus, Leporis	Lep	Hare	6	−20
Carina, Carinae	Car	Ship's Keel	9	−60	Libra, Librae	Lib	Scales	15	−15
Cassiopeia, Cassiopeiae	Cas	Cassiopeia	1	+60	Lupus, Lupi	Lup	Wolf	15	−45
Centaurus, Centauri	Cen	Centaur	13	−50	Lynx, Lyncis	Lyn	Lynx	8	+45
Cepheus, Cephei	Cep	Cepheus	22	+70	Lyra, Lyrae	Lyr	Lyre	19	+40
Cetus, Ceti	Cet	Whale	2	−10	Mensa, Mensae	Men	Table	5	−80
Chamaeleon, Chamaeleontis	Cha	Chameleon	11	−80	Microscopium, Microscopii	Mic	Microscope	21	−35
Circinus, Circini	Cir	Compass	15	−60	Monoceros, Monocerotis	Mon	Unicorn	7	−5
Columba, Columbae	Col	Dove	6	−35	Musca, Muscae	Mus	Fly	12	−70
Coma Berenices, Comae Berenices	Com	Berenice's Hair	13	+20	Norma, Normae	Nor	Level	16	−50
Corona Australis, Coronae Australis	CrA	Southern Crown	19	−40	Octans, Octantis	Oct	Octant	22	−85
					Ophiuchus, Ophiuchi	Oph	Serpent Bearer	17	0
Corona Borealis, Coronae Borealis	CrB	Northern Crown	16	+30	Orion, Orionis	Ori	Orion	5	+5
					Pavo, Pavonis	Pav	Peacock	20	−65
Corvus, Corvi	Crv	Crow	12	−20	Pegasus, Pegasi	Peg	Winged Horse	22	+20
Crater, Crateris	Crt	Cup	11	−15	Perseus, Persei	Per	Perseus	3	+45
Crux, Crucis	Cru	Southern Cross	12	−60	Pheonix, Phoenicis	Phe	Phoenix	1	−50
Cygnus, Cygni	Cyg	Swan	21	+40	Pictor, Pictoris	Pic	Easel	6	−55
Delphinus, Delphini	Del	Dolphin	21	+10	Pisces, Piscium	Psc	Fishes	1	+15
					Piscis Austrinus, Piscis Austrini	PsA	Southern Fish	22	−30

Appendix 2: The 88 Constellations

Latin name, Genitive	Abbr.	English	RA	Dec	Latin name, Genitive	Abbr.	English	RA	Dec
Puppis, Puppis	Pup	Ship's Stern	8	−40	Taurus, Tauri	Tau	Bull	4	+15
Pyxis, Pyxidis	Pyx	Ship's Compass	9	−30	Telescopium, Telescopii	Tel	Telescope	19	−50
Reticulum, Reticuli	Ret	Net	4	−60	Triangulum, Trianguli	Tri	Triangle	2	+30
Sagitta, Sagittae	Sge	Arrow	20	+10	Triangulum Australe	TrA	Southern Triangle	16	−65
Sagittarius, Sagittarii	Sgr	Archer	19	−25	Tucana, Tucanae	Tuc	Toucan	0	−65
Scorpius, Scorpii	Sco	Scorpion	17	−40	Ursa Major, Ursae Majoris	UMa	Great Bear	11	+50
Sculptor, Sculptoris	Scl	Scuptor	0	−30	Ursa Minor, Ursae Minoris	UMi	Little Bear	15	+70
Scutum, Scuti	Sct	Shield	19	−10	Vela, Velorum	Vel	Ship's Sails	9	−50
Serpens, Serpentis	Ser	Serpent			Virgo, Virginis	Vir	Virgin	13	0
Caput		Head	16	+10	Volans, Volantis	Vol	Flying Fish	8	−70
Cauda		Tail	18	−5	Vulpecula, Vulpeculae	Vul	Little Fox	20	+25
Sextans, Sextantis	Sex	Sextant	10	0					

Appendix 3

Stars for the Hertzsprung–Russell Diagram (page 104)

(Up to visual magnitude 8.05, from *Sky Catalogue 2000.0* by A. Hirshfeld and R.W. Sinnott; dwarf stars from South African Observatory and other sources.)
Sp.t./C,L = spectral type, class and luminosity; M = absolute magnitude; V = visual magnitude; $B-V$ = blue – visual colour index; pc = distance in parsec; t = distance measured trigonometrically; ts = adjusted according to spectrum. Other distances from absolute magnitudes.

Sp.t./C,L	Star	M	V	$B-V$	pc	Sp.t./C,L	Star	M	V	$B-V$	pc
WC8	γ^2 Vel	−4.1	1.78	−0.19	370	B 2 V	ϕ Cen	−2.5	3.83	−0.21	190
O 5	ζ Pup	−7.1	2.25	−0.26	700	B 3Ia	o^2 CMa	−6.8	3.03	−0.09	860
O 8	λ Ori	−5.0	3.66	−0.18		I	90706	−6.3	7.06	0.47	3400
O 8 I	149404	−6.2	5.47	0.4	1500	Ib	v Sco	−5.7	2.69	−0.22	480
O 9 III	σ Ori	−6.0	2.76	−0.23		II	α^2 Cru	−3.2	1.88	−0.26	110
O 9.5 Ia	α Cam	−6.2	4.29	0.03	860	III	α Tel	−2.9	3.51	−0.17	180
O 9.5 Ib	ζ Ori	−5.9	2.05	−0.21	340	IV	α Pav	−2.3	1.94	−0.20	71
O 9.5 II	δ Ori	−4.4	2.23	−0.22		IV-V	σ Sgr	−2.0	2.02	−0.22	64
O 9.5	ζ Oph	−3.4	2.56	0.02		V	δ Cae	−1.7	5.07	−0.19	230
B 0 Ia	ε Ori	−6.2	1.70	−0.19	370	V	δ^1 Lyr	−1.6	5.58	−0.15	270
Ib	224868	−5.8	7.26	0.13	2600	B 4 IVe	ε Aps	−2.0	5.06	−0.1	240
III	β Cru	−5.0	1.25	−0.23	130−	V	ζ Cir	−1.4	6.09	−0.06	590
IVe	γ Cas	−4.6	2.47	−0.15	24	B 5 Ia	η CMA	−7.0	2.04	−0.07	760
V	δ Sco	−4.1	2.82	0.14	170	Iab	212545	−6.3	7.70	0.38	6300
B 1 Ia	κ Cas	−6.6	4.16	0.14	930	I-II	ϕ Vel	−6.0	3.54	−0.08	770
Ib	ζ Per	−5.7	2.85	0.12	340	Ib	4768	−5.7	7.57	0.38	2300
Ib	97522	−5.4	7.73	0.31	2100	III	ι Aql	−2.2	4.36	−0.08	180
II	β Cen	−5.1	0.61	−0.23	140	IV	σ Cas	−3.5	4.88	−0.07	400
II-III	β CMa	−4.8	1.98	−0.23	220	IV	α Eri	−1.6	0.46	−0.16	26
III	σ Sco	−4.4	2.89	0.13	180	V	α Gru	−1.1	1.74	−0.13	21
IV	α^1 Cru	−3.9	1.41	−0.26	110	B 6 Ia	15497	−7.1	7.02	0.78	2100
V	α Vir	−3.5	0.98	−0.24	79	Ib	7902	−5.7	6.96	0.41	1800
V	μ^1 Sco	−3.0	3.04	−0.20	160	III	ζ Dra	−1.9	3.17	−0.12	97
B 2 Ia	χ^2 Ori	−6.8	4.63	0.28	1100	IV	π Ari	−1.3	5.22	−0.06	190
I	99939	−6.3	7.23	0.04	3500	V	ψ^2 Lup	−0.9	4.75	−0.15	140
Ib	138101	−5.7	7.39	0.23	2400	V	ζ^1 CrB	0.4	6.0	−0.1	130
II	ε CMa	−4.4	1.50	−0.21	150	Vp	o And	1.4	3.6	−0.09	35 +
II	ι CMa	−3.9	4.38	−0.07	410	B 7 Iae	187399	−7.1	7.01	0.18	3400
III	γ Ori	−3.6	1.64	−0.22	110	Iab	80558	−6.4	5.87	0.54	1400
V	λ Sco	−3.0	1.63	−0.22	84	III	β Tau	−1.6	1.65	−0.13	40

Appendix 3: Stars for the Hertzsprung–Russell Diagram

Sp.t./C,L	Star	M	V	B–V	pc
B 7 IV	ι Lyr	–1.0	5.28	–0.11	180
V	α Leo	–0.6	1.35	–0.11	26ts
B 8 Ia	β Ori	–7.1	0.12	–0.03	280
Ia–Iab	105071	–6.8	6.33	0.22	300–
Iab	91024	–6.5	7.61	0.27	4600
Ib	14322	–5.6	6.79	0.32	2000
II	γ CMa	–3.4	4.11	–0.12	320
III	γ Crc	–1.2	2.59	–0.11	70
V	β Per	–0.2	2.12	–0.05	29ts
8.5 V	ζ Peg	0.0	3.40	–0.09	48
B 9 Ia	σ Cyg	–7.1	4.23	0.12	1600
Iab	40589	–6.5	6.05	0.25	2400
Ib	99316	–5.5	7.40	0.33	2500
III	γ Lyr	–0.8	3.24	–0.05	59
IVe	ε Sgr	–0.3	1.85	–0.03	26
V	α Peg	0.2	2.49	–0.04	31ts
9.5 V	ω Aqr	0.4	4.49	–0.04	43–
A 0 Ia	92207	–7.1	5.45	0.50	180–
I	93737	–6.6	6.00	0.27	2500
Ib	η Leo	–5.2	3.52	–0.03	560
Ib.II	91054	–4.0	7.67	0.63	880
II	213050	–2.8	7.30	0.10	940
III	α Dra	–0.6	3.65	–0.05	71
IV	γ Gem	0.0	1.93	0.00	26ts
IV–V	μ Ari	0.3	5.70	0.00	120
V	α Lyr	0.5	0.03	0.00	8.1t
V	α CrB	0.6	2.23	–0.02	24ts
p	ω Her	2.0	4.57	0.00	33t
A 1 Ia	149019	–8.1	7.40	0.89	5300
Ia	α Cyg	–7.5	1.25	0.09	560
Ia	12953	–7.3	5.67	0.61	1900
V	α Gem	1.2	1.99	0.04	14ts
V	α CMa	1.4	–1.46	0.00	2.7t
A 2 Ia	o² Cen	–7.2	5.15	0.49	2100
Iab	91533	–6.7	6.00	0.32	2500
Ib	58439	–5.0	6.3	0.1	1600
III	δ Aqr	–0.2	3.27	–0.2	30ts
A 2 III	β Crt	0.2	4.48	0.03	72
IV	β Aur	0.6	1.90	0.03	22ts
V	ζ UMa	1.4	2.27	0.02	18ts
p	192640	2.4	4.97	0.14	33t
p	Psc	2.6	4.96	0.03	30t
A 3 Ia	223358	–7.6	5.13	0.67	2000
Iab	13476	–6.8	6.44	0.6	2200
Ib	104035	–4.8	5.61	0.18	1100
II–III	γ UMi	–1.1	3.05	0.05	69
III	β Eri	0.0	2.79	0.13	28ts
IV	γ Cam	0.9	4.63	0.03	56
V	κ Phe	1.7	3.94	0.17	19ts
V	α PsA	2.0	1.16	0.09	6.7t
V	ξ Vir	2.2	4.85	0.18	34t
A 4 Iab	223767	–6.9	7.23	0.63	150–
Ib	110786	–4.8	7.68	1.25	1200
IV	ζ Men	1.0	5.64	0.20	76
V	δ Leo	1.9	2.56	0.12	16ts
A 5 Ia	164514	–7.7	7.42	1.09	2700
Iab	198288	–6.9	6.95	0.71	2500
II	o Sco	–2.1	4.55	0.84	92
III	β Pic	0.3	3.85	0.17	24ts
IV	β Pav	1.2	3.42	0.16	28
V	β Ari	2.1	2.64	0.13	14ts
A 6 III	ε Cnc	0.4	6.30	0.17	150
IV	ψ Oct	1.3	5.50	1.3	160
d	205924	2.2	5.67	0.25	38ts
A 7 Ib	148743	–4.8	6.50	0.37	1300
III	γ Boo	0.5	3.03	0.19	32
IV	κ Boo	1.5	4.40	0.2	38
IV–V	α Cep	1.9	2.44	0.22	14ts
IV–V	α Aql	2.2	0.77	0.22	5.1t
V	ι UMa	2.4	3.14	0.19	15ts
A 8 Ib	υ Car	–2.0	2.97	0.27	99
A 9 I	842	–6.7	7.95	0.51	5500
III	γ Her	0.6	3.75	0.27	42
IV	ω Eri	1.6	4.39	0.25	36
V	157792	2.5	4.17	0.28	25ts
F 0 Ia	α Car	–8.5	–0.72	0.15	360

Appendix 3: Stars for the Hertzsprung-Russell Diagram

Sp.t./C,L	Star	M	V	B–V	pc	Sp.t./C,L	Star	M	V	B–V	pc
F 0 I	1457	–6.6	7.82	0.62	4600	F 5 IV	α Cmi	2.6	0.38	0.42	3.5t
I–II	θ Sco	–5.6	1.87	0.40	280	IV–V	ε Cet	2.8	4.84	0.45	32ts
Ib	α Lep	–4.7	2.58	0.21	296	IV–V	μ Dra	2.9	4.92	0.48	26t
III	ζ Leo	0.6	3.44	0.31	36	V	ι Peg	3.4	3.76	0.44	13ts
III–IV	υ Tau	1.1	4.29	0.26	42	F 6 I–II	201078	–5.5	5.82	0.56	1600
IV	μ Cet	1.7	4.27	0.31	30ts	Ib	180028	–4.6	6.97	0.81	1600
IV–V	δ Aql	2.1	3.26	0.32	16ts	II	ρ Pup	–2.0	2.81	0.43	92
V	112429	2.4	5.24	0.28	37t	III	ι Vir	0.7	4.08	0.52	22
V	γ Vir	2.6	2.75	0.36	11ts	IV	ξ Sco	2.2	4.16	0.45	26ts
V	204153	3.4	5.60	0.32	28t	V	κ Tuc	3.7	4.86	0.47	18ts
F 1 dIII	ε Sex	0.6	5.26	0.31	85	V	γ Lep	4.1	3.60	0.47	8.1t
V	ε Scl	2.8	5.31	0.39	30	Vp	γ Pav	4.5	4.22	0.49	8.6t
d	77370	2.8	5.16	0.42	24ts	F 7 Ib	29260	–4.6	6.80	0.50	1000
F 2 Ia	ι¹ Sco	–8.4	3.03	0.51	1700	IV	ο Peg	2.3	5.16	0.48	26–
Iab	75276	–6.6	5.75	0.56	2300	IV–V	σ² UMa	3.1	4.80	0.49	22ts
Ib	ν Aql	–4.6	4.66	0.60	530	V	θ Per	3.8	4.12	0.49	13ts
II	ν Her	–2.0	4.41	0.39	190	V	χ Dra	4.1	3.57	0.49	7.86
II–III	ο¹ Eri	–0.7	4.04	0.33	85–	F 8 Ia	δ CMa	–8.0	1.86	0.65	940
F 2 III	η Sco	0.6	3.33	0.41	21ts	Iab	γ¹ Nor	–6.3	4.99	0.80	1700
III–IV	χ Leo	1.3	4.63	0.33	47	I–II	δ Vol	–5.4	3.98	0.79	730
IV	β Cas	1.9	2.27	0.34	13t	Ib	α UMi	–4.6	2.02	0.60	209
V	ρ Psc	3.0	5.38	0.39	30ts	III	ζ Pic	0.6	5.45	0.51	61–
F 3 Ib	8906	–4.6	7.11	0.74	1400	IV	123999	2.4	4.83	1.1	29ts
III	ο Gem	0.6	4.90	0.40	70	IV	α For	3.3	3.87	0.52	14t
IV	μ Vir	1.9	3.88	0.38	26ts	V	β Vir	3.6	3.61	0.55	10t
IV–V	79940	2.5	4.62	0.45	22ts	V	β Cae	4.0	5.05	0.37	17ts
V	τ⁶ Eri	3.1	4.23	0.42	17ts	F 9 Ia	β Dor	–8.0	3.8	0.80	2300
F 4 Iab	61715	–6.5	5.68	0.65	2200	V	χ Her	4.2	4.62	0.56	15ts
III	ο Oct	0.7	5.15	0.49	70	d	13612	4.2	5.54	0.57	22ts
IV	ω And	2.0	4.83	0.42	36	VI	201891	5.2	7.38	0.51	27t
V	101606	3.3	5.73	0.43	34ts	G 0 Ia	ο¹ Cen	–8.0	5.13	1.08	3000
F 5 Ia	170764	–8.2	6.5	0.5	3900	I	58526	–6.3	5.97	0.92	2600
I	178359	–6.4	7.8	0.5	36+	Ib	β Aqr	–4.5	2.91	0.83	300
Ib	α Per	–4.6	1.80	0.48	190	II	ε Leo	–2.0	2.98	0.80	95
Ib–II	725	–3.3	7.09	0.63	1000	III	ε Hya	0.6	3.38	0.68	34
II	υ² Cen	–2.0	4.34	0.60	160	IV	η Boo	2.7	2.68	0.58	9.8t
II–III	π Peg	–0.6	4.29	0.46	96	IV	ζ Her	3.0	2.81	0.65	9.8t
III	ξ Gem	0.7	3.36	0.43	23ts	IV–V	1461	3.6	6.46	0.68	27ts
IV	θ Cyg	2.1	4.48	0.38	19ts	V	λ Aur	4.4	4.71	0.63	13ts

Appendix 3: Stars for the Hertzsprung–Russell Diagram

Sp.t./C,L	Star	M	V	B–V	pc
G 0 V	η Cas	4.6	5.44	0.57	5.9t
V	β Com	4.7	4.26	0.57	8.3t
V	ξ UMa	4.9	3.79	0.59	7.7t
V	ζ Tuc	5.0	4.23	0.58	7.1t
G 1 Ib	59890	−4.5	4.65	0.93	640
IV	β Hyi	3.8	2.80	0.62	6.3t
d	κ For	4.5	5.20	0.60	14ts
G 2 I	175580	−6.2	6.8	0.7	2000
Ib	α Aqr	−4.5	2.96	0.98	290
II	β Lep	−2.1	2.84	0.82	97
II–III	η Peg	−0.9	2.94	0.86	53
III	φ Vir	0.4	4.81	0.70	29ts
IV	μ Cnc	3.0	5.30	0.63	31ts
V	α¹ Cen	4.4	0.001	0.71	1.3t
V	ζ¹ Ret	4.7	5.54	0.64	13ts
V	Sun	4.85	−26.72	0.62	4.85 ×10⁻⁶
V	224930	5.4	5.75	0.67	12t
G 3 I	199997	−6.2	7.66	0.84	5900
Ib	ξ Pup	−4.5	3.34	1.24	230
IV	ι Hor	3.0	5.41	0.56	17ts
d	3795	4.9	6.14	0.70	21ts
V	8262	4.9	6.96	0.62	27ts
V	153631	5.5	7.14	0.6	21t
G 4 Ia	187921	−8.0	7.60	0.8	13 000
Ibp	ζ Cap	−4.5	3.74	1.00	450
II–III	o UMa	−0.9	3.36	0.84	71
gIV	ζ Pyx	0.3	4.89	0.9	75
IV–V	67458	4.1	6.81	0.6	26ts
V	11112	5.0	7.13	0.65	23ts
d	ψ Cnc	5.3	5.73	0.81	17+
G 5 Ia	12399	−8.0	7.49	0.90	4000
I	96566	−6.2	4.61	1.03	160−
Ib	186442	−4.5	6.52	1.23	1300
II	β Crv	−2.1	2.65	0.89	89
III	μ Vel	0.3	2.69	0.90	30
III–IV	δ CrB	1.8	4.63	0.80	38
IV CN	χ Eri	3.2	3.70	0.85	15ts
IV	μ Her	3.9	3.42	0.75	8.1t
G 5 IV	δ Pav	4.8	3.56	0.76	5.7t
V	κ Cet	5.0	4.83	0.68	9.3t
d	χ Ser	5.2	5.85	0.68	19ts
V	α Men	5.4	5.09	0.72	8.7t
V	38392	6.6	6.15	0.94	8.1t
G 6 I	κ TrA	−6.2	5.09	1.13	1700
II	α Ret	−2.1	3.35	0.91	120
g	υ¹ Eri	0.3	4.51	0.98	63
V	115617	5.1	4.74	0.71	8.4t
d	ρ Eri	5.3	5.57	1.05	12
V	152391	5.6	6.64	0.76	16t
G 7 Iab	95109	−6.1	6.11	1.10	1000
III	λ Pyx	0.3	4.69	0.92	76
d	114260	5.5	7.37	0.73	17ts
G 8 Ia	119796	−8.0	6.51	1.98	2900
Iab	63302	−6.1	6.35	1.78	1200
Ib	ε Gem	−4.5	2.98	1.40	210
II	ι Cnc	−2.1	4.02	1.01	130
II–III	λ Peg	−0.9	3.95	1.07	33ts
III	α Aur	0.3	0.08	0.80	13ts
G 8 III–IV	υ² Cas	1.8	4.63	1.21	29ts
II CN	ε Cet	2.8	4.84	0.89	32ts
IV	β Aql	3.2	3.71	0.86	11
V	ξ Boo A	5.5	4.55	0.76	6.8t
IV–V	82885	5.6	5.41	0.77	9.2t
Vp	τ Cet	5.7	3.50	0.72	3.64t
Vp	GRB 1830	6.7	6.45	0.75	8.8t
G 9 III	ε Vir	0.2	2.83	0.94	32ts
d	22468	5.7	5.71	0.92	10ts
K 0 Iape	214369	−8.0	7.76	1.89	6200
Iab	45829	−6.1	6.63	1.58	2100
Ib	1397	−4.4	6.9	1.1	91+
II	ε Car	−2.1	1.86	1.28	62
II–III	α Cas	−0.9	2.23	1.17	37
III	β Gem	0.2	1.14	1.00	11ts
III–IV	β Cen	1.7	2.06	1.01	14ts
III	ψ Aqr	1.8	4.21	1.11	30ts
IV	β Ret	3.2	3.85	1.13	17ts
IV	δ Eri	3.8	3.54	0.92	9.0t

Appendix 3: Stars for the Hertzsprung-Russell Diagram

Sp.t./C,L	Star	M	V	B–V	pc
K 0 d	121056	4.6	6.19	1.02	21
V	180161	4.8	7.03	0.79	27t
V	188088	5.3	6.18	1.02	15t
V	70 Oph A	5.7	4.03	0.86	5.1t
V	3651	5.8	5.87	0.85	11t
V	σ Dra	5.9	4.65	0.79	5.7ts
V	105417	6.6	7.53		15t
V	194215	6.9	5.85	1.10	14t
K 1 I	174947	–6.1	5.7	1.2	2300
Ib	ζ Cep	–4.4	3.35	1.57	220
II	ζ And	–2.2	4.06	1.12	48ts
III	ν^2 CMa	0.0	3.95	1.06	25ts
IV	γ Cep	2.2	3.21	1.03	16t
V	α^2 Cen	5.7	1.39	0.88	1.3t
V	o^2 Eri	6.0	4.43	0.82	4.9t
V	190404	6.1	7.28	0.82	19ts
K 2 C	ν^2 Sgr	–6.0	4.83	1.41	1400
Ib	ε Peg	–4.4	2.38	1.53	160
Ib–II	3147	–3.3	6.91	2.05	350
II	π^6 Ori	–2.2	4.47	1.40	190
II–III	ω Hya	–1.1	4.97	1.22	170
IIIp	α Boo	–0.3	–0.04	1.23	11t
III	α Ari	–0.1	2.00	1.15	26ts
IV	τ Sgr	0.0	3.32	1.19	40
V	ε Eri	6.1	3.73	0.88	3.3t
V	13579	6.3	7.18	0.90	22ts
V	3765	6.6	7.36	0.94	14t
V	156384	7.0	5.91	1.04	7.1t
V	101998	7.8	7.14	1.44	10
K 3 Ia	4817	–8.0	6.07	1.88	3100
Iab	σ^1 CMa	–6.0	3.86	1.73	520
Iab–Ib	187238	–5.2	7.7	1.1	2300
Ib	β Ara	–4.4	2.85	1.46	240
Ib–II	o^2 Cyg	–3.4	3.98	1.52	280
II	β Cyg	–2.4	3.08	1.13	120
III	ν Psc	–0.2	4.44	1.36	42ts
V	219134	6.4	5.56	1.01	6.8t
V	191408	6.6	5.32	0.87	5.6t
V	125072	6.7	6.66	1.03	9.6t
K 3 V	5133	6.9	7.16	0.94	13ts
V	111180	8.0	7.73	1.49	11s
K 4 C	34255	–5.9	5.61	1.75	1100
II	ζ Aur	–2.3	3.75	1.22	160
III	β UMi	–0.3	2.08	1.47	29ts
V	29883	6.2	8.61	0.91	23t
V	131977	7.1	5.64	1.4	5.6t
d	97584	7.4	7.63	1.03	13ts
V	217580	7.4	7.46	0.95	15ts
V	110296	8.2	7.78	1.55	7.8
K 5 Ib	λ Vel	–4.4	2.21	1.66	150
III	α Tau	–0.3	0.85	1.54	21ts
d	137778	6.6	7.58	0.86	16t
V	ε Ind	7.0	4.67	1.06	3.5t
V	12208	7.4	7.44	1.69	10
V	61Cyg A	7.6	5.22	1.26	3.4t
V	156026	7.7	6.34	1.16	5.4t
V	4378	8.0	7.92	1.4	11ts
K 6 III	η Men	–0.3	5.47	1.52	140
g	u Hyi	–0.4	3.24	1.33	49
d	70 Oph B	7.5	5.9	0.86	5.1ts
d	64468	8.2	7.79	0.95	16ts
K 7 V	99279	7.3	7.22	1.26	7.3t
V	157881	8.2	7.54	1.36	7.5t
V	GRB 1618	8.3	6.61	1.37	4.5t
V	61 Cyg B	8.4	6.04	1.37	3.4t
K 9 III	ν Gru	–0.4	5.47	0.95	76–
M 0 Iab	σ Cma	–5.7	3.46	1.73	460
III	β And	–0.4	2.06	1.58	27ts
V	79210	8.7	7.64	1.43	6.0t
V	Lac 8760	8.8	6.68	1.45	3.9t
V	191849	9.0	7.97	1.46	6.1t
M 1 Ia	42543	–7.2	6.2	1.6	3100
Ib	α Sco	–4.7	0.96	1.83	100
III	λ Hyi	–0.5	5.07	1.37	130
V	36395	9.1	7.97	1.47	5.9t
M 2 Iab	α Ori	–6.0	0.50	1.85	200
Ibp	101712	–4.8	7.86	1.78	2600
II	δ Sge	–2.4	3.82	1.41	170

Appendix 3: Stars for the Hertzsprung–Russell Diagram

Sp.t./C,L	Star	M	V	B–V	pc	Sp.t./C,L	Star	M	V	B–V	pc
M 2 II–III	β Peg	–1.4	2.42	1.67	54		v Maanen				
III	ψ Leo	–0.5	5.35	1.63	140	dG 5	HL 4	15.4	14.5	1.0	6.7t
Ve	Lac 9352	9.6	7.34	1.5	3.7t	dM 0	Kapteyn	10.9	8.9	1.56	4.0t
V	LaI21185	10.5	7.49	1.51	2.5t	dM 1	Σ 2269	8.5	15.0		200
M 3 Iab	14270	–5.6	7.84	2.27	1600	dM 2e	GRB 34A	10.3	8.1	1.55	3.6t
II	η Sgr	–2.4	3.11	1.56	130	dM 4	Krg 60A	11.8	9.8	1.63	4.0t
III	γ Cru	–0.5	1.63	1.59	27	dM 4e	Krg 60B	13.4	11.4	1.8	4.0t
V	153336	10.6	5.86	1.62		dM 4	Ross 614	13.2	11.2	1.74	3.98t
M 4 II	δ Lyr	–2.4	4.30	1.68	220	dM 4e	GRB 34B	13.3	11.05	1.78	3.6t
II–III	1364	–1.4	7.30	1.6	560	dM 4e	40 Eri C	12.6	11.05	1.68	4.9t
III	θ Aps	–0.5	6.0	1.6	200	dM 4	Σ 2398 A	11.2	8.9	1.54	3.53t
V	217580	7.4	7.46	0.95	15ts	dM 5	Barnard	13.2	9.5	1.75	1.83t
M 5 III	β Gru	–2.4	2.11	1.62	53–	dM 5	Σ 2398 B	12.0	9.7	1.58	3.57t
M 6 III	ε Oct	–2.13	5.10	1.47	280	dM 5e	Prox Cen	15.1	10.7	1.97	1.3ts
DA 0	40Eri B	11.2	9.7	0.03	4.9t	dM 6e	UV Cet A	15.2	12.4	1.9	2.76t
DA 3	LB 1497	11.0	16.5		125	dM 6e	UV Cet B	15.9	12.9	1.9	2.76t
DA 5	α² CMa	11.4	8.7	0.04	2.7t	dM 6e	Wolf 359	16.8	13.6	2.01	2.33t
DA 5	α² CMi	13.1	10.8	0.5	3.5t	dM 6e	v Biesbroeck	19.3	17.5	2.3	5.5t
dF 5	Wolf 28	14.2	12.9	0.55	5.5t						

In the H–R diagram (Figure 10.9, page 169) the absolute magnitudes are plotted against the spectral types, classes and luminosities (Sp.t./C,L). In Figure 10.10, page 171, the absolute magnitudes are plotted against the B–V colour indices.

Appendix 4

Masses and Luminosities of Double Stars

Masses and luminosities in multiples of those of the Sun (Periods: d = days; y = years)

Double star	Co-ordinates RA	Dec	Masses		Luminosities		Periods of revolution	Spectral type
Plaskett	06 37.0	+06 08	60	40	2000	1000	14.414 d	O8e, O
AO Cas	00 17.7	+51 26	32	30	2000	2000	3.523 55 d	O9 III, O9 III
V444 Cyg	20 19.5	+38 44	32	18	6900	1450	4.212 38 d	WN5, B1
UW CMa	07 18.7	−24 57	23	19	8000	8000	4.393 51 d	O7, O7
Y Cyg	20 52.1	+34 39	17.3	17.1	5000	5000	2.996 33 d	B0 IV, B0 IV
V Pup	07 58.1	−49 14	16.5	9.7	1100	580	1.454 49 d	B3 IV, B1 V
μ Sco	16 51.5	−38 02	14	9	1000	700	1.440 27 d	B2 V, B6 V
S Aur	05 00.4	+39 39	8.3	6.8	2100	400	2.661 70 y	B7 V, K4 II
36 And	00 52.3	+23 22	6.7	6.2	7	6	165 y	K1, K
σ Aql	19 39.2	+05 24	6.9	5.5	440	275	1.950 27 d	B3, B4
68 Her	17 15.5	+33 09	7.5	2.9	260	100	2.051 03 d	B3 III, B5
U Oph	17 16.3	+01 13	5.4	4.7	480	480	1.677 35 d	B5 V, B8
λ Tau	04 00.4	−12 29	6.1	1.6	525	70	3.953 02 d	B3 V, A4
γ Per	03 04.8	+53 30	4.7	2.7	100	25	14.647 y	A3, F5
U Cep	01 02.3	+81 53	4.7	1.9	130	25	2.493 d	B8 V, gG8
U Sge	19 18.5	+19 36	3.5	1.4	90	40	3.380 62 d	B8 V, G2 III
β Per	03 08.2	+40 57	3.7	0.8	90	3.5	2.867 31 d	B8, K0 IV
β Aur	05 59.5	+44 57	2.35	2.25	55	55	8.78 d	A0, A2
α Aur	05 16.3	+45 53	2.14	2.04	90	70	104.022 d	F6, G8 III
δ Lib	15 00.7	−08 30	2.7	1.2	46	3	2.327 35 d	A0, G
α CMa	06 46.8	−16 42	2.35	0.98	23	0.002	49.98 y	A1 V, DA5
α Gem A	07 34.6	+31 53	1.6	1.6	12	12	9.212 8 d	A1 V, A2 V
RW Tau	04 03.9	+28 08	2.55	0.55	130	3.5	2.768 85 d	B8 Ve, K0 IV
α Oph	17 34.7	+12 35	2.40	0.6	40	?	8.5 y	G0 III, ?
α Com	13 10.0	+17 32	1.45	1.41	3	3	25.85 y	F5 V, F5 V
α CrB	15 34.5	+26 44	2.5	0.9	45	0.5	17.359 91 d	G0, dG6
ε Hya	08 46.8	+06 25	1.75	1.6	70	?	15.3 y	F8, G0 III
α CMi	07 39.0	+00 58	1.7	0.65	5.8	0.0005	40.65 y	G5, ?
13 Cet	00 35.2	−03 36	1.27	1.05	1	1	6.91 y	F8 V, G0
α Gem B	07 34.6	+31 53	1.2	1.1	6	6	2.928 3 d	A5, A5
γ Vir	12 41.7	−01 27	1.18	1.12	0.7	0.7	171 y	F0 V, F0 V

Appendix 4: Masses and Luminosities of Double Stars

Double star	Co-ordinates		Masses		Luminosities		Periods of revolution	Spectral type
	RA	Dec						
α Cen	14 39.6	−60 50	1.1	0.86	1.5	0.4	80.089 y	G2 V, dK1 V
ζ Her	16 39.4	+31 41	1.12	0.78	6	0.5	34.38 y	G0 IV, dK0
ξ Boo	14 51.4	+19 06	0.87	0.76	0.5	0.07	149.9 y	G8 V, K4 V
η Cas	00 49.1	+57 49	0.85	0.54	1.2	0.04	480 y	G0 V, dM0
α Gem C	07 34.6	+31 53	0.63	0.57	0.025	0.025	0.814 3 d	dK6, dK6
o^2 Eri	04 15.2	−07 39	0.75	0.44	0.3	0.0027	8000 y	G5, DA
61 Cyg	21 04.4	+38 28	0.6	0.5	0.065	0.038	653 y	K5 V, K7 V
85 Peg	00 02.2	+27 05	0.82	0.8	70	?	26.27 y	G0 V, ?
70 Oph	18 05.5	+02 30	0.82	0.6	0.4	0.08	87.85 y	K0 V, dK6
Σ 2398	18 42.8	+59 38	0.27	0.25	0.0027	0.0013	346 y ?	dM4, dM5
Krg 60	22 28.1	+57 42	0.26	0.14	0.0013	0.0004	44.46 y	dM3, dM4e
Ross 614	06 29.4	−02 45	0.14	0.08	0.00048	0.000016	165 y	dM6e, ?
UV Cet	01 38.8	−17 58	0.044	0.035	0.00004	0.00003	200 y	dM6e, dM6e

The data in the table above, as well as in Appendices 5, 6, 7, 8.1, 9 and 10, collected from: *Burnham's Celestial Handbook* by R. Burnham Jr. (Dover Publications Inc., New York); *Variable Stars* by J.S. Glasby (Constable, London); *Veränderliche Sterne* by C. Hofmeister, G. Richter and W. Wenzel (Springer-Verlag, Berlin); *Russian Catalogue of Variable Stars* by P.N. Kholopov and N.P. Kukarkin; *Sky Catalogue 2000.0, Volumes I and II* by A. Hirshfeld and R.W. Sinnott (Sky Publishing Corporation and Cambridge University Press, Cambridge, London and New York); and from the South African Astronomical Observatory.

Appendix 5

Galactic or Open Clusters

NGC Number	Mes.	Co-ordinates RA	Dec	Const.	Width	Magnitudes	No. of stars	Dist. (pc)	Spectrum	Notes
752		1 57.8	+37 41	And	50'	8...	70	400	A2	Loose
1039	34	2 42.0	+42 47	Per	35'	8...	80	440	B8	Loose, pretty
	45	3 47.0	+24 07	Tau	110'	3...	200	125	B5	Pleiades
		4 27	+16	Tau	330'	3...	25	46	A2	Hyades, A, F stars
1746		5 03.6	+23 41	Tau	45'	8...	50	420	B5	Beautiful
1912	38	5 28.4	+35 50	Aur	20'	8...	100	1200	B5	Loose
1960	36	5 35.8	+34 08	Aur	12'	9...	60	1200	B3	Dense
2099	37	5 52.1	+32 33	Aur	20'	9...	150	1300	B9	Dense, beautiful
2168	35	6 08.4	+24 21	Gem	30'	8...	120	850	B4	Magnificent
2244		6 32.1	+4 52	Mon	40'	6–9	15	1500	O5	In Rosette nebula
2264		6 41.0	+9 53	Mon	26'	4.6–9	20	790	O8	In Cone nebula
2287		6 46.9	–20 40	CMa	38'	7...	50	640	B5	Bright
2323	50	7 03.0	–8 20	Mon	16'	9–14	100	1000	B8	Loose, beautiful
2360		7 17.6	–15 37	CMa	12'	9–12	50	1100	B8	Rich
2422	47	7 36.4	–14 29	Pup	29'	6...	25	470	B3	Loose, bright
2437	46	7 41.6	–14 48	Pup	27'	10...	150	1600	B9	Beautiful, Pl. neb. 2438
2447	93	7 44.4	–23 52	Pup	22'	8–14	50	1100	B9	Very dense
2477		7 52.1	–38 32	Pup	27'	11...	300	1200	B8	Very dense
2516		7 58.3	–60 51	Car	60'	7–13	100	410	B3	Very dense
2539		8 11.5	–12 49	Pup	20'	11...	100	910	A0	Dense
2547		8 10.5	–49 15	Vel	20'	7–15	50	420	B3	Scattered
2548	48	8 13.5	–5 47	Hya	54'	9–13	50	630	A0	Pretty
2632	44	8 39.8	+20 00	Cnc	95'	7–7.5	55	160	A0	Praesepe, scattered
2682	67	8 50.2	+11 50	Cnc	29'	10–16	40	790	B8	Main Sequence., some giants
3114		10 02.6	–60 05	Car	40'	9–13	100	940	B9	Scattered
I 2602		10 43.0	–64 22	Car	70'	5...	30	150	B0	Bright, scattered
3532		11 06.2	–58 39	Car	60'	8–12	150	480	B5	Very rich
3766		11 35.2	–61 35	Cen	12'	8–13	60	1930	B0	Rich
4755		12 53.3	–60 19	Cru	10'	6–10	50	1500	B3	κ Crucis, Jewel Box
5460		14 07.3	–48 18	Cen	30'	8...	30	770	B8	Bright
5617		14 29.4	–60 42	Cen	15'	8...	50	1600	B0	Rich, bright
6025		16 03.3	–60 30	TrA	10'	7...	30	770	B3	Bright, has Cepheids

Appendix 5: Galactic or Open Clusters

NGC Mes. Number	Co-ordinates RA	Dec	Const.	Width	Magnitudes	No. of stars	Dist. (pc)	Spectrum	Notes
6087	16 18.5	−57 54	Nor	20'	7–10	40	900	B5	Loose, Cepheids
6124	16 25.3	−40 39	Sco	25'	9–12	100	560	B8	Scattered
6152	16 32.7	−52 37	Nor	30'	9…	60	1030	B5	Scattered
6067	16 12.9	−54 12	Nor	15'	10…	100	1700	B2	Dense, fine field
6193	16 40.9	−48 45	Ara	20'	6…	30	1300	O7	Nebulous
6208	16 49.1	−53 49	Ara	20'	9–12	50	990		Scattered
6231	16 53.7	−41 47	Sco	15'	7–13	100	2000	O9	Has Wolf–Rayet stars
Tr 24	16 57.0	−40 40	Sco	60'	8–16	200	1600	O7	Scattered
6405 6	17 39.8	−32 12	Sco	25'	7–10	50	490	B5	Beautiful
6475 7	17 53.6	−34 49	Sco	80'	7–11	50	240	B5	Magnificent, naked eye
6494 23	17 56.6	−19 01	Sgr	25'	9…	100	640	B9	Dense, beautiful
6611 16	18 18.6	−13 47	Ser	9'	8–12	50	1200	O6	Some B and K stars
I 4725 25	18 31.4	−19 15	Sgr	35'	6–10	50	710	B4	Loose, Cepheids
6633	18 27.5	+6 34	Oph	27'	7…	65	310	B6	Scattered
I 4756	18 39.0	+5 27	Ser	70'	7…	35	390	B7	Scattered
6694 26	18 45.0	−9 24	Sct	14'	11–14	25	1500	B8	Dense
6705 11	18 50.8	−6 17	Sct	12'	11–14	200	2000	B8	Richest cluster, dense
6709	18 51.5	+10 21	Aql	12'	8…	40	950	B5	Scattered
6913 29	20 23.8	+38 31	Cyg	7'	8…	20	1300	B0	Scattered
7092 39	21 32.0	+48 25	Cyg	30'	7…	25	290	A0	Scattered
7209	22 05.0	+46 28	Lac	25'	9–12	50	900	A0	Beautiful, Scattered

Appendix 6

Selected Eclipsing Binaries

Type	Star	Co-ordinates RA	Dec	Magnitudes Max.	Min.	Period (days)	Spectral type
6.1 Algol	β Per	03 07.9	+40 56	2.12	3.4	2.493 d	B8 V; G5 IV; Am
	λ Tau	04 00.4	+12 29	3.3	3.8	3.952 d	B3 V; A4 V
	R Z Eri	04 43.8	−10 41	7.79	8.71	39.282 44	A5; F5 Vm; gG8
	AR Aur	05 18.0	+33 46	6.15	6.8	4.134 d	Ap (Hg, Mn); B9 V
	δ Ori	05 32.0	−00 18	1.94	2.13	5.732 476	B0 III; O9 V Mintaka
	TW And	00 03.3	+32 51	8.8	10.86	4.122 774	F0 V; K0
	WW Aur	06 32.2	+32 28	5.79	6.5	2.525 d	A3m; A3m
	R CMa	07 19.3	−16 23	5.70	6.3	1.135 d	F IV
	V Pup	07 58.0	−49 14	4.7	5.2	1.454 d	B1 Vp; B3 IV
	CV Vel	09 00.5	−51 32	6.5	7.3	6.889	B2 V; B2 V
	δ Lib	15 00.7	−08 30	4.92	5.9	2.327 d	B9.5 V
	U Oph	17 16.3	+01 13	5.88	6.5	1.677 d	B5 Vnn; B5 V
	U Sge	19 18.6	+19 36	6.58	9.1	3.380 d	B8 III; K
	V 505 Sgr	19 52.9	−14 37	6.48	7.5	1.182 d	A0 V; F8 IV
6.2 Beta Lyrae	UW CMa	07 18.7	−24 34	4.84	5.33	4.393 41	O7 Ia; O
	TT Aur	05 09.7	+39 35	8.59	9.5	1.332 733	B2 Vn; B5
	YY CMi	08 06.6	+01 56	8.33	9.13	1.094 02	F6 Vn
	GG Lup	15 18.9	−40 47	5.4	6.0	2.164 175	B5; A0
	μ^1 Sco	16 51.9	−38 03	2.80	3.08	1.440 27	B1.5 V; B6.5 V
	V 1010 Oph	16 49.2	−15 40	6.1	7.0	0.661 d	A5 V
	V 861 Sco	16 56.3	−40 49	6.07	6.6	7.848 d	B0.5 Iae
	U Her	17 17.2	+33 06	4.6	5.3	2.051 d	B1.5 Vp; B5 III
	β Lyr	18 49.9	+33 22	3.34	4.3	12.935 d	B7 Ve; A8p
	V 599 Aql	19 02.6	−10 43	6.67	6.75	1.849 08	B2 V; B8
	V 525 Sgr	19 07.2	−30 10	7.9	8.8	0.705 122	A2
	V 337 Aql	19 04.2	−02 02	8.57	9.27	2.733 83	B0.5 Vp; B2 V

Appendices 6.3, 6.4

Type	Star	Co-ordinates RA	Dec	Magnitudes Max.	Min.	Period (days)	Spectral type
6.3 W Ursae Majoris	YY Eri	04 12.2	−10 28	8.80	9.50	0.321 495	G5; G5
	S Ant	09 52.3	−28 38	6.4	6.92	0.648 345	A9 Vn
	W UMa	09 43.8	+55 57	7.9	8.63	0.333 637	dF8p; F8p
	AH Vir	12 14.3	+11 49	9.0	9.61	0.407 522	K0 V; K0 V
	SX Crv	12 40.2	−18 48	8.99	9.25	0.316 639	F8
	ι Boo	15 03.8	+47 39	6.5	7.1	0.267 816	B2 V; G2 V
	V 502 Oph	16 41.4	+00 30	8.34	8.84	0.453 393	G2 V; F9 V
	AK Her	17 14.0	+16 21	8.83	9.32	0.421 523	F2; F6
	V 566 Oph	17 56.9	+04 59	7.5	7.96	0.409 647	F4 V
6.4 Other Eclipsing Binaries	X Tri	02 00.6	+27 53	8.9	11.8	0.971 531	A3; G3
	RW Tau	04 03.9	+28 08	7.98	11.14	2.768 840	B8 Ve; K0 IV
	ε Aur	05 02.0	+43 49	2.92	3.83	9890 d	F0 Ia
	TZ Men	05 30.2	−84 47	6.2	6.9	8.659 d	B9.5 IV–V
	RY Gem	07 27.4	+15 40	8.5	11.3	9.300 525	A2 Ve; K2
	RZ Cnc	08 39.1	+31 48	8.67	10.03	21.642 998	K2 III; K4 III
	S Vel	09 33.2	−45 13	7.74	9.50	5.933 666	A5 Ve; K5 IIIe
	α Vir	13 25.2	−11 10	0.97	1.04	4.014 54	B1 III; B2 V
	ZZ Boo	13 56.2	+25 55	5.8	6.40	4.991 77	G2 V; G2 V
	U CrB	15 18.2	+31 39	7.66	8.79	3.452 2	B6 V; F8 III–IV
	RW Ara	17 34.8	−57 09	8.85	11.45	4.367 21	A1 IV; K3 III
	W Ser	18 05.0	−29 35	4.30	5.08	7.594 71	F4; G1 Ib
	BL Tel	19 06.6	−51 25	7.72	9.82	778.1 d	F8; M
	Z Vul	19 21.7	+25 34	7.38	9.20	2.454 93	B4 V; A2–A3 III
	V 695 Cyg	20 13.6	+46 44	3.77	3.88	3784.3 d	K2 II; B3 V
	RU Cap	20 32.6	−21 41	9.2	15.2	347.37 d	M9e
	Y Psc	23 34.4	+07 55	9.0	12.0	3.765 86	A3; K0

Appendix 7

Selected Variable Stars

Type		Star	Co-ordinates		Magnitudes		Period (days)	Spectral type
			RA	Dec	Max.	Min.		
7.1	Delta Cephei	X Pup	07 32.8	−20 55	7.82	9.24	25.961 d	F6–G2
		SV Mon	06 21.4	+06 28	7.61	8.88	15 232 1 d	F6–G4
		RT Aur	06 28.3	+30 30	5.00	5.8	3.728	F4 Ib–G IIb
		ζ Gem	07 03.9	+20 35	3.66	4.1	10.150	F7 Ib–G3 Ib
		VX Pup	07 32.6	−21 56	7.73	8.51	3.017 7/2.136	F5–F8
		AX Vel	08 10.8	−47 42	7.9	8.5	3.671 3/2.592 8	F8
		Y Car	10 32.2	−58 30	7.53	8.48	3.639 8/2.559	F3
		U TrA	16 07.3	−62 55	7.47	8.25	2.568 4/1.824 9	F5–F7
		ZZ Car	09 45.2	−62 30	3.28	4.18	35.535 84	F6 Ib–K0 Ib
		VY Car	10 44.6	−57 34	6.87	8.05	18.990	F6–G4 Iab–Ib
		U Car	10 57.6	−59 43	5.72	7.0	38.768	F6–G7 Iab
		S Cru	12 54.4	−58 26	6.22	6.92	4.689 97	F6–G1 Ib–II
		Y Sgr	18 21.0	−18 52	5.40	6.1	5.773	F8 I
		U Sgr	18 31.9	−19 07	6.34	7.08	6.744 925	F5 Ib–G1.5 Ib
		TT Aql	19 08.2	+01 18	6.46	7.70	13.754 6	F6–G5
		η Aql	19 52.5	+01 00	3.48	4.3	7.176	F6 Ib–G4 Ib
		S Sge	19 55.8	+16 37	5.28	6.0	8.382	F6 Ib–G5 Ibv
		δ Cephei	22 29.0	+58 24	3.48	4.3	5.366	F5 Ib–G IIb
7.2	RR Lyrae	RR Cet	01 32.1	+01 21	9.10	10.10	0.553 028 14	A7–F5
		SS For	02 07.9	−26 52	9.45	10.60	0.495 432	A3–G0
		X Ari	03 08.5	+10 27	8.97	9.95	0.651 142 6	A8–F4
		FO Vir	13 29.8	+01 06	6.5	6.8	0.60	A2
		UV Oct	16 32.5	−83 55	8.92	9.79	0.542 625	A6–F6
		MT Tel	19 01.9	−46 39	8.68	9.28	0.316 897	A0
		RR Lyr	19 25.3	+42 47	7.06	8.12	0.566 867	A8–F7
		XZ Cyg	19 32.5	+56 23	9.00	10.16	0.466 473 1	A6–F6
		V Ind	21 11.5	−45 04	9.12	10.48	0.479 603 0	A5–G3

Appendices 7.3, 7.4, 7.5

	Type	Star	Co-ordinates RA	Dec.	Magnitudes Max.	Min.	Period (days)	Spectral type
7.3	Delta Scuti	θ Tuc	00 33.4	−71 16	6.06	6.15	0.052	A7 IV
		UV Ari	02 45.0	+12 27	5.18	5.22	0.035 5	A7 IV
		KW Aur	05 15.4	+32 41	4.95	5.08	0.088 088	A9 IV
		V 356 Aur	05 42.5	+29 00	8.01	8.12	0.189 16	F4 IIIp
		AZ CMi	07 44.1	+02 24	6.44	6.51	0.095 26	F0 III
		VZ Cnc	08 40.6	+09 51	7.18	7.91	0.17836415	A7 III–F2 III
		γ Boo	14 32.1	+38 19	3.02	3.07	0.2903137	A7 III
		TU Crv	12 36.0	−20 32	6.53	6.55	0.082	F0 III
		δ Sct	18 42.3	−09 03	4.98	5.16	0.193770	F2 IIIp
		α Lyr	18 36.9	+38 47	0.02	0.03	0.07	A0 Va
		ρ¹ Sgr	19 21.7	−17 51	3.90	3.93	0.050	F0 IV–V
		δ Del	20 43.5	+15 04	4.39	4.49	0.158	A7 IIIp
7.4	W Virginis	SU Col	05 07.8	−33 52	11.13	12.30	21.55	A2 II
		ST Tau	05 45.1	+13 35	7.79	8.56	4.034269	F5–G5
		ST Pup	06 48.7	−37 16	9.28	10.68	19.0	F2 Ie–G2 I
		RU Cam	07 21.7	+69 40	8.1	9.79	22 d	C0.1–C3.2e (R0)
		W Vir	13 25.1	−03 13	9.46	10.75	17.2736	F0 Ib–G0 Ib
		BL Her	18 01.2	+19 15	9.70	10.62	1.30745	F0–F6 II–III
		RS Pav	18 07.4	−58 58	10.10	11.25	19.954	Ke
		AP Her	18 50.3	+15 57	10.19	11.18	10.4156	F2 Ib–II–G0
		PZ Aql	18 55.6	−02 53	11.25	11.97	8.7530	F5–G3
		κ Pav	18 56.9	−67 14	3.94	4.75	9.088	F5 I–II
7.5	Long Period	R And	00 24.0	+38 35	5.8	14.9	409.33 d	S3.5e–S8.8e (M7e)
		R Ari	02 16.1	+25 03	7.4	13.7	186.78 d	M3e–M6e
		o Cet	02 19.1	−02 59	2.0	10.1	331.96 d	M5e–M9e
		R Hor	02 53.7	−49 55	4.7	14.3	403.97 d	M7 IIIe
		R Gem	07 07.4	+22 42	6.0	14.0	369.81 d	S2.9e–S8.9e
		R Leo	09 47.3	+11 27	4.4	11.3	312.43 d	M8 IIIe
		T Cen	13 41.5	−33 35	5.5	9.0	90.44 d	K0e–M4 IIe
		R Cen	14 16.3	−59 54	5.3	11.8	546.2 d	M4e–M8 IIe
		V CrB	15 49.5	+39 34	6.9	12.6	357.63 d	C6.2e (N2e)
		RR Sco	16 56.6	−30 35	5.0	12.4	279.42 d	M6 II–IIIe–M8 IIe
		TV Her	18 14.7	+31 49	9.0	14.6	303.72 d	M4e
		χ Cyg	19 50.4	+32 54	3.3	14.2	406.93 d	S6.2e–S10.4e

	Type	Star	Co-ordinates RA	Dec	Magnitudes Max.	Min.	Period (days)	Spectral type
7.6	Semi-Regular	R Pic SRa	04 46.0	−49 15	6.7	10.0	164.2 d	M IIIe–M4 IIe
		V Cvn SRa	13 19.3	+45 33	6.52	8.5	191.89 d	M4e–M6e III
		T Cen SRa	13 41.5	−33 35	5.5	9.0	90.44 d	K0e–M4 IIe
		R Dor SRb	04 36.7	−62 05	4.8	6.6	338 d	M8 IIIe
		η Gem SRb	06 14.6	+22 31	3.2	3.9	232.9 d	M3 III
		L² Pup SRb	07 13.4	−44 38	2.6	6.2	140.42 d	M5 IIIe
		AK Hya SRb	08 39.7	−17 17	6.33	6.9	112 d	M4 III
		FH Vir SRb	13 16.2	+06 32	6.92	7.4	70 d	M6 III
		T Cet SRc	00 21.5	−20 05	5.0	6.9	158.9 d	M5–M6SIIe
		α Ori SRc	05 54.9	+07 24	0.4	1.3	2110 d	M1–M2 Ia–Ib
		α Sco SRc	16 29.0	−26 25	0.88	1.8	1733 d	M1.5 Iab–Ib + B4 Ve
		α¹ Her SRc	17 14.4	+14 24	3.0	4.0	?	M5 Ib–II
		VX Sgr SRc	18 07.8	−22 13	6.5	12.5	732 d	M4 Iae–M9.5
		IS Gem SRd	06 49.4	+32 37	6.6	7.3	47 d	K3 II
		BM Sco SRd	17 40.7	−32 13	6.8	8.7	850 d	K2.5 Ib
		TV Psc SR	00 28.0	+17 54	4.65	5.42	70 d	M3 IIIv
		SX Her SRd	16 07.5	+24 55	8.6	10.9	102.90 d	G3ep–K0 (M3)
		β Peg SRd	23 03.8	+28 05	2.31	2.74	?	M2.5 II–III
7.7	Irregular	V Ari	02 15.0	+12 14	7.7	8.3		R4
		TV Gem	06 11.8	+21 52	8.7	9.5	182 d	M1 Iab
		BL Ori	06 25.5	+14 43	8.5	9.7		Nb (C6.2)
		W CMa	07 08.1	−11 55	6.35	7.9		C6.3 (N)
		η Car	10 45.1	−59 41	−0.8	7.9	(S Dor type)	pe
		SW CrB	15 40.8	+38 43	7.8	8.5	100 d	M0
		FQ Ser	16 08.6	+08 37	6.31	6.60		gM4
		V Aql	19 04.4	−05 41	6.6	8.4	353 d	C5.4–C6.4 (N6)
		U Del	20 45.5	+18 05	7.6	8.9	110 d	M5 II–III
7.8	RW Aurigae	RW Aur	05 11.2	+30 20	9.6	13.6		dG5en
		CO Ori	05 27.8	+11 25	10.0	13.0		dK0n
		R Mon	06 38.7	+08 44	10.0	14.3		G5n
		T Cha	11 57.5	−79 22	10.0	13.2		dG5n
		RU Lup	15 56.3	−37 49	9.3	13.2		dG5en
7.9	T Orionis	LP Ori	05 35.2	−05 28	7.8	9.2		B1.5p
		NU Ori	05 35.5	−05 16	6.83	6.93		B1 V
		U Ori	05 55.6	+20 11	4.8	12.6	372.40 d	M6.5 IIIe
		RY Ori	05 35.6	−04 50	7.25	7.36		B3 Vp
		T Ori	05 35.8	−05 29	9.5	12.6		B8–A3ep V
		V 380 Ori	05 36.4	−06 43	8.7	10.81		B8–A2e (T)

Appendices 7.10, 7.11, 7.12, 7.13, 7.14, 7.15

	Type	Star	Co-ordinates RA	Dec	Magnitudes Max.	Min.	Period (days)	Spectral type
7.10	T Tauri	T Tau	04 22.0	+19 32	8.4	13.5		dGe–K1e
		RY Tau	04 22.0	+28 27	9.3	13.0		dF8e–dG5e
		SU Aur	04 56.0	+30 34	9.3	11.8		F5–G2 IIIne
		V 380 Ori	05 36.4	–06 43	8.2	10.81		B8–A2e
7.11	R Coronae Borealis	SU Tau	05 48.8	+19 04	9.1	16.0		G0ep–(C I, O)
		UW Cen	12 43.0	–54 30	9.1	14.5		K
		R CrB	15 48.6	+28 09	5.71	14.8		C0.0 (F8pe)
		RS Tel	18 18.5	–46 33	9.3	13.0		R8
		RY Sgr	19 16.3	–33 32	6.0	15.0		G0 Ipe (C I, O)
7.12	RV Tauri	RV Tau	04 45.0	+26 06	9.8	13.3	78.731 d	G2e Ia–M2 Ia
		U Mon	07 30.6	–09 46	6.1	8.1	92.26	F8e–K0 Ibp
		AC Her	18 30.0	+21 52	7.43	9.7	75.461	F2 Ibp–K4e
		R Sct	18 47.2	–05 43	4.45	8.2	140.05	G0 Iae–K0 Ibp
		AR Sgr	18 59.7	–23 42	9.1	13.5	87.87	F5e–G6
		R Sge	20 14.1	+16 44	9.46	11.46	70.594	G0 Ib–G8 Ib
		V Vul	20 36.5	+26 36	8.06	9.35	75.72	G4e–K3 (M2)
		AD Aql	18 59.1	–08 20	8.83	13.42	65.4	Fp (R)
7.13	Beta Cephei	γ Peg	00 13.2	+15 11	2.80	2.87	0.157 495	B2 IV
		ν Eri	04 36.3	–03 21	3.4	3.6	0.177 904	B2 III
		β CMa	06 22.7	–17 57	1.93	2.09	0.250 03	B1 II–III
		δ Cru	12 15.1	–58 45	2.78	2.84	0.151 038	B2 IV
		β Cru	12 47.7	–59 41	1.23	1.31	0.236 507	B0.5 III–IV
		α Vir	13 25.2	–11 10	0.97	1.04	4.014 54	B1 III–IV + B2 V
		β Cen	14 03.8	–60 22	0.61	0.7	0.157	B1 III
		θ Oph	17 22.0	–25 00	3.25	3.29	0.140 531	B2 IV
		λ Sco	17 33.6	–37 06	1.59	1.65	0.213 701	B2 IV + B
		β Cep	21 28.7	+70 33	3.16	3.27	0.190 488	B2 IIIev
7.14	UV Ceti	UV Cet	01 38.8	–17 58	6.8	12.95		M5.5 Ve
		DY Eri	04 15.3	–07 39	4.43	9.43		K IV + M4 Ve
		Wolf 359	07 44.7	+03 34	8.6	12.93	2.780 964	M4.5 Ve
		AD Leo	10 19.7	+19 52	9.41	10.94		M4.5 Ve
		EV Lac	22 46.9	+44 20	8.28	11.83		dM4.5e
7.15	Rotating	SX Ari	03 12.2	+27 15	5.67	5.81	0.727 892	B7p (Si, HeI)
		θ Aur	05 59.7	+37 13	2.62	2.70	1.373 5	B9.5p (Si)
		BM Cnc	08 13.1	+29 39	5.53	5.65	4.116	B9p (Si, Cr)
		V 815 Cen	11 07.3	–42 38	5.14	5.17	2.443	A3p (Sr)
		α^2 CVn	12 56.0	+38 19	2.84	2.98	5.469 39	A0p (Si, Eu, Hg)

Type		Star	Co-ordinates		Magnitudes		Period (days)	Spectral type
			RA	Dec	Max.	Min.		
7.16	Symbiotic (Z And)	BF Cyg	19 23.7	+29 40	9.3	13.4		Bep + M5 III
		AG Peg	21 51.0	+12 38	6.0	9.4	830.14 d	WN6 + M1–M3 III
		Z And	22 23.4	+48 48	8.0	12.4		M2 III + B1e
		R Aqr	23 43.6	–15 19	5.8	12.4	386.96	M5e–M8.5e
7.17	Cataclysmic (U Geminorum)	U Gem	07 54.8	+22 01	8.2	14.9	103 d	M4.5 + WD
		BV Cen	13 31.0	–54 57	10.7	13.6	149.4	
		AE Aqr	20 39.9	–00 53	10.4	12.0		
		SS Cyg	21 42.5	+43 34	8.2	12.4	50.1	A1–dGep
		RU Peg	22 13.8	+12 41	9.0	13.1	67.8	sdBe + G8 IV
	Z Cam	WW Cet	00 11.2	–11 30	9.3	16.8	31.2 d	
		RX And	01 04.3	+41 17	10.3	14.0	14.3 d	
		Z Cam	08 24.7	+73 08	10.2	14.5	22 d	
		SY Cnc	09 00.8	+17 55	10.6	13.7	27.3 d	
		AH Her	16 44.0	+25 16	10.2	14.7	19.8 d	

7.18 Novae

Names and dates	Co-ordinates RA	Dec	Magnitudes	
Na (Rapid)				
V476 Cyg 1920	19 58.4	+ 05 37	2.0	16.2
XX Tau 1927	05 19.4	+ 16 43	6.0	16.5
QZ Aur 1964	05 28.7	+ 33 19	6	18
BT Mon 1939	06 43.8	− 02 01	4.5	17+
RW UMi 1956	16 47.7	+ 77 02	6	21
V603 Aql 1918	18 48.9	+ 00 35	− 1.4	12
V476 Cyg 1920	19 58.4	+ 53 37	2	16.2
Nb (Medium)				
T Aur 1891	05 32.0	+ 30 27	4.1	15.5
RR Pic 1925	06 35.6	− 62 38	1.2	12.4
V732 Sgr 1936	17 56.1	− 27 22	6	16
DQ Her 1934	18 07.5	+ 45 51	1.3	15.6
HR Del 1967	20 42.3	+ 19 10	3.7	12.4
Nc (Slow)				
EU Sct 1949	18 56.2	− 04 12	8	17
V928 Sgr 1947	18 19.0	− 28 06	8.5	16.5
Nr (Repeating)				
T Pyx 1890, 1902, 1920, 1944, 1967	09 04.7	− 32 23	6.3	14.0
T CrB 1866, 1946	15 59.5	+ 25 55	2	10.8
U Sco 1863, 1906, 1936, 1979	16 22.5	− 17.53	8.8	19
RS Oph 1898, 1933, 1958, 1967	17 50.2	− 06 43	5.3	12.3
V101 Sgr 1901, 1973	18 32.1	− 29 24	6.2	14.7
WZ Sge 1913, 1946, 1979	20 07.6	+ 17 42	7	15.5
VY Aqr 1907, 1962	19 32.7	+ 17 45	4.5	19.5

7.19 Pulsars

PSR	Co-ordinates RA	Dec.	Period (seconds)
0031 − 07	00 34.0	− 07 22	0.942 950 784 86
0138 + 59	01 41.7	+ 60 09	1.222 948 267 23
0525 + 21	05 28.9	+ 22 00	3.745 497 029 02
M1 0531 + 21	05 34.5	+ 22 01	0.033 134 040 75
0538 − 75	05 36.5	− 75 44	1.245 855 434 9
0628 − 28	06 30.8	− 28 35	1.244 417 072 6
0655 + 64*	07 00.6	+ 64 18	0.195 670 944 8
0740 − 28	07 42.8	− 28 23	0.166 752 446 61
0820 + 02*	08 23.0	+ 01 59	0.864 872 75
0833 − 45	08 35.3	− 45 11	0.089 247 268 25
0919 + 06	09 22.2	+ 06 38	0.430 614 311 65
0950 + 08	09 53.0	+ 07 56	0.253 065 068 19
1323 − 62	13 27.3	− 62 23	0.529 906 294 3
1356 − 60	14 00.0	− 60 38	0.127 500 776 85
1530 − 53	15 34.0	− 53 34	1.368 880 509 0
1738 − 08	17 41.4	− 08 41	2.043 081 510 6
1804 − 08	18 07.6	− 08 48	0.163 727 360 83
1900 + 01	19 03.5	+ 01 36	0.729 301 632 74
1913+16*	19 15.5	+ 16 06	0.059 029 995 26
1920 + 21	19 22.9	+ 21 11	1.077 919 155 14
1933 + 16	19 35.8	+ 16 17	0.358 736 248 27
1937 + 21	19 39.6	+ 21 35	0.001 557 806 49
1952 + 29	19 54.4	+ 29 23	0.426 676 785 59
1953 + 29*	19 55.5	+ 29 09	0.006 133 17
2045 − 16	20 48.6	− 16 17	1.961 566 879 85
2303 + 30	23 06.0	+ 31 00	1.575 884 744 27
2319 + 60	23 21.9	+ 60 24	2.256 483 704 9

* = Member of binary system

7.20 Planetary Nebulae

NGC	M No.	Const.	Co-ordinates		Magnitudes		Diam.	Size
			RA	Dec	Visual	Central star		
246		Lyr	00 47.0	−11 53	8.0	11.94	225″	4′ × 3.5′
1535		Eri	04 14.2	−12 44	9.6	12.24	18/44	20″ × 17″
2392		Gem	07 29.2	+20 55	9.9	10.47	13/44	40″ × 10″ (Eskimo)
3132		Vel	10 07.7	−40 26	8.2	10.07	47	84″ × 52″
3242		Hya	10 24.8	−18 38	8.6	12.0	16/1250	40″ × 35″
6210		Her	16 44.5	+23 49	9.3	12.90	14	20″ × 16″
6572		Oph	18 21.1	+06 51	9.0	13.6	8	15″ × 12″
6720	M57	Lyr	18 53.6	+33 02	9	14.8	70/150	80″ × 60″ (Ring)
6853	M27	Vul	19 59.6	+22 43	7.6	13.94	350/910	8′ × 5′
7009		Aqr	21 04.2	−11 22	8.3	11.5	25/100	25″ × 25″ (Saturn)
7293		Aqr	22 29.6	−20 48	6.5	13.47	769	12′ × 12′

Appendix 8

8.1 Nebulae in the Milky Way

Constellation	Name	Number		Co-ordinates		Appearance
Aquila			B 133	19 03.8	+ 02 19	Dark patch
Ara		6188		16 40.5	− 48 47	Dark and bright
Auriga			I 405	05 16.2	+ 34 16	Bright 18′ × 30′
Carina	Eta, Keyhole	3372		10 43.8	− 59 52	Bright 85′ × 80′
Crux	Coal Sack			12 53	− 63	Dark 3° × 3°
Cygnus	Veil	6990/6992		20 56.4	+ 31 43	Bright, gossamer
Cygnus	Pelican	7000	I 5076	20 50.8	+ 44 21	Bright 80′ × 80′
Cygnus	North America	7000		20 58.8	+ 44 20	Bright 100′
Dorado	Tarantula	2070		05 38.7	− 69 06	Bright 20′
Monoceros	Hubble's Variable	2261		06 39.2	+ 08 44	Bright, small
Monoceros	Cone	2264		06 40.9	+ 09 54	Dark and bright
Musca		5189		13 30.4	− 65 45	185′ × 130′
Ophiuchus	Rho Ophiuchi		I 4604	16 25.6	− 23 26	Starry
Ophiuchus	Barnard's S neb.		B 72	17 22.6	− 23 44	Snake, dark
Orion	Great nebula	1976	M 42	05 35.4	− 05 27	Bright 65′ × 65′
Orion	42, 54 Orionis	1977		05 35.5	− 04 52	Bright 40′ × 25′
Orion		1982	M 43	05 35.6	− 05 16	Next to M42
Orion	Horse's Head		B 33	05 41.0	− 02 24	Dark and bright
Orion	Zeta Orionis	2024		05 40.7	− 02 27	Bright
Orion		2068	M 78	05 46.7	+ 00 03	Bright 6′ × 4′
Perseus	California	1499		04 00.7	+ 36 37	Bright 145′ × 40′
Sagittarius	Trifid	6514	M 20	18 02.6	− 23 02	Bright 25′ × 25′
Sagittarius	Lagoon	6523	M 8	18 03.8	− 24 23	Bright 80′ × 40′
Sagittarius			B 86	18 02.7	− 27 50	Dark 5′ × 3′
Sagittarius	Omega; Swan	6618	M 17	18 20.8	− 16 11	Bright 45′ × 35′
Scorpius			I 4592	16 12.0	− 19 28	Bright 150′ × 60′
Scorpius	Lambda	6334		17 20.5	− 35 43	Bright 30′
Serpens	Eagle	6611	M 16	18 18.8	− 13 47	Bright, also cluster
Taurus	Crab	1952	M 1	05 34.5	+ 22 01	Bright 6′ × 4′
Taurus	SN remnant	5147	Sh2-240	05 39.1	+ 28 00	Filaments

8.2 Molecules in Interstellar Nebulae

Yr	Molecule	Formula	Wavelength	Yr	Molecule	Formula	Wavelength
19–				19–			
37	methylidyne	CH	4300 Å	74	dimethyl ether	CH_3OCH_3	3.47/9.6 mm
40	cyanogen	CN	3875 Å	74	ethynyl radical	$-CCH$	3.43 mm
41	methylidyne ion	CH^+	3745 Å	74	methylamine	CH_3NH_2	3.48/4.1 mm
51	hydrogen	H	21 cm	75	acrylonitride	CH_2CHCN	21.86 cm
63	hydroxyl radical	$-OH$	2.2/5/6.3/18 cm	75	cyanamide	NH_2CN	2.98/3.43 mm
68	ammonia	NH_3	1.2,/1.26 cm	75	diazenyl ion	N_2H^+	3.22 mm
68	water vapour	H_2O	1.35 cm	75	ethanol	CH_3CH_2OH	2.8/3.5 mm
69	formaldehyde	$HCHO$	1/2.2/6.2 cm	75	methyl formate	HCO_2CH_3	18.6 cm
70	carbon monoxide	CO	2.6 mm	75	sulphur dioxide	SO_2	3.46/3.58 mm
70	cyanoacetylene	HC_2CN	3.3 cm	75	sulphur nitride	SN	2.60 mm
70	cyanogen radical	$-CN$	2.64 mm	76	cyanodiacetylene	HC_4CN	2.8/11.28 cm
70	formic acid	$HCOOH$	18.3 cm	76	formyl radical	HCO^+	3.46 mm
70	formyl ion	HCO^+	3.36 mm	76	methyl cyanoacetylene	CH_3C_2CN	3.46 mm
70	hydrogen	H_2	1100 Å	77	cyanoethynyl	C_2CN	3.37 mm
70	hydrogen cyanide	HCN	3.38 mm	77	cyanohexatri-yne	$HC_2C_2C_2CN$	2.95 cm
70	methanol	CH_3OH	0.3/1.2/36 cm	77	ketene	CH_2CO	2.94/3.67 mm
71	acetaldehyde	CH_3CHO	28.1 cm	77	methylamine	CH_3NH_2	3.48/4.1 mm
71	carbon monosulphide	CS	2.04 mm	77	nitrosyl hydride	HNO	3.68 mm
71	carbonyl sulphide	OCS	2.74 mm	78	ethyl cyanide	CH_3CH_2CN	2.58 mm
71	formamide	$HCONH_2$	6.49 cm	78	nitric oxide	NO	1.99 mm
71	hydrogen isocyanide	HNC	3.31 mm	79	methyl mercaptan	CH_3SH	2.93/3.95 mm
71	isocyanic acid	$HNCO$	0.34,/1.36 cm	80	thiocyanic acid	$HNCS$	2.13/3.65 cm
71	methyl cyanide	CH_3CN	2.72 mm	81	carbon monoxide ion	CO^+	2.56 mm
71	propyne	CH_3C_2H	3.51 mm	81	proton-carbonic acid	$HOCO^+$	2.34 mm
71	silicon monoxide	SiO	3.49 mm	81	thioformyl ion	HCS^+	1.2/3.5 mm
71	thioformaldehyde	H_2CS	9.5 cm	84	methyldiacetylene	CH_3C_4H	1.23/1.47 cm
72	hydrogen sulphide	H_2S	1.78 mm	84	silicon dicarbide	SiC_2	1.76/3.22 mm
72	methyleneimine	CH_2NH	5.67 cm	84	tricarbon monoxide	C_3O	1.56 cm
73	sulphur monoxide	SO	3.49 mm	85	hydrochloric acid	HCl	4.79 mm

From: P de la Cotardière, *Larousse Astronomy* (Hamlyn, London), and Colin Ronan, *Encyclopedia of Astronomy* (Hamlyn, London).

Appendix 9
Globular Clusters

Galactic longitudes ℓ and latitudes b (for Figure 12.21, page 234) kindly calculated by G.C. Jacobs, as well as by J.J. Franken.

Values A' and $k°$ for the projections in Figure 12.22, page 235, on the plane SMP through the Sun and the mid-point and pole of the Milky Way, deduced from ℓ and b.

Number, constellation and size	Co-ordinates RA α	Dec δ	Figure 12.21 Galactic longtitude ℓ	latitude b	Distance modulus $m - M$	Distance (1000s of light years = A)	Figure 12.22 Projection on plane Dist. A'	SMP $k°$
N0104 47 Tuc 25'	00 23.9	−72 06	305.8	−45.2	13.16	14	11.5	−60
N0288 Scl 10'	00 52.6	−26 37	167.6	−89	14.7	28.4	28	+89
N0362 Tuc 10'	01 03.1	−70 52	301.4	−46	14.9	31.1	25	−64
SW 1 Phe	01 51.0	−44 31	271.3	−69				
N1261 Hor 3'	03 12.1	−55 14	270.2	−52.1	16.05	52.9	42	−90
PAL 1 Cep	03 33.7	+79 34	130.2	+19.2	20.1	341	236	−28
ESO 1 Hor	03 54.9	−49 37	258.1	−48.4		378		
SW 2 Eri	02 24.6	−21 12	218.2	−41	19.2	226	200	+48
Sersic Dor	04 36.1	−58 51	268.5	−40.3	18.57	169	109	+88
PAL 2 Aur	04 45.8	+31 22	170.8	−8.7	20.9	493	487	+9
N1851 Col	05 14.0	−40 03	244.5	−34.8	15.55	42	28	+58
N1904 M79 Lep 7'	05 24.0	−24 32	227.3	−29	15.6	43	33	+39
N2298 Pup 3'	06 48.8	−36 00	245.8	−15.8	15.6	43	20.6	+35
N2419 Lyn 2'	07 37.8	+38 53	180.6	+25.7	19.9	311	311	−26
N2808 Car 7'	09 12.6	−64 51	282.4	−11.3	15.6	43	12	−43
ESO 3 Cha	09 21.0	−77 16	292.4	−19.2	15.45	40	19.5	−42
PAL 3 Sex	10 05.3	+60 06	240.8	+42.1	20.0	326	248	−62
N3201 Vel 10'	10 17.4	−46 23	277.6	+8.6	14.15	22	4.4	+49
PAL 4 UMa	11 29.0	+29 00	203	+72.2	19.85	304	302	−74
N4147 Com 4'	12 09.9	+18 34	255.1	+77.3	16.25	58	56.7	−87
N4372 Mus 18'	12 25.5	−72 38	301.2	−10.1	14.9	31	16.7	−19
N4590 M68 Hya 9'	12 39.2	−26.43	300.2	+35.8	15.0	32.6	23	+55
N4833 Mus 6'	12 59.3	−70 51	303.8	−8.3	14.85	30.4	17	−15
N5024 M53 Com 10'	13 12.7	+18 12	334.7	+79.3	16.34	60.4	60	+80
N5053 Com	13 16.2	+17 43	337.2	+78.5	16.03	52.4	52	+79
N5139 ω Cen 30'	13 26.5	−47 27	309.5	+14.7	13.92	20	13	+22
N5272 M3 CVn 18'	13 42.0	+28 24	41.1	+78.3	15.0	32.6	32	+81
N5286 Cen 4'	13 46.2	−51 21	311.9	+10.2	15.6	43	29	+15
AM 4 Hya	13 56.1	−27 09	320.6	+33.2				

Appendix 9: Globular Clusters

Number, constellation and size	Co-ordinates RA α	Dec δ	Figure 12.21 Galactic longtitude ℓ	latitude b	Distance modulus $m - M$	Distance (1000s of light years = A)	Figure 12.22 Projection on plane Dist. A'	SMP $k°$
N5466 Boo 5'	14 05.2	+ 28 33	41.5	+ 73.2	15.96	51	50	+ 77
N5634 Vir	14 29.4	− 05 57	342.7	+ 48.8	17.15	88	86	+ 50
N5694 Hya 2'	14 39.4	− 26 31	331.5	+ 29.9	17.8	118	107	+ 33
I 4499 Aps	14 59.6	− 82 12	307.5	− 20.8	17.05	84	56	− 32
N5824 Lup 3'	15 03.7	− 33 03	332.9	+ 21.6	17.4	98.5	89	+ 24
PAL 5 Ser	15 15.9	− 00 06	1.1	+ 45.3	16.75	73	73	+ 45
N5897 Lib 8.5'	15 17.1	− 21 00	343.3	+ 29.8	15.65	44	42.6	+ 31
N5904 M5 Ser 13'	15 18.3	+ 02 06	4.1	+ 46.3	14.51	26	26	+ 46
N5927 Lup 6'	15 27.7	− 50 39	326.9	+ 4.4	15.8	47	39	+ 5
N5946 Nor 2'	15 35.2	− 50 39	327.9	+ 3.8	16.6	68	58	+ 5
N5986 Lup 5'	15 45.8	− 37 46	337.3	+ 12.8	15.9	49	45	+ 14
PAL 14 Her	16 10.9	+ 14 58	28.9	+ 41.7	19.2	225	210	+ 46
N6093 M80 Sco 7'	16 16.8	− 22 58	353.0	+ 19.0	15.22	36	35.7	+ 19
N6121 M4 Sco 20'	16 23.3	− 26 31	351.3	+ 15.5	12.75	11.6	11.5	+ 16
N6101 Aps 10'	16 25.3	− 72 11	317.9	− 16.2	15.7	45	34	− 21
N6144 Sco 3'	16 27.0	− 26 01	352.2	+ 15.2	15.9	49	48.6	+ 15
N6139 Sco 5'	16 27.4	− 38 50	342.7	+ 6.5	16.9	78	74.5	+ 7
N6171 Oph 4'	16 32.3	− 13 03	3.6	+ 22.5	14.73	29	29	+ 23
N6205 M13 Her 23'	16 41.5	+ 36 28	59.0	+ 40.6	14.35	23.6	17.9	+ 59
N6218 M12 Oph 10'	16 46.8	− 01 56	15.9	+ 25.8	14.3	23.6	22.9	+ 27
N6229 Her 3.5'	16 47.0	+ 47 32	73.5	+ 40.1	17.5	103	70	+ 71
N6235 Oph 2'	16 53.2	− 22 10	359.2	+ 13.0	16.15	55	55	+ 13
N6254 M10 Oph 8'	16 56.9	− 04 06	154.0	+ 22.6	14.05	21	19	− 25
N6256 Sco	16 59.2	− 37 07	348.1	+ 2.8	16.3	59	57.7	+ 3
PAL 15 Oph 4'	16 59.8	− 00 33	19.2	+ 23.7		228		
N6266 M62 Sco 6'	17 00.9	− 30 06	353.9	+ 6.8	15.35	38	38	+ 7
N6273 M19 Oph 6'	17 02.3	− 26 16	357.1	+ 8.9	16.35	61	60.9	+ 9
N6284 Oph 2'	17 04.2	− 24 46	358.6	+ 9.4	15.95	51	51	+ 9
N6287 Oph 2'	17 04.9	− 22 42	0.4	+ 10.5	15.62	43	43	+ 11
N6293 Oph 2'	17 09.9	− 26 35	357.9	+ 7.3	15.54	42	42	+ 7
N6304 Oph 2'	17 14.3	− 29 27	356.1	+ 4.9	15.2	36	36	+ 5
N6316 Oph 1'	17 16.3	− 28 08	357.4	+ 5.3	16.9	78	78	+ 5
N6341 M92 Her 8'	17 17.0	+ 43 09	68.3	+ 34.6	14.5	26	16.7	+ 62
N6325 Oph 1'	17 17.7	− 23 46	1.2	+ 7.5	16.7	71	71	+ 8
N6333 M9 Oph 4'	17 18.9	− 18 31	5.8	+ 10.2	15.49	41	40.8	+ 10
N6342 Oph 1'	17 20.9	− 19 35	5.2	+ 9.2	17.1	86	85.6	+ 9

Appendix 9: Globular Clusters

Number, constellation and size	Co-ordinates RA α	Dec δ	Figure 12.21 Galactic longtitude ℓ	latitude b	Distance modulus $m - M$	Distance (1000s of light years = A)	Figure 12.22 Projection on plane Dist. A'	SMP $k°$
N6356 Oph 2'	17 23.3	− 17 49	7.0	+ 9.7	17.77	74	73	+ 10
N6355 Oph 1'	17 23.7	− 26 21	359.9	+ 4.9	16.6	68	68	+ 5
N6352 Ara 8'	17 25.1	− 48 25	341.7	− 7.6	14.25	23	22	− 8
TER 2 Sco 1.5'	17 27.2	− 30 48	356.6	+ 1.8				
N6366 Oph 4'	17 27.5	− 05 04	18.6	+ 15.5	14.8	30	28.5	+ 16
TER 4 Sco	17 30.3	− 31 36	356.3	+ 0.8				
HP 1 Oph 2.9'	17 30.8	− 29 59	357.7	+ 1.6				
N6362 Ara 9'	17 31.4	− 67 03	325.7	− 18.0	14.7	28	23.6	− 22
GR 1 Sco	17 31.7	− 33 50	354.6	− 0.6		32.6		
LIL 1 Sco	17 33.1	− 33 23	355.1	− 0.7		29.3		
N6380 Sco 2'	17 34.1	− 39 04	350.6	− 4.1		28.3		
TER 1 Sco 2.8'	17 35.5	− 30 29	357.9	+ 0.5				
TON 1 Sco 3.4'	17 35.8	− 38 33	345.8	− 7.2				
N6388 Sco 4'	17 36.0	− 44 44	351.1	− 3.9	16.40	62	61	− 4
N6402 M14 Oph 6'	17 37.4	− 03 15	21.6	+ 14.3	16.9	78	72.9	+ 15
N6401 Oph 1'	17 38.3	− 23 54	3.7	+ 3.4	16.4	62	62	+ 3
N6397 Ara 19'	17 40.3	− 53 40	338.4	− 12.4	12.3	9.4	8.77	− 14
PAL 6 Sgr 7'.2	17 43.4	− 26 13	2.4	+ 1.3	18.5	163	163	+ 1
N6426 Oph 2'	17 44.7	+ 03 10	28.3	+ 15.8	17.4	99	88	+ 18
TER 5 Sgr 2.1'	17 47.9	− 24 47	4.1	+ 1.2				
N6440 Sgr 1'	17 48.6	− 20 22	8.0	+ 3.3	18.00	130	129	+ 3
N6441 Sco 3'	17 49.9	− 37 03	353.8	− 5.5	16.2	57	56.7	− 6
TER 6 Sco 1.2'	17 50.4	− 31 17	558.8	− 2.7				
N6453 Sco 3.5'	17 50.6	− 34 36	356.0	− 4.4	17.1	86	85.8	− 4
N6496 Sco 3'	17 58.7	− 44 16	348.3	− 10.5	14.0	21	20.6	− 11
TER 9 Sgr	18 01.5	− 26 51	3.9	− 2.6				
N6517 Oph 1'	18 01.6	08 58	19.5	+ 6.3	17.4	98	92	+ 7
N6522 Sgr 2'	18 03.0	− 30 02	1.3	− 4.5	15.65	44	44	− 5
N6535 Sgr 1.5'	18 03.6	00 18	27.4	+ 9.9	15.2	36	32	+ 11
N6528 Sgr 1'	18 04.5	− 30 03	1.4	− 4.7	15.85	48	48	− 5
N6539 Sgr 2.5'	18 04.6	− 07 35	27.1	+ 6.3	16.0	52	48.6	+ 7
N6544 Sgr 8.9'	18 07.1	− 25 00	6.1	− 2.7	15.2	36	35.8	− 3
N6541 CrA 6'	18 07.7	− 43 42	349.5	− 11.7	14.60	27	26.6	− 12
N6553 Sgr 2'	18 09.0	− 25 54	5.5	− 3.6	16.05	53	52.8	− 4
N6558 Sgr 2'	18 10.0	− 31 46	0.5	− 6.5	15.8	47	47	− 7
I1276 Sgr 7'	18 10.5	− 07 12	22.1	+ 5.2	17.6	108	100	+ 6

Appendix 9: Globular Clusters

Number, constellation and size	Co-ordinates RA α	Dec. δ	Figure 12.21 Galactic longtitude ℓ	latitude b	Distance modulus $m - M$	Distance (1000s of light years = A)	Figure 12.22 Projection on plane Dist. A'	SMP $k°$
TER 11	18 12.2	−22 45	8.7	−2.7				
N6569 Sgr 2'	18 13.4	−31 50	0.8	−7.2	16.2	57	57	−7
N6584 Tel 6'	18 18.3	−52 13	342.4	−16.9	16.2	57	54.6	−18
N6624 Sgr 3'	18 23.4	−30 22	3.1	−8.4	15.15	35	35	−8
N6626 M28 Sgr 6'	18 24.3	−24 52	8.1	−6.1	15.01	33	32.7	−6
N6638 Sgr 2'	18 30.7	−25 30	8.2	−7.7	15.3	37	36.6	−8
N6637 M69 Sgr 4'	18 31.1	−32 21	2.0	−10.8	15.3	37	37	−11
N6642 Sgr 2'	18 31.6	−23 29	10.1	−7.0	14.6	27	26.6	−7
N6652 Sgr 2'	18 35.5	−33 00	1.8	−11.9	15.8	47	47	−12
N6656 M22 Sgr 18'	18 36.1	−25 54	10.2	−8.1	13.60	17	16.7	−8
PAL 8 Sgr 4.7'	18 41.2	−19 50	14.4	−7.3	18.6	171	166	−8
N6681 Sgr 7.8'	18 42.9	−32 18	3.1	−11.0	15.40	39	38.9	−13
N6712 Sct 3'	18 52.8	−08 43	25.6	−4.8	15.21	36	32.5	−5
N6715 M54 Sgr 9'	18 54.8	−30 29	5.9	−14.6	17.11	86	85.6	−15
N6717 Sgr 3.9'	18 54.8	−22 42	13.2	−11.4	15.1	34	33	−12
N6723 Sgr 7'	18 59.2	−36 38	0.3	−17.8	14.58	27	27	−18
N6749 Aql 6.3'	19 05.0	+01 52	36.3	−2.7				
N6752 Pav 15'	19 10.5	−59 59	336.7	−26.1	13.20	14	13	−28
N6760 Aql 6.6'	19 11.0	+01 01	36.4	−4.4	15.9	49	39.5	−6
N6779 M56 Sgr 5'	19 16.4	+30 11	62.9	+8.0	15.60	43	20.3	+17
TER 7 Sgr	19 17.4	−34 40	3.6	−20.6				
PAL 10 Sge 3.5'	19 17.9	+18 33	52.7	+2.3		32.6		
ARP 2 Sgr 3.7'	19 28.4	−30 22	8.8	−21.3				
N6809 M55 Sgr 15'	19 39.7	−30 58	9.1	−23.8	13.80	19	18.8	−24
PAL 11 Aql 3.2'	19 45.0	−08 02	32.1	−16.1	16.1	54	46	−19
N6838 M71 Sge 6'	19 53.6	+18 46	57.0	−4.9	13.55	17	9.3	−9
N6864 M75 Sgr 3'	20 05.8	−21 56	20.6	−26.2	16.85	76	72	−28
N6934 Del 2'	20 34.0	+07 23	52.5	−19.3	16.20	57	37.8	−30
N6981 M72 Aqr 3'	20 53.2	−12 33	35.6	−33.1	16.25	58	50.6	−39
N7006 Del 1'	21 01.3	+16 10	64.2	−19.7	18.12	137	72.7	−39
N7078 M15 Peg 10'	21 29.7	+12 09	65.5	−27.6	15.26	37	21.9	−52
N7089 M2 Aqr 7'	21 33.3	−00 51	53.9	−36.1	15.45	40	30	−51
N7099 M30 Cap 6'	21 40.1	−23 12	27.6	−47.3	14.60	27	25.6	−51
PAL 12 Cap 2.9'	21 46.4	−21 16	31.0	−48.1	16.20	57	53.5	−52
PAL 13 Peg 1.8'	23 06.5	+12 45	87.8	−42.8	17.10	86	58.5	−88
N7492 Aqr 6.2'	23 08.2	−15 38	54.4	−63.8	16.40	62	58	−74

Appendix 10

Bright Galaxies

(From: *Astronomical Almanac* and Burnham's *Celestial Handbook*.)
SA = ordinary spiral; SB = barred spiral; SAB = in-between type.

(s) = spiral arms originate directly from nucleus of galaxy.
(r) = spiral arms originate in ring around nucleus of galaxy.
a, b, c, d = degree of spread of spiral arms; from a = tight to d = very loose.
S = lens-shaped (O$^-$, O$^\circ$, O$^+$); I = irregular in shape; m = faint spiral arms.
E0 to E7 = elliptical galaxy (from spherical to flattened). p = peculiar.

Number and constellation		Co-ordinates RA	Dec.	Size (minutes)		Magnitude	Morphological type	Radial velocity
N0045	Cet	00 14	− 23 12	8 ×	5.5	12.1	SA(s)dm	+ 468
N0055	Scl	00 15	− 39 12	25	4	7.8	SB(s)m9	+ 124
N0134	Scl	00 30	− 33 17	5	1	11.4	SAB(s)bc	+ 1579
N0185	Cas	00 39	+ 48 19	3.5	2.8	11.8	E3 p	− 251
N0205	And	00 40	+ 41 40	8	3	10.8	E5 p	
N0221 M32	And	00 43	+ 40 50	3.6	3.1	8.7	E2	− 205
N0224 M31	And	00 43	+ 41 15	160	40	5	SA(s)b	− 61
N0247	Cet	00 47	− 20 47	18	5	10.7	SAB(s)d	+ 159
N0253	Scl	00 47	− 25 19	22	6	7	SAB(s)c	+ 250
Small M C	Tuc	00 53	− 72 51	210		1.5	SB(s)m p	+ 175
N0300	Scl	00 55	− 37 43	21	14	11.3	SA(s)d	+ 141
N0488	Psc	01 22	+ 05 14	3.5	3	11.2	SA(r)b	+ 2267
N0598 M33	Tri	01 34	+ 30 38	60	40	6.7	SA(s)cd	− 179
N0628 M74	Psc	01 37	+ 15 46	9	9	10.2	SA(s)c	+ 655
N0891	And	02 22	+ 42 20	12	1	12.2	SA(s)b p	+ 528
Fornax	For	02 40	− 34 33	65	30	12.5	dE4	+ 47
N1068 M77	Cet	02 43	− 00 02	2.5	1.7	8.9	SA(rs)b	+ 1153
N1097	For	02 46	− 30 18	9	5.5	10.6	SB(s)b	+ 1274
N1232	Eri	03 10	− 20 36	7	6	10.7	SAB(rs)c	+ 1683
N1300	Eri	03 20	− 19 26	6	3.2	11.3	SB(rs)bc	+ 1568
N1365	For	03 34	− 36 09	8	3.5	11.2	SB(s)b	+ 1663
N1398	For	03 39	− 26 21	4.5	3.8	10.7	SB(r)ab	+ 1407
N1433	Hor	03 42	− 47 14	7	6	11.4	SB(r)ab	+ 1067
N1566	Dor	04 20	− 54 57	5	4	10.5	SAB(s)bc	+ 1492
Large M C	Dor	05 24	− 69 46	360		1.0	SB(s)m	+ 313
N2683	Lyn	08 52	+ 33 26	9	1.3	10.6	SA(rs)b	+ 405
N2903	Leo	09 32	+ 21 31	11 ×	4.7	9.7	SAB(rs)bc	+ 556
N2997	Ant	09 45	− 31 10	6	5	11.0	SAB(rs)c	+ 1087
N3031 M81	UMa	09 55	+ 69 05	18	10	8.0	SA(s)ab	− 36
N3034 M82	UMa	09 55	+ 69 42	8	3	9.2	I0 p	+ 216
N3115	Sex	10 05	− 07 42	4	1	10.0	S0$^-$	+ 661

Appendix 10 (continued)

Number and constellation		Co-ordinates		Size (minutes)		Magnitude	Morphological type	Radial velocity
		RA	Dec					
N3184	UMa	10 18	+ 41 26	5.5	5.5	10.5	SAB(rs)cd	+ 591
N3198	UMa	10 20	+ 45 34	9	3	11.0	SB(rs)c	+ 663
N3344	LMi	10 43	+ 24 57	6	5.4	11.0	SAB(r)bc	+ 585
N3351 M95	Leo	10 44	+ 11 44	4	3	10.4	SB(r)b	+ 777
N3368 M96	Leo	10 47	+ 11 51	6	4	9.1	SAB(rs)ab	+ 897
N3379 M105	Leo	10 48	+ 12 36	2.1	2	10.6	E1	+ 889
N3486	LMi	11 00	+ 29 00	5.5	4.2	11.2	SAB(r)c	+ 681
N3521	Leo	11 06	− 00 01	6	4	10.2	SAB(rs)bc	+ 804
N3623 M65	Leo	11 19	+ 13 07	7.8	1.6	9.5	SAB(rs)a	+ 806
N3627 M66	Leo	11 20	+ 13 01	8 ×	2.5	8.8	SAB(s)b	+ 726
N3628	Leo	11 20	+ 13 37	12	2	12	Sb p	+ 846
N3938	UMa	11 53	+ 44 09	4.5	4	11.2	SA(s)c	+ 808
N4192 M98	Com	12 14	+ 14 56	8.2	2	10.7	SAB(s)ab	− 141
N4214	CVn	12 15	+ 36 21	7	4.5	10.5	IAB(s)m	+ 291
N4216	Vir	12 16	+ 13 10	7.2	1	10.9	SAB(s)b	+ 129
N4254 M99	Com	12 19	+ 14 27	4.5	4	10.1	SA(s)c	+ 2407
N4258	CVn	12 19	+ 47 20	19.5	6.5	9.0	SAB(s)bc	+ 449
N4274	Com	12 20	+ 29 38	5	1.2	11.5	SB(r)ab	+ 929
N4303 M61	Vir	12 22	+ 04 30	5.7	5.5	9.6	SAB(rs)bc	+ 1569
N4321 M100	Com	12 23	+ 15 51	5.2	5	10.6	SAB(s)bc	+ 1585
N4374 M84	Vir	12 25	+ 12 55	2	1.8	9.1	E1	+ 951
N4382 M85	Com	12 25	+ 18 13	3	2	9.3	SA(s)O$^+$p	+ 722
N4395	CVn	12 26	+ 33 34	10	8	11.0	SA(s)m	+ 319
N4406 M86	Vir	12 26	+ 12 58	3	2	9.7	E3	− 248
N4472 M49	Vir	12 30	+ 08 02	4	3.4	8.6	E2	+ 912
N4486 M87	Vir	12 31	+ 12 25	3	3	9.2	E0 °p	+ 1282
N4501 M88	Com	12 32	+ 14 27	5.7	2.5	10.2	SA(rs)b	+ 2279
N4517	Vir	12 33	+ 00 08	9	1	11.4	SA(s)cd	+ 1121
N4535	Vir	12 34	+ 08 14	6	4	10.7	SAB(s)c	+ 1957
N4536	Vir	12 34	+ 12 13	7	2	11.0	SAB(rs)bc	+ 1804
N4565	Com	12 36	+ 26 01	15	1.1	10.5	SA(s)b p	+ 1225
N4569 M90	Vir	12 37	+ 13 11	7	2.5	11.1	SAB(rs)ab	− 236
N4579 M58	Vir	12 38	+ 11 51	4	3.5	8.2	SAB(rs)b	+ 1521
N4594 M104	Vir	12 40	− 11 36	7	1.5	8.7	SA(s)a p	+ 1089
N4621 M59	Vir	12 42	+ 11 40	2	1.5	9.3	E5	+ 430
N4631	CVn	12 42	+ 32 34	12.5	1.2	9.7	SB(s)d p	+ 608

Appendix 10: Bright Galaxies

Appendix 10 (continued)

Number and constellation		Co-ordinates RA	Dec	Size (minutes)		Magnitude	Morphological type	Radial velocity
N4649 M60	Vir	12 43	+ 11 35	3	2.5	9.2	E2	+ 1114
N4656	CVn	12 44	+ 32 12	19.5	2	11.0	SB(s)m p	+ 640
N4725	Com	12 50	+ 25 32	7.5	4.8	10.5	SAB(r)ab p	+ 1205
N4736 M94	CVn	12 51	+ 41 09	5	3.5	7.9	SA(r)ab	+ 308
N4826 M64	Com	12 57	+ 21 42	7.5	3.5	6.6	SA(rs)ab	+ 411
N5033	CVn	13 13	+ 36 37	8	4	11.0	SA(s)c	+ 877
N5055 M63	CVn	13 16	+ 42 03	9	4	10.1	SA(rs)bc	+ 504
N5102	Cen	13 22	− 36 36	6	3.5	10.8	SA0$^-$	+ 468
N5128	Cen A	13 25	− 43 00	10	8	7.2	E1/S0$^+$s p	+ 559
N5194 M51	CVn	13 30	+ 47 13	10	5.5	8.1	SA(s)bc p	+ 463
N5236 M83	Hya	13 37	− 29 51	10	8	10.1	SAB(s)c	+ 514
N5247	Vir	13 38	− 17 52	4.5	4	11.9	SA(s)bc	+ 1357
N5364	Vir	13 56	+ 05 02	5	4	11.5	SA(rs)bc p	+ 1241
N6744	Pav	19 09	− 63 52	9	9	10.6	SAB(r)bc	+ 838
N6822	Sgr	19 45	− 14 49	20	10	11.0	IB8(s)m	− 54
N7331	Peg	22 37	+ 34 24	10	2.4	10.4	SA(s)b	+ 821
N7424	Gru	22 57	− 41 06	6	6	12	SAB(rs)cd	+ 941
N7793	Scl	23 58	− 32 37	6	4	9.7	SA(s)d	+ 228

Bibliography

The author thanks the authors and publishers of the following works which he consulted.

Abell GO, *Drama of the Universe*, Holt, Rinehart and Winston, New York, 1978
Asimov I, *Birth and Death of Stars*, Gareth Stevens, Milwaukee, 1989
Asimov I, *Exploding Suns*, Michael Joseph, London, 1985
Asimov I, *The Collapsing Universe*, Walker & Co., New York, 1977
Asimov I, *The Universe*, Penguin Books, Middlesex, UK, 1971
Baker PHD, *History of Manned Space Flight*, New Cavendish Books, London, 1980
Ball RS, *Elements of Astronomy*, Longmans Green & Co., London, 1896
Barlow CWC and Bryan GH, *Elementary Mathematical Astronomy*, University Tutorial Press Ltd, Oxford
Beatty J et al., *The New Solar System*, Sky Publishing Corp., Cambridge, Massachusetts, 1981
Berlage P, *Het Ontstaan en Vergaan der Werelden*, Wereld Bibliotheek, The Netherlands, 1929
Bethe H, Energy Production in the Sun, from *Origins of the Solar System*, Sky Publishing Corp., Cambridge, Massachusetts, 1940
Bok BJ, The Birth of Stars, from *New Frontiers in Astronomy, Scientific American*, W.H. Freeman & Co., San Francisco, 1977
Burnham R Jr., *Burnham's Celestial Handbook*, Dover Publications, New York, 1978
Cillié G and Wargau WF, *Halley se Komeet 1985/86*, UNISA, Pretoria, 1985
Davidson, *Elements of Mathematical Astronomy*, Hutchinson's Scientific and Technical Publications, London
de la Cotardière P and Morris MR, *Larousse Astronomy*, Hamlyn, London
de Klerk JH, *Sterrekunde-Woordeboek*, JH de Klerk, Potchefstroomse Universiteit vir C H O, Potchefstroom, South Africa, 1990
de Villiers CW, *Wonderlike Heelal*, Kaap en Transvaalse Drukkers, Kaapstad, South Africa
Dixon RT, *Dynamic Astronomy*, Prentice-Hall Inc., New Jersey, 1975
Dole SH, *Habitable Planets for Man*, Blaisdell Publishing Co., New York, 1964
Eddington AS, *The Nature of the Physical World*, J.M. Dent & Sons, London, 1935
Farmer G and Hamblin DJ, *First on the Moon*, Michael Joseph, London, 1970
Ferris T, *Galaxies*, Stewart, Tabori & Chang Publishers, New York, 1982
Gamow G, The Evolutionary Universe, from *New Frontiers in Astronomy, Scientific American*, W.H. Freeman & Co., San Francisco, 1956
Glasby JS, *The Nebular Variables*, Pergamon Press, Oxford
Glasby JS, *Variable Stars*, Constable, London
Graham-Smith F, *Radio Astronomy*, Penguin Books, Middlesex, UK, 1974
Gray DE, *American Institute of Physics Handbook*, McGraw-Hill, New York, 1972
Haber H, *Our Blue Planet*, Angus and Robertson Ltd, London, 1969
Heiserman D, *Radio Astronomy for the Amateur*, T A B Books, Pennsylvania, 1977
Henbest N, *The Mysterious Universe*, Marshall Cavendish Books, London, 1981
Herzberg G, *Atomic Spectra and Atomic Structure*, Dover Publications, New York, 1944
Hey JS, *The Evolution of Radio Astronomy*, Elek Science, London, 1973
Hirshfeld A and Sinnott RW, *Sky Catalogue 2000.0*, Cambridge University Press and Sky Publications Corp., Cambridge, London and New York, 1982
Hodge PW, *Galaxies and Cosmology*, McGraw-Hill Book Co., New York, 1966
Hoffmeister C, Richter G and Wenzel W, *Veränderliche Sterne*, Springer-Verlag, Berlin
Hoyle F, *Astronomy*, Macdonald, London, 1962
Hoyle F, *Astronomy and Cosmology*, W.H. Freeman & Co., San Francisco, 1975
Hoyle F, *Frontiers of Astronomy*, Heinemann, London, 1961
Hoyle F, *The Intelligent Universe*, Michael Joseph, London, 1983
Hunt G and Moore P, *Jupiter*, Mitchell Beazley, London, 1981
Iben I, Globular Cluster Stars, from *New Frontiers in Astronomy, Scientific American*, W.H. Freeman & Co., San Francisco, 1970
Jastrow R, *Stars, Planets and Life*, Heinemann, London, 1968
Jennison RC, *Introduction to Radio Astronomy*, Newnes, London, 1966
Kholopov J, Kukarkin NP et al., *Russian General Catalogue of Variable Stars*, NAUKA, Moscow
Klepesta J and Rükl A, *Constellations*, Paul Hamlyn London, 1960
Koestler A, *The Sleepwalkers*, Penguin Books, Middlesex, UK, 1964
Kraus JD, *Radio Astronomy*, McGraw-Hill, New York, 1966
Maddison RA, *Dictionary of Astronomy*, Hamlyn, London, 1980
Moller C and Rasmussen E, *The World of the Atom*, George Allen & Unwin, London, 1939
Moore P, *The Craters of the Moon*, Lutterworth Press, London, 1967
Moore P, *The Guinness Book of Astronomical Facts and Feats*, Guinness Superlatives Ltd, Enfield, UK, 1979
Moore P, *The Southern Stars*, Howard B. Timmins, Cape Town/Mitchell Beazley, London, 1972

Moore P, *Survey of the Moon*, Eyre and Spottiswoode, London, 1963
Moore P, *Watchers of the Stars*, Michael Joseph, London, 1974
Nicholson I, *The Sun*, Mitchell Beazley, London, 1982
Norton AP, and Inglis JG, *Norton's Star Atlas*, Gall and Inglis, Edinburgh, 1969
Payne Gaposchkin O, *Introduction to Astronomy*, Methuen & Co. Ltd, London, 1961
F Reddy, *Halley's Comet 1985*, Pan Books, London
Ronan CE, *Encyclopaedia of Astronomy*, Hamlyn, London, 1979
Roth GD, *Astronomy Handbook*, Springer-Verlag, Berlin, 1975
Sandage AR, The Redshift, from *New Frontiers in Astronomy, Scientific American*, W.H. Freeman & Co., San Francisco, 1956
Satterthwaite GE, *Encyclopaedia of Astronomy*, Hamlyn, London, 1977
Schramm DN, The Age of the Elements, from *New Fronteirs in Astronomy, Scientific American*, W.H. Freeman & Co., San Francisco, 1974
Schwarzschild M, *Structure and Evolution of the Stars*, Dover Publications, New York, 1958
Sciama DW, *Modern Cosmology*, Cambridge University Press, London, 1971
Sky and Telescope, Sky Publications, Cambridge, Massachusetts
Stranathan JD, *The Particles of Modern Physics*, The Blakiston Co., Philadelphia, 1942
Stratton FJ, *Astronomical Physics*, Methuen & Co. Ltd, London
Strong J, *Flight to the Stars*, Temple Press Books, London, 1965
Thackeray AD, *Astronomical Spectroscopy*, Macmillan, New York, 1961
Thackeray AD, *Cosmic Abundance of the Elements*, Macmillan, New York, 1961
The Astronomical Journal
The Astrophysical Journal
Tolansky S, *Introduction to Atomic Physics*, Longmans Green, London, 1945
van de Kamp P, Barnard's Star as an Astrometric Binary, from *Origins of the Solar System*, Sky Publishing Corp., Cambridge, Massachusetts, 1975
van de Kamp P, *Encyclopaedia of Astronomy*, McGraw-Hill, New York
Vehrenberg H, *Atlas of Deep-Sky Spendors*, Treugesell-Verlag, Düsseldorf, 1983
Weinberg S, *The First Three Minutes*, Bantam Books Inc., New York
Whipple FL, *Earth, Moon and Planets*, Harvard University Press, Cambridge, Massachusetts, 1970
Whipple FL, The Nature of Comets, from *New Frontiers in Astronomy, Scientific American*, W.H. Freeman & Co., San Francisco, 1974
Wilson A, *Space Shuttle Story*, Viscount Books, London, 1986
Yenne B, *Encyclopaedia of United States' Spacecraft*, Hamlyn, New York

Index

aberration constant, 56
aberration of light, 55b, 56a
absolute magnitude, 162b–164, 168a, 169, 171
absorption spectrum, 40b
accretion, 75b, 99a, 107–109, 186, 228, 231, 249, 250, 263, 267, 268
accretion disc, 207b, 219b
Achernar, 4, 9, 164
active optics, 45
Adams JC, 78a
Adhara, 164
Adonis, 74b, 75b, 76
Adrastea, 122b, 127
Agena, 98b, 164
ages of clusters, 232b, 233a
ages of elements, 263a
Air Pump, 3a, 284
Airy disc, 172a, 174
Airy G.B., 78a
Akira Fujii, 115b, 116a
albedo 68a, 84, 85
Albert Einstein, 68a
Alcock's comet, 91a
Alcor, 177a
Alcyone, 186
Aldebaran, 2a, 7, 164, 167a, 187a, 266
Aldrin E, 100a, 102b, 103
Alexander C, 131a
Algol, 3b, 7, 195b, 196a
Algol variables, 195b, 196a, 296
Almagest, 18a
Alnilam, 2a
Alnitak, 2a
Alpha Aurigae, 167a, 168b
Alpha Boötis, 165b, 167a, 190a, 191b
Alpha Canis Majoris, 163a, 165b, 266
Alpha Canis Minoris, 167a, 266
Alpha Canum Venaticorum, 207a
Alpha Carinae, 10a, 166b
Alpha Centauri, 9, 159a, 160a, 161, 162a, 164, 177, 178, 179, 180, 181a, 266
Alpha Crucis, 162a, 164, 177
Alpha Draconis, 17a
Alpha Geminorum, 166b, 181b
Alpha Herculis, 203b
Alpha Leonis, 3a
Alpha Lyrae, 162a, 165b, 166b, 170b, 267a
Alpha Orionis, 2a, 167a, 203b, 266
Alpha Piscis Austrini, 166b
Alpha Scorpii (see Antares)
Alpha Tauri (see Aldebaran)

Alpher BA, 257
Alpheus (river), 3a
Alphonsine Tables, 18a
Altair, 5, 39a, 164
Altar, 3a, 284
Amalthea, 69b, 78b, 126b, 127a
AM Cr A, 203a
Amguid crater, 93b
amino acids, 269b, 270a
Ananke, 127a
Andromeda, 3b, 4, 284
Andromeda galaxy, 226b, 240
Andromeda nebula, 210a, 223b, 253b
Angel R, 45b
Ångstrom A.J., 41a
Ångstrom units, 41a
angular diameters, 62b
angular momentum, 216b
annular eclipse, 139
Antarctic meteorites, 92b, 94b
antenna (radio), 251b
antineutrinos, 261b
Antlia, 3a, 6, 243a, 284
apastron, 178a
Apennine Mts., 51b
Apex of Sun's Way, 226b
aphelion, 24a, 31b, 58a, 139a
Aphrodite, 113b
apocynthion, 102b
apogee, 50b, 97a, 139a
Apollo, 74b, 76
Apollo asteroids, 75b
Apollo missions, 100a, 104a, 105a
Apollo-8, 100a
Apollo-11, 100a, 101a
Apollo-15, 29a, 106b
apparent magnitude, 162b
Appendices, 273–313
Apus, 9, 10a, 284
Aquarius, 3b, 4, 10, 17b, 39b, 137b, 138, 284
Aquila, 3b, 5, 218a, 221a, 284
Ara, 3a, 5, 9, 235a, 284
Araki, comet, 91a
Archer, 3b, 285
Arcturus, 6, 39a, 164, 165b, 167a, 190, 191a
Arecibo, 111b, 251
Ariel, 71b, 132b, 134
Aries, 4, 8a, 10b, 11, 17b, 137b, 284
Aristarchos, 13, 14, 18
Aristotle, 29a
Armstrong N., 100a, 102b, 103

Arrow, 3a, 285
Arsia Mons, 117a
ascending node, 58a, 83b
Ascraeus Mons, 117a
Asgard, 107a, 126a
astrometric binaries, 181b
Astronomical Journal, 111b, 187a
astronomical unit, 24a, 48a, 54a, 56b, 57, 188b, 206b, 259b
Astrophysical Journal, 188b, 207b, 224b, 257, 260b
Atlas, 98b, 130, 131b
Atlas-Agena, 98b
Atlas-Centaur, 99b
atmospheres, planetary, 69b, 113a, 84, 85
Audouze J, 230a
Augeas, 3a
Auriga, 2b, 7, 8, 9a, 284
Aurora Australis, 115, 116a, 151
Aurora Borealis, 115, 116a, 151b
autumnal equinox, 38b, 138
azimuth, 38

Baade WHW, 65a, 223b
Babcock HW, 146a
Baily F, beads, 140a
Baldwin B, 216b
Ballick B, 228a
Balmer JJ, 42b
Balmer series, 40b, 43a, 142b, 166b
Barlow lens, 36b
Barnard EE, 60b, 61a, 69b, 78b, 126b
Barnard's star, 158b, 161, 182b, 183, 266
Barnes–Evans formula, 176
Barnes TG, 175, 176
barred spirals, 242, 243
Barringer crater, 93b
Bartholdi P, 175b
barycentre, 44a, 177a, 179, 196a
Bayer J, 165a
BD + 5°16, 68, 161
BD — 12°45, 23, 161
Becklin EE, 228a
Belinda, 134b
Bell J, 253b, 357a
Bennett J, comet, 89
Bessel FW, 62b, 64b, 158b, 159, 181b, 182
Beta Aurigae, 184a
Beta Cephei, 206a, 301
Beta Crucis, 162a, 164
Beta Leonis Minoris, 184a

Beta Lyrae, 197b, 198a, 296
Beta Orionis, 2a, 166b, 266
Beta Persei, 133b, 195b, 196a
Beta Pictoris, 2a, 88b, 204b, 267a
Beta Trianguli, 168a
Betelgeuse, 2a, 7, 164, 174, 187a, 191a, 266
Bethe H, 154a
Bianca, 134b
Biela , comet, 90a
Bielids, 91a
Biermann LF, 87b
big bang theory, 257b, 259, 261
binaries, 44a, 213b
Bird of Paradise, 10a, 284
black body, 151b, 152a, 190
black body curve, 257b
black hole, 198a, 214a, 228a, 250b, 262b, 264
blink comparator, 76b, 77a
Blitz L, 226b, 227
blueshift, 219a
blue-white super-giants, 167b
BM Orionis, 184b
Bode J, 73
Bok B, 231a
Bok globules, 231a
bolides, 92a
bolometer, 68b
bolometric correction, 171a, 176a, 190
bolometric magnitude, 168a, 171a, 176a, 188b, 190
Bond GP, 78b
Bondi H, 257b, 260b
Boötes, 3a, 6, 8, 284
Boötes cluster, 248a
Boss L, 187b
Bouguer, photometer, 44a
bow shock, 121b, 150, 151a
Bradley J, 55b, 62b
Brahe T, 21b, 22
brightest stars, 239a, 247
bright galaxies, 311, 312, 313
bright giants, 167b
brightness of stars, 160b
Brookes' comet, 88b
Brown RH, 175a
Brown RL, 228a
brown stars, 262b
Bruno Giordano, 20b
Bruston P, 230a
Bull, 2a, 187a, 285
Bunsen R, 40b, 44a
Bunsen photometer, 40b, 44a

317

Burbidge EM, 212b, 254a
Burbidge GR, 212b, 254a
Burg Prof, 181a
Burke BF, 253a
Burnham R, 184b, 229a, 241b, 247a
Burnham SW, 165b
butterfly diagram, 145b
BY Draconis, 207a

Caelum, 2a, 7, 284
caesium clock, 39a
Callisto, 28, 122b, 123a, 126, 268a
Caloris Basin, 107a, 110b, 111a, 268a
Calypso, 130, 131b
Camelopardalis, 8, 9a, 284
canali, 58b
Cancer, 2b, 7, 10b, 11, 137b, 185b, 284
Cancer, tropic of, 16
Canes Venatici, 3a, 6, 241b, 284
Canis Major, 2, 7, 284
Canis Majoris stars, 206a
Canis Minor, 2b, 7, 284
Canopus, 2b, 7, 9, 10a, 121a, 162a, 163–166, 184a, 206a
Capella, 2b, 7, 164, 167a
Cape of Good Hope, 115
Cape Town Observatory, 52a, 79a, 159a, 226a
Capricorn, Tropic of, 16
Capricornus, 3b, 4, 5, 10, 137b, 138a, 284
carbon–nitrogen–oxygen cycle, 205
Carina, 2b, 6, 7, 9, 10a, 284
Carme, 127a
Carswell RF, 256a
Cassegrain telescope, 34a
Cassini gap, 47a, 61a, 128b
Cassini GD, 28a, 47a, 64b, 69a
Cassiopeia, 3b, 4, 8, 10a, 251b, 284
Cassiopeia A, 252b
Castor, 2b, 7, 164, 166b, 180b, 181b
cataclysmic variables, 207, 208a, 302
Celaeno, 186
celestial pole, 37b
celestial sphere, 38b
Centaur, 3a
Centaurus, 3a, 6, 9, 10a, 284
Centaurus A, 244, 252a
Centripetal force, 29, 30, 31a
Cepheid variables, 222–225, 232, 238b, 239b
Cepheus, 8, 10a, 17a, 284
Ceres, 73b–75a, 76
Cetus, 3b, 4, 8a, 284
Chamaeleon, 9, 10a, 284
Chameleon, 10a, 284
Chandrasekhar S, 213b, 214a, 218a
charge coupled device, 45a, 80b, 170a
Charioteer, 2b, 284
Charon, 80b, 81a
Chi Cygni, 203a
Chiron, 76b
chlorophyll, 270a
Christy J, 80b
chromatic aberration, 35a
chromosphere, 140b
Churms J, 168a
Cincinnati, 74b
Circinus, 5, 10a, 284
circular measure, 25a, 276
Clark AC, 182a
Clark DH, 212a, 218b
clusters, 185–187, 294, 307
C–N–O cycle, 205
cobalt-56, 213, 214a
COBE, 257
Cocke WJ, 253b

coefficient of correlation, 188b, 277, 278
Colette volcano, 113b
collimator, 39b, 142b
Collins M, 100a, 103b
colour, 170
colour index, 170, 171, 176, 190
colour index diagram, 170b, 171
Columba, 2a, 7, 284
Coma Berenices, 3a, 6, 243b, 284
Coma Berenices cluster, 3a, 246b
Comet, 1682, 82a
cometary orbits, 86b
comet nucleus, 87a
comets, 1b, 81b
comets, discovery of, 86b
comets' spin, 88
command module, 100a, 104a
Compass, 2b, 284
compound lens, 35b
Cone nebula, 229a, 230a
conjunction, 70a
constant of gravitation, 30b, 64b
Constellation - I, 239b
Constellations, 10a, 284, 285
continuous creation, 261a
Copernicus crater, 51, 98b, 99a, 108b
Copernicus N, 18a, 19
Corbin TE, 188a
Cordelia, 134b
Corona, 1b, 149
Corona Australis, 3b, 5, 284
Corona Borealis, 3a, 5, 284
Corona Borealis cluster, 248a
coronagraph, 149a
coronal voids, 149b
coronium, 149b
correcting plate, 34b
correlation coefficient, 188b, 277, 278
Corvus, 3a, 6, 244a, 284
cosine, formula, 275
Cosmic Background Explorer, 257
cosmic egg, 261a, 262a
cosmo-genesis, 261a, 262a
cosmological principle, 260b, 263a
CP 1919 + 21, 253b
Crab, 2b, 284
Crab nebula, 211, 212, 216, 217a, 252a, 253
Crane, 3b, 284
Crater, 3a, 6, 284
craters, 108, 267b
Crawford AB, 257
crepe ring, 61b
Cressida, 134b
Crow, 3a, 284
Crux, 6, 9, 10a, 284
Cup, 3a, 284
Curtis RH, 167b
cyanogen, 87b, 230b
cyclic universe, 261b, 262a
Cygnus, 3a, 5, 8, 17a, 218a, 221, 251a, 284
Cygnus A, 253, 254a
Cygnus X1, 253b

Danielson E, 123
dark matter, 227a, 262b
D'Arrest H, 78a
Davis J, 175a
declination, 38, 39a
declination axis, 37b
declination circle, 37b, 38a
degenerate matter, 261b
Deimos, 116b, 268a
de Lacaille NL, 178a
Delisle L, 54b
Delphinus, 3a, 5, 284

Delta Canis Majoris, 167a
Delta Cephei, 4, 8, 201a, 210, 225, 226a, 298
Delta Cepheids, 201 –202a
Delta Geminorum, 168a
Delta Orionis, 201a
Delta Pavonis, 9, 10a, 270b
Delta Scuti, 202b, 210, 299
Deneb, 5, 164
Denebola, 6
de-oxyribonucleic acid, 269b
De Revolutionibus Orbium Coelestium, 20a
descending node, 58a, 83b
Desdemona, 134b
Deslandres H, 142b
Despina, 136a
deuterium, 153b, 216b
deuteron, 153b, 154
de Vaucouleurs G, 243
DHM 0054-284, 255
Dicke RH, 257
diffraction grating, 40
dimethyl hydrazine, 102b
dinosaurs, 75b
Dione, 47a, 130
Dione B, 130, 132a
Dipper, 2b
Disney MJ, 253b
distance determination, 247, 268af
distance modulus, 163a, 235a
docking manoeuvres, 102a, 103b
Dole SH, 270
Dolland J, 55a
Dollfus A, 78b, 128b, 131b
Dolphin, 3a, 284
Donati's comet, 87b
Doppler CJ, effect, 43b, 61b, 62a, 112a, 182b, 201b, 212a
Doppler shift, 218b
Dorado, 2b, 7, 9, 10a, 238, 250a, 284
double stars, 44a, 177, 184
Dove, 2a, 284
DQ Herculis, 208b
Draco, 3a, 5, 8, 9b, 10a, 17a, 284
Dragon, 3a, 284
Draper H, 165a, 228b
dwarf novae, 207a
dwarfs, red, 160b, 182b, 210, 266b
dwarfs, white, 182b, 191b, 207b, 209b, 210, 213a, 217–219, 262, 266a
Dyce RB, 111, 112

Eagle, 3b, 284
early stars, 166a
Earth, 10b, 18b, 26, 27, 67, 82, 114–116
  atmosphere, 84, 113a, 268b, 269
  circumference, 15b, 18
  core (nucleus), 65a
  craters on, 93
  data not listed here, 84
  density, 30b, 65a, 84
  distance from Sun, 57b, 84
  eccentricity of orbit, 18a, 84, 139a
  equatorial radius, 50a, 56b, 84
  escape velocity, 65b, 84
  flattening, 63a
  from space, 106a
  grazers, 75b, 76
  magnetic field, 115
  magnetopause, 150
  magnetosheath, 150
  magnetosphere, 150
  mass, 30b, 64a, 84
  Moon not a moon, 66
  speed in orbit, 56a, 84
  size to scale, 81b, 82

eastern elongation, 20b
eccentricity, 17b, 25a, 57b, 75a, 83b
eclipse of Moon, 138–140
eclipse of Sun, 138–140
eclipsing binaries, 195–200
Ecliptic, 3a, 4, 9a, 11, 15b, 18b, 38, 137a, 138, 139b, 226
Ecliptic plane, 58a
ecosphere, 270a
Eddington AS, 188, 190a, 201b
Einstein A, 31, 154a, 196b, 198a, 205b, 254b, 256
Einstein ring, 256b
Elara, 127a
electron, 42a, 153b
electro-magnetic spectrum, 41
elements, abundance, 141b
elevation, 38
ellipse, 31b, 86a
elliptic galaxies, 263b
elongation, greatest, 21a
emission spectrum, 40
Encke JF, comet, 87b, 89
Encke gap, 61a
Endeavour shuttle, 46a
energy levels, 42
Epimetheus, 128b, 130, 131b
Epsilon Aurigae, 199a
Epsilon Eridani, 160, 161, 167a, 176, 270b
Epsilon Hydrae, 170b
Epsilon Indi, 161
Epsilon Orionis, 166b
Epsilon Sculptoris, 168b
equator, 4, 16, 18b, 38
equatorial diameters, 62, 84, 85
equatorial horizontal parallax, 49, 50
equinoxes, 17b, 39b, 137a, 138a
Equuleus, 3b, 4, 284
Eratosthenes, 14b, 15, 18
Eridanus, 4, 7, 8a, 9, 243a, 284
Eros, 54a, 74b, 76
eruptive variables, 207, 208
escapement, 39a, 107b
escape velocity, 65b, 66a, 98b, 198a, 214a, 218b
Eta Aquilae, 202a
Eta Carinae, 10a, 206
Eta Coronae Borealis, 181a
Eta Tauri, 186a
ethanol, 306b
ether, 56b
Eunomia, 74a
Euphrosyne, 74b, 75a
Europa, 28, 121b, 122b, 123a, 124b, 125
European Southern Observatory, 37a, 45a, 206b, 217
European Space Agency, 87a
Evans DS, 175, 176
Evershed J, 145a
expanding universe, 260b, 261
exploding galactic nuclei, 249b, 250a
Explorer satellite, 98a, 115a
E0 galaxy, 242a
E5 galaxy, 242a
E7 galaxy, 241b
$E=mc^2$, 154a, 205b

Fabricius J, 144a
false disc, 172a
Feltz KA, 199b
Ferris T, 241b
Fesen RA, 211a
filaments, 145a, 146a, 147b
filar micrometer, 36b, 62b, 157a, 177b
filters, 170b

Index

Finsen WS, 74b, 181a
First Point of Aries, 17b, 37b, 38, 39b, 53a, 58a
First Point of Libra, 39b, 58a
Fishes, 8a, 284
fixed stars, 2a
Fizeau AHL, 43b, 56b
FK Comae Berenices, 207a
Flamsteed Numbers, 165a, 186
flares, solar, 151a
flare star, 206b
flat earth, 15a
Flather EF, 188b
flattening, 52b
flexus, 125a
fluorescence, 87b
Fly, 10a, 284
Flying Fish, 10, 285
Flying Horse, 3b, 284
Foal, 3b, 284
Fomalhaut, 4, 164, 166b
forbidden lines, 149b, 229a,
formaldehyde, 231a
formic acid, 231a
Fornax, 4, 7, 8a, 284
Fornax system, 240, 241
Foucault JBL, 56b
Fountain JW, 128b
Fowler WA, 212b, 254a
Fox, 3a, 285
Franken JJ, 235b
Fraunhofer J, 40b
Fraunhofer lines, 40b, 42a, 149a
frequency, 40a
Fresnel A–J, 172b
F-ring, 128b, 131b
fundamental data, 56b
Fujii Akira, 115b, 116a
FU Orionis, 209a
Furnace, 8a, 284
Fusion, 154, 213a

Gagarin J, 98b
galactic clusters, 184b, 294, 295
Galatea, 136a
galaxies, 221–236
 classification of, 241b, 242
 clusters of, 246b, 247a
 elliptical, 241b, 263b
 evolution of, 263
 exploding, 264b
 formation of, 263a
 irregular, 244a
 recession of, 245
 shape of, 227b, 228
 speeds in, 227
 spiral, 242
Galilean moons, 123b
Galileo G, 27a, 144a, 221a
Galileo probe, 76b
Galileo Regio, 125b
Galle JG, 78a
Gamma Andromedae, 38, 39a, 98
Gamma Arietis, 207a
Gamma Crucis, 164, 167a
Gamma Lyrae, 198a
Gamma Orionis, 166b
gamma ray bursters, 239b
gamma rays, 41, 154, 205b, 261a
Gamma Velorum, 166a
Gamow G, 257, 261a
Ganymede, 28, 122b, 123a, 125, 126a
Garradd G, 214b
Gascoigne W, 36b
Gatewood G, 226a
Gemini, 2b, 7, 10b, 11, 137b, 138a, 284

Gemini capsule, 100a
Geminids, 91b
Gezare DY, 175a
Giacobinids, 252a
Giant planets, 69b, 85
Giant stars, 167b, 170a
Gibeon meteorites, 92a
Gill D, 54a, 79a, 226a
Giordano Bruno, 20b
Giotto probe, 87a
Giver LP, 61b, 62a
GK persei, 208b
Glasby JS, 204a
Glenn J, 99b
globular clusters, 231–236, 264a
Goat, 3b, 284
Goldfish, 2b, 10a, 284
Gold T, 253b, 257b, 260b
Goodricke J, 195b, 196a, 197b, 201a
Gorenstein P, 218a
Gorgons, 3b
graduated circles, 39a
Grand Tour, 120
grating, 40a
Graving Tool, 2a
gravitational lenses, 256
gravitational vortex, 198a, 214a, 218a, 219b, 228a, 250b, 264
gravitation, law of, 30b, 179b, 180a
gravity, 65b
grazing occultation, 70b
Great Bear, 2b, 177a, 285
great crunch, 262a
Great Dog, 2a, 284
greatest elongation, 70a
Great Nebula in Orion, 2a, 184b, 228a, 231b
Great Red Spot, 121b, 122a
Greenwich Observatory, 52a
Groombridge-34, 161
Groth EJ, 259b
Grus, 3b, 4, 9, 284

habitable planets, 171b, 270a
Hale GE, 142b
Hale Observatories, 143, 248a
Hale telescope, 37a, 45b
Hall RG, 188b
Halley E, 54a, 82a
Halley's comet, 82b, 83, 86a, 87b
halo, 231b
Hanson R, 188a
Harding KL, 74a
Hare, 2a, 284
Harriot T, 47a
Hartmann JF, 230b
Harvard College Observatory System, 165b, 222b
Hayashi C, 261b
Haywain, 2b, 285
HD Catalogue, 165a
HD 33793, 161
HD 77370, 168b
HEAO-2, 217a
Hebe, 74b
Helene, 130, 132a
heliocentric annual parallax, 158b
heliometer, 55a
heliopause, 136b
Helios-2, 239b
helium, 41a, 154, 261b, 263a
helium flash, 212b, 266b
helium, ionised, 166, 202
Helix nebula, 209b
Henderson T, 159a
Herbst T, 90b

Hercules, 3a, 5, 17a, 233b, 284
Hercules cluster, 247a
Herdsman, 3a, 284
Herman H, 257
Hermes, 75b
Herschel FW (William), 41a, 62a, 71, 73b
Herschel wedge, 140b
hertz, 42b
Hertzsprung E, 168, 169
Hertzsprung–Russell diagram, 168, 169
Hewish A, 252a, 253b
Hewitt J, 257
Hey JS, 230b, 252a, 254a
Hidalgo, 74b, 75a
Hilton AS, 78a
Himalia, 127a
Hipparchus, 16b, 17b, 18, 160b, 211a
Hirshfeld A, 168a
HK lines, 42a, 246a, 248a
Hoba meteorite, 92b
Hodge P, 187a
Hoffmeister C, 202a
Hofman KH, 206b
Holwarda, 203a
homogeneous universe, 260b
Hooke R, 29b
Hooker telescope, 37a
horizon, 38
Horologium, 7, 8a, 284
Horsehead nebula, 229
hour angle, 39a
hour circle, 38b
Howell TF, 257b
Hoyle F, 212b, 257b, 260b
HR Dephini, 209a
H–R diagram, 168, 169, 185b, 186, 203a, 210, 232, 263b
Hubble EP, 216b, 223, 241b, 242, 245
Hubble's constant, 245b, 246a, 249a, 254b, 255, 264a
Hubble's law, 245b, 249a
Hubble Space Telescope, 45b, 81a, 90b
Humason ML, 223a, 245a, 262b
Hunting Dogs, 3a, 284
Hunting Scene, 2a
Huygens C., eyepiece, 35b, 39a, 47a, 56b, 69a, 131a
Hyades, 187, 188b
Hydra, 2b, 3a, 6, 7, 242b, 284
Hydra cluster, 246b, 248a
hydrocyanic acid, 231a
hydrogen, 261b, 263a
hydrogen spectrum, 42
hydroxyl radical, 231a
Hydrus, 9, 10a, 284
Hygeia, 74
hyperbola, 86a
Hyperion, 78b, 130, 131b

Iapetus, 47a, 130, 131
Iben I, 233
Icarus, 74b, 75b, 76
Ida, 76
IDS Catalogue, 181a
impacts on Jupiter, 90b, 91a
inclination of equator, 58b
inclination of orbit, 21b, 58a, 74
Indian, 3b, 284
Indus, 3b, 4, 5, 284
inferior conjunction, 57a, 62b
infrared, 41, 244a
Innes RTA, 159b, 165b, 181a
instability region, 210
intelligent life, 270b
interference, 172a, 173

interference fringes, 173, 174, 254b
interferometry, 174, 191a, 251b
International Astronomical Union, 48a
International Geophysical Year, 97
interstellar molecules, 230b, 231
intrinsic brightness, 162b
intrinsic variables, 201–219
inverse squares, 30b
Io, 28, 121b, 122b, 123a, 124, 268b
ionised atoms, 149b, 166, 193
ionosphere, 151b
ions, 87a
Iota Boötis, 199a
Iota Librae, 184a
IRAS satellite, 91b
Iris, 74b
irregular galaxies, 244a
irregular variables, 240a, 300
Irvine–Michigan–Brookhaven, 215a
Ishtar Terra, 113b
isotropic universe, 245a, 260b

Jacobs GC, 235b
Jansky K, 230b, 250b
Janssen J, 142a
Janus, 78b, 128b, 130, 131b
Japanese Observatory, 45b
Jausa LF, 94b
Jet Propulsion Laboratory, 98b, 116b
Jewel Box, 185a
Jewitt D, 123
Johannesburg Observatory, 74b, 159, 181a
Johannes Kepler, 23, 211a
Johnson HL, 170b
Joly's photometer, 44a
Joy AN, 199b
Juliet, 134b
Juno, 74, 67, 76
Jupiter, 9, 10b, 11a, 18b, 19, 20, 26, 27, 82, 85, 120–124
 angular diameter, 62b
 atmosphere, 69b, 85, 122b
 bow shock, 121b
 –C Booster, 98b
 cross-section, 137b
 data not listed here, 85
 density, 65a, 85
 diameter, 62b, 85
 distance from Sun, 27, 85
 distant moons, 127
 family of comets, 88b
 flattening, 63b
 gravity, 65b, 77b, 83, 88b
 Great Red Spot, 59b
 inclination of equator, 58b, 85
 magnetic field, 120b, 122
 mass, 64b, 85
 moons, 28b, 123–127
 period of rotation, 122a
 perturbation of Halley's comet, 83b
 pull on Uranus, 78a
 radio emission, 122a
 ring, 127b
 sidereal period, 16a, 57b, 85
 size to scale, 81b, 82
 slingshot acceleration, 120, 122
 synodic period, 57b, 85
 velocity in orbit, 67, 85
 winds in atmosphere, 122b

Kamiokande, 215a
Kappa Ceti, 170b
Kappa Crucis, 185a
Kappa Geminorum, 168a

Kapteyn JC, 158b, 226b, 227b
Kapteyn's star, 161
Keck WM, telescope, 45a
Keel, 2b, 284
Keeler JE, 60a, 61b
Kellner eyepiece, 36a
Kepler, Johannes, 23–27, 211a
Kepler's laws, 25b, 27a, 30b, 61b, 180a
KH lines, 42a, 142, 143a
Kholopov PN, 195a
Kids, 2b
kinetic energy, 149a, 150a
Kirchhoff G, 40b, 41a
Kirkwood D, gaps, 75a
Kirzenberg JD, 48a
Kopal Z, 188b
Kowal C, 76b
Krüger-60, 161
Kuiper Airborne Observatory, 72b, 215a
Kuiper GP, 71b, 78b, 110b, 179a
Kukarkin P, 195a

Labeyrie A, 175a
Lacaille-8760, 161, 165b
Lacaille-9352, 161
Lacerta, 3b, 4, 284
Lacy CH, 175, 176
Lagoon nebula, 229b
Lagrange points, 77b, 131b
Lakshmi Planum, 113b
Lalande-211851, 161
Landau LD, 214a
Landsat satellite, 94a, 114b, 115
Laplace PS, 64b, 264b
Laques P, 130
Large Magellanic Cloud, 10a, 214b, 215, 216a, 243b
Larissa, 136a
Larson S, 128b
laser reflectors, 107a
Lassell W, 71b, 78b
late stars, 166a
Laurent C, 230a
Law of gravitation, 30, 64a
Leavitt HS, 160a, 222, 223
Lecasheux J, 130
Leda, 127a
Leibnitz GW, 28b
Leiden Observatory, 252b, 253a
Leighton R, 146a
Lemaitre G, 257, 261a
Leo, 2b, 6, 10b, 11, 137b, 138a, 284
Leo Minor, 2b, 6, 284
Leonids, 91b
Leo systems, 239b, 240, 241
Leo I and II, 240, 241a
Lepus, 2a, 284
Lesser Dog, 2b, 284
Lesser Lion, 2b, 284
Level, 3a, 284
Leverrier UJJ, 62b, 78a
Levy D, comet, 90b
Libra, 3a, 10, 137b, 284
light year, 159
limb darkening, 141a, 195b
Lin C, 231b
line of apsides, 196b
line of best fit, 225, 281, 282, 283
line of nodes, 83a, 138
linea, 125a
Lion, 2b, 284
Lipperscheij H, 27b, 28a
Lippincott SL, 188b
Lizard, 3b, 284
LMC, 9, 239, 240

Local Group, 241
Lockyer N, 41a, 142a
Lodge O, 149b
logarithms, 276b, 277
longitude of ascending node, 86a
longitude of perihelion, 86a
long period Cepheids, 203a, 223b
long period variables, 203a, 299
Lorentz AA, 145a
Lowell Observatory, 22b, 185a
Lowell P, 58b, 116a
luminosities of double stars, 188, 189, 292, 293
Luna I and III, 98
Lunar Excursion Module, 100a
Lunar Lander, 102a
Lunar Orbiter, 98b
Lupus, 2a, 3a, 5, 6, 7, 9, 284
Luyten L-789-6, 161
Lyman T, series, 42b, 43a
Lynx, 2b, 7, 8, 9a, 284
Lyot BF, 149a
Lyra, 3a, 5, 17a, 284
Lyre, 3a, 284
Lyrids, 91b
Lysithea, 127a

macula, 125a
Maffei P, 244a
Maffei-1, 244a
Magellanic Clouds, 9, 222, 223a, 226a, 238, 239
Magellan space probe, 114a
magnetic storm, 151
magnetogram, 151
magnetohydrodynamic waves, 149b
magnetopause, 150, 151a
magnetosheath, 151a
magnetosphere, 151a
magneto tail, 150
magnitude absolute, 161, 162b, 163
magnitude apparent, 160b
magnitude of stars, 160b, 161, 162b, 163
magnitude visual, 161, 162b, 163
Maia, 186
Main Sequence, 169, 170, 185b, 186, 188b, 233a, 263b
Maksutov telescope, 34b
manned orbital flight, 99b
Mare Imbrium, 99b, 107a
Mare Orientale, 107a, 268a
Mare Tranquillitatis, 103a
Margon B, 219a
maria, 98b, 109b, 267b
Marietta Aerospace, 182a
Mariner-4, -6, -7, -9, 116
Mariner-10, 110a, 267b
Marius crater, 99a
Marius Von Gunzenhausen, 27b, 123b
Mars, 10b, 11a, 18b, 19–27, 53, 67, 82, 116–119
  atmosphere, 69b, 84, 113a
  canals, 116
  craters, 116a, 118
  data not listed here, 84
  density, 65a, 84
  diameter, 62b, 84
  distance, 23, 24, 27a, 84
  elliptic orbit, 25
  flattening, 63
  floods on, 118
  ice caps, 58b
  inclination of equator, 58b, 84
  moons, 116b, 117a
  parallax, 53b

retrograde motion, 21
rim, 69b
seasons, 58b
sidereal period, 16a, 19b, 23a, 27a, 59a, 84
size to scale, 81, 82a
surface temperature, 68a, 84
synodic period, 57a, 84
velocity in orbit, 67, 84
mascons, 98b
maser, 253a
masses of double stars, 188, 189, 292, 293
mass–luminosity law, 188–190, 233a, 278, 279
mass spectrometer, 155b, 263a
mathematical principles, 273–283
Matthews JA, 254a
Mauna Kea-Loa, 117b
Maunder E., minimum, 145b
Max Planck Institute, 87a, 90b
Maxwell Montes, 113b
McClure B, 188a
McDonald Observatory, 175b
McGraw-Hill telescope, 257a
McKinley DWR, 253a
McKinnon W, 81a
McMillan RS, 48a
McNamara DH, 199b
Medusa, 3b
Melotte PJ, 123
Mensa, 9, 10a, 284
Mercury, 10b, 11a, 16a, 18b, 19, 20, 26, 27, 67, 82, 107a, 110–112, 253a, 267b
  atmosphere, 69b, 84
  Caloris basin, 110b, 111a, 126a
  data not listed here, 84
  density, 65a, 84, 110b
  diameter, 62b, 84
  eccentricity of orbit, 26a, 58a, 84
  escarpments, 110b
  inclination of orbit, 21b, 58a, 84
  period of rotation, 59a, 84, 111, 112
  shift of perihelion, 196b
  sidereal period, 20a, 27a, 59a, 84
  size to scale, 81b, 82
  synodic period, 84
Mercury capsule, 99b
Mercury, transit of, 54a
meridian, 38
Merope, 186
Messier CJ, 231b
Messier objects, 231b
  M1, 7, 211, 216, 237,
  M3, 6, 237
  M6, 5, 185
  M7, 5, 185
  M8, 5, 229
  M13, 5, 233
  M16, 5
  M17, 5
  M20, 5, 230, 237
  M22, 5, 237
  M31, 4, 210, 223
  M33, 4, 240,
  M42, 2a, 7, 184, 228, 231, 237
  M44, 7
  M46, 7, 185
  M51, 3a, 6, 241, 244
  M57, 5, 166, 209, 237
  M67, 185
  M72, 5, 236
  M74, 4, 226
  M81, 242
  M82, 249, 264, 265
  M83, 6, 243
  M87, 6, 241, 242, 244, 252, 256
Meteor Crater, 93b

meteors, 90–94, 267b
methane, 268b, 269
methyl alcohol, 231a
methylidene, 230b
Metis, 122b, 127
MG, 1131+0456, 256b
Michelson AH and Morley EW, 56b, 260b
Microscope, 3b, 284
Microscopium, 3b, 5, 284
micro waves, 41a, 263b
Milky Way, 2a, 3b, 27b, 184b, 218b, 221–236
Milky Way neighbours, 239b, 240, 241
Miller S, 269b
Mimas, 61a, 78b, 129a, 130
Minnette HC, 252a
Minor Planets, 73b–77
Mintaka, 2a
Minton RB, 182a
Mira, 4, 203, 210
Miranda, 71b, 132b, 134, 268b
missing mass, 264b
Mitton S, 254a
Mizar, 6, 8, 177a, 184a
molecules, interstellar, 306
Monoceros, 2b , 7, 229, 284
Moon, 3a, 11, 18, 104–109
  angular diameter, 50b, 51a, 139a
  as a planet, 65b, 66
  average distance, 56b
  craters, 108
  data not listed here, 84
  density, 65a, 84
  distance, 48, 49, 84
  eccentricity of orbit, 18, 50, 84, 139a
  elliptical orbit, 50b
  escape velocity, 65b, 84
  far side, 106
  heights of peaks, 51
  inclination of equator, 58b, 84
  inclination of orbit, 58a, 84, 138b
  magnitude, 162a
  maria, 98b, 106b, 267b
  occultations, 70, 175
  parallax, 48, 49
  phases, 13a
  quakes, 103, 107b
  research on, 104b
  rocks, 265b, 267b
  sidereal period, 16a, 84
  surface temperature, 69a, 84
  synodic month, 140a
  synodic period, 16a, 57a, 84
  width of craters, 51
moons of planets, 69
Morabito L, 124a
Morgan WW, 170b, 226b
Morley EW, 56b, 260b
morphological types of galaxies, 241a, 242, 243
Morrison P, 250a, 265a
Mount Palomar, 37a, 45b, 76b, 223b
Mount Wilson, 37a, 174a, 210a, 223a
Mueller S, 81a
Mu Geminorum, 175b, 207, 208a
Muller PM, 98b
multiple stars, 184b–188a
Murdin P, 218b
Murray BC, 116b
mylar filter, 140b

nadir, 38
Naiads , 136a
Napier (Neper) J, 24a
Narratio Prima, 20a
NASA, 46a, 72b, 106b, 120, 121

Index

National Optical Astronomical Observatories, 216b, 217, 233b
nearest stars, 161
nebulae, 228b, 229, 230a
nebulae in Milky Way, 228b, 229, 230a
nebular variables, 204a
Neptune, 26, 27a, 67, 78, 82, 134b, 135a
   black spot, 135a
   data not listed here, 85
   density, 65a, 85
   diameter, 78b, 85, 135a
   discovery of, 78
   distance from Sun, 27a, 78b, 85
   inclination of equator, 58b, 78b, 85
   magnetic field, 134b
   mass, 78b, 85
   polar diameter, 85
   rings, 134b
   satellites, 135b, 136
   sidereal period, 27a, 78b, 85
   size to scale, 81b, 82
   velocity in orbit, 67, 85
Nereid, 78b, 135a, 136
Net, 8a, 285
neutrino, 153b, 154, 155b, 205, 212a, 215a
neutron, 153b, 154
neutron star, 212b, 213b, 216b, 219b
Newcomb S, 55a
new moons, 78b
Newton I, 28, 29, 34a
   law of gravitation, 30b, 179b, 180a
   laws of motion, 29, 179b
   light theory, 56b
   mass of Jupiter, 64b
   mass of Sun, 64b
   spectrum, 35a
   telescope, 34
   third law of motion, 29b
Neugebauer G, 228a
New Technology Telescope, 45a
Nicholson SB, 123
Nicollier C, 89
Nix Olympica, 117b
NGC objects
   147, 240
   185, 240
   205, 240, 242
   221, 240
   224, 223
   253, 4, 250
   598, 240
   628, 226
   1275, 250
   1300, 243
   1566, 250
   1866, 293
   1910, 239
   1936, 239
   1976, 228
   1978, 239
   2070, 239
   2237, 229
   2244, 7, 229
   2362, 185
   2682, 185
   2997, 243
   3077, 244
   4038/9, 244
   4486, 241, 242, 252
   4565, 243
   4755, 185
   4826, 6
   5128, 244, 252
   5139, 232
   5195, 241, 244
   5236, 243
   6121, 5
   6397, 235, 237
   6405, 185
   6474, 185
   6514, 230
   6822, 240
   6981, 236
   6990, 237
   7293, 4, 209
   7317/20, 244
nodes, 11b, 58a, 138b
non-thermal radiation, 251b
Nordlingen craters, 94b
Norma, 3a, 5, 9, 284
North Celestial Pole, 9b, 38
Northern Crown, 3a, 284
Northern Lights, 115, 116a, 151b
North Polar Branch, 231b
North Polar Series, 162a
Nova Delphini, 209a, 253b
novae, 207a, 208–210, 253b, 266b, 303
Nova Serpentis, 253b
NP 0532, 253b
nuclear reactions, 154, 205
N9 nebula, 239b

Oberon, 71b, 132b, 133, 134b
observable limit, 260a
observable universe, 260a
occultations, 72b
Oceanus Procellarum, 99
Octans, 9, 10a, 284
Octant, 10a, 284
oculars, 35, 36
Olbers HWM, 73b, 74a, 260
Olbers' Paradox, 260
Olympus Mons, 117
Omega Centauri, 232, 235, 236
Omega nebula, 229b
Omega Piscium, 184a
Omicron Ceti, 203a
Omicron Cygni, 199a
Omicron-2 Eridani, 181
onion-like core, 213
Oort cloud, 88a, 268b
Oort JH, 88a, 226a, 228a, 253a
open clusters(galactic), 185–187, 294, 295
open ellipse, 196b, 197a
Ophelia, 134b
Ophiuchus, 3a, 5, 10, 11a, 284
Oppenheimer JR, 214a
opposition, 19b, 23, 57a, 70a
optical doubles, 177a
ordinary giants, 167b, 172
Orion, 2a, 7, 138a, 187a, 204a, 221a, 229, 284
Orion arm, 226a
Orionids, 91b
Orion's Belt, 2a
orthoscopic eyepiece, 36a
Osiander, 20a
Osterbroek DE, 226b
outer planets' opposition, 21b, 70a
Overbeek MD, 74a, 151
oxygen in atmosphere, 269
ozone, 151b

Pair of Compasses, 284
Pallas, 74, 76
PAL-I and II, 235, 238a
Pandora, 130, 131b
Papadopoulos C, 79b, 232a
parabola, 86a
parallax, 22, 158, 161, 162b
parallax of Mars, 53b
parallax of Moon, 48, 49
parallax of nearby star, 22a
parallax of Sun, 53, 54
parallelogram of velocities, 183b
Parker EN, 149b
parsec, 159b, 160a
Parsons W, 62a, 241b
particles or waves, 56b
Paschen–Ritz series, 42b, 43a
Pasiphae, 127a
Pavo, 5, 9, 10a, 284
Pavonis Mons, 117a
Peacock, 10a, 284
peculiar variables, 206a
Peebles PJE, 257, 259b
Pegasus, 3b, 4, 244a, 284
Pendulum Clock, 8a, 284
penumbra, 139, 140a, 145a
Penzias AA, 257
periastron, 178a
pericynthion, 102b
perigee, 50b, 97a, 139a
perihelion, 25a, 31b, 58a, 70b, 83b, 139a
perijove, 91a
period–luminosity law, 222–225, 232a
period of rotation, 59a
Perrine CD, 123
Perseids, 91
Perseus, 3b, 4, 7, 8, 9a, 284
Perseus arm, 226a
perturbation of cometary orbits, 88
Peters CHF, 181b
Petrejus, 20a
Pettersen BR, 206b
Pettingill CH, 111, 112
Pfund–Brackett series, 43a
PG, 0804+761, 255
PHL, 957, 255
PHL, 1127-14, 255
Phobos, 116a, 117b, 268a
Phoebe, 78b, 130, 131b
Phoenix, 4, 8a, 9, 284
photo-electric effect, 44b
photo-electric photometer, 44b, 170a
photographic magnitude, 170
photography, 79
photometers, 44a
photomultiplier, 44b
photosphere, 141a
photosynthesis, 119b, 270a
Piazzi G, 74
Pickering EC, 78b, 184a
Pickering W, 130
Pictor, 2a, 7, 9, 284
Piddington JH, 252a
Pioneer ring, 128b
Pioneer space probe, 59b, 98b
Pioneer Venus, 113
Pioneer-10 and, 11 -11, 120, 128
Piper P, 94b
Pisces, 4, 8a, 10, 11, 17b, 39b, 137b, 138, 284
Piscis Australis, 3b, 4, 284
PKS, 1127-14, 255, 284
PKS, 1610-77, 255
plage, 144
Planck, Max, 43b, 190, 191
Planck MKEL, 190, 191
Planck's constant, 191b, 261a
Planck's laws, 190
planck time, 261a
planetary nebula, 166a, 209b, 266b, 304
planetary periods, 26, 84, 85
planetary positions, 70a
planetary systems, 159a
planetesimals, 75b, 267a
planetoids, 75b, 267a
planets, 10b, 57–81, 97–136
   atmospheres, 69b, 268b, 269, 270a
   densities, 65a, 84, 85
   distances, 23b, 24, 84, 85
   formation of, 267
   masses, 64, 84, 85
   probability of, 267a
   relative distances, 23b, 24, 84, 85
   volumes, 63a, 84, 85
Plaskett JS, star, 180b
plasma, 151a, 153a
Pleiades, 7, 8b, 185b, 186
Pleione, 186
Plough, 2b, 285
Pluto, 26, 27a, 67, 80, 81a, 82, 85
   albedo 81b, 85
   data not listed here, 85
   density, 65a, 85
   diameter, 81a, 85
   discovery of, 80
   distance, 27a, 80b, 85
   eccentricity of orbit, 80b, 85
   inclination of equator, 58b, 85
   inclination of orbit, 58a, 80b, 85
   mass, 81a, 85
   retrograde spin, 58b
   size to scale, 81b, 82
Pogson NR, 160b, 162a, 207a
polar axis, 37b, 38a
Polaris, 8, 9b, 10a, 17a
Polestar, 9b
Pollux, 2b, 7, 164
Popper DM, 188b
Population I, II stars, 202b, 206a, 223b, 231a, 233b, 239a, 263b, 264a, 265b, 266
Portia, 134b
position angle, 177b, 178
positron, 153b, 154
positron–electron pairs, 261a
Potsdam Observatory, 201a, 230b
precession, 17, 18b, 39b, 58a
Preston GW, 188a
primary eclipse, 196
primeval atmosphere, 269b
primeval nebula, 88a, 93b, 99a, 268a
prism, 39b
Proctor crater, 118a
Prometheus, 130, 131b
prominences, 147a, 148
prominences, eruptive, 148
proper motion, 181b, 182
Proteus, 135a, 136a
proton, 42a, 153b, 154
proton–proton reaction, 154
proto-star, 231a, 263b
Proxima Centauri, 159b, 161, 170a, 181a, 206b
Ptolemy, 12b
Puck, 134b
Pulik G, 191a
pulsar, 211b, 216a–218a, 219, 253b, 303
Puppis, 2b, 7, 185a, 285
Pythagoras' theorem, 13a, 112a, 275
Pyxis, 2b, 7, 285

quadrature, 23, 70a
quantum, 42b
quantum theory, 191b
Quadrantids, 91b
quartz clock, 39a
quasars, 250b, 254, 255, 256a, 265a

Index

radar, 53
Radcliffe Observatory, 232b
radial velocity, 43b, 106b, 182b, 183, 239a
radians, 276
radioactive dating, 108a, 155b
radioactive elements, 263a
radio astronomy, 230b, 250b, 253a
radio galaxies, 253
radio interferometry, 254b
radio telescope, 231b
radio telescope, resolving power of, 251b
Ram, 8a, 284
Ramsay W, 41a
Ramsden eyepiece, 35b, 36a
R Andromedae, 167b
Ranger probe, 98b
rapid novae, 208a, 303a
Rateldraai meteorite, 92a
Rayet GAP, 166a
Rayleigh–Jeans law, 251a
Rayleigh limit, 174a
R Cor Bor, 169b, 195a, 204b, 205a, 210, 301
R Doradus, 203b
Reber G, 230b, 251a
recession velocity, 218b, 245, 248a
red dwarfs, 160b, 182b, 210, 266b
red giants, 213a, 219b, 263b, 266a
redshift, 219a, 245–249, 254a, 255, 260b
refraction, 35a, 39b, 52
regolith, 107b
Regulus, 3a, 6, 39a, 164
Reifenstein EC, 253b
relative diameters, 63a, 84, 85
relative distances of planets, 57b, 84, 85
relativistic speeds, 149a, 152b, 248b, 249a
repetitive novae, 208a, 209a, 303
resonance periods, 59, 75, 112b
Reticulum, 8a, 9, 285
retrograde motion, 19b, 58b, 59b, 83b
retrograde spin, 58b, 113a, 114
Reynolds R, 131a
Rhaeticus, 20a
Rhea, 47a, 130
rhenium-187, 264b
Rho Persei, 168a
Riccioli GB, 177a
Richter G, 202a
Rigel, 2a, 7, 59a, 164, 166b, 266
right ascension, 38, 39a
right ascension circle, 37b, 38a
Ring nebula M57, 166a, 209
Ringtail galaxies, 244a
Ritchey GW, 210a
River, 8a, 284
R Leonis, 203b
Robinson EL, 207b
Roche EA, orbit limit, 60b, 91a
rocket, principle of, 97
Roll PG, 257
Rømer OC, 47a, 53b
Rosalind, 134b
Rosette nebula, 229
Ross −128, −154, −248, −164, 161
Rosse, Lord, 62a, 241
rotating variables, 206b, 301
Rougoor GW, 228a, 253a
Rover car, 104a, 105b
Rowland HA, 40a
Royal Observatory Edinburgh, 216a
RR Lyrae stars, 202, 210, 223b, 232b, 239b, 298
RT Serpentis, 209a
Rufus WC, 167b
Rumford photometer, 44a
Russell HN, 168, 169
RV Tauri, 199a, 206a, 210, 301

RW Aurigae, 204a, 210, 300
RW Tauri, 201a
RX Andromedae, 208a
Ryle M, 252a, 254a
RY Sagittarii, 205a

Sagitta, 3a, 5, 285
Sagittarius, 3b, 5, 10, 137b, 138a, 195a, 227b, 229b, 230, 250b, 251a, 285
Sagittarius A, 230b, 234, 250b, 265a
Sagittarius Arm, 226a
Sagittarius B2, 231b
Sails, 2b, 285
Sandage AR, 224a, 2432a, 246, 254a, 262a
S Andromedae, 210, 211, 223
Sanduleak −69°, 202, 215a, 218a
SAO, 158687, 72b
Saros, 11b, 140a
SA, SB, SAB galaxies, 242
satellite, first, 97a
satellites, 84, 85, 124–136
Satterthwaite GE, 171a
Saturn, 10b, 11a, 18b, 19, 20, 26–28, 67, 82, 128
  clouds, 132a
  data not listed here, 85
  density, 65a, 85
  diameter, 62b, 85
  flattening, 63a
  inclination of equator, 58b, 60b, 85
  inclination of orbit, 21b, 60b, 85
  length of day, 85, 132a
  magnetic field, 128b
  period of rotation, 85
  rings, 60–62, 128
  satellites, 128b–131b
  sidereal period, 16a, 19b, 27a, 60a, 85
  size to scale, 81b, 82
  synodic period, 85
  tiny moons, 131b
Saturn-V, 100a
Scales, 3a, 284
Schiaparelli GV, 58b, 116a
Schmidt–Cassegrain telescope, 34b
Schmidt M, 226b, 227, 254a
Schmitt H, 105b
Schramm DN, 263a, 264b, 265b
Schroter JH, 62b
Schwarzschild M, 188b
Scorpion, 3a, 3b, 285
Scorpius, 3a, 5, 10, 137b, 138a, 181a, 185a, 221a, 285
Scott J, 29a
Sco X1, 253b
Sculptor, 4, 8a, 250a, 285
Sculptor system, 239b, 240, 241
Scutum, 3b, 5, 285
S Doradus, 210, 239a
Sea of Storms, 99b
seasons, 16, 137b, 138a
Secchi A, 41a, 165b
secondary eclipse, 196
second law of thermodynamics, 149a
seismic waves, 65b, 107b
seismometers, 102a, 103, 108a
Seldner M, 259b
semi-conductors, 44b
semi-major axis, 50a, 178b
semi-regular variables, 203b, 300
separation between stars, 177
Serpens, 5, 285
Serpent, 5, 285
Serpent Holder, 3a, 284
Serra da Congalha, 94a
Service Module, 100a

setting of circles, 39a
Seven Sisters, 2a, 8b, 185b, 186
Seward FD, 218a
Sextans, 3a, 6, 285
Sextant, 3a, 285
Seyfert galaxy, 250
Shakeshaft JR, 257b
Shapiro II, 111, 112
Shapley H, 199a, 201b, 223a, 232a, 239b, 245a
Sharpless SL, 226b
Shaula, 164
Shelton I, 214b
Shield, 3b, 285
Ship Argo, 2b
Shklovsky IS, 254a
Shoemaker E and C, 90b
Shoemaker–Levy comet, 90b, 91a
Shreiner C, 144a
Shu FH, 231b
Shwabe SH, 144a
sidereal day, 39a
sidereal period, 19b, 57a
sidereal time, 39a
Siebers B, 259b
Sigler R, 35a
Sigma Orionis, 166b
Sigma-2398, 161
SIMA-SIAL, 114b
similar triangles, 274b, 275a
sine, sine rule, 274
Sinnott RW, 168a
Sinope, 127a
Sinton WM, 69a, 72a
Sirius, 2a, 7, 11b, 160a, 161, 162a, 163a, 164, 165, 181b, 182, 190b, 266
Sirius B, 182
Sithylemenkat crater, 94a
Sjogren WL, 98b
Skylab, 147b, 149b
Slipher VM, 186b, 245a
slow novae, 203a, 209a, 303
Small Magellanic Cloud, 10a, 243b
SMC, 239, 240
Smith BA, 130
Smith FG, 252a
Smith TE, 90b
soft landing on Moon, 99b
Sol (Mars pro), 59a
solar constant, 152b
Solar System scale model, 81b, 82
Solar System size, 57b
solar wind, 87b, 88, 115a, 121b, 149b, 150, 151, 266a, 268a
solstices, 137b, 138a
Somerville W, 245b, 247
South African Astronomical Observatory, 83a, 168a, 215
Southern Cross, 3a, 10a, 79a, 80, 221a, 284
Southern Crown, 3b, 284
Southern Fish, 3b, 284
Southern Lights, 115b, 116a, 151
Southern Triangle, 10a, 285
SouthWestern Cape, 114b, 115
Space Age, 98a
Space Colony, 271b
space exploration, 98a
space–time continuum, 31, 263b
space walk, 100a
speckle interferometry, 175a
spectral lines, 41b, 42
spectral types, 160a, 161, 164, 166, 167
spectra, stellar, 142a, 165b
spectroheliograph, 142b, 143a
spectroscope, 39b, 141b

spectroscopic binaries, 44a, 184a, 207a
spectrum, 40
speed of light, 40a, 43a, 47b, 56, 219a
spherical aberration, 35a
Spica, 6, 164
spicules, 143b, 150a
Spinrad H, 61b, 62a
spiral arms, 226, 227, 231b, 265
spiral galaxies, 242
Sproule Observatory, 183b
Sputnik, 97a
SS Cygni, 207b
SS433, 218b, 219
Stachnik RV, 175a
Staelin DH, 253b
star diagonal, 140b
star formation, 265b
Stars, 157–219
  angular diameters, 175, 176
  birth of, 263b
  brightness, 157, 160b
  diameters, 171b–177, 190, 191
  distances, 157–160, 164
  double, 44a, 177–184,
  early, 166a
  eclipsing binaries, 196, 197
  first generation, 265b
  fixed, 10b
  flare, 159b
  for H–R diagram, 286–291
  late, 166a
  luminosities, 169, 170a, 171, 189, 292, 293
  magnitude, 161–164
  masses, 179, 180, 292, 293
  massive, 264a
  names, 165a, 195a
  neutron, 212b, 213b, 216b, 219b
  parallax of, 158, 161, 162b
  position angles, 177b, 178
  proper motions, 158b
  red dwarfs, 160b, 266b
  red giants, 219b, 266a
  second generation, 228a, 231b, 265b
  separation, 177
  super-giants, 167
  temperatures, 166, 167, 191b, 192, 193
  third generation, 231a, 265b
  variable (see under Variable stars)
  white dwarfs (see under letter W)
star streaming, 226a
ST Carinae, 201a
steady–state theory, 257b, 260b, 261a
Stefan–Boltzmann law, 192b
stellar diameters, 171b–177, 190, 191
stellar spectra, 142a, 165b
stellar types F-2-K-1, 270a
Stephan's Quintet, 244a
Stephenson CB, 218b
Stephenson FR, 212a
Stern, 2b, 284
Strand AA, 188b
Strand KA, 184b
Strong J, 69a
Struve FGW, 165b, 181a
Struve O, 165b
Struve-2398, 161
sub-giants, 167b
Sudbury structure, 93b
sulci, 125b
Sun, 18, 137–155, 252a
  absolute magnitude, 163b, 164
  atmosphere, 140a, 141b
  chromosphere, 140b, 147b, 148a, 153
  collapse of, 212b
  composition, 153, 155b
  convection zone, 153

Index

corona, 140, 147b, 149, 153
density, 153a
diameter, 14b
distance, 54, 137a
energy, 153b, 154
flares, 148b
equatorial horizontal parallax, 54a, 56b
gamma rays, 154, 155
gravitation on Moon, 31a, 64a
magnitude, 162a
mass, 64a, 152b
neutrinos, 153b, 154, 155b, 205, 212a, 215a
nuclear reactions, 154, 205
nucleus, 153
parallax, 53, 54
period of rotation, 144
photosphere, 141a, 153
plages, 147b, 148b
plasma, 151a, 153a
position in Galaxy, 235
proton–proton reaction, 154
radiation energy, 151b, 152, 155a
radiation zone, 153, 155
radio waves, 251b
size, 153
size to scale, 81b, 82
spectrum, 141b, 142
spicules, 143b, 147a
spots, 143b, 144, 145a
standard model, 153
sunspot cycle, 145b, 146a, 151b, 252a
temperature, 145a, 149a, 152a, 153a, 193
third generation star, 265b
volume, 63a
Sun–Moon relative distances, 13a,
  relative sizes, 13a, 14a
Sun's Way, 226b, 227a
superior conjuncntion, 62b
super-giants, 167
super massive object, 228a
supernova, 79b, 210–216a, 261b, 262a, 263a
  remnants, 212, 216b, 239b, 250b, 254b
  historic, 212
  Type I, Type II, 211b, 212a
Surveyor probe, 98b, 99b
S Vulpeculae, 203b
Swan, 3a, 229b, 284
Swan nebula, 305
SX Herculis, 203b
Sydney Observatory, 253a
symbiotic stars, 302
synchronous rotation, 59a, 135b
synchrotron, 152b
synchrotron radiation, 131b, 152b, 214a, 216b, 217b, 219b
Synnott SP, 123
synodic period, 57a

Table (Mountain), 10a, 284
Tachard G, 177a
Talemzane crater, 93b
Tammann GA, 224a, 246
tangent, 274
Tarantula nebula, 214b, 239a
Tau Ceti, 160, 161, 270b
Taurids, 91b
Taurus, 2a, 7, 10b, 11, 137b, 138a, 187a, 211a, 216a, 285
Taurus A, 252a
Taurus Moving Cluster, 188a
Taurus X-1, 216b
Tau Sagittarii, 184
Tau Tauri, 184

Taylor DJ, 253b
Taylor G, 74a
T Centauri, 203a
tectites, 92b, 93a
telemetry, 119b
Telescope, 3b, 285
telescope mountings, 37, 38a
telescopes, 33–39
Telescopium, 3b, 5, 285
Telesto, 130, 131b
telluric lines, 41b, 142a
Temple–Tuttle comet, 91b
terrestrial planets, 84
Tethys, 47a, 130
Thalassa, 136a
Thales, 11b
Tharsis Ridge, 117
Thebe, 122b
theory of relativity, 31, 56b, 58b, 154a, 196b, 198a, 219, 248b, 256, 257a, 246b
thermic radiation, 251b
Theta Orionis, 184b, 229a
thorium-232, 264b
Thorpe AM, 153b
Three Kings, 2a
Thuban, 8, 9b, 17a
time dilation, 219
Titan, 47a, 128b, 129a, 130, 131a
Titania, 71b, 132b, 133b, 134b
titanium, 42a
titanium oxide, 167a
Titan-2 rocket, 100a
Titius–Bode rule, 73, 78b
Titius J, 73
Tombaugh CW, 80a
T Orionis, 204a, 300
Toro, 76
transit of Mercury, 54a
transit of Venus, 54b, 55a
transverse velocity, 183
Trapezium, 184b, 229a
Triangle, 8a, 285
  Southern, 285
  Winter, 2b
triangle, area of, 274b
Triangulum T, 4, 8a, 240, 285
Triangulum Australe, 9, 10a, 285
Trifid Nebula, 229b, 230a
Trigonometric relationships, 273
tritium, 211b
Triton, 78b, 135, 136, 268a
Trojans, 77b
Tropic of Cancer, 16, 138a
Tropic of Capricorn, 16, 138a
T Tauri variables, 204b, 231b, 266b, 268a, 301
Tucana, 4, 9, 10a, 233b, 285
Tucker WH, 218a
Tunguska, 94b
T Vulpeculae, 202a
TW Ceti, 199a
Twins, 2b, 284
TX Piscium, 175b
Tycho Brahe, 21b, 22, 81b, 157a, 211a, 216a
Tycho crater, 93a, 108b
Tycho's supernova, 216, 217a, 252b
types of stars and spectra:
  W, O, B, A, F, G, K, M, R, N, C, S 166, 167

UBVRI-system, 170b
UBV-system, 170b
U Cephei, 199b
U Delphini, 204a
U Germinorum variables, 195a, 207, 208a, 210, 302

ultraviolet, 41, 209b
ultraviolet camera, 104b
ultraviolet spectrograph, 104b
Ulugh Beg, 18a
umbra, 11b, 139, 140a, 145a
Umbriël, 71b, 132b, 133b, 134b
Unicorn, 2b, 284
Universe, 223a, 260a
  age of, 246a,
  closed, 262b
  expanding, 246, 249, 262b
  open, 262b
  rate of expansion, 262a
uranium-238, 264b
Uranus, 26, 27a, 67, 71, 82, 132, 133a
  data not listed here, 85
  density, 65a, 85
  diameter, 85
  distance, 71b, 85
  flattening, 63a
  inclination of equator, 58b, 71b, 85
  intensity of sunlight, 72a
  magnetic field, 133a
  mass, 71b, 85, 132b
  moons, 71b
  occultation-1977, 72b, 73
  period of rotation, 85, 133a
  perturbations of, 77b
  rings, 133a
  satellites, 134b
  sidereal period, 27a, 71a, 72a, 85
  size to scale, 81b, 82
  temperature, 85
  velocity in orbit, 85
Ursa Major, 2b, 6, 8, 9b, 242b, 249b, 285
Ursa Major cluster, 248a
Ursa Minor, 8, 10a, 17a, 285
U Sagittae, 199b, 200, 296
U Scorpii, 209a
US Naval Observatory, 80b
UV Ceti, 161, 180b, 206b, 210, 253a, 301
UV Leonis, 199a
UW Canis Majoris, 198b
Uwe Keller, Dr. H, 87a

Valhalla, 107a, 126a
Valles Marineris, 117b
Van Allen JA, belts, 98a, 115, 151a
van Biesbroeck's star, 170a
van Bueren HG, 187b
van de Hulst HC, 252b
van de Kamp P, 183b, 188b
van den Bergh S, 243b
van den Bos WH, 74b, 165b, 181a
van Gent H, 232b
V Aquilae, 204a
Variable stars, 195–219, 296–302
  Algol type, 195b, 196a, 296
  Beta Cephei type, 206a, 301
  Beta Lyrae type, 197b, 198a, 296
  cataclismic, 107–108a, 302
  Delta Cephei type, 201, 202a, 210, 298
  Delta Scuti, 202b, 210, 299
  irregular, 204a, 300
  long period, 203a, 299
  other eclipsing, 297
  peculiar, 206a
  R Cor Bor, 195a, 204b, 205a, 210, 301
  rotating, 206b, 301
  RR Lyrae, 202, 210, 298
  RV Tauri, 199a, 206a, 210, 301
  RW Aurigae, 204a, 210, 300
  semi-regular, 203b, 300
  symbiotic, 207a, 302

T Orionis, 204a, 300
T Tauri, 204b, 301
U Geminorum, 195a, 207, 210, 302
U Sagittae, 199b, 200, 296
UV Ceti, 206b, 210, 301
W Ursae Majoris, 198b, 199a, 297
W Virginis, 202b, 203a, 210, 299
Z Andromedae, 302
Z camelopardalis, 302
V Arietis, 167b
Vaughan AH, 188a
Vega, 5, 17a, 88b, 162a, 164, 165b, 166b, 170b
Vega- I, 87
Vehrenberg H, 217a, 229b, 233b
Veil nebula, 218a
Vela, 2b, 6, 7, 9, 217a, 285
Vela pulsar, 217a
Vela X, 253b
velocity in orbit, 66a, 67
velocity of light, 47b, 48a, 56
velocity of recession, 218b, 245, 248a
Venera probes, 98b, 113a, 239b
Venus, 10b, 11a, 16a, 18b, 19–23, 26, 27, 67, 113, 114, 253a
  angular diameter, 62b
  atmosphere, 69b, 84, 113a
  brightest, 62b
  clouds, 59b
  data not listed here, 84
  density, 65a, 84
  diameter, 62b, 84
  distance from Sun, 25–27, 84
  eccentricity of orbit, 26a, 58a, 84
  inclination of equator, 58b, 84
  inclination of orbit, 21b, 84
  length of day, 113
  magnitude, 162a
  period of rotation, 59b, 84
  phases, 22b
  retrograde motion, 113a
  sidereal period, 20a, 27a, 84, 113a
  size to scale, 81b, 82
  synodic period, 57a, 84
  transit, 54, 55a
vernal equinox, 38
vertical circle, 38
Very Large Array, 252, 256a, 257a
Very Large Telescope, 45a
Vesta, 74, 76
Vidal-Madjar A, 230a
Viking-1 and, 2, 48b, 119
Virgo, 3a, 6, 10, 11a, 127b, 138a, 241b
Virgo A, 252
Virgo cluster, 241b, 246b, 248a
visible spectrum, 41
visual magnitude, 162b
visual surface brightness parameter, 176
Vogel HC, 44a
Volans 9, 10a, 285
Von Braun W, 100a
von Weizsäcker C, 154a
Voyager-1, 98b, 122b, 128b, 136b
Voyager-2, 98b, 120, 123a, 130a, 132a, 134b, 135a, 136b
Vulpecula, 3a, 5, 285
VV Cephei, 199a
VY Aquarii, 209a
V407 Cygni, 209a
V444 Cygni, 199a
V476 Cygni, 208b
V603 Aquilae, 208a
V732 Sagittarii, 208b, 209a
V939 Sagittarii, 209a
V941 Sagittarii, 209a

Wallerstein G, 187b
Walsh D, 256a
Ware J, 228b
Water Bearer, 3b, 284
wavelength, 40a
W Canis Majoris, 167b
W Cygni, 168a
Weaver HA, 90b
Weigelt G, 206b
Weinberg S, 264b
Wenzel W, 202a
Wesson PS, 260b
West R, 217
West's comet, 87b, 90a
Wetherill GW, 267b
Weymann AJ, 256a
Wezen, 167a
Whale, 3b, 8a, 284
Wheatstone bridge, 68b
Whipple FL, 87a
Whirlpool Galaxy, M51, 241, 242
white dwarfs, 182b, 191b, 207b, 209b, 210, 213a, 217a, 218a, 219b, 262b, 266a
Wien W, law, 151b, 152a, 191b, 192
Wilkinson DT, 257
William Herschel, 41a, 62a, 71, 78b, 178a, 180b, 221

Williams, Dr CN, 92
Wilson-Bappu effect, 188a
Wilson RW, 257
Winter G, 94a
Winter Triangle, 2b
WM Keck telescope, 45a
Wolf, 3a, 284
Wolf R, 144a
Wolf CJE, 166a
Wolf–Rayet bands, 209b
Wolf–Rayet stars, 166a
Wolf–relative numbers, 145b
Wolf–359, 161
Wood HE, 181a
Wostok, 98b
Wright KO, 188b
W Ursa Majoris variables, 198b, 199a, 208a, 297
W Virginis stars, 202b, 203a, 210, 299

Xi Aquarii, 184a
Xi Boötis, 181a
Xi Scorpii, 184a
X-rays, 41, 216b, 218b, 244a, 264b
X Sagittarii, 175b
XX Tauri, 209a

year, length of, 11b, 17b, 138b
Yerkes Observatory, 33b
Young T, 173
Yuty crater, 118
YY Geminorum, 181b

Z Andromedae, 302
Z Aquarii, 203b
Z Camelopardalis, 207a, 208a, 302
Zeeman effect, 145a
Zeiss, 241b
zenith, 38
zenith angle, 48b, 49
Zeta Aurigae, 184a
Zeta Geminorum, 202a
Zeta Orionis, 229b
Zeta Puppis, 166a
Zeta Ursae Majoris, 177a, 184a
zirconium oxide, 167a
Zodiac, 10b, 11
zones of avoidance, 75a
Zuben Elakrab, 3a
Zuben Elchemali, 3a
Zuben Elgenubi, 3a
Zurich number, 145b
Zwicky F, 211b, 214a
ZZ Ceti, 206b

ZZ Piscis, 206b, 207a

2.5-metre telescope, 174, 210a, 223b, 245a
3C-9, 255
3C-41, 216b
3C-48, 254, 255
3C-58, 212a
3C-120, 250b
3C-273, 250b, 254, 255
3C-295, 255
3C-345, 255
3C-446, 255
3° background radiation, 257, 263a
5-metre telescope, 45b, 69a, 223b, 238a, 251b, 254a, 256
6-metre telescope, 37a
8C-1435 +635, 255b
19 Tauri, 168a
25 Tauri, 186
30 Doradus, 214b, 239a
40 Eridani, 181
42 Librae, 175b
47 Tucanae, 9, 233b, 235, 236, 237
61 Cygni, 158b, 161, 177, 180b
70 Ophiuchi, 177b, 178, 181b
1973NA, 74b
1976 UA, 75b